Water Policy and Planning

in a Variable and Changing Climate

Insights from the Western United States

Drought and Water Crises

Series Editor: Donald A. Wilhite

Published Titles:

Water Policy and Planning in a Variable and Changing Climate
Editors: Kathleen A. Miller, Alan F. Hamlet, Douglas S. Kenney, and Kelly T. Redmond

Drought, Risk Management, and Policy: Decision Making under Uncertainty
Editors: Linda Courtenay Botterill and Geoff Cockfield

Remote Sensing of Drought: Innovative Monitoring Approaches
Editors: Brian D. Wardlow, Martha C. Anderson, and James P. Verdin

Water Policy and Planning
in a Variable and Changing Climate

Insights from the Western United States

Edited by

Kathleen A. Miller • Alan F. Hamlet
Douglas S. Kenney • Kelly T. Redmond

CRC Press
Taylor & Francis Group
Boca Raton London New York

CRC Press is an imprint of the
Taylor & Francis Group, an **informa** business

CRC Press
Taylor & Francis Group
6000 Broken Sound Parkway NW, Suite 300
Boca Raton, FL 33487-2742

First issued in paperback 2017

ISBN 13: 978-1-138-49086-4 (pbk)
ISBN 13: 978-1-4822-2797-0 (hbk)

Library of Congress Cataloging-in-Publication Data

Names: Miller, Kathleen A., editor.
Title: Water policy and planning in a variable and changing climate /
editors, Kathleen A. Miller [and three others].
Description: Boca Raton : Taylor & Francis, 2016. | "A CRC title." | Includes
bibliographical references and index.
Identifiers: LCCN 2015040346 | ISBN 9781482227970 (alk. paper)
Subjects: LCSH: Municipal water supply. | Water-supply. | Climatic changes.
Classification: LCC TD390 .W36 2016 | DDC 333.91--dc23
LC record available at http://lccn.loc.gov/2015040346

Visit the Taylor & Francis Web site at
http://www.taylorandfrancis.com

and the CRC Press Web site at
http://www.crcpress.com

In memory of George James Miller, whose love and encouragement inspired and nurtured the lead editor of this book.

Contents

Preface

The editors and authors of this book were motivated to undertake this project because we see a perfect storm brewing as a changing climate complicates the already difficult policy issues facing the western US water management community. Challenges include the need to balance the goals of providing reliable, affordable, and high-quality water to a rapidly growing population; protecting that population from loss of life and property during floods; restoring aquatic ecosystems long damaged by blocked and dewatered streams; and preserving the open landscapes, free-flowing river reaches, and productive farms that are valued by western citizens and visitors alike.

The West's extensive system of water storage, conveyance, and distribution infrastructure was built over the course of a century and a half to provide secure livelihoods for an ever-growing population despite the region's scarce and highly variable water supplies. In many parts of the West, these projects were designed to work in concert with the regular seasonal accumulation and melting of mountain snowpacks. Climate change threatens to disrupt that synergy.

Over at least the past three decades, members of the scientific community have warned the West's water managers about the potential impacts of greenhouse gas–induced climate change on western snowpacks (Revelle and Waggoner 1983). In particular, warmer temperatures are likely to increase the fraction of cool-season precipitation falling as rain rather than as snow, trigger earlier melting of snowpacks, and consequently reduce late summer streamflows and soil moisture in mountainous watersheds. A large body of scientific research now supports those projections and confirms that changes in the timing of spring runoff consistent with the long-term projections have already been occurring in many areas (Cayan et al., Chapter 2 this volume).

Historically, the development of water infrastructure and institutions has been directed at serving out-of-stream water uses, primarily irrigated agriculture, and has given short shrift to the preservation of natural aquatic ecosystems and the services they provide for fisheries, recreation, natural flow regulation, and water quality protection. Belated recognition of these values is leading to transformative changes in western water policy, with new federal and state laws and programs, and environmental advocates now firmly at the table in water planning forums. Federal agencies, once charged with damming and diverting western rivers, are now playing important roles in efforts to protect and restore their ecological resources. The change in policy focus has spawned conflicts, as those formerly favored by public water policies feel threatened (e.g., the conflicts between farmers and federal management agencies in the Klamath Basin in 2001 and 2002); but as westerners grow weary of expensive and unproductive litigation, new ways of resolving conflicts and developing cooperative approaches to water planning are being pioneered.

The material presented here describes some of those efforts and provides assessments of the implications of climate variability and change for key policy issues surrounding western water resources. The book also provides essential background information to enable readers to comprehend the challenges ahead, as well as the physical and institutional factors shaping options for addressing those challenges. The authors bring a wealth of experience and expertise related to western water resources, and the material presented is intended to be of value to practicing water policy and management professionals as well as to students in graduate and advanced undergraduate courses who aspire to work in this field.

Many members of the West's water management community are well aware of emerging climate change but are struggling to develop planning approaches to allow effective adaptation to climate conditions that will be changing in ways that are, as yet, only partly predictable (WUCA 2010). However, others have been skeptical of climate science and slow to take seriously the need to consider the possible impacts of a changing climate on their long-term planning or policy positions. Regardless of the credence one gives to the extensive body of scientific evidence surrounding the issue of anthropogenic climate change, recent events demonstrate the value of developing water management strategies that will be resilient to extreme climatic conditions.

As we conclude this project at the beginning of 2016, it seems as if the western region has received a foretaste of changes to come. Indeed, record warm temperatures coupled with extreme drought allowed little snowpack accumulation in California during winter 2014–2015. The following summer saw significant reductions in water deliveries to California's farm sector, restricted water supplies for urban areas, accelerating declines in groundwater levels, massive wildfires, and further stress on environmental resources. In the Pacific Northwest, conditions were little better. There, too, warm winter temperatures dramatically reduced the winter snowpack, and continued warmth and worsening drought fostered an early and violent wildfire season.

These events are providing a wake-up call for many western citizens and fueling concerns for what the future might hold. But this is not a time for panic or rash decisions. It is a time to energize progress toward building more effective policy processes focused on responding to climate-related risks and achieving a shared vision for resilient, equitable, and environmentally sound management of the West's water resources. In writing this book, our chief objective has been to promote the development of a community of water resource professionals and informed citizens to lead that task.

Kathleen A. Miller

References

Revelle, R.R. and P.E. Waggoner. 1983. Effects of a Carbon Dioxide-Induced Climatic Change on Water Supplies in the Western United States. In *Changing Climate: Report of the Carbon Dioxide Assessment Committee*. pp. 419–432. Washington D.C.: National Academy Press.

WUCA. 2010. Decision Support Planning Methods: Incorporating Climate Change Uncertainties into Water Planning. Report prepared for Water Utility Climate Alliance by Edward Means III, Maryline Laugier, Jennifer Daw, Marc Waage and Laurna Kaatz, January 2010. Available at http://www.wucaonline.org/assets/pdf/actions_whitepaper_012110.pdf.

Acknowledgement

Work on this volume was partially supported by the National Center for Atmospheric Research. The National Center for Atmospheric Research is sponsored by the National Science Foundation. Any opinions, findings, and conclusions or recommendations expressed in this publication are those of the authors and do not necessarily reflect the views of the National Science Foundation.

Editors

Dr. Kathleen A. Miller (Lead Editor) is an economist working at the National Center for Atmospheric Research, in the Climate Science and Applications Program. She conducts research on climate impacts, vulnerability, and adaptation. Her work focuses especially on natural resource governance and adaptation planning under uncertainty and on modeling interactions between human strategic behavior and dynamic natural systems. She collaborates with scientists from other disciplines to understand how natural resource systems will respond to the combined impacts of climate variability, direct human exploitation, other anthropogenic stressors, and management actions. She is the author of numerous papers on the management of water, fisheries, and other natural resources in the context of climate variability and prospective climate change. Dr. Miller's work includes serving as a coordinating lead author and lead author of chapters for the Intergovernmental Panel on Climate Change (IPCC) Working Group II, Third and Fourth Assessment Reports, and as a lead author for the IPCC Technical Paper on Climate Change and Water. Other publications include *Climate Change and Water Resources: A Primer for Municipal Water Providers* (AWWA Research Foundation 2006).

Dr. Douglas S. Kenney has been with the Natural Resources Law Center since 1996. He researches and writes extensively on several water-related issues, including law and policy reform, river basin- and watershed-level planning, the design of institutional arrangements, water resource economics, and alternative strategies for solving complex resource issues. Dr. Kenney has served as a consultant to a variety of local, state, multistate, and federal agencies, including several Interior Department agencies, Environmental Protection Agency (EPA), the US Forest Service, and special commissions (e.g., the Western Water Policy Review Advisory Commission); and national governments and nongovernmental organizations in Asia and Africa. Additionally, he has made presentations in (at least) 19 states (and the District of Columbia), 7 nations, and 4 continents.

Dr. Alan F. Hamlet is an assistant professor in the Department of Civil & Environmental Engineering & Earth Sciences, College of Engineering, at University of Notre Dame. Dr. Hamlet's research is focused on the integrated modeling of climate variability and change, surface water hydrology, water resource systems, the built environment, and aquatic and terrestrial ecosystems. He has been actively involved in stakeholder education and outreach programs in the Pacific Northwest for many years and is a leader in the development of decision support systems and sustainable climate change adaptation strategies in the water sector. Dr. Hamlet also has a long-term interest in the impacts of climate on renewable energy systems.

Dr. Kelly T. Redmond is the deputy director and regional climatologist at the Western Regional Climate Center at the Desert Research Institute, Reno, Nevada. He maintains an interest in all facets of climate and climate behavior; its temporal variability; spatial characteristics and physical causes; how climate interacts with other human and natural processes; and how such information is acquired, used, communicated, and perceived. As regional climatologist for the western United States, Dr. Redmond has played an active role nationally in development of the climate services sector. He has taught graduate and undergraduate classes in climatology, forecasting and synoptics, atmospheric dynamics, and hydrology. Interdisciplinary interactions have encompassed topics such as fisheries and wildlife, forestry, water resources and hydrology, and western land management, with much emphasis on observational and data management systems, and National Oceanic and Atmospheric Administration (NOAA) coop and reference networks. He is currently working on several projects for the National Integrated Drought Information System (NIDIS). He is closely involved in the NOAA Regional Integrated Sciences and Assessment (RISA) Program and the Department of Interior Climate Science Center Program. He has served on and contributed to approximately a dozen committees for the National Academy of Sciences and National Research Council. He interacts daily with members of the public and with print, radio, and television media across the West.

Contributors

Chapter Lead Authors

Celeste Cantú is general manager for the Santa Ana Watershed Project Authority (SAWPA). Since 2006, she has been working on the crest-to-coast, corner-to-corner Integrated Regional Watershed Management Plan, called One Water One Watershed (OWOW), that addresses all water-related issues and joins all entities and hundreds of stakeholders seeking to create a new vision of resiliency for the Santa Ana River Watershed. Previously, Celeste served as the executive director for the California State Water Resources Control Board and as the US Department of Agriculture rural development state director for California.

Daniel R. Cayan is a research meteorologist at the Scripps Institution of Oceanography, University of California, San Diego, and a researcher at the US Geological Survey. His research is aimed at understanding climate variability and changes over the Pacific Ocean and North America and how they affect the water cycle and related sectors, with specific interests in regional climate phenomena and impacts in California.

Thomas V. Cech is director of the One World One Water (OWOW) Center for Urban Water Education and Stewardship at the Metropolitan State University of Denver. He was executive director of the Central Colorado Water Conservancy District in Greeley and taught undergraduate- and graduate-level water resources courses at the University of Northern Colorado and Colorado State University. Tom has published extensively on water resource history, management, and policy, including work on Colorado water law and administration.

Heather Cooley is director of the Pacific Institute's Water Program. She conducts and oversees research on an array of water issues, such as the connections between water and energy, sustainable water use and management, and the hydrologic impacts of climate change. Ms. Cooley has authored numerous scientific papers and coauthored five books, including *A 21st Century U.S. Water Policy, The Water–Energy Nexus in the American West,* and *The World's Water.*

Barbara Cosens is a professor in the College of Law and affiliate of the Waters of the West Program at the University of Idaho. Her research interests include the integration of law and science in education, water governance, and dispute resolution; adaptive water governance and resilience; and the recognition and settlement of Native American water rights. She is a member of the Universities Consortium on Columbia River Governance.

David Lewis Feldman is professor and chair of the Department of Planning, Policy and Design, and professor of political science at the University of California, Irvine, and director of Water UCI—an interschool initiative focused on grand challenges facing water worldwide. His most recent books include *Water* (Polity Books 2012), *The Politics of Environmental Policy in Russia* (with Ivan Blokov of Greenpeace Russia, Elgar Books 2012), *The Geopolitics of Natural Resources* (Elgar 2011), and *Water Policy for Sustainable Development* (Johns Hopkins 2007).

Denise D. Fort is an environmental lawyer and research professor of law at the University of New Mexico School of Law. She recently resigned her position as a professor of law and director of the school's Utton Center to focus on climate change advocacy. She chaired, by appointment of President Clinton, the Western Water Policy Review Advisory Commission, which prepared a seminal report on western water policy. Her research and publications address environmental law, water policy, river restoration, and climate policy.

Jeanine Jones is the California Department of Water Resources' interstate resources manager. She serves on the Western States Water Council, Colorado River Board of California, and the Climate Adaptation Workgroup of the federal Advisory Committee on Water Information. She is a registered professional engineer in California and Nevada and has a BS and MS in civil engineering.

Ronald Kaiser is a professor of water law and policy and chair of the Water Management and Hydrological Science graduate degree program at Texas A&M University. His water research focuses on (1) conservation, (2) marketing, (3) planning, and (4) groundwater management. A number of his marketing and conservation research recommendations are incorporated in the Texas Water Code. He has published over 100 articles and papers, 2 books, and 6 book chapters.

Theodore S. Melis is currently deputy center director for the US Geological Survey's Southwest Biological Science Center in Flagstaff, Arizona. He was also formerly physical sciences program manager, deputy chief, and acting chief of the Grand Canyon Monitoring and Research Center between 1996 and 2012, a period in which he worked to develop long-term monitoring protocols to evaluate large-scale flow and nonflow experiments tied to the operation of Glen Canyon Dam on the Colorado River. Besides his early career interests in hillslope processes and how they influence fluvial settings, his interests also include adaptive management and use of decision support tools to assist resource managers in long-term experimental strategies associated with freshwater ecosystems in complex social settings.

Richard B. Norgaard is currently professor emeritus, Energy and Resources Group, University of California, Berkeley. Generally recognized as an ecological economist, he served on the CALFED Independent Science Board, then as chair of the Delta Independent Science Board (DISB), and currently as one of DISB's 10 members. He also was a chapter lead author and chapter review editor for the Millennium Ecosystem Assessment and a chapter lead author on the Fifth Assessment of the Intergovernmental Panel on Climate Change.

Andrea J. Ray is a scientist at the National Oceanic and Atmospheric Administration Earth System Research Lab in Boulder, Colorado. She does interdisciplinary work to connect climate research to applications, especially in natural resource management, and often participates in planning and policy teams, working to transition research knowledge.

Brian D. Richter is the chief scientist of the Global Water Program of The Nature Conservancy, an international conservation organization, where he promotes sustainable water use and management with governments, corporations, investors, and local communities. He is also the president of Sustainable Waters, a global water education organization, and teaches a course on water sustainability at the University of Virginia.

Jason Anthony Robison is an assistant professor at the University of Wyoming (UW) College of Law. Professor Robison has held water law– and water policy–related positions with the Colorado River Governance Initiative, Harvard Kennedy School, Harvard Water Security Initiative, Oregon Department of Justice, and Oregon Supreme Court. His research interests primarily concern relations among federal, state, and tribal sovereigns over water resources within the American West, including climate change's far-reaching implications for existing allocation and governance institutions.

Edella Schlager is a professor in the School of Government and Public Policy at the University of Arizona; senior research fellow with the Ostrom Workshop in Political Theory and Policy Analysis at Indiana University; and editor of the *Policy Studies Journal*. Her research interests focus on the governance of common-pool resources, particularly water in the western United States. She is currently studying the design and performance of intergovernmental agreements for managing transboundary watersheds.

Chadwin B. Smith is director of Natural Resources Decision Support for Headwaters Corporation. He is leading implementation of adaptive management on the Platte River and has provided independent expert advice on adaptive management to programs on the Middle Rio Grande, Everglades, Trinity River, and Missouri River. He also is a member of the Collaborative Adaptive Management Network's (CAMNet's) Core Advisory Group.

Kelly Helm Smith is the communication and planning specialist at the National Drought Mitigation Center, based at the University of Nebraska–Lincoln. Smith is particularly interested in ways to connect drought planning with other planning processes.

Robert C. Wilkinson is adjunct professor at the Bren School of Environmental Science and Management, and senior lecturer emeritus in the Environmental Studies Program, at the University of California, Santa Barbara. His teaching, research, and consulting focus is on water, energy, climate change, and environmental policy issues. Dr. Wilkinson cochairs the US Sustainable Water Resources Roundtable and has served as an advisor to government agencies, nongovernmental organizations, foundations, and companies in the United States and abroad.

Connie A. Woodhouse is a professor in the School of Geography and Development with joint appointments in the Laboratory of Tree-Ring Research and the Department of Geosciences at the University of Arizona. Her primary research focuses on climatic and hydrologic conditions of the last 2000 years in western North America, using tree rings to develop reconstructions of past rainfall, streamflow, and drought. Another area of research is on the connections between science and decision making. Research projects have included reconstructions of past streamflow in the upper Colorado River and Rio Grande basins, reconstructions of the North American monsoon in the Southwestern US, and past fire and climate relationships in northern New Mexico.

Sandra B. Zellmer is the Robert Daugherty Professor at the University of Nebraska College of Law, where she teaches torts, environmental law, natural resources, water law, and related courses. She has published numerous articles and commentary on water conservation and use, environmental law, and related topics, as well as several books: *Natural Resources Law* (with Laitos, West 2015); *Water Law in a Nutshell* (with Getches and Amos, West 2015); *A Century of Unnatural Disasters: Mississippi River Stories* (with Klein, NYU 2014); and *Comparative Environmental Law* (Carolina 2013).

Other Contributing Authors

John Abatzoglou is an associate professor in the Department of Geography at the University of Idaho.

David M. Baasch is director of biological and ecological services at the Platte River Recovery Implementation Program, Office of the Executive Director, Headwaters Corporation.

Nigel Bankes is a professor of law at the University of Calgary, where he holds the chair in Natural Resources Law, and an adjunct professor of law at the University of Tromsø.

Christopher J. Carparelli is a Big Sky Watershed Corps member with the Beaverhead Watershed Committee in Dillon, Montana.

Tapash Das is a technologist professional at CH2M, and previously was a postdoctoral researcher and project scientist at Scripps Institution of Oceanography, University of California, San Diego.

Michael D. Dettinger is a research hydrologist with the US Geological Survey.

Michelle Faggert is a student at the University of Virginia, studying environmental science with a specialization in biological and environmental conservation.

Jason M. Farnsworth is director of Habitat Management and Rehabilitation Services at the Platte River Recovery Implementation Program, Office of the Executive Director, Headwaters Corporation.

Alexander Fremier is an associate professor in the School of the Environment at Washington State University.

Michael J. Hayes is the director for the National Drought Mitigation Center (NDMC) located within the School of Natural Resources at the University of Nebraska–Lincoln.

Katherine K. Hirschboeck is an associate professor of climatology, Laboratory of Tree-Ring Research, University of Arizona, with joint appointments in Hydrology and Water Resources, Atmospheric Sciences, Arid Lands Resources Sciences, and the School of Geography and Development.

Shaleen Jain is an associate professor of civil and environmental engineering (CIE) at the University of Maine (UM).

Jerry F. Kenny is executive director of the Platte River Recovery Implementation Program, Office of the Executive Director, Headwaters Corporation.

Christine A. Klein is Chesterfield Smith Professor of Law at the University of Florida Levin College of Law and the director of the LLM program in environmental and land use law.

Noah Knowles is a research hydrologist with the US Geological Survey.

Josh Korman is the principal of Ecometric Research, a small Vancouver-based consulting firm that does contract work for the US Geological Survey, BC Hydro, Fisheries and Oceans Canada, and the Ministry of Environment.

Jeffrey J. Lukas is a research integration specialist with the Cooperative Institute for Research in Environmental Sciences (CIRES) Western Water Assessment (WWA), a National Oceanic and Atmospheric Administration–supported applied research program at the University of Colorado, Boulder.

Tyler Lystash is a recent graduate of the University of Virginia. He held an internship sponsored by The Nature Conservancy and the National Fish and Wildlife Foundation in 2014 and is now employed by the Peace Corps in Paraguay.

David M. Meko is a research professor in the Laboratory of Tree-Ring Research at the University of Arizona.

Kiyomi Morino is a research associate at the Laboratory of Tree-Ring Research at the University of Arizona.

David Pierce works at the Division of Climate, Atmospheric Sciences, and Physical Oceanography at the Scripps Institution of Oceanography.

William E. Pine, III is an associate professor at the University of Florida, where he teaches courses in riverine and quantitative ecology.

Emily Maynard Powell is the global water analyst in the Global Water Program of The Nature Conservancy, an international conservation organization.

Roger S. Pulwarty is a physical scientist and is currently director of the National Integrated Drought Information System (NIDIS) located in Boulder, Colorado.

F. Martin Ralph is the director of the Center for Western Weather and Water Extremes at Scripps Institution of Oceanography, University of California, San Diego, and formerly a division chief, program manager, and program developer in the National Oceanic and Atmospheric Administration.

Crystal J. Stiles is an applied climatologist with the High Plains Regional Climate Center, located at the University of Nebraska–Lincoln.

Edwin Sumargo is a graduate student at the Scripps Institution of Oceanography, University of California, San Diego.

Michael D. Yard is an aquatic ecologist at the US Geological Survey Grand Canyon Monitoring and Research Center, Flagstaff, Arizona.

Section I

Overview and Background

1

Introduction: The Context for Western Water Policy and Planning

Kathleen A. Miller, Alan F. Hamlet, and Douglas S. Kenney

CONTENTS

ABSTRACT Here we describe the context for water policy and planning in western United States and the implications of natural climate variability and anthropogenic climate change for the region's water resources. The questions related to water resource planning and policy development that face western water managers, policy makers, and citizens are made more challenging by extreme droughts, floods, and the likelihood of further changes in climate. This chapter discusses the purpose and focus of the book and provides an overview of its content and key messages.

1.1 Rationale and Purpose

As we wrapped up work on this book, drought and record warm temperatures had left California and the Pacific Northwest states grappling with the water resource impacts of paltry winter snowpacks, parched watersheds, and raging wildfires (California State Water Resources Control Board 2015; NIDIS 2015; NOAA 2015a). In California, the water content of the snowpack stood at 5% of normal on April 1, 2015 (California Department of Water Resources 2015) and low streamflow forecasts for the summer irrigation season prompted the governor to declare a drought emergency. Shortly thereafter, the State Water Resources Control Board began mandating curtailment of surface water withdrawals from the state's major rivers. As of June 12, curtailment orders on the Sacramento and San Joaquin Rivers extended to all water rights with priority dates junior to 1903, leaving those rights holders unable to draw upon their usual sources of supply. Elsewhere in 2015, a multiyear drought in Texas and Oklahoma gave way to torrential downpours and deadly flash flooding.

These extreme conditions provide stark reminders of the inherently variable climate of the western region of the United States and the challenges it poses for water policy and planning. Water resource professionals also are becoming increasingly aware of the fact that the range of high and low streamflows represented in the historical record provides an incomplete picture of the full range of natural variability and an inadequate guide to future hydroclimatic conditions. For example, it is now widely recognized that the Colorado River Compact of 1922 allocated fixed amounts of water to the Upper and Lower Basins based on relatively short streamflow records that were not representative of long-term streamflows as revealed by paleoclimatic reconstructions, nor of likely future conditions.

Global climate change is already underway, and there is strong evidence that it is likely to alter patterns of water availability worldwide. The fact that nearly all major rivers in the western United States arise in snow-fed mountainous watersheds presents a distinct source of vulnerability due to temperature impacts alone (Cayan et al., Chapter 2 of this volume). Warmer temperatures will inevitably alter the dynamics of snow accumulation and melting, leading to impairment of the natural water storage capacity of high-elevation catchments and alteration of seasonal river flow regimes (Adam et al. 2009; Stewart 2009).

Although this volume documents the fact that a number of prominent water resource management agencies in the West are currently moving toward consideration of climate projections in formal planning, there is also abundant evidence that many present-day water managers are repeating the fundamental mistakes made in the Colorado River Compact by assuming that twentieth century climate variability is a reasonable proxy for twenty-first-century conditions. The US Federal Energy Regulatory Commission (FERC) licensing process is a notable example (Ray, Chapter 8 of this volume), and Federal Emergency Management Agency (FEMA) flood maps that give information on eligibility for the federal flood insurance program are another (Zellmer and Klein, Chapter 19 of this volume). Despite forward motion, there is a long way to go.

We have chosen to focus on the western United States in this book because water is generally scarce relative to human uses in this region, and the control of water has played an important role in shaping its historical development and continued economic vitality. In defining the region, we rely on the tradition established by the 1902 federal Reclamation Act (National Reclamation Act of 1902), which authorized federal support for the construction of irrigation projects in the 17 contiguous western states. In other words, our region of analysis includes the Great Plains states from the Dakotas southward to Texas, and the states westward to the Pacific coast.

The impacts of climate change are expected to compound the many existing pressures on western water resources. These have grown as the region's population has swelled—leading to transformation of the landscape, new water demands, and shifting public views on how the resource should be managed. In many cases, growing demands are encountering diminishing or degraded sources of supply. For example, surface water supplies may be polluted, and groundwater resources may be pumped at unsustainable rates, leading to declining aquifer levels, increasing pumping costs, and depletion of baseflows in interconnected surface streams. Furthermore, the cumulative ecological impacts of substantial water withdrawals and the region's history of river basin engineering have left many species in a fragile condition and less able to withstand episodes of drought and extreme high temperatures. A compelling argument can be made that the West's major river basins and aquifers are now best understood as human-dominated systems (Richter et al., Chapter 5 of this volume).

Assessment of the implications of climate variability and change for such systems requires attention to both their human and natural dimensions. More particularly, the dynamic interactions between these elements will determine the outcomes of any proposed changes in policy or infrastructure. Important points to recognize when considering options for improving the climate resilience of the West's water resource systems include the spatially complex patterns of water availability and demands; the large number of independent, but interconnected, decision makers; and the multitude of values at stake. There is no central water authority in any western state (and certainly none for the West as a whole), but rather, a decentralized multilayered system of water management by multiple entities including individuals, irrigation organizations, federal and state project operators, municipal water authorities, regional water management bodies, and drainage districts.

There is a wealth of available information on climate change and its implications for western hydrology and water management (e.g., Lewis 2003; Climatic Change 2004; Brekke et al. 2009). There also is a growing body of literature providing advice on planning for adaptation to climate change in the face of considerable uncertainty (e.g., Lempert et al. 2006; Wilby and Dessai 2010; WUCA 2010; Stakhiv 2011; Yates and Miller 2011). Much of this work focuses on planning problems relevant to small sets of well-defined decision makers and thus fails to explore the complexity of real-world planning processes. The broader policy issues that climate change presents for balancing competing interests, sustaining societal benefits from water resources, and protecting ecological resources have been the subject of a separate body of literature, largely in law and policy science journals (e.g., Craig 2010; Adler 2012). Integration of these different spheres of knowledge will be important for the tasks of developing sound policies and effective planning processes.

Our purpose in writing this book is to help readers understand not only the challenges ahead but also possible strategies for managing water-related risks, securing human well-being, and protecting environmental resources in a variable and changing climate. This requires an accurate understanding of the starting point for adaptive actions—in other words, the existing physical, social, and institutional features that will shape available options for responding to the combined impacts of human activities and climate change on water resources and flooding risks. It also requires a sound and well-informed approach for incorporating science information in water policy and planning deliberations.

Such an approach recognizes that science is not a static repository of "truth" but, rather, an evolving process of systematic inquiry focused on gathering evidence to shed light on how systems work and what causes them to change. As such, uncertainties are an inherent part of the process. Rather than seeing uncertainty as a reason for rejecting scientific input, citizens, policy makers, and water resource professionals would do well to embrace uncertainty as a central aspect of the problems they seek to address. On the other side, science professionals can benefit by acquiring a realistic understanding of the motivations and considerations, apart from their own areas of expertise, that are important in driving policy and planning processes. Mutual understanding is needed between all parties engaged in making or informing water-related decisions in order to meet the challenges posed by an increasingly variable and changing climate. In this spirit, the topics covered in this volume were selected to provide a foundation for building a holistic and well-grounded understanding of the issues and context for future decision-making regarding the West's water resources.

1.2 Climate and Western Water

The West's climate is highly variable in both time and space (Cayan et al., Chapter 2 of this volume). Large-scale moisture transport into the region is driven by ever-changing patterns in the exchange of heat, moisture, and momentum between the atmosphere and oceans. Where that moisture falls as rain or snow depends heavily on the West's complex topography. Mountaintops and westward-facing slopes capture much of the available moisture, leaving dry rain-shadow landscapes on the lee-side of the major mountain ranges. Within a short drive across any of the coastal and mountain states, one can witness the dramatic impacts of differences in temperature and precipitation regimes on vegetation and water availability.

The rivers that provide most of the West's water arise in those well-watered mountainous areas. Their seasonal patterns of flow are shaped by the accumulation of snow over the winter season and its rapid release as meltwater over the course of the spring and early summer. Western water management infrastructure and water use practices are tuned to that regular seasonal cycle. In essence, mountain snowpacks represent key water storage reservoirs—holding flows back in the winter when they might otherwise pose a flood risk and the water is not wanted for human uses, and releasing it to flow into the valleys just at the start of the irrigation season.

The warming anticipated from global climate change is likely to disrupt this long-established pattern. The loss of snowpack storage could spell reduced late summer streamflows, reduced water availability for irrigation and other human uses, and higher water temperatures. Those impacts would be especially likely in stream basins where there is little available natural or human-constructed storage capacity to capture and regulate earlier and possibly larger spring runoff.

Throughout history, human communities have adapted to differences in regional climates, and their cultures, technologies, and settlement patterns have been shaped by struggles to cope with climate variability, including extremes of droughts and floods. At present, we are entering an uncharted era in human experience with the global climate system. Rising atmospheric concentrations of carbon dioxide (CO_2) and other greenhouse gases are well documented (IPCC 2013; WMO 2014). As of 2015, the concentration of CO_2 in the global atmosphere stood in the neighborhood of 400 parts per million by volume (ppm), a level unprecedented in human history. Using ice core records to extend modern measurements back to 800,000 years before present, researchers have found that CO_2 concentrations over that period fluctuated between roughly 170 and 300 ppm, with low and high values corresponding, respectively, to glacial and interglacial periods (Jouzel et al. 2007; Lüthi et al. 2008; Scripps Institute of Oceanography 2015). To put this record in perspective, recall that there were no modern humans 800,000 years ago—that was the era of ancestral humans *Homo erectus*. *Homo sapiens* did not emerge until about 200,000 years ago (Smithsonian 2015).

At the beginning of the industrial revolution in the mid-eighteenth century, the concentration of CO_2 in the atmosphere was about 280 ppm, close to the mean level experienced during previous interglacial periods. Since 1750, human activities have released approximately 555 (±85) gigatons of carbon (GtC) into the atmosphere, of which about two-thirds was from fossil fuel burning and cement production and the remainder from deforestation and other land-use changes. As a result, atmospheric CO_2 concentrations have increased by about 43% since preindustrial times (IPCC 2013). The sharp spike to modern levels occurred in the geologically very short period of two and a half centuries.

Continuing growth in CO_2 concentrations, together with rising concentrations of methane (CH_4), nitrous oxide (N_2O), and more powerful manufactured greenhouse gases such as halocarbons, has altered the Earth's energy balance, leading to ongoing and expected future warming of ocean and land-surface temperatures. Changing large-scale atmospheric circulation patterns and acceleration of the hydrologic cycle are projected outcomes of global climate change.

There is abundant evidence that the measured changes in radiative forcing have already caused climate changes around the globe. Multiple independent research teams have recorded a 0.8°C increase in global average annual temperatures since the middle of the nineteenth century. As documented by the Intergovernmental Panel on Climate Change, each of the last three decades has been successively warmer at the Earth's surface than any preceding decade since 1850. The oceans also have warmed, accounting for more than 90% of the net energy storage in the climate system between 1971 and 2010 (IPCC 2013).

Rising greenhouse gas concentrations will lead to changes throughout the climate system. For example, there may be a positive feedback to initial warming as shrinking snow and ice cover reduce the reflection of energy back to space. Another positive feedback comes from the fact that warm air can hold more moisture than can cold air, and atmospheric water vapor is itself a powerful greenhouse gas (NOAA 2015b). Global evaporation and precipitation will likely increase (Trenberth 2011), but there will be strong regional differences in the impacts on water availability as storm tracks shift in response to warming. Most climate model simulations project "...a poleward shift of the midlatitude storm tracks and equatorward contraction of convergence zones..." (Seager et al. 2012, p. 3355). In the Northern Hemisphere, such shifts are expected to concentrate greater precipitation in already wet far-northern areas and along the equator while fostering drier conditions in subtropical dry zones (around ±30° latitude) like northern Mexico and the US Southwest. Many parts of the western United States fall in the transition zone between those projected trends, making projections of precipitation changes especially difficult. However, even with limited predictability of precipitation trends, the projected impacts of warmer temperatures on evaporation and transpiration provide considerable information about future water stress across the western United States.

The important role of temperature in driving hydrologic change is illustrated in Figure 1.1, which shows projected changes by the mid-twenty-first century in annual runoff and climatic water deficit (a measure of evaporative demand in excess of available soil moisture) in a scenario of high greenhouse gas emissions. The mean changes as simulated

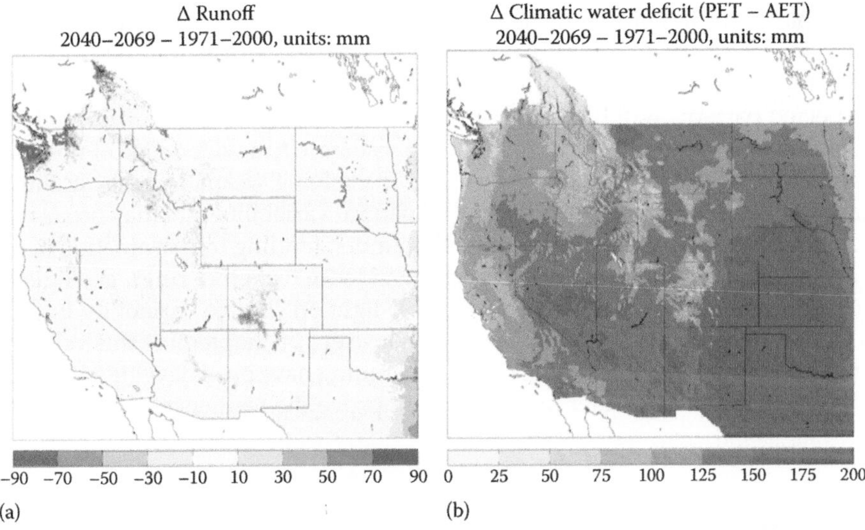

FIGURE 1.1
Projected change in annual runoff (a) and climatic water deficit as measured by the difference between potential evapotranspiration (PET) and actual evapotranspiration (AET) (b) between the late-twentieth-century (1971–2000) climate and the mid-twenty-first-century (2040–2069) climate under a high-emissions (RCP 8.5) scenario and as simulated by 20 climate models. RCP denotes a Representative Concentration Pathway as defined in the Fifth Assessment Report of the Intergovernmental Panel on Climate Change (IPCC 2013). (Figure courtesy of John Abatzoglou, a contributor to this volume.)

by 20 climate models are shown only in areas where at least two-thirds of models agree on the sign of change. Note that annual runoff is projected to increase in a few areas where greater winter runoff would play a dominant role. However, even in those locations, moisture stress is projected to increase in response to the impacts of warming on evaporation and plant water use (Stephenson 1998).

The projected widespread increases in climatic water deficit are posited to contribute to an increased likelihood of wildfires and forest disturbance due to insect outbreaks and drought-induced mortality (e.g., Allen et al. 2010). Such disturbances would further affect watershed hydrology, potentially leading to flashier runoff, higher stream temperatures, and water quality degradation (Adams et al. 2012; Luce et al. 2012; Thompson et al. 2013).

A point that is inadequately understood by the general public is the fact that both the near-term climate and the climate of future decades will reflect the combined influences of natural variability and anthropogenic climate change. Seager and colleagues make the following argument: "There is a growing sense that a purely natural (i.e., uninfluenced by human activity) climate system no longer exists, and it is widely assumed that climate events like heat waves, stormy winters, droughts, and floods bear at least some imprint of human-induced climate change, rendering the term 'natural climate variability' a relic of the preindustrial age" (Seager et al. 2012, p. 3356). This statement points to the difficulty of disentangling the separate influences of natural and human-forced processes in determining the observed sequence of climate conditions, a topic also discussed by Cayan et al. (Chapter 2 of this volume).

In the western United States, where climate is innately highly variable, trends in weather patterns over typical planning horizons will be especially difficult to foresee. In short, the sequence of climate events over the next couple of decades is likely to be dominated by natural variability but shaped by the underlying influence of anthropogenic climate change (Deser et al. 2012a,b).

For example, natural temperature variability superimposed on increasing temperature trends due to climate change will likely result in periods of plateau (no trend) and rapid increases as the cycles of natural variability reinforce or oppose the long-term trends. These expected patterns will likely be a source of confusion for the uninformed.

These facts suggest a need to focus on building resilience to a wide range of possible conditions together with flexibility to respond quickly to evolving weather events. It will be especially important to understand the full range of natural variability that may occur regardless of global climate change, and to incorporate that understanding into evaluation of planning options. Paleoclimate reconstructions, based on evidence from tree rings, geologic features, and other information, are shedding considerable light on climate variability in this region prior to modern records. As described by Woodhouse et al. (Chapter 9 of this volume), methods for incorporating that information in water planning have been developed and are being used by some of the West's water managers to improve the resilience of their systems.

1.3 Institutional Context

Fickle and unevenly distributed water resources have long posed challenges for human efforts to secure sustainable livelihoods in the western United States. Long before the nineteenth-century influx of Euro-American settlers into the territories west of the Mississippi River, Native and Hispanic communities in the Southwest developed irrigation systems to allow farming in areas where crops otherwise could not succeed. Irrigation continues to be vitally important for western US agriculture, accounting for most of the water consumed by human activities in the West.

The story of water planning and policy in the western United States has revolved around efforts to solve problems posed by limited and highly variable water availability in a physically challenging environment. The West's extensive present-day system of dams, levees, and infrastructure for the transport and distribution of water is the outgrowth of more than a century and a half of such problem solving. These efforts were often uncoordinated, involving decisions made at different points in time by a multitude of entities ranging from individual farmers and miners to state and federal government agencies.

Early water development projects were typically driven by narrowly defined objectives: for example, securing a reliable source of irrigation water to allow a family or larger community to survive, or shunting floodwaters away from an established community to prevent its destruction. Uncoordinated local solutions to specific local problems, however, frequently engendered new problems for other people, places, and times. For example, levees to protect one community from flooding typically function simply to deposit the floodwaters farther downstream (Mount 1995).

The West's water institutions grew up as a parallel effort to resolve numerous conflicts between parties seeking to rely on the same water source or to resolve incompatibilities between their proposed development projects. Competition for scarce water supplies and the adverse impacts on other parties of water diversions and stream channel modifications quickly led to disputes. These, in turn, laid the early groundwork for the development of western water law and the evolution of administrative arrangements that now govern the management, use, and development of western water resources. Prior appropriation was widely, although not exclusively, adopted as the principle for resolving conflicts over allocation of scarce and fleeting water supplies. It brought stability by clearly identifying who would or would not have the right to withdraw water from a river when flows were low. It also provided a strong incentive for latecomers to build reservoirs to secure their access to water when seasonal streamflows declined.

In the modern era, prior appropriation has many critics, who point to it as having historically ignored the ecological and aesthetic value of natural streamflows and as now locking water use into archaic patterns dominated by irrigated agriculture (Wilkinson 1991; Kenney 2005). However, the doctrine has been modified in practice, and there is evidence of flexibility (Tarlock 2001; Benson 2012). For example, the reservation of water for environmental purposes through instream flow provisions is a common, albeit fairly recent, addition to most western water codes (Gillilan and Brown 1997). Likewise, the movement of water to new and "higher" uses is authorized through the legal treatment of water rights as transferrable property. In both contexts, however, the promise is often greater than the reality.

The stunted evolution of water markets is particularly significant, as transaction costs can make otherwise viable transfers prohibitive. Transfers of water rights are regulated in various ways to prevent harm to other rights holders, for example, those relying on the return flows from an existing irrigation right. Such protections serve a useful purpose. However, the efficiency with which they are implemented varies dramatically from state to state depending on the quality of state record keeping and the design of the review process. In some cases, documentation of actual water use is so poor that enforcement of water rights has occurred only when necessitated by drought, and then awkwardly (Wilson 2012; Lund et al. 2014; Hanemann et al. 2015). A proposed water transfer can provide a reason to review past uses and to update records, an essential step to marketing but a potentially costly and dangerous one for a water user whose rights have heretofore escaped serious governmental scrutiny.

In other settings, up-to-date records and well-developed administrative infrastructure already exist, allowing relatively efficient enforcement of water rights and providing adequate documentation to allow water markets to function without undue constraints (e.g., MacDonnell and Rice 2008; Jones and Cech 2009; Western Governors Association 2012; Hansen et al. 2014). But even in these situations, transfers can be discouraged by their high political costs, as the socioeconomic impacts to areas losing water can be significant (National Research Council 1992).

Water management infrastructure and institutional arrangements for water allocation and mitigation of flood risks developed jointly over time in response to the region's growth and changing water demands. In recent decades, this complementary relationship between infrastructure expansion and the rules of management has become strained by concerns about the adverse impacts of dams, levees, and water diversions on aquatic ecosystems and associated efforts to preserve the values that unimpaired watersheds can provide to the human community. The range of interests, values, and disciplines of participants in water decisions has expanded accordingly, increasingly transforming water management from a technical, engineering-driven exercise to a much more complicated endeavor.

In the Columbia River Basin in the Pacific Northwest, for example, the number of water management objectives that must be considered in developing new water management policies has expanded dramatically since the 1960s, when the basin's operating policies were first developed. Despite these expanding management concerns, the primary basis of the Columbia's operating policies remains the Columbia River Treaty (CRT; 1964), which was (and remains) focused almost entirely on the conjunctive management of hydropower and flood control. Although the CRT is often cited as one of the most successful international water treaties in the world, it does not address important impact pathways associated with climate change, such as changes in seasonal hydropower generation (Hamlet et al. 2010), evolving flood risks (Lee et al. 2009; Tohver et al. 2014), or the effects of intensifying low flows and warmer water temperatures on salmon populations (Mantua et al. 2010). Although the need for change is readily apparent in the Columbia Basin (Miles et al. 2000), the social, legal, and political challenges that are tied to potential changes in the CRT are formidable, and the way forward is far from clear (Cosens et al., Chapter 10 of this volume).

Legislation and formal administrative mandates and protocols are key aspects of the West's water institutions, but the term *institutions* is a much broader concept. One definition was provided by Nobel Laureate Douglass North: "Institutions are the humanly devised constraints that structure political, economic and social interaction. They consist of both informal constraints (sanctions, taboos, customs, traditions, and codes of conduct), and formal rules (constitutions, laws, property rights)" (North 1991, p. 97). Thus, to understand the context for present-day water resource decision making, one may need to consider the informal customs and expectations of water users and managers in addition to the formal legal definitions of their rights and obligations.

Sound but generic advice on climate adaptation planning is available from several sources (e.g., National Research Council 2009; Major and O'Grady 2010; NDWAC 2011). Key messages include the need to approach climate change adaptation as a risk management problem and the need to create flexibility. Somewhat less attention has been given to specifics about how to achieve these objectives in the context of ongoing planning processes that are often marked by conflicts between different stakeholder groups whose members are reluctant to let go of perceived entitlements and cherished preferred solutions.

Even among water management professionals, it is common to find strong preferences for adherence to known strategies, reliance on observed records, and suspicion of model projections outside of the range of recent experience (Hamlet 2011). The prospect of continuing climate change and the growing likelihood of extreme droughts and floods belies the presumed security of such an approach. Following old strategies under rapidly changing conditions is likely to yield disappointing—and even dangerous—results.

A very different planning perspective is needed to secure reliable water supplies, protect ecological resources, and mitigate flood hazards when the underlying probability distribution of extreme climate conditions is not stable. The requisite perspective is one that seeks to understand the implications of different plausible trajectories of changing water-related risks and growing pressures on resources, and one that fosters creative and proactive problem solving while avoiding unnecessary conflict and hardship. Also required are (1) a willingness to embrace uncertainty by acknowledging our incomplete understanding of complicated water-dependent socioecological systems and our imperfect ability to predict their responses to climate-driven perturbations and (2) a purposeful strategy of learning from the inevitable surprises that will come from limited understanding and predictability.

Armed with such a perspective, those engaged in water resource planning and flood hazard mitigation would be better able to participate in a long-term ongoing program of evaluating vulnerabilities, identifying robust response options, and taking action. A supportive policy environment is needed to facilitate such exploration and to ensure that planning processes are open but structured to facilitate constructive engagement of all relevant stakeholder communities. Examples of such processes, as well as the consequences of their absence, can be found in the material presented in this volume.

1.4 Summary and Overview

We are confronting a reality of growing pressures on the West's water resources. Potentially large physical changes in the region's climate and hydrology coupled with limited predictability create new challenges for the ongoing need to make water policy and planning decisions. It is important to move beyond the near-term sectoral interests that have long dominated water planning and to adopt a planning perspective focused on the long-term well-being of all citizens and the health of the environment on which that well-being depends. That will require taking into account the rapidly evolving climate system and its impacts on water resource systems. Decisions made now are likely to cast long shadows. That is, they will have enduring impacts arising from the long life of water infrastructure and the lasting legacy of both water allocation decisions and institutional innovations. In this setting, it is easy enough to caution policy makers and water planners to choose wisely, and to adopt strategies to respond to changing climate conditions. It is much more difficult to define what that means or how to make it happen.

We do not pretend that we have a recipe for efficient adaptation to prospective climate change. Although there have been success stories in many areas of the West (a number of them documented in this volume), there are no silver bullets or one-size-fits-all solutions. Rather, we argue that it is important for anyone engaged in water policy discussions or analysis of water planning options to start with a realistic understanding of the values and interests at stake; how they relate to one another; and what we do, do not, and cannot know about the effects of any proposed action as affected by the West's highly variable and changing climate.

The chapters in this volume identify many of the issues that will require policy attention or planning decisions in coming decades. It is clear that significant problems exist for western water management, even in the absence of anthropogenic climate change. It also is clear that anticipated climate change will worsen many of these problems and may foreclose some solution options. For example, past water development and land-use decisions have massively altered watersheds, streamflow regimes, and aquatic habitats, leading many species to the brink of extinction (Richter et al., Chapter 5 of this volume). Recognition of the ecological values that have been lost has spurred restoration efforts, but it has proved difficult and expensive to undo the damage, and only limited stream-recovery success stories have been achieved thus far. The task will become increasingly daunting as summer streamflows decline and water temperatures rise in response to anthropogenic climate change. Furthermore, thresholds may be crossed beyond which even massive expenditures to acquire water for stream restoration may fail to preserve species in their native habitats. However, that does not imply that preservation and restoration goals should be abandoned. It does imply that creative thinking and thorny negotiations will be needed to maintain a desirable balance between healthy aquatic ecosystems and out-of-stream water uses.

Groundwater management is another subject that will become more challenging in a changing climate. The practice of turning to groundwater use when surface water supplies are unavailable can be a sensible strategy for mitigating the impacts of short-duration droughts. Indeed, the economic impacts of the 2014–2015 California drought would have been more immediate and painful were it not for the ability of agricultural water users to sustain production by pumping groundwater (Howitt et al. 2014). But when groundwater rights are poorly constrained, aquifer depletion may lead to long-term consequences for future water availability. That risk will escalate if climate change reduces summer streamflows. Attention to groundwater policies will be required, but the issues are complicated. For example, Fort (Chapter 3 of this volume) notes that groundwater mining is an especially difficult problem because regulation has been historically lacking, and it is hard to define an appropriate balance between sustaining groundwater levels and allowing use when some aquifers are essentially nonrenewable. Interactions between surface water and groundwater also present challenging issues for policy development, especially when a changing climate may alter the impacts of groundwater pumping on nearby streams, as discussed by Cech (Chapter 18 of this volume).

Despite the magnitude of the challenges, however, we identify many reasons for optimism. The following chapters paint a picture of a policy environment that is complex and resistant to change but not immutable. Old habits and ways of thinking are giving way to newer concerns about sustainability and protection of the public interest in healthy aquatic environments. New processes are being tried that give voice to a broader range of interests and values in the development of policy recommendations and proposals for infrastructure projects and management strategies. While periods of water stress still inflame old conflicts, they also are increasingly engendering new collaborative efforts to secure a sustainable and desirable water future for communities, states, and river basins. Adopting needed reforms, and at a pace sufficient to meet rising challenges, is far from easy—but very few aspects of western water management have ever been easy. The following pages offer a wealth of examples, insights, and considerations to guide those efforts.

References

Adam, J.C., A.F. Hamlet, and D.P. Lettenmaier, 2009. Implications of global climate change for snowmelt hydrology in the twenty-first century. *Hydrological Processes* 23:962–972.

Adams, H.D., C.H. Luce, D.D. Breshears, C.D. Allen, M. Weiler, V.C. Hale, A.M.S. Smith, and T.E. Huxman, 2012. Ecohydrological consequences of drought- and infestation-triggered tree die-off: Insights and hypotheses. *Ecohydrology* 5:145–159.

Adler, R.W., 2012. Balancing compassion and risk in climate adaptation: U.S. water, drought and agricultural law. *Florida Law Review* 64:201–267.

Allen, C.D., A.K. Macalady, H. Chenchouni, D. Bachelet, N. McDowell, M. Vennetier, T. Kitzberger et al., 2010. A global overview of drought and heat-induced tree mortality reveals emerging climate change risks for forests. *Forest Ecology and Management* 259(4):660–684.

Benson, R.D., 2012. Alive but irrelevant: The prior appropriation doctrine in today's western water law. *University of Colorado Law Review* 83(3):675–714.

Brekke, L.D., J.E. Kiang, J.R. Olsen, R.S. Pulwarty, D.A. Raff, D.P. Turnipseed, R.S. Webb, and K.D. White, 2009. Climate Change and Water Resources Management: A Federal Perspective. U.S. Geological Survey Circular 1331. Available at http://pubs.usgs.gov/circ/1331/.

California Department of Water Resources, 2015. News for Immediate Release, April 1, 2015: Sierra Nevada Snowpack Is Virtually Gone; Water Content Now Is Only 5 Percent of Historic Average, Lowest Since 1950. Available at http://www.water.ca.gov/news/newsreleases/2015/040115snowsurvey.pdf.

California State Water Resources Control Board, 2015. Available at http://www.waterboards.ca.gov/waterrights/water_issues/programs/drought/analysis/.

Climatic Change, 2004. Special issue on the effects of climate change on water resources in the West. *Climatic Change* 62(1–3):1–388.

Craig, R.K., 2010. "Stationarity is dead"—Long live transformation: Five principles for climate change adaptation law. *Harvard Environmental Law Review* 34:9–74.

Deser, C., R. Knutti, S. Solomon, and A.S. Phillips, 2012a. Communication of the role of natural variability in future North American climate. *Nature Climate Change* 2:775–779. doi: 10.1038/nclimate1562.

Deser, C., A.S. Phillips, V. Bourdette, and H. Teng, 2012b. Uncertainty in climate change projections: The role of internal variability. *Climate Dynamics* 38:527–546. doi: 10.1007/s00382-010-0977-x.

Gillilan, D.M. and T.C. Brown, 1997. *Instream Flow Protection: Seeking a Balance in Western Water Use.* Washington, D.C.: Island Press.

Hamlet, A.F., 2011. Assessing water resources adaptive capacity to climate change impacts in the Pacific Northwest region of North America. *HESS* 15:1427–1443. doi:10.5194/hess-15-1427-201. http://www.hydrol-earth-syst-sci.net/15/1427/2011/hess-15-1427-2011.pdf.

Hamlet, A.F., S.-Y. Lee, K.E.B. Mickelson, and M.M. Elsner, 2010. Effects of projected climate change on energy supply and demand in the Pacific Northwest and Washington State. *Climatic Change* 102(1–2):103–128.

Hanemann, M., C. Dykman, and D. Park, 2015. California's flawed surface water rights. In *Sustainable Water Challenges and Solutions from California.* Allison Lassiter (ed.), pp. 52–81. Berkeley: University of California Press.

Hansen, K., R. Howitt, and J. Williams, 2014. Econometric test of water market structure in the western United States. *Natural Resources Journal* 55:127–152.

Howitt, R., J. Medellín-Azuara, D. MacEwan, J. Lund, and D. Sumner, 2014. Economic analysis of the 2014 drought for California agriculture. Center for Watershed Sciences at University of California Davis, UC Agricultural Issues Center, and ERA Economics, Davis, CA. Available at https://watershed.ucdavis.edu/files/biblio/DroughtReport_23July2014_0.pdf.

IPCC, 2013: *Climate Change 2013: The Physical Science Basis.* Contribution of Working Group I to the Fifth Assessment Report of the Intergovernmental Panel on Climate Change [Stocker, T.F., D. Qin, G.-K. Plattner, M. Tignor, S.K. Allen, J. Boschung, A. Nauels, Y. Xia, V. Bex and P.M. Midgley (eds.)]. Cambridge, UK and New York: Cambridge University Press, 1535 pp. Available at: https://www.ipcc.ch/report/ar5/wg1/.

Jones, P.A. and T. Cech, 2009. *Colorado Water Law for Non-Lawyers.* Boulder, CO: The University Press of Colorado.

Jouzel, J., V. Masson-Delmotte, O. Cattani, G. Dreyfus, S. Falourd, G. Hoffmann, B. Minster et al., 2007. Orbital and millennial Antarctic climate variability over the last 800,000 years. *Science* 317:793–796.

Kenney, D.S., 2005. Prior appropriation and water rights reform in the western United States. In *Water Rights Reform: Lessons for Institutional Design.* B.R. Bruns, C. Ringler, and R. Meinzen-Dick (eds.), pp. 167–182. Washington, D.C.: International Food Policy Research Institute.

Lee, S.-Y., A.F. Hamlet, C.J. Fitzgerald, and S.J. Burges, 2009. Optimized flood control in the Columbia River Basin for a global warming scenario. *ASCE Journal of Water Resources Planning and Management* 135(6):440–450. doi: 10.1061/(ASCE)0733-9496(2009).

Lempert, R.J., D.G. Groves, S.W. Popper, and S.C. Bankes, 2006. A general, analytic method for generating robust strategies and narrative scenarios. *Management Science* 52(4):514–528.

Lewis, W.M. Jr. (ed.), 2003. *Water and Climate in the Western United States,* Boulder, CO: University Press of Colorado, 294 pp.

Luce, C., P. Morgan, K. Dwire, D. Isaak, Z. Holden, and B. Rieman, 2012. *Climate Change, Forests, Fire, Water, and Fish: Building Resilient Landscapes, Streams, and Managers.* Gen. Tech. Rep. RMRS-GTR-290. Fort Collins, CO: U.S. Department of Agriculture, Forest Service, Rocky Mountain Research Station.

Lund, J., B. Lord, W. Fleenor, and A. Willis, 2014. Drought Curtailment of Water Rights—Problems and Technical Solutions. Center for Watershed Sciences Report, University of California, Davis.

Lüthi, D., M. Le Floch, B. Bereiter, T. Blunier, J.M. Barnola, U. Siegenthaler, D. Raynaud et al., 2008. High-resolution carbon dioxide concentration record 650,000–800,000 years before present. *Nature 453*(7193):379–382.

MacDonnell, L.J. and T.A. Rice, 2008. Moving agricultural water to cities: The search for smarter approaches. *Hastings West–North West Journal of Environmental Law and Policy* 14:105–158.

Major, D.C. and M. O'Grady, 2010. Adaptation assessment guidebook. *Annals of the New York Academy of Sciences* 1196:229–292.

Mantua, N., I. Tohver, and A. Hamlet, 2010. Climate change impacts on streamflow extremes and summertime stream temperature and their possible consequences for freshwater salmon habitat in Washington State. *Climatic Change* 102(1–2):187–223.

Miles, E.L., A.K. Snover, A.F. Hamlet, B. Callahan, and D. Fluharty, 2000. Pacific Northwest regional assessment: The impacts of climate variability and climate change on the water resources of the Columbia River Basin. *Journal of the American Water Resources Association* 36(2):399–420.

Mount, J.F., 1995. *California Rivers and Streams: The Conflict between Fluvial Process and Land Use.* Berkeley: University of California Press.

National Reclamation Act, 1902. P.L. 57-161, 32 Stat. 388.

National Research Council, 1992. *Water Transfers in the West: Efficiency, Equity, and the Environment.* Washington, D.C.: National Academy Press.

National Research Council, 2009. *Informing Decisions in a Changing Climate.* Washington, D.C.: National Academy Press.

NDWAC (National Drinking Water Advisory Council), 2011. *Climate Ready Water Utilities.* Final Report to the U.S. Environmental Protection Agency. Available at http://water.epa.gov/drink/ndwac/climatechange/upload/CRWU-NDWAC-Final-Report-12-09-10-2.pdf.

NIDIS (National Integrated Drought Information System), 2015. US Drought Portal. Available at http://www.drought.gov.

NOAA (National Oceanographic and Atmospheric Administration), 2015a. Statewide average temperature ranks July 2014–June 2015. Available at http://www.ncdc.noaa.gov/sotc/service/national/statewidetavgrank/201407-201506.gif.

NOAA (National Oceanographic and Atmospheric Administration), 2015b. Greenhouse gases. Available at https://www.ncdc.noaa.gov/monitoring-references/faq/greenhouse-gases.php.
North, D.C., 1991. Institutions. *Journal of Economic Perspectives* 5(1):97–112.
Scripps Institute of Oceanography, 2015. The Keeling Curve. Available at https://scripps.ucsd.edu/programs/keelingcurve/
Seager, R.M., N. Naik, and L. Vogel, 2012. Does global warming cause intensified interannual hydroclimate variability? *Journal of Climate* 25:3355–3372.
Smithsonian, 2015. Human evolution timeline interactive. Available at http://humanorigins.si.edu/evidence/human-evolution-timeline-interactive.
Stakhiv, E.Z., 2011. Pragmatic approaches for water management under climate change uncertainty. *Journal of the American Water Resources Association*. 47:1183–1196.
Stephenson, N., 1998. Actual evapotranspiration and deficit: Biologically meaningful correlates of vegetation distribution across spatial scales. *Journal of Biogeography* 25:855–870.
Stewart, I.T., 2009. Changes in snowpack and snowmelt runoff for key mountain regions. *Hydrological Processes* 23:78–94.
Tarlock, A.D., 2001. The future of prior appropriation in the new West. *Natural Resources Journal*. 41:769–793.
Thompson, M.P., J. Scott, P.G. Langowski, J.W. Gilbertson-Day, J.R. Haas, and E.M. Bowne, 2013. Assessing watershed-wildfire risks on national forest system lands in the Rocky Mountain region of the United States. *Water* 5(3):945–971.
Tohver, I.M., A.F. Hamlet, and S.-Y. Lee, 2014. Impacts of 21st-century climate change on hydrologic extremes in the Pacific northwest region of North America. *JAWRA Journal of the American Water Resources Association* 50(6):1461–1476.
Trenberth, K.E., 2011. Changes in precipitation with climate change. *Climate Research* 47:123–138.
Western Governors Association, 2012. *Water Transfers in the West: Projects, Trends, and Leading Practices in Voluntary Water Trading*. Denver, CO: Western Governors Association. Available at http://www.westgov.org/component/content/article/102-articles/initiatives/373-water-transfers.
Wilby, R.L. and S. Dessai, 2010. Robust adaptation to climate change. *Weather* 67:180–185.
Wilkinson, C.F., 1991. In memoriam: Prior appropriation 1848–1991. *Environmental Law* 21:v–xviii.
Wilson, C.M., 2012. Improving Water Right Enforcement Authority. Report to the State Water Resources Control Board and the Delta Stewardship Council, September 19. Sacramento, CA.
WMO (World Meteorological Organization), 2014. WMO Greenhouse Gas Bulletin No. 10: The State of Greenhouse Gases in the Atmosphere Based on Global Observations through 2013. Geneva: World Meteorological Organization ISSN 2078-0796. Available at http://www.wmo.int/pages/prog/arep/gaw/ghg/documents/GHG_Bulletin_10_Nov2014_EN.pdf.
WUCA, 2010. Decision Support Planning Methods: Incorporating Climate Change Uncertainties into Water Planning. Report prepared for Water Utility Climate Alliance by E. Means III, M. Laugier, J. Daw, M. Waage and L. Kaatz, January 2010. Available at http://www.wucaonline.org/assets/pdf/actions_whitepaper_012110.pdf
Yates, D. and K. Miller. 2011. *Climate Change in Water Utility Planning: Decision Analytic Approaches*. Denver: The Water Research Foundation.

2

Natural Variability, Anthropogenic Climate Change, and Impacts on Water Availability and Flood Extremes in the Western United States

Daniel R. Cayan, Michael D. Dettinger, David Pierce, Tapash Das, Noah Knowles, F. Martin Ralph, and Edwin Sumargo

CONTENTS

ABSTRACT The western United States (*the West*) undergoes considerable hydrologic variability in response to regional climate fluctuations that are termed *anomalous* by climate scientists because they depart from long-term average conditions. Regional climate fluctuations persist for seasonal to multidecadal durations, usually in association with larger-scale climate patterns. They play a crucial role in determining regional hydrologic variability by affecting trends of important drivers such as precipitation and temperature, sometimes by promoting particular blends of influential weather events. In California and other regions of the West, much of the annual precipitation is delivered by relatively few very large storms, which are usually atmospheric river events. Besides providing its water supply, these storms also drive year-to-year differences in annual precipitation totals, and cause most of the region's floods. During years or multiyear periods when these very large storms are absent, the region may fall into drought. Historically, droughts have had a strong presence in the West, but recent droughts have exhibited unusually warm temperatures, likely a harbinger of dry events in future decades when climate change threatens to make overall conditions even warmer. Other early signs of climate change that have been observed include declines of mountain snowpacks, which supply spring and summer runoff for the region. Along with warmer surface temperatures have come higher elevation freezing levels, more rain and less snow, and earlier snowmelt and earlier snowmelt runoff. Anthropogenic climate changes, which are projected to build as greenhouse gas concentrations rise, would result in further warming and amplified hydrologic changes. Global climate models suggest that precipitation may shift toward fewer overall wet days but somewhat increased extreme storm events. Further shifts in snowpack, runoff, and increased moisture loss to the atmosphere would reduce soil moisture and streamflows in summer. Annual discharge in arid western watersheds may decline, which would exacerbate dry spells. Heavier winter precipitation events and higher elevation rain/snow transition zones would cause greater flood volumes in some mountain catchments by the latter half of the twenty-first century.

2.1 The Highly Variable Precipitation Regime

Shifting climate patterns have long been recognized as influencing seasonal, annual, and longer-term precipitation and other time-aggregated hydrologic measures. Increasingly, studies are revealing how certain patterns of climate variability may condition events on weather timescales, and how individual weather events can influence longer-term climate aggregates. In either case, there are associated effects on hydrologic variability. Thus, in this chapter, we emphasize linkages between climate patterns and weather events, with attendant impacts on terrestrial hydrology. Our overall geographic focus is the western conterminous United States (hereafter "the West"), a region noted for its extreme weather and hydroclimatic variability. This variability is structured around seasonal precipitation regimes. The heaviest precipitation season varies across the western United States, peaking in the winter season along the West Coast, shifting to a spring or summer maximum in the lee of the Rocky Mountains and on the High Plains and to a double-peaked pattern in areas strongly affected by the southwestern summer monsoon. It is important, however, that cool-season precipitation plays an unusually important role in determining hydrologic variability in most of the West, because in most locations, it provides a considerable fraction of the annual total and because much of the precipitation in the warm season simply evaporates. Cool-season precipitation in the Sacramento/San Joaquin watershed explains about 95% of the variance in annual flow. In the Columbia River Basin, it explains about 82% of the variance in annual flow, and even in the Colorado River Basin, it explains 57% of the variability in annual flow (Westerling et al. 2008).

Weather and climate phenomena also cause substantial irregular precipitation variability across all timescales (Figure 2.1). Strong seasonality and interannual and longer-term

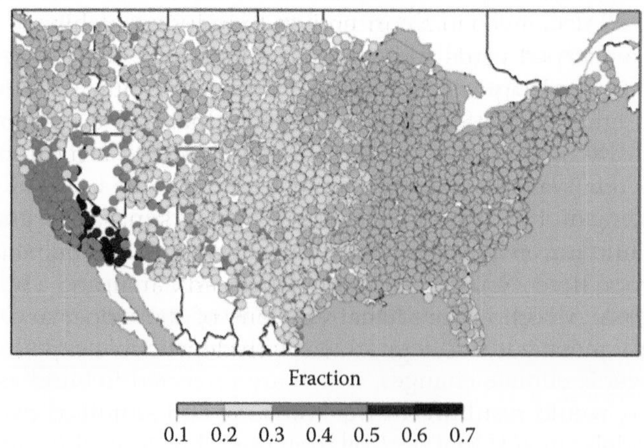

FIGURE 2.1
Coefficients of variation of water year precipitation total at long-term monitoring stations across the conterminous United States, water years 1951–2008. (From Dettinger, M.D. et al., *Water*, 3(2), 445–478, 2011.)

variability are especially pronounced in California, a region whose fluctuating hydroclimate has been the subject of many research investigations, in particular, several by the authors. Thus, a number of this chapter's examples are drawn from the California region.

The coastal states' strongly seasonal precipitation regime derives from the annual cycle of atmospheric circulation over the North Pacific. In late spring, North Pacific storminess subsides and shifts poleward so that conditions along the West Coast are relatively dry throughout summer and early fall. In fall, westerly winds and storminess begin to intensify over the North Pacific, and by winter, these features have migrated southward. These Pacific storms propagate into western North America, albeit sporadically and to varying degrees in different years.

Climate patterns associated with anomalously low or high precipitation over the region have Pacific Ocean roots but extend into North America (e.g., Klein and Bloom 1987; Cayan and Peterson 1989). Changing patterns of anomalous precipitation over the West Coast are modulated seasonally and also quite strongly affected by several important modes of Pacific and North American climate variability that operate at different timescales (Barnston and Livezey 1987). Some of these modes have a strong tropical Pacific influence. These include intraseasonal fluctuations associated with the Madden–Julian Oscillation (MJO, the most important mode of tropical intraseasonal variability); the El Niño/Southern Oscillation (ENSO), Earth's dominant seasonal-to-interannual climate variability mode; and longer-term shifts in sea surface temperature (SST) and atmospheric circulation patterns associated with the Pacific Decadal Oscillation (PDO).

- The MJO is an equatorial, propagating pattern of anomalous rainfall, cloud cover, pressure, and wind, occurring mainly in the Indian and Pacific Oceans (e.g., Wheeler and Hendon 2004). It is an eastward propagating coupled ocean–atmosphere mode, and when it reaches certain locations along the tropical corridor, it produces remote atmospheric patterns that can reinforce or divert regional storminess at particular locations along the West Coast (Mo and Higgins 1998a,b).
- ENSO is a coupled ocean–atmosphere pattern of climate variability that features variations in SST over the tropical eastern Pacific Ocean (Sarachik and Cane 2010). ENSO oscillates irregularly at timescales of a few to several years between a warm phase, called El Niño and a cool phase, called La Niña. Although it is seated in the tropical Pacific, ENSO extremes are known to produce anomalous weather and climate patterns in many regions of the globe. When in its warm (El Niño) phase, it affects the winter storm track and intensity over the North Pacific and tends to produce drier-than-normal conditions in the Pacific Northwest and wetter-than-normal conditions in the Southwest (Redmond and Koch 1991; Cayan et al. 1999).
- The PDO is an irregular oscillation of broad-scale North Pacific SSTs, whose warm phase has positive (warmer-than-average, El Niño–like) anomalies in the eastern North Pacific and negative (cooler-than-average, La Niña–like) anomalies in the central and western North Pacific (Mantua et al. 1997). The oscillation from warm to cool and back to warm phases is irregular but happens over time periods of 10–40 years, and there may be shorter-period excursions in between.

ENSO teleconnections to North American and West Coast weather appear to be conditioned by lower-frequency variability (Gershunov and Barnett 1998). For example, greater and broader-scale precipitation anomalies along the West Coast are observed in Los Niños when the PDO is also in its warm phase, having anomalously warm SSTs in the eastern North Pacific and anomalously cool SSTs in the central and western North Pacific. However, MJO, ENSO, and PDO are not the only contributions to a broad mix of western US precipitation patterns. Precipitation occurs in response to many synoptic patterns, which may be favored or discouraged by different climate patterns (e.g., Weaver 1962; Mo and Higgins 1998b; Cayan et al. 1999; Ralph and Dettinger 2011; Guan et al. 2013; Jones and Carvalho 2014). Shifts, even subtle ones, in the center of action of weather and climate patterns produce important changes in the distribution of precipitation over the coast and interior western United States (Klein and Bloom 1987; Cayan and Peterson 1989; Dettinger et al. 1998; Mo and Higgins 1998b).

In California, the narrow window of storminess (typically between November and March) that supplies most of the year's precipitation is heavily affected by climate fluctuations. In other areas, such as the eastern part of the United States, each season has the potential to contribute significantly to the annual total at that location. But in California, the warm-season months are generally so dry that there is little chance to compensate for a dry winter. On the other hand, if storm conditions are very active during the winter season, the annual supply and generally other hydrologic measures such as annual discharge will be in excess of the long-term average. California's annual precipitation totals routinely vary from as little as 50% to greater than 200% of long-term averages, greater interannual variability than at than most other locations in the United States (Figure 2.1).

Among the winter storms that occur in a given year, the presence or absence of very large storms is a strong determinant of that year's overall precipitation and is a major source of the year-to-year variability observed at each location. Considering the entire region from the Rocky Mountains to the West Coast, large storms occur throughout the year, but in the far West, the largest storms are heavily weighted toward the winter season (Figure 2.2a). Somewhat surprisingly, the heaviest precipitation totals during the largest winter storms, in favored moist locations on windward slopes in the far West, are comparable to most of the heaviest tropical storm precipitation totals from US Gulf Coast locations (Figure 2.2b) (Ralph and Dettinger 2012). The disproportionately large contribution of a few large storms to annual precipitation is especially pronounced in California, where an impressive fraction of the year's precipitation and also its interannual variation arises from the relatively small number of large storms (Figure 2.3). Many of the floods and much of the water supply in the far western states are attributable to "atmospheric river" (AR) storms (e.g., Ralph et al. 2006, 2013; Neiman et al. 2008, 2011, 2013; Leung and Qian 2009; Dettinger et al. 2011; Ralph and Dettinger 2012; Cordeira et al. 2013; Guan et al. 2013). If particularly unfavorable large-scale patterns persist, and a few large storms happen to bypass California in a given winter, precipitation totals are much reduced, often leading to drought (Dettinger and Cayan 2014).

As underscored by recent dry spells, drought is a familiar occurrence in the West. In the Southwest United States, several intermittent dry spells have been described, both in the instrumental and in the preinstrumental record (e.g., Woodhouse et al. 2010; Cook et al. 2014). The areal extent of drought over the Southwest during 2001–2010 was the second largest observed for any decade from 1901 to 2010 (Hoerling et al. 2013). Streamflow totals in the four major drainage basins of the Southwest were 5–37% lower during 2001–2010 than their average flows in the twentieth century. Persistent dryness in the Colorado Basin and more recently in California and Nevada has occurred during the last two decades (MacDonald 2010; Borsa et al. 2014; California Department of Water Resources 2015).

Shukla et al. (2015) demonstrate that the high variability and unusual probability distribution of precipitation in California in comparison to other regions of the conterminous United States translates to some important drought-related properties: California has

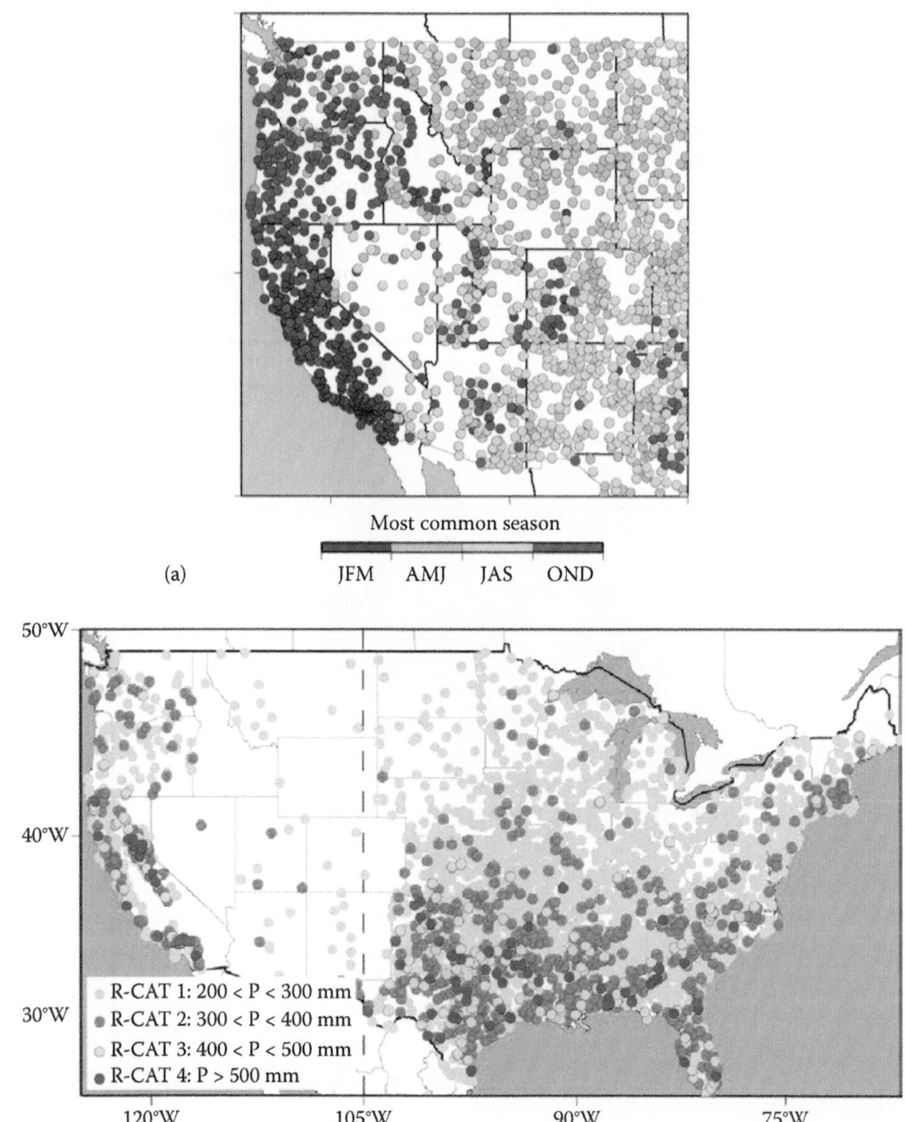

FIGURE 2.2
(a) Seasonality of extreme precipitation events based on daily precipitation totals from long-term monitoring stations (dots) with records for 30 years or longer. The dots are color coded, corresponding to the season (JFM = January–March; AMJ = April–June; JAS = July–September; OND = October–December) when more of the top 10 daily precipitation events occurred than any other season. (From Ralph, F.M. et al., *J. Contemp. Water Res. Educ.*, 153(1), 16–32, 2014.) (b) Maximum 3-day precipitation totals at 5877 cooperative observer network (COOP) stations in the conterminous United States during 1950–2008. Each site used here had to have at least 30 years of records. (From Ralph, F.M., and Dettinger, M.D., *Bull. Am. Meteorol. Soc.*, 93, 783–790, 2012.)

lower precipitation amounts (relative to their overall mean values) for a given low percentile level yet higher percentile levels for a fixed 75% of normal precipitation amount, for example.

Drought indicators describing the beginning of drought or the recovery from drought are needed by decision makers and the public to detect and assess drought conditions

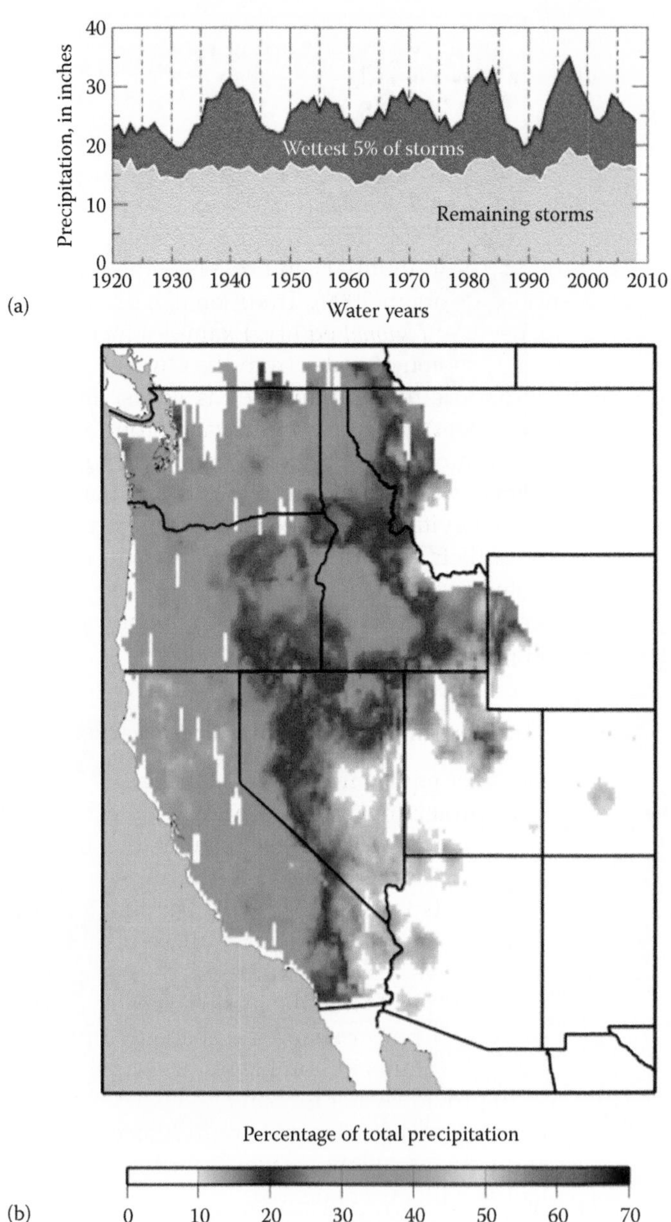

FIGURE 2.3
(a) Five-year moving averages of contributions to water year precipitation from upper 5% (dark gray) and remaining 95% (gray) daily precipitation events in San Francisco Bay delta catchment. (After Dettinger, M., and Cayan, D.R., *San Francisco Estuary Watershed Sci.*, 12(2), 7 pp., 2014.) (b) Contributions to yearly total precipitation from wet-season (November–April) days on which atmospheric rivers made landfall on the West Coast. From 1/8° gridded precipitation based on cooperative weather stations, water years 1998 through 2008. (From Dettinger, M.D. et al., *Water*, 3(2), 445–478, 2011.)

and take action to reduce impacts (Steinemann et al. 2015). Dettinger (2013) found that droughts often end more abruptly than they begin; these sharp endings result from the arrival of an especially wet month via a few very large storms. A survey of the storm types that occurred during "drought-busting" months along the West Coast revealed that a major portion of the heavy precipitation events were produced by landfalling AR events, with the remainder resulting mostly from other forms of persistent low-pressure systems.

Characteristics of individual storms matter greatly in determining the amount and form of the annual supply of precipitation in the West. Ralph et al. (2013) studied landfalling AR storms striking windward slopes of California's coastal mountains and found that the amount of precipitation is governed largely by one primary measure—the amount of moisture transport in the upslope direction. Further, the duration of AR storms has a disproportionate impact on runoff—a doubling of AR duration produced nearly six times greater peak streamflow and more than seven times the storm total runoff volume. Storm tracks and topographic structure in the far West have a strong effect on precipitation downstream, whereby the moisture transport that fuels heavy precipitation events in the Intermountain West is fed through notches in coastal mountains including the Cascade, Sierra Nevada, and Peninsular Mountains of the West Coast (e.g., Alexander et al. 2015; Rutz et al. 2015). Antecedent conditions also matter—for example, when antecedent soil moisture was less than 20%, even heavy rainfall did not lead to significant streamflow (Ralph et al. 2013)—another way in which longer-period patterns are involved in governing hydrologic responses.

Historical records indicate that the heaviest daily precipitation in Sierra Nevada locations occurs during relatively warm storms (Cayan and Riddle 1992; Pandey et al. 1999), and the highest rainfall rates occur when snow lines are highest (White et al. 2010). Warmer storms generate higher runoff—in assessing snow levels and runoff in four different watersheds in California, White et al. (2002) found that a 600 m rise in the freezing level tripled the peak outflow in three of the four basins. Along the West Coast, many of these relatively warm heavy precipitation events occur during AR events (Ralph and Dettinger 2011).

A general pattern across the West is that most of the water supply is derived from precipitation falling in mountainous higher-elevation terrain. In comparison to low-lying upwind locations, precipitation amounts in higher-elevation windward slopes are enhanced by topographic lifting of moist air (Pandey et al. 1999), but due to variations in meteorological conditions such as wind speed and direction, humidity, and stability, the effect of this mechanism differs from case to case. Dettinger et al. (2004), studying orographic enhancement of winter storm precipitation in the Sierra Nevada, found that the orographic ratio (OR) between higher- and lower-elevation precipitation gauges ranged from nearly equal amounts of precipitation to 10 or more times as much precipitation at the higher altitudes. Strongly orographic storms in the Sierra Nevada were found to most commonly have winds that transport water vapor across the range from a nearly westerly direction, which contrasts with wind directions associated with the overall wettest storms, whose wind directions were somewhat more southerly. High-OR storms were found to be somewhat warmer than storms with very low OR values, yielding storm-time snow lines 150–300 m higher during high-OR storms. In the Sierra Nevada, La Niña winters have produced more storms with high ORs than have El Niño winters. Winters during negative (La Niña–like) PDO conditions tend to yield slightly more storms with large ORs than do positive-PDO winters. One important atmospheric dynamics phenomenon that affects the OR is the presence or absence of a barrier jet, either coastal (e.g., Neiman et al. 2002) or the Sierra barrier jet (e.g., Lundquist et al. 2010; Neiman et al. 2014). This occurs because the barrier jet is a virtual obstacle that displaces upward air motion (and thus condensation and precipitation) at a position "upwind" from the actual terrain (e.g., Neiman et al. 2002; Kingsmill et al. 2013).

Recently, Luce et al. (2013) found that, in the Pacific Northwest, a spate of diminished westerly winds in recent decades has likely reduced orographic precipitation enhancement in higher elevations and thus contributed to reduced snowpack and declining streamflow. This mechanism has not previously been emphasized but warrants consideration, along with regional changes in temperature and other mechanisms, in explaining observed hydrologic changes as well as anticipating future changes.

2.2 Temperature-Related Changes

Climate warming in recent decades has affected multiple aspects of the hydrologic system in the West (e.g., Mote 2006; Barnett et al. 2008; Hoerling et al. 2013). Although temperature instruments, especially in mountains, have generally not been installed for purposes of tracking climate changes, a widespread network of stations indicate a broad footprint of warming surface temperatures over the West (e.g., Bonfils et al. 2008). The upward trend in the region has roughly paralleled the warming of global average surface temperatures (Figure 2.4b), which have risen steadily since the late 1970s at a rate of +0.15°C to +0.20°C per decade (Hansen et al. 2010). Hoerling et al. (2013) reported that annual averaged temperatures over the Southwest for 2001–2010 were 0.8°C warmer than the 1901–2000 average, noting, "Key features of a warming Southwest appear robustly across various data sets and methods of analysis" (p. 76), and "The period since 1950 has been warmer in the Southwest than any comparable period in at least 600 years, based on paleoclimatic reconstructions of past temperatures" (p. 75).

Greater warming has occurred in the nighttime than in the daytime (Bonfils et al. 2008), and temperatures during California heat waves have become increasingly expressed in warmer nighttime temperatures than in the past (Gershunov et al. 2009). Looking throughout the West, warming has occurred in each season (e.g., Cordero et al. 2011). Importantly, wet days have warmed as much or more than dry days during the interval from 1949 to 2004, which helps to explain a trend toward lower fractions of snowfall relative to rainfall in mountain settings across the West (Knowles et al. 2006). One key aspect of the warming in recent years can be described as a shift toward earlier warm weather—Regonda et al. (2005) found that, over 1950–1999, the date of earliest occurrence of a persistent (7 days or longer) warm spell has advanced by 5 to more than 15 days earlier in many parts of the West.

Warming at the surface has been accompanied by changes in the altitude of freezing in the atmosphere, which in recent years has been about 200 m higher than during 1950–1975 (Figure 2.4a). The freezing altitude influences climate and hydrologic structure in multiple ways, including the elevation of the rain/snow transition, frozen versus thawed ground, the duration of snowpack, and various ecological functions.

Drought has been a prominent feature of climate in the West in the past two decades, and there is also plentiful evidence of dry spells in the historical instrumental record and well before that. Paleoclimatologists find widespread evidence of drought in sources ranging from tree-ring widths (Woodhouse et al. 2010) to lacustrine and riverine deposits to submerged trees or buried stumps. This includes several-decades-long and extremely severe "megadroughts" during the past 2000 years (Stine 1994). However, a distinguishing feature in recent droughts has been the occurrence of unusually warm conditions during a period

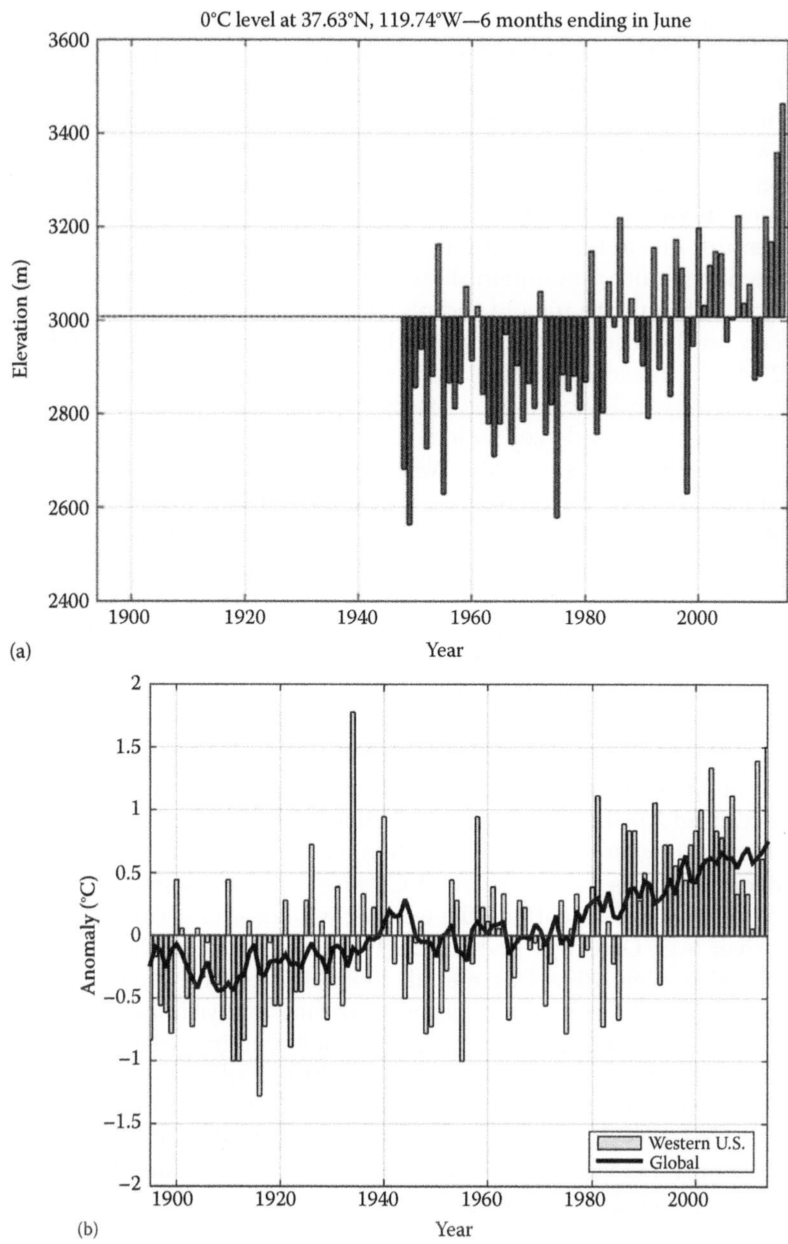

FIGURE 2.4
(a) January–June average freezing level (meters) in the central Sierra Nevada region. (From North American Freezing Level Tracker, Western Regional Climate Center, http://www.wrcc.dri.edu/cwd/products/.) The baseline indicates 1981–2010 mean. (b) Global (black) and western US (light gray) annual surface temperature anomalies between 1895 and 2014. The anomalies are based on the respective 1901–2000 averages. The global anomaly data set is obtained from NOAA National Climatic Data Center (NCDC). The western US anomaly data set is from Desert Research Institute Western Regional Climate Center's (DRI WRCC) WestMap tool.

that is also unusually dry, e.g., in the Southwest (MacDonald 2010) and in California (Seager et al. 2014; California Department of Water Resources 2015). Griffin and Anchukaitis (2014) determined that the 2012–2015 California drought registers as the strongest in the last 1000 years, to a great degree because of the precipitation deficit but also because of the exceptional warmth during recent winters (Figure 2.4) (Seager et al. 2014; Bond et al. 2015).

Other processes also have contributed to hydrologic changes. For example, wind-borne dust deposition in western watersheds has seen a several-fold increase since before the nineteenth century due to human activities (Neff et al. 2008). This darkens snow surfaces and may hasten melt-out of mountain snowpack by several days to weeks (e.g., Painter et al. 2010). Aerosols, from both remote and regional sources, may also be involved through varying effects on cloud seeding and cloud droplet concentrations (Rosenfeld et al. 2008; Ault et al. 2011; Creamean et al. 2013). Although many different mechanisms are likely contributing to the hydrologic changes observed in recent decades, increased winter and spring temperatures are clearly key factors.

Recent hydroclimatic changes in the West can be seen as part of a larger shift in climate taking place over North America as a whole. Since 1950, North America has warmed considerably (Hansen et al. 2010), and remotely sensed observations reveal a large-scale decline in winter and spring snowpack, especially in Canada (Gan et al. 2013). Consistent with this continental decline, the West has experienced a marked increase in temperature, and substantial loss of spring snowpack has occurred in most of its mountainous terrain. These changes, seen from a series of manual snow course observations, shorter series of automated snow sensors, and hydrologic model estimates, have been described in a growing body of research (e.g., Roos 1987; Cayan et al. 2001; Mote et al. 2005; Regonda et al. 2005; Peterson et al. 2008; Pierce et al. 2008; Clow 2010; Kapnick and Hall 2012). Using 1950–1997 data, Mote et al. (2005) removed effects of variable precipitation and found that rising temperatures caused April 1 snow water equivalent (SWE) to decline by more than 30% at sites across the West, where average winter temperatures are relatively warm (December to February temperatures greater than –5°C). Most studies of the reductions of spring snowpack in the West have identified the warmer winters and springs after the mid-1970s as the key driver (e.g., Dettinger and Cayan 1995; Mote et al. 2005; Regonda et al. 2005; Knowles et al. 2006; Mote 2006; Pierce et al. 2008; McCabe and Wolock 2009; Clow 2010; Kapnick and Hall 2012). Associated effects of warming on hydrologic variables include a reduction in the fraction of precipitation falling as snow, an increase in the fraction falling as rain (Figure 2.5) (Knowles et al. 2006), and a shift to earlier flows in snow-dominated rivers from western Canada southward to the southern Rocky Mountains, with streamflow shifted several days earlier than was observed the 1950s and 1960s, as illustrated in Figure 2.6 (Dettinger and Cayan 1995; Cayan et al. 2001; McCabe and Clark 2005; Regonda et al. 2005; Stewart et al. 2005). Furthermore, the fraction of streamflow occurring during winter and early spring has increased (Dettinger and Cayan 1995; Stewart et al. 2005), and the date of peak SWE has shifted to earlier in the season (Hamlet et al. 2005; Kapnick and Hall 2012).

McCabe and Wolock (2009) demonstrated that a Westwide decline in spring SWE is related to temperature variation over the region, at both interannual and decadal timescales, and that reductions in SWE since the 1980s have only partially been counteracted by increases in precipitation. These winter and spring temperature changes are tied, partly, to large-scale atmospheric circulation shifts in the Pacific and North America region (Dettinger and Cayan 1995; Abatzoglou 2011; Johnstone and Mantua 2014), but it is unclear to what extent such circulation shifts may also reflect anthropogenic climate change mechanisms (Hartmann 2015).

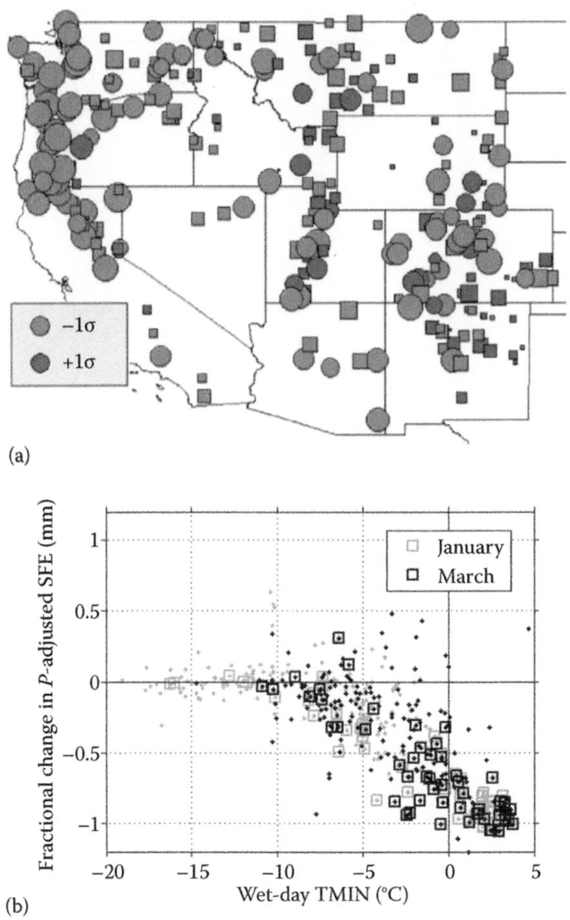

FIGURE 2.5
(a) Linear trends in precipitation falling on snowy days as a fraction of total precipitation (snowfall equivalent/precipitation total [SFE/P]) during November–March 1949–2004: red symbols indicate negative trends (decreasing snowfall fraction), and blue indicates increasing fractions; symbol radius is proportional to magnitude of change over record period, measured in standard deviations of the detrended time series as indicated; circles indicate high trend significance ($P < .05$), and squares indicate lower trend significance. (b) SFE changes (1949–2004) versus average minimum temperature on wet days (TMINw) for January and March. SFE change has been adjusted to remove changes that result from precipitation change over the period. The greater number of very cold (TMINw < −10°C) stations in January compared to March results in less widespread SFE/P declines in January. Statistically significant ($P < .05$) trends are highlighted with squares. (From Knowles, N. et al., *J. Clim.*, 19, 4545–4559, 2006.)

The warmer parts of mountain catchments, primarily low and mid elevations, have exhibited the greatest losses of spring snowpack (e.g., Mote 2006; Kapnick and Hall 2012). However, even the colder, higher elevations of Colorado have exhibited earlier snowmelt, by 2–3 weeks, as shown by Clow (2010), who determined snowmelt directly from a network of snow sensor records from 1978 to 2007. Reductions in accumulated spring snowpack that have occurred were found to be caused, mostly, by losses in the mid to latter portion of the snow accumulation season, when daytime temperatures rise above freezing in most elevations (Kapnick and Hall 2012; Pederson et al. 2013). Hamlet et al. 2005 used model simulations to demonstrate that most of the region-wide decline of spring SWE in the West

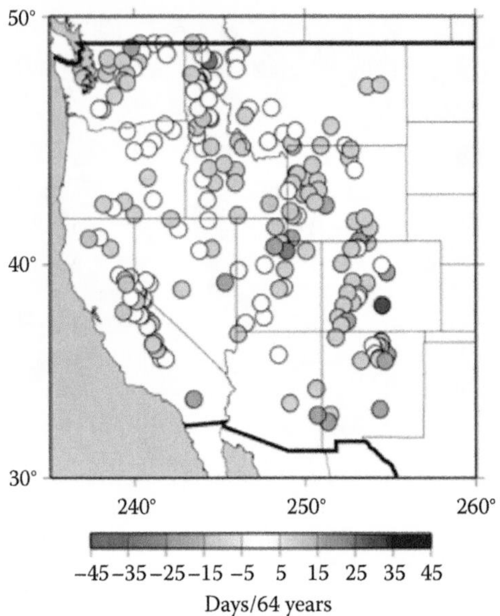

FIGURE 2.6
Change in streamflow timing (days/64 years) as measured by linear trend over 1950–2013 of center of mass (date when half of the annual streamflow has been discharged) for snowmelt-dominated streams in the West. (After Stewart, I.T. et al., *J. Clim.*, 18, 1136–1155, 2005.)

in recent decades has been associated with warming temperatures, rather than changes in precipitation. McCabe and Wollock (2009), in a different modeling exercise, showed that anomalously warm temperatures since about 1980 have produced lower spring snowpack than would otherwise have occurred if temperature had been closer to long-term averages.

If recent warming-related "changes" are simply multidecade fluctuations, they could be expected to revert to cooler conditions as in previous decades, but evidence is mounting that indicates that anthropogenic effects are playing a role. To determine the extent to which recent changes are caused by anthropogenic climate change, efforts have focused on questions of *detection* and *attribution*—in other words, detection of changes that are unusually large in comparison to historical variation and attribution of those changes to natural or man-made sources. Recent studies have employed multicentury natural climate "control" simulations along with observational time histories to address these questions. Results indicate that the recent warming in North America has been caused, to some degree, by the continuing accumulation of greenhouse gases (GHGs) and other human impacts on the climate system (Karoly et al. 2003; IPCC 2013).

In a regional detection and attribution effort that focused on water resource–related changes specific to the West, Bonfils et al. (2008) assessed observed (1950–1999) temperature trends, Pierce et al. (2008) studied changes in spring SWE as a fraction of precipitation (SWE/P) over nine mountainous regions in the West, and Hidalgo et al. (2009) investigated shifts in streamflow timing in the combined flow of the Sacramento and San Joaquin Rivers, the Colorado River at Lees Ferry, and the Columbia River at the Dalles. Combining these measures, Barnett et al. (2008) considered coincident changes in the temperature, spring snowpack, and river discharge timing. These studies indicated that warming-associated changes, including hydrological measures, are *unlikely*, at a high statistical confidence, to have occurred due to natural variations. Furthermore, they concluded that changes in the climate due to anthropogenic GHGs, ozone, and aerosols are causing part of the recent changes. Importantly, precipitation variations tend to be dominated by natural variability (e.g., Hoerling et al. 2010), so detection of long-term changes that might have anthropogenic drivers is unlikely to emerge for many decades (e.g., Pierce and Cayan 2013).

Detection and attribution results were reinforced and broadened by Das et al. (2009), who investigated changes in recent decades across the West, finding that the observed winter temperature and several hydrologic measures have undergone significant trends over considerable parts (37–89%, depending on measure) of the snow-dominated landscape. These observed trends are not likely to have resulted from natural variability alone, as gauged from the distribution of trends produced from a long control simulation. Significant trends toward lesser snow accumulation and earlier runoff were found in a relatively large portion of the Columbia River Basin and to a lesser extent in the California Sierra Nevada and in the Colorado River Basin. The greatest trends occurred in regions with a mean spring temperature close to freezing, where warming might be most effective in changing snow to rain and in causing earlier melt of accumulated snowpack. Das et al. (2009) found that nearly all of the changes that registered as statistically significant were in the sense that is consistent with warming (not cooling) conditions—e.g., earlier runoff and diminished spring snow accumulation.

2.3 Projected Climate Changes

Climate changes caused by projected greenhouse-forced warming would have growing impacts on western US water resource management and distribution (e.g., Barnett et al. 2005; Christensen and Lettenmaier 2007; Udall and Bates 2007; Cayan et al. 2008). A major concern is that more of the annual flow will occur in winter and much of the water in the West that is stored as snow in winter and spring will melt earlier. Limited reservoir capacity and the need for flood control storage makes it difficult to store increased winter runoff, and earlier snowmelt in spring reduces summer inflows to reservoirs. Another concern is that climate change may exacerbate various forms of extreme events, including both droughts and flood flows.

In fact, increasing hydrologic changes over many parts of the West will probably occur (e.g., Lettenmaier and Gan 1990; Barnett et al. 2005; Overpeck et al. 2013; Lukas et al. 2014) because GHGs are almost certain to continue to accumulate in the atmosphere, making further warming highly likely (IPCC 2013). Under both moderate and relatively high (Representative Concentration Pathway [RCP] 8.5) emissions,* surface temperatures projected by an ensemble of downscaled global climate model (GCM) simulations rise by 1°C or more over recent historical averages, by the middle of the twenty-first century (Figure 2.7). Emissions scenarios, of course, matter greatly, but in the first half of the twenty-first century, the warming produced by the high-emissions scenario is not much greater than that of the lower-emissions scenario. By the latter half of the twenty-first century, considerably higher GHG concentrations under the higher-emissions scenario are projected to lead to

* For its Fifth Assessment Report, the Intergovernmental Panel on Climate Change (IPCC) used multiple climate models to estimate the effects of four distinct RCPs. These were constructed by varying assumptions about demographic, technological, economic, and land-use trends to achieve different projected levels of radiative forcing by the year 2100 (about 2.6, 4.5, 6.0, and 8.5 W/m^2).

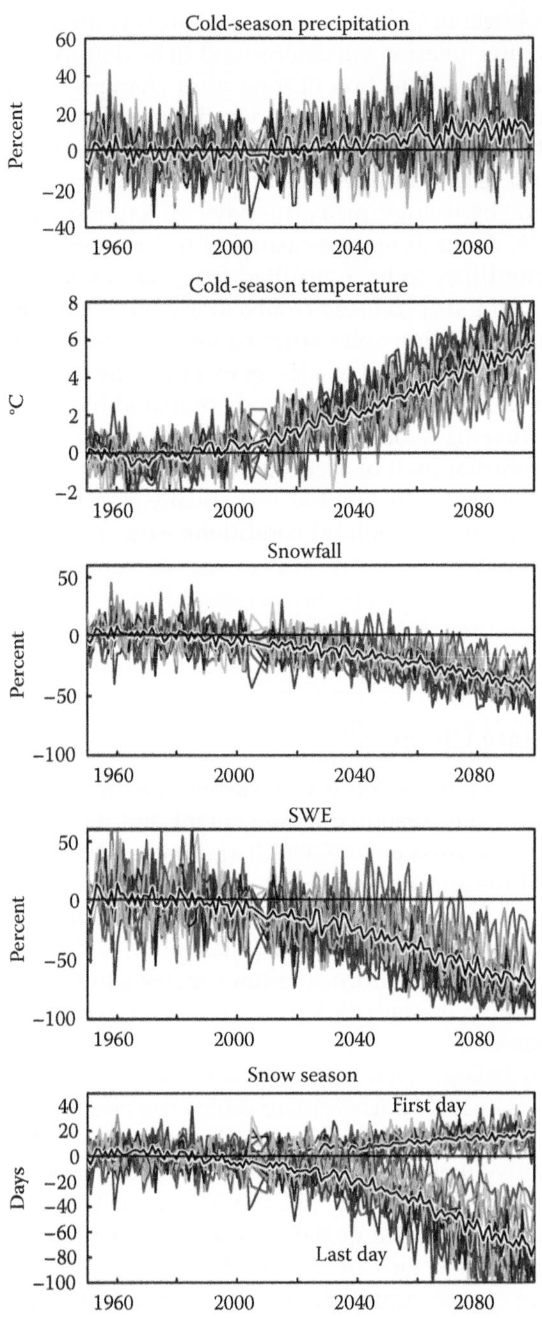

FIGURE 2.7
Observed and projected changes in cold-season (October–March) precipitation and temperature, snowfall, SWE, and the length of snow season from 1950 to 2100, for mountain snow regions of the West. From VIC hydrologic model simulations based upon an ensemble of 13 downscaled RCP 8.5 Coupled Model Intercomparison Project Phase 5 (CMIP5) GCM projections. The bottom figure shows the changes of both the first day (getting later in the fall or early winter) and the last day (getting earlier in the spring) of the snow season. Colors indicate each of the 13 GCMs included in the ensemble so that color envelope shows intermodel spreads. Black contours indicate the multimodel ensemble means. (From Pierce, D.W., and Cayan, D.R., *J. Clim.*, 26, 4148–4167, 2013.)

increasingly greater warming than would occur in the low-emissions scenario. Although temperature projections are broadly consistent across GCMs, precipitation changes seen in simulations by a suite of GCMs are dominated by shorter-timescale natural variability (Deser et al. 2012), considerable fluctuations in regional climate structure, and great diversity across individual GCMs. This leads to considerable differences in precipitation projections. Nonetheless, there is a 70% consensus toward drier conditions in Mexico, extending into the southern portions of the western United States (IPCC 2013; Polade et al. 2014).

Pierce and Cayan (2013) investigated the time required for climate change–driven trends of snow-related variables to emerge from the "noise" of natural variation. They used variable infiltration capacity (VIC) hydrologic model (Liang et al. 1994; Hamlet et al. 2005) simulations of 13 GCMs forced with two representative GHG concentration pathways (RCP 4.5 and RCP 8.5) and calculated linear trends in snow-related hydroclimate measures over the twenty-first century. In addition to rising temperatures, the model projections showed the earliest significant downward trends in the fraction of precipitation that falls as snow (snowfall water equivalent/precipitation total [SFE/P]) and the fraction of snow water equivalent (SWE) retained in the snowpack as of April 1 (SWE/P). In comparison, snowfall, next to precipitation, was the noisiest variable and took the longest time to detectably change. Of the model simulations, 80% showed a significant downward trend in the primary snow indicators by 2030 (Figure 2.7), when averaged over the snow-dominated regions of the West. The RCP 8.5 simulations produced stronger declines and earlier emergence of detectable statistically significant trends compared to those for RCP 4.5. Declining trends in SWE and snowfall were found to emerge earlier and more strongly in regions with warmer cold-season climate (e.g., the Oregon Cascades, Sierra Nevada, and Washington Cascades) and to emerge later in cooler climates (e.g., the Colorado Rockies and Wasatch). The season during which snow cover persists became shorter over the twenty-first century, with the end of the snow season changing (becoming earlier) more quickly than the start of the snow season. According to the simulations, all of these regions will still build a snowpack during some years by the end of the twenty-first century, but in most years, the snowpack will be diminished considerably from present-day levels. Concerning mechanisms involved in snow reduction, the model simulations indicated that as the climate warms, the transition of precipitation from snow to rain plays a more important role in the decline in April 1 SWE than does earlier snowmelt (Figure 2.8) during the historical snow accumulation seasons. In the Sacramento River and San Joaquin River watersheds that feed the San Francisco Bay, runoff changes and associated estuarine salinity effects from climate warming were most strongly driven by shifts in runoff from low to middle snowmelt-dominated elevations, ranging from 1300 to 2700 m (Knowles and Cayan 2004).

In the southwestern United States, several studies have combined GCM and hydrologic models and concluded that streamflow in the Colorado Basin and some other catchments will likely decline in response to climate change (Milly et al. 2005; Christensen and Lettenmaier 2007; Seager et al. 2007; Cayan et al. 2010). As reported by Vano et al. (2013), estimated reductions from climate change impacts range from about –5% to about –20% by midcentury (Hoerling et al. 2010; Das et al. 2011; Vano et al. 2012). Models suggest responses of annual Colorado River discharge to changes in precipitation and temperature that range from approximately 1% to 2% change in flow per 1% increase in precipitation and –5% to –28% change in flow per 1°C increase in temperature (Hoerling et al. 2010; Vano et al. 2012).

The extent to which runoff would decline as climate warms appears to vary considerably across the West. Using estimates from VIC hydrological model experiments, Das et al. (2011) found that runoff and streamflow are more sensitive to warming in the Colorado Basin than in the Columbia River or west-slope Sierra Nevada drainages in California.

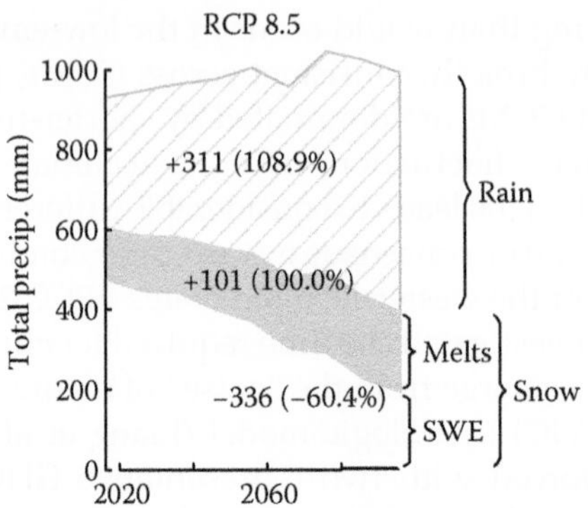

FIGURE 2.8
Projected change in the surface water budget by the end of the twenty-first century in Sierra Nevada based on ensemble mean of VIC model results from 13 downscaled RCP 8.5 GCM simulations. Note that total water (sum of all components) increases over the projected period due to incremental increase in multimodel mean precipitation over the period. Surface water components, accumulated from October 1 to March 31, from rain and snowmelt and stored in snowpacks, are indicated by hatched, gray, and white wedges. (From Pierce, D.W., and Cayan, D.R., *J. Clim.*, 26, 4148–4167, 2013.)

The stronger response of the Colorado Basin to climate warming is associated with the low runoff in proportion to the precipitation that is delivered to this generally arid watershed. Thus, any increase in losses due to increased evapotranspiration and sublimation will have a relatively large impact on runoff and streamflow. In addition to warming-driven reductions in runoff, parts of the West, especially in the south near the Mexican border, could experience diminished precipitation, which would cause further reductions in runoff.

The GCM projections indicate that climate warming will occur throughout all months of the year (IPCC 2007, 2013), but the projections consistently exhibit greater warming in summer than in winter. From a set of downscaled Special Report on Emissions Scenarios (SRES) A2* simulations, Das et al. (2011) found, as an average across several GCMs, about 3°C warming in winter and nearly 5°C warming in summer. The same study used VIC hydrological model experiments to investigate the effect of warming on runoff and streamflow, and found that warming throughout the year produced reductions in annual flows in major western watersheds but that warming in the warmer months had stronger impacts than did warming in the cooler months of the year.

During droughts, when persistent reductions in precipitation occur, the effects of warming may come into even sharper focus. The added adverse impacts of warming add to the concern that, in parts of the West, water supplies may not be able to meet even current levels of demand (Barnett and Pierce 2009a,b). Warming has compounded the effects of precipitation deficits during the recent 2000s drought in the Southwest, that was more or less focused on the Colorado River Basin, and during the 2012–2015 drought in California and neighboring states (California Department of Water Resources 2015). Comparing twenty-first-century Southwest drought characteristics from GCM climate change projections with those in the observed and modeled historical era, Cayan et al. (2010) found that projected future drought episodes became more extreme. As in the historical period, the driest years in the projections almost always occurred in the midst of longer dry periods.

* The IPCC (2000) Special Report on Emissions Scenarios (SRES) defined a set of emission scenarios that were used in GCM simulations for the Third and Fourth IPCC Assessment Reports. The A2 scenario is a fairly high-emissions scenario (https://www.ipcc.ch/pdf/special-reports/spm/sres-en.pdf).

VIC hydrologic model calculations indicated that persistence of depleted soil moisture over the historical record ranged from 4 to 10 years, but in the twenty-first century projections, some of the dry events persisted for more than 10 years. Moreover, summers in several of the projected droughts are even warmer than the already-warm adjacent years that were not in drought—i.e., warming is compounded when the land surface dries out.

Although climate change impacts on annual total precipitation are quite uncertain in much of western North America, the manner in which the precipitation is delivered at shorter timescales is very likely to shift. For California, ensemble mean projected changes in precipitation for the mid- to late-twenty-first century have been shown to favor somewhat wetter winters and drier springs (Pierce et al. 2012, 2013). These winter precipitation increases are largely driven by increases in daily precipitation intensity. In spring, any increases in intensity were overwhelmed by a diminished number of days with precipitation. Polade et al. (2014), in a study using 28 GCMs from the Intergovernmental Panel on Climate Change (IPCC) Fifth Assessment analysis, found that the occurrence of dry days increased by 5–15 per year in California by the end of the twenty-first century under RCP 8.5 emissions, a result that was repeated over each of the Earth's Mediterranean climate regions. On the other hand, although it is projected that the overall frequency of wet days may decrease in many areas of California, there may be increases in the largest precipitation events (Pierce et al. 2013). These shifting distributions of daily precipitation affect projections of annual total amounts. Pierce et al. (2013) found that GCM-to-model disagreement regarding projected changes in California's annual precipitation were mostly attributed to the varying degrees of change in the frequency or intensity of the relatively few heavy precipitation events each year.

An important consideration about climate warming and changes in event-scale weather characteristics is how they might conspire to impact flood flows (Hamlet and Lettenmaier 2007). Mantua et al. (2010) investigated watershed changes in the Pacific Northwest under scenarios of climate change. In the latter half of the twenty-first century, they found that watersheds in the current climate that are *transitional* (both rain and snow runoff) would become more purely rainfall runoff landscapes, resulting in increased winter flood frequency and magnitudes. Tohver et al. (2014) found similar patterns using a more sophisticated downscaling technique and further highlighted the role of rising snow lines and increasing contributing basin area in basins with the largest increases in flood risk. Using dynamic downscaling techniques, Salathé et al. (2014) showed that flood impacts in rain-dominant basins in the Pacific Northwest may be much larger than predicted by previous studies due to increasing intensity of ARs and orographic effects on the west slopes of the Cascades not captured by statistical downscaling. Das et al. (2013) used outputs from 16 historic and projected twenty-first-century conditions under the SRES A2 emissions scenario, downscaled to the Sierra Nevada, to investigate possible climate change effects on flooding and found that the number of days of precipitation did not change over the twenty-first century. However, there was an increase in the most intense precipitation events. Additionally, warming projected over this period produced a greater proportion of precipitation falling as rain instead of snow, amplifying observed changes already occurring during the last few decades, as shown by Knowles et al. (2006). Using VIC hydrological model simulations whose input was the downscaled GCM simulations, Das et al. (2013) found that by the end of the twenty-first century, all 16 climate projections yielded larger floods (return periods ranging from 2 to 50 years) for both the Northern Sierra Nevada and Southern Sierra Nevada. The importance of shorter-period phenomena is underscored by the fact that there was a consensus of increasing flood magnitudes produced by the model runs, despite approximately half of the projections having reduced mean precipitation amounts, relative to the twentieth-century historic period (Figure 2.9).

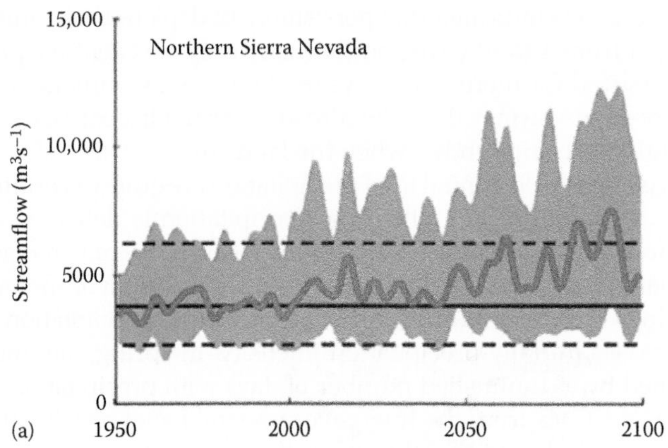

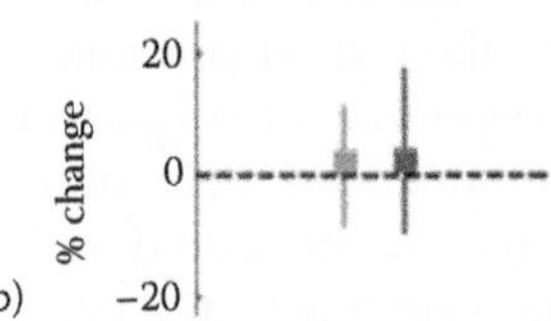

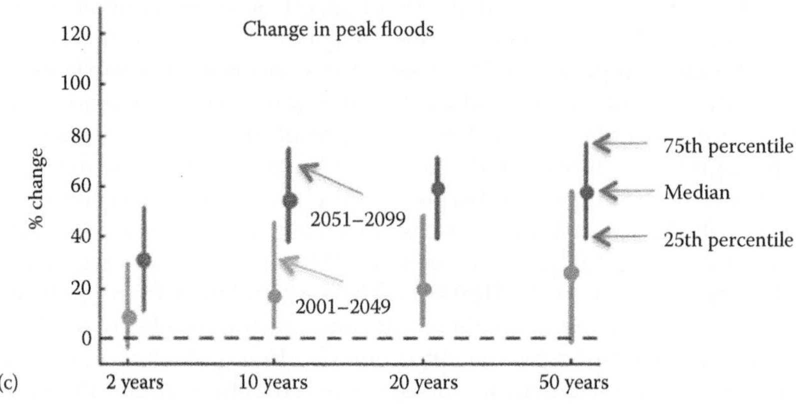

FIGURE 2.9
(a) VIC simulated northern Sierra Nevada annual maximum 3-day streamflow increase found in downscaled output from 16 GCMs, run under SRES A2 emissions scenario. Multimodel median is shown by thick gray curved line; envelope of 3-day maximum flows between 25th and 75th percentiles is shown by gray shading. Horizontal lines represent historical median (solid black line) and 25th and 75th percentiles (dotted black lines) from 1951 to 1999. Results are smoothed using low-pass filter. (b) Very little change in annual total discharge is shown by percentage changes (relative to 1951–1999) in mean annual streamflow from VIC simulations as simulated by 16 downscaled GCMs. (c) Increases (in percent) of flood magnitudes for selected return periods from same model simulations as in upper panels. Changes in the period 2001–2049 (gray) and 2051–2099 (black) are shown side by side. For each of the return periods, filled squares show ensemble medians, and vertical whiskers extend from 25th to the 75th percentile of the 16 GCM simulations. (From Das, T. et al., *J. Hydrol.*, 501, 101–110, 2013.)

2.4 Conclusions

The western United States (the West) has a remarkably varied hydroclimate, in terms of temporal variability, spatial diversity, and the range of projected futures that it faces. Recent research demonstrates how a surprisingly large portion of this variation is due to the timing and high intensity of a relatively small number of extreme weather events. Other mechanisms may be more important than is presently recognized, including changes in wind flows that drive orographic precipitation and the varying effects of aerosols in the atmosphere and dust and soot deposited at the surface. These emerging findings add incentive to better understand processes that operate at the confluence of weather and climate, and sharpen focus upon forecasting at few-day to several-day lead times.

In addition to natural variation from synoptic to multidecadal timescales, conditions in the West appear to be undergoing long-term changes. An interconnected set of hydrologic shifts, including more rain and less snow, diminished spring snowpack, and earlier mountain runoff, has been observed in snow-dominated watersheds in response to warmer winters and springs since the mid-1970s. Although the region exhibits natural climate variability on decadal timescales and long droughts have occurred in the past, there is a striking similarity of observed trends to those that are projected under climate warming. This, combined with evidence from a series of detection and attribution studies, indicates that the hydrologic changes are, to some degree, the early phase of a response to anthropogenic climate change. Exceptionally warm dry spells in the Colorado Basin and over the West Coast during the past decade add to concerns about changing climate.

The projected effects of increasing concentrations of atmospheric GHGs on western US hydrology are substantial, even under moderate scenarios of climate change. Under higher (SRES A2 or RCP 8.5) scenarios, these changes and impacts would be extremely challenging. The region's water supply and its vulnerability to flood hazards depend on high-volume precipitation events, so understanding the disposition of the region's major storm events under climate change is vital in preparing for future impacts.

Many of the observed changes in the region's hydroclimate have been discovered from time series data of opportunity developed from strings of measurements that were not designed for climate purposes. Sustaining these traditional observations is vital to track and evaluate further changes. Additionally, to understand processes driving fluctuating and changing hydroclimate and to track long-term changes, a new cohort of carefully designed observations and monitoring networks is needed. Clearly, besides better surface observations, an evolving description of the 3-D atmospheric structure is required to explain how climate varies and may change. The Cal Water program of field studies (Ralph et al. 2015) has explored both the AR component and also emerging aerosol–cloud–precipitation dimensions of this challenge, and has led to some of the findings presented herein. Field campaigns like Cal Water, along with other institutionalized monitoring efforts to collect major new data sets, will be critical for evaluating climate model representation of these phenomena and for interpreting the underlying physical processes. Elucidating changes in specific physical processes will make detection and attribution of overall system changes that much more certain and will make those detections possible earlier.

Acknowledgments

The lead author was supported by the California Energy Commission, the US Geological Survey (USGS)–Sponsored Southwest Climate Science Center, and the National Oceanic and Atmospheric Administration (NOAA) Regional Integrated Science Applications (RISA) program through the California Nevada Climate Applications Program (CNAP).

References

Abatzoglou, J. T. 2011. Influence of the PNA on declining mountain snowpack in the western United States. *Int. J. Climatol.* 31(8):1135–1142, doi:10.1002/joc.2137.

Alexander, M. A., J. D. Scott, D. Swales, M. Hughes, K. Mahoney, and C. A. Smith. 2015. Moisture pathways into the U.S. intermountain west associated with heavy winter precipitation events. *J. Hydrometeor.* 16:1184–1206.

Ault, A., C. Williams, A. White, P. Neiman, J. Creamean, C. Gaston, M. Ralph, and K. Prather. 2011. Detection of Asian dust in California orographic precipitation. *J. Geophys. Res.—Atmospheres* 116:D16205, doi:10.1029/2010JD015351.

Barnett, T. P., and D. W. Pierce. 2009a. Sustainable water deliveries from the Colorado River in a changing climate. *Proc. Natl. Acad. Sci.* 106(18):7334–7338, doi:10.1073/pnas.0812762106.

Barnett, T. P., and D. W. Pierce. 2009b. Reply to comment by J. J. Barsugli et al. on When will Lake Mead go dry? *Water Resour. Res.* 45:W09602, doi:10.1029/2009WR008219.

Barnett, T. P., J. C. Adam, and D. P. Lettenmaier. 2005. Potential impacts of a warming climate on water availability in snow-dominated regions. *Nature* 438:303–309.

Barnett, T. P., D. W. Pierce, H. Hidalgo, C. Bonfils, B. Santer, T. Das, G. Bala et al. 2008. Human-induced changes in the hydrology of the western United States. *Science* 316:1080–1083.

Barnston, A. G., and R. E. Livezey. 1987. Classification, seasonality and persistence of low-frequency atmospheric circulation patterns. *Mon. Wea. Rev.* 115:1083–1126.

Bond, N. A., M. F. Cronin, H. Freeland, and N. Mantua. 2015. Causes and impacts of the 2015 warm anomaly in the NE Pacific. *Geophys. Res. Lett.* 42(9):3414–3420, doi: 10.1002/2015GL063306.

Bonfils, C., D. W. Pierce, B. D. Santer, H. Hidalgo, G. Bala, T. Das, T. Barnett et al. 2008. Detection and attribution of temperature changes in the mountainous western United States. *J. Climate* 21:6404–6424, doi:10.1175/2008JCLI2397.1.

Borsa, A. A., D. C. Agnew, and D. R. Cayan. 2014. Ongoing drought-induced uplift in the western United States. *Science* 345(6204):1587–1590, doi:10.1126/science.1260279.

California Department of Water Resources. 2015. California's Most Significant Droughts: Comparing Historical and Recent Conditions. State of California, February 2015. Available at http://www.water.ca.gov/waterconditions/docs/California_Signficant_Droughts_2015_small.pdf.

Cayan, D. R., and D. H. Peterson. 1989. The influence of North Pacific atmospheric circulation on streamflow in the West. *Geophys. Monograph 55: Aspects of Climate Variability in the Pacific and the Western Americas*, December 1989, 375–397.

Cayan, D. R., and L. Riddle. 1992. Atmospheric Circulation and Precipitation in the Sierra Nevada: Proceedings, International Symposium on Managing Water Resources during Global Change. American Water Resources Association, Reno, Nevada, November 1–5.

Cayan, D. R., K. T. Redmond, and L. G. Riddle. 1999. ENSO and hydrologic extremes in the western United States. *J. Clim.* 12:2881–2893.

Cayan, D. R., S. Kammerdiener, M. D. Dettinger, J. M. Caprio, and D. H. Peterson, 2001. Changes in the onset of spring in the western United States. *Bull. Am. Met Soc.* 82(3):399–415.

Cayan, D. R., E. P. Maurer, M. D. Dettinger, M. Tyree, and K. Hayhoe. 2008. Climate change scenarios for the California region. *Clim. Change* 87:S21–S42.

Cayan, D. R., T. Das, D. W. Pierce, T. P. Barnett, M. Tyree, and A. Gershunov. 2010. Future dryness in the southwest US and the hydrology of the early 21st century drought. *PNAS* 107(50):21271–21276. http://www.pnas.org/cgi/doi/10.1073/pnas.0912391107.

Christensen, N. S., and D. P. Lettenmaier. 2007. A multimodel ensemble approach to assessment of climate change impacts on the hydrology and water resources of the Colorado River basin. *Hydrol. Earth Syst. Sci.* 3:1–44.

Clow, D. W. 2010. Changes in the timing of snowmelt and streamflow in Colorado: A response to recent warming. *J. Clim.* 23:2293–2306.

Cook, B. I., J. E. Smerdon, R. Seager, and E. R. Cook. 2014. Pan-continental droughts in North America over the last millennium*. *J. Clim.* 27:383–397.

Cordeira, J. M., F. M. Ralph, and B. J. Moore. 2013. The development and evolution of two atmospheric rivers in proximity to Western North Pacific tropical cyclones in October 2010. *Mon. Wea. Rev.* 141:4234–4255.

Cordero, E. C., W. Kessomkiat, J. T. Abatzoglou, and S. A. Mauget. 2011. The identification of distinct patterns in California temperature trends. *Clim. Change* 108:357–382.

Creamean, J. M., K. J. Suski, D. Rosenfeld, A. Cazorla, P. J. DeMott, R. C. Sullivan, A. B. White et al. 2013. Dust and biological aerosols from the Sahara and Asia influence precipitation in the western U.S. *Science* 339:1572–1578, doi:10.1126/science.1227279.

Das, T., H. Hidalgo, D. Cayan, M. Dettinger, D. Pierce, C. Bonfils, T. P. Barnett, G. Bala, and A. Mirin. 2009. Structure and origins of trends in hydrological measures over the western United States. *J. Hydrometeorol.* 10:871–892, doi:10.1175/2009JHM1095.1.

Das, T., D. W. Pierce, D. R. Cayan, J. A. Vano, and D. P. Lettenmaier. 2011. The importance of warm season warming to western U.S. streamflow changes. *Geophys. Res. Lett.* 38:L23403, doi:10.1029/2011GL049660.

Das, T., E. P. Maurer, D. W. Pierce, M. D. Dettinger, and D. R. Cayan. 2013. Increases in flood magnitudes in California under warming climates. *J. Hydrol.* 501:101–110.

Deser, C., A. Phillips, V. Bourdette, and H. Teng. 2012. Uncertainty in climate change projections: The role of internal variability. *Clim. Dynam.* 38:527–547.

Dettinger, M. 2011. Climate change, atmospheric rivers, and floods in California—A multimodel analysis of storm frequency and magnitude changes, *J. Am. Water Resour. Assoc.* 47(3):514–523.

Dettinger, M. D. 2013. Atmospheric rivers as drought busters on the US west coast. *J. Hydrometeorol.* 14:1721–1732, doi:10.1175/JHM-D-13-02.1.

Dettinger, M. D., and D. R. Cayan. 1995. Large-scale atmospheric forcing of recent trends toward early snowmelt runoff in California. *J. Clim.* 8:606–623.

Dettinger, M., and D. R. Cayan. 2014. Drought and the California delta—A matter of extremes. *San Francisco Estuary and Watershed Science* 12(2), 7 pp.

Dettinger, M. D., D. R. Cayan, H. F. Diaz, and D. M. Meko. 1998. North–south precipitation patterns in western North America on interannual-to-decadal time scales. *J. Clim.* 11(12):3095–3111.

Dettinger, M., K. Redmond, and D. Cayan. 2004. Winter orographic precipitation ratios in the Sierra Nevada—Large-scale atmospheric circulations and hydrologic consequences. *J. Hydrometeorol.* 5:1102–1116.

Dettinger, M. D., F. M. Ralph, T. Das, P. J. Neiman, and D. R. Cayan. 2011. Atmospheric rivers, floods and the water resources of California. *Water* 3(2):445–478, doi:10.3390/w3020445.

Gan, T. Y., R. G. Barry, M. Gizaw, A. Gobena, and R. Balaji. 2013. Changes in North American snowpacks for 1979–2007 detected from the snow water equivalent data of SMMR and SSM/I passive microwave and related climatic factors. *J. Geophys. Res. Atmos.* 118:7682–7697, doi:10.1002/jgrd.50507.

Gershunov, A., and T. P. Barnett. 1998. Interdecadal modulation of ENSO teleconnections. *Bull. Am. Meteorol. Soc.* 79:2715–2725, 10.1175/1520-0477(1998)079<2715:imoet>2.0.co;2

Gershunov, A., D. R. Cayan, and S. F. Iacobellis. 2009. The Great 2006 heat wave over California and Nevada: Signal of an increasing trend. *J. Clim.* Early online release, doi:10.1175/2009JCLI2465.1.

Griffin, D., and K. J. Anchukaitis. 2014. How unusual is the 2012–2014 California drought? *Geophys. Res. Lett.* doi:10.1002/2014GL062433.

Guan, B., N. P. Molotch, D. E. Waliser, E. J. Fetzer, and P. J. Neiman. 2013. The 2010/2011 snow season in California's Sierra Nevada: Role of atmospheric rivers and modes of large-scale variability. *Water Resour. Res.* 49:6731–6743.

Hamlet, A. F., and D. P. Lettenmaier. 2007. Effects of 20th century warming and climate variability on flood risk in the western U.S. *Water Resour. Res.* 43:W06427.

Hamlet, A. F., P. W. Mote, M. P. Clark, and D. P. Lettenmaier. 2005. Effects of temperature and precipitation variability on snowpack trends in the western United States. *J. Clim.* 18: 4545–4561.

Hansen, J., R. Ruedy, M. Sato, and K. Lo. 2010. Global surface temperature change. *Rev. Geophys.* 48:RG4004, doi:10.1029/2010RG000345.

Hartmann, D. L. 2015. Pacific sea surface temperature and the winter of 2014. *Geophys. Res. Lett.* 42(6):1894–1902, doi:10.1002/2015GL063083.

Hidalgo, H. G., T. Das, M. D. Dettinger, D. R. Cayan, D. W. Pierce, T. P. Barnett, G. Bala et al. 2009. Detection and attribution of streamflow timing changes to climate change in the western United States. *J Clim.* 22:3838–3855.

Hoerling, M. P., J. Eischeid, and J. Perlwitz. 2010. Regional precipitation trends: Distinguishing natural variability from anthropogenic forcing. *J. Clim.* 23:2131–2145.

Hoerling, M., M. Dettinger, K. Wolter, J. Lukas, J. Eischeid, R. Nemani, B. Liebmann, and K. Kunkel. 2013. Present and climate—Evolving conditions. Chapter 5 in Garfin, G., Jardine, A., Merideth, R., Black, M., and LeRoy, S. (eds.), *Assessment of Climate Change in the Southwest United States.* Washington, DC: Island Press, 74–100.

IPCC. 2007. *Climate Change 2007: The Physical Science Basis.* Working group I contribution to the fourth assessment report of the Intergovernmental Panel on Climate Change. Cambridge University Press, Cambridge, United Kingdom and New York, USA. 996 pp.

IPCC. 2013. *Climate Change 2013: The Physical Science Basis.* Contribution of Working Group I to the Fifth Assessment Report of the Intergovernmental Panel on Climate Change [Stocker, T. F., D. Qin, G.-K. Plattner, M. Tignor, S. K. Allen, J. Boschung, A. Nauels, Y. Xia, V. Bex and P. M. Midgley (eds.)]. Cambridge University Press, Cambridge, United Kingdom and New York, 1535 pp., doi:10.1017/CBO9781107415324.

Johnstone, J. A., and N. J. Mantua. 2014. Atmospheric controls on northeast Pacific temperature variability and change, 1900–2012. *Proc. Natl. Acad. Sci. USA* 111(40):14360–14365.

Jones, C., and L. M. V. Carvalho. 2014. Sensitivity to Madden-Julian Oscillation variations on heavy precipitation over the contiguous United States. *Atmos. Res.* 147–148:10–26, doi:10.1016/j.atmosres.2014.05.002.

Kapnick, S., and A. Hall. 2012. Causes of recent changes in western North American snowpack. *Clim. Dyn.* 38:1885–1899.

Karoly, D. J., K. Braganza, P. A. Stott, J. M. Arblaster, G. A. Meehl, A. J. Broccoli, and K. W. Dixon. 2003. Detection of human influence on North American climate. *Science* 302:1200–1203.

Kingsmill, D. K., P. J. Neiman, B. J. Moore, M. Hughes, S. E. Yuter, and F. M. Ralph. 2013. Kinematic and thermodynamic structures of Sierra barrier jets and overrunning atmospheric rivers during a land-falling winter storm in Northern California. *Mon. Wea. Rev.* 141:2015–2036.

Klein, W. H., and H. J. Bloom. 1987. Specification of monthly precipitation over the United States from the surrounding 700mb height field. *Mon. Wea. Rev.* 115:2118–2132.

Knowles, N., and D. R. Cayan. 2004. Elevational dependence of projected hydrologic changes in the San Francisco estuary and watershed. *Clim. Change* 62:319–336.

Knowles, N., M. D. Dettinger, and D. R. Cayan. 2006. Trends in snowfall versus rainfall in the western United States. *J. Clim.* 19:4545–4559.

Lettenmaier, D. P., and T. Y. Gan. 1990. Hydrologic sensitivities of the Sacramento–San Joaquin River Basin, California, to global warming. *Water Resour. Res.* 26:69–86.

Leung, L. R., and Y. Qian. 2009. Atmospheric rivers induced heavy precipitation and flooding in the western U.S. simulated by the WRF regional climate model. *Geophys. Res. Lett.* 36:L03820, doi:10.1029/2008GL036445.

Liang, X., D. P. Lettenmaier, E. F. Wood, and S. J. Burges. 1994. A simple hydrologically based model of land surface water and energy fluxes for GSMs. *J. Geophys. Res.* 99(D7):14415–14428.

Luce, C. H., J. T. Abatzoglou, and Z. A. Holden. 2013. The missing mountain water: lower westerlies decrease orographic enhancement in the Pacific northwest USA. *Science* 342(6164):1360–1364, doi:10.1126/science.1242335.

Lukas, J., J. Barsugli, N. Doesken, I. Rangwala, and K. Wolter. 2014. Climate Change in Colorado: A Synthesis to Support Water Resources Management and Adaptation. A Report for the Colorado Water Conservation Board. Boulder, CO: Western Water Assessment, p. 114.

Lundquist, J. D., J. R. Minder, P. J. Neiman, and E. M. Sukovich. 2010. Relationships between barrier jet heights, orographic precipitation gradients, and streamflow in the northern Sierra Nevada. *J. Hydrometeor.* 11(5):1141–1156.

MacDonald, G. 2010. Water, climate change, and sustainability in the Southwest. *Proc. Natl. Acad. Sci.* 107:21256–21262.

Mantua, N. J., S. R. Hare, Y. Zhang, J. M. Wallace, and R. C. Francis. 1997. A Pacific interdecadal climate oscillation with impacts on salmon production. *Bull. Am. Meteorol. Soc.* 78:1069–1079.

Mantua, N., I. Tohver, and A. F. Hamlet. 2010. Climate change impacts on streamflow extremes and summertime stream temperature and their possible consequences for freshwater salmon habitat in Washington State. *Clim. Change* 102(1–2):187–223, doi:10.1007/s10584-010-9845-2.

McCabe, G. J., and M. P. Clark. 2005. Trends and variability in snowmelt runoff in the western United States. *J. Hydrometeorol.* 6:476–482.

McCabe, G. J., and D. M. Wolock. 2009. Recent declines in Western U.S. snowpack in the context of twentieth-century climate variability. *Earth Interact.* 13:1–15.

Milly, P. C. D., K. A. Dunne, and A. V. Vecchia. 2005. Global pattern of trends in streamflow and water availability in a changing climate. *Nature* 438:347–350.

Mo, K. C., and R. W. Higgins. 1998a. Tropical influences on California Precipitation. *J. Clim.* 11:412–430.

Mo, K. C., and R. W. Higgins. 1998b. Tropical convection and precipitation regimes in the western United States. *J. Clim.* 11:2404–2423.

Mote, P. W. 2006. Climate-driven variability and trends in mountain snowpack in western North America, *J. Clim.* 19:6209–6220.

Mote, P. W., A. F. Hamlet, M. P. Clark, and D. P. Lettenmaier. 2005. Declining mountain snowpack in western North America. *Bull. Am. Meteorolog. Soc.* 86:39–49.

Neiman, P. J., F. M. Ralph, A. B. White, D. A. Kingsmill, and P. O. G. Persson. 2002. The statistical relationship between upslope flow and rainfall in California's coastal mountains: Observations during CALJET. *Mon. Wea. Rev.* 130:1468–1492.

Neiman, P. J., F. M. Ralph, G. A. Wick, J. Lundquist, and M. D. Dettinger. 2008. Meteorological characteristics and overland precipitation impacts of atmospheric rivers affecting the West Coast of North America based on eight years of SSM/I satellite observations. *J. Hydrometeorol.* 9(1):22–47, doi:10.1175/2007JHM855.1.

Neiman, P. J., L. J. Schick, F. M. Ralph, M. Hughes, and G. A. Wick. 2011. Flooding in western Washington: The connection to atmospheric rivers. *J. Hydrometeorol.* 12:1337–1358, doi:10.1175/2011JHM1358.1.

Neiman, P. J., F. M. Ralph, B. J. Moore, M. Hughes, K. M. Mahoney, J. M. Cordeira, and M. D. Dettinger. 2013. The landfall and inland penetration of a flood-producing atmospheric river in Arizona. Part 1: Observed synoptic-scale, orographic and hydrometeorological characteristics. *J. Hydrometeorol.* 14:460–484.

Neiman, P. J., F. M. Ralph, B. J. Moore, and R. J. Zamora. 2014. The regional influence of an intense Sierra Barrier Jet and landfalling atmospheric river on orographic precipitation in northern California: A case study. *J. Hydrometeorol.* 15:1419–1439.

Neff, J. C., A. P. Ballantyne, G. L. Farmer, N. M. Mahowald, J. L. Conroy, C. C. Landry, J. T. Overpeck, T. H. Painter, C. R. Lawrence, and R. L. Reynolds. 2008. Increasing eolian dust deposition in the western United States linked to human activity. *Nature—Geosciences*. 1(3):189–195, doi:10.1038/ngeo133.

Overpeck, J., G. Garfin, A. Jardine, D. E. Busch, D. Cayan, M. Dettinger, E. Fleishman, A. Gershunov et al. 2013. Summary for decision makers. In Garfin, G., Jardine, A., Merideth, R., Black, M., and LeRoy, S. (eds.), *Assessment of Climate Change in the Southwest United States: A Report Prepared for the National Climate Assessment*, 1–20. A report by the Southwest Climate Alliance. Washington, DC: Island Press.

Painter, T. H., J. S. Deems, J. Belnap, A. F. Hamlet, C. C. Landry, and B. Udall. 2010. Response of Colorado River runoff to dust radiative forcing in snow. *Proc. Natl. Acad. Sci.* 107(40):17125–17130.

Pandey, G. R., D. R. Cayan, and K. P. Georgakakos. 1999. Precipitation structure in the Sierra Nevada of California during winter. *J. Geophys. Res.* 104(D10):12019–12030.

Pederson, G. T., J. L. Betancourt, and G. J. McCabe. 2013. Regional patterns and proximal causes of the recent snowpack decline in the Rocky Mountains, U.S. *Geophys. Res. Lett.* 40:1811–1816, doi:10.1002/grl.50424.

Peterson, D. H., I. Stewart, and F. Murphy. 2008. Principal hydrologic responses to climatic and geologic variability in the Sierra Nevada, California. *San Franc. Estuary Watershed Sci.* 6(1), jmie_sfews_10996. Retrieved from http://escholarship.org/uc/item/2743f2n3.

Pierce, D. W., and D. R. Cayan, 2013. The uneven response of different snow measures to human-induced climate warming. *J. Clim.* 26:4148–4167, doi:10.1175/JCLI-D- 12-00534.1.

Pierce, D. W., T. P. Barnett, H. G. Hidalgo, T. Das, C. Bonfils, B. D. Santer, G. Bala et al. 2008. Attribution of declining western U.S. snowpack to human effects. *J. Clim.* 21:6425–6444, doi:10.1175/2008JCLI2405.1.

Pierce, D. W., T. Das, D. R. Cayan, E. P. Maurer, and N. L. Miller et al. 2012. Probabilistic estimates of future changes in California temperature and precipitation using statistical and dynamical downscaling. *Clim. Dyn.*, published online March 30 2012, doi:10.1007/s00382-012-1337-9.

Pierce, D. W., D. R. Cayan, T. Das, E. P. Maurer, N. L. Miller, Y. Bao, M. Kanamitsu et al. 2013. The key role of heavy precipitation events in climate model disagreements of future annual precipitation changes in California. *J. Clim.* 26(16):5879–5896.

Polade, S. D., D. W. Pierce, D. R. Cayan, A. Gershunov, and M. D. Dettinger 2014. The key role of dry days in changing regional climate and precipitation regimes. *Sci. Rep.* 4:4364, 8 pp., doi:10.1038/srep04364.

Ralph, F. M., and M. D. Dettinger. 2011. Storms, floods and the science of atmospheric rivers. *EOS Trans. Am. Geophys. Union*. 92:265–266.

Ralph, F. M., and M. D. Dettinger. 2012. Historical and national perspectives on extreme west coast precipitation associated with atmospheric rivers during December 2010. *Bull. Am. Meteorol. Soc.* 93:783–790.

Ralph, F. M., P. J. Neiman, G. A. Wick, S. I. Gutman, M. D. Dettinger, D. R. Cayan, and A. B. White. 2006. Flooding on California's Russian River: Role of atmospheric rivers. *Geophys. Res. Lett.* 33:L13801, doi:10.1029/2006GL026689.

Ralph, F. M., T. Coleman, P. J. Neiman, R. J. Zamora, and M. D. Dettinger. 2013. Observed impacts of duration and seasonality of atmospheric-river landfalls on soil moisture and runoff in coastal Northern California. *J. Hydrometeorol.* 14:443–459, 10.1175/jhm-d-12-076.1.

Ralph, F. M., M. Dettinger, A. White, D. Reynolds, D. Cayan, T. Schneider, R. Cifelli et al. 2014. A vision for future observations for western U.S. extreme precipitation and flooding. *J. Contemporary Water Res. Educ.* 153(1):16–32.

Ralph, F. M., K. A. Prather, D. Cayan, J. R. Spackman, P. DeMott, M. Dettinger, C. Fairall et al. 2015. CalWater field studies designed to quantify the roles of atmospheric rivers and aerosols in modulating U.S. West Coast precipitation in a changing climate. *Bull. Am. Meteorol. Soc.* (in press December 2015).

Redmond, K. T., and R. W. Koch. 1991. Surface climate and streamflow variability in the western United States and their relationship to large scale circulation indices. *Water Resour. Res.* 27(9):2381–2399.

Regonda, S. K., B. Rajagopalan, M. Clark, and J. Pitlick. 2005. Seasonal cycle shifts in hydroclimatology over the western United States. *J. Clim.* 18:372–384.

Roos, M. 1987. Possible changes in California snowmelt patterns, *Proc., 4th Pacific Climate Workshop*, Pacific Grove, California, 22–31.

Rosenfeld, D., W. L. Woodley, D. Axisa, E. Freud, J. G. Hudson, and A. Givati. 2008. Aircraft measurements of the impacts of pollution aerosols on clouds and precipitation over the Sierra Nevada, *J. Geophys. Res.* 113:D15203, doi:10.1029/2007JD009544.

Rutz, J. J., J. W. Steenburgh, and F. M. Ralph. 2015. The inland penetration of atmospheric rivers over western North America: A Lagrangian analysis. *Mon. Wea. Rev.* 143:1924–1944.

Salathé, E. P. Jr., A. F. Hamlet, C. F. Mass, S.-Y. Lee, M. Stumbaugh, and R. Steed. 2014. Estimates of 21st century flood risk in the Pacific northwest based on regional climate model simulations. *J. Hydrometeorol.* 15(5):1881–1899, doi: 10.1175/JHM-D-13-0137.1.

Sarachik, E. S., and M. A. Cane. 2010. *The El Niño–Southern Oscillation Phenomenon*. Cambridge, UK: Cambridge Univ Press.

Seager, R. M. Ting, I. Held, Y. Kushnir, J. Lu, G. Vecchi, and H. P. Huang et al. 2007. Model projections of an imminent transition to a more arid climate in southwestern North America. *Science* 316:1181–1184.

Seager, R., M. Hoerling, S. Schubert, H. Wang, B. Lyon, A. Kumar, J. Nakamura, and N. Henderson. 2014. Causes and predictability of the 2011–14 California drought, Rep., 40 pp., doi:10.7289/V7258K7771F.

Shukla, S., A. Steinemann, S. F. Iacobellis, and D. R. Cayan. 2015. Annual drought in California: Association with monthly precipitation and climate phases. *J. Appl. Meteorol. Climatol.* 54:2273–2281.

Steinemann, A. C., S. F. Iacobellis, and D. R. Cayan. 2015. Developing and evaluating drought indicators for decision-making. *J. Hydrometeorol.* 16:1793–1803.

Stewart, I. T., D. R. Cayan, and M. D. Dettinger. 2005. Changes towards earlier streamflow timing across western North America. *J. Clim.* 18:1136–1155.

Stine, S. 1994. Extreme and persistent drought in California and Patagonia during medieval time. *Nature* 369:546–549.

Tohver, I., A. F. Hamlet, and S.-Y. Lee. 2014. Impacts of 21st century climate change on hydrologic extremes in the Pacific Northwest region of North America. *J. Am. Water Resour. Assoc.* 50(6):1461–1476, doi: 10.1111/jawr.12199.

Udall, B., and G. Bates. 2007. Climatic and Hydrologic Trends in the Western U.S.: A Review of Recent Peer-Reviewed Research, Intermountain Climate Summary, January 2007. Available at http://wwa.colorado.edu/climate/iwcs/archive/IWCS_2007_Jan_feature.pdf.

Vano, J. A., T. Das, and D. P. Lettenmaier. 2012. Hydrologic sensitivities of Colorado River runoff to changes in precipitation and temperature. *J. Hydrometeorol.* 13:3932–949, doi:10.1175/JHM-D-11-069.1.

Vano, J. A., B. Udall, D. R. Cayan, J. T. Overpeck, L. D. Brekke, T. Das, H. C. Hartmann et al. 2013. Understanding uncertainties in future Colorado River streamflow. *Bull. Am. Meteorol. Soc.* 95:59–78, doi:10.1175/BAMS-D-12-00228.1.

Weaver, R. L. 1962. Meteorology of hydrologically critical storms in California. U.S. Weather Bureau, Hydrol. Rep. No. 37, 207 pp.

Westerling, A., T. Barnett, A. Gershunov, A. F. Hamlet, D. Lettenmaier, N. Lu, E. Rosenberg, and A. Steinemann. 2008. Climate Forecasts for Improving Management of Energy and Hydropower Resources in the Western U.S., Final Report prepared for the California Energy Commission, Public Interest Energy Research (PIER) Program, 125 pp., June. Available at http://uc-ciee.org/downloads/ClimateForecasts.Westerling.pdf.

Wheeler, M. C., and H. H. Hendon. 2004. An all-season real-time multivariate MJO index: Development of an index for monitoring and prediction. *Mon. Wea. Rev.* 132:1917–1932.

White, A. B., D. J. Gottas, E. Strem, F. M. Ralph, and P. J. Neiman. 2002. An automated brightband height detection algorithm for use with Doppler radar vertical spectral moments. *J. Atmos. Oceanic Technol.* 19:687–697.

White, A. B., D. J. Gottas, A. F. Henkel, P. J. Neiman, F. M. Ralph, and S. I. Gutman. 2010. Developing a performance measure for snow-level forecasts. *J. Hydrometeorol.* 11:739–753.

Woodhouse, C. A., D. M. Meko, G. M., MacDonald, D. W. Stahle, and E. R. Cook. 2010. A 1200-year perspective on the 21st century drought in southwestern North America. *Proc. Natl. Acad. Sci.* 107:21283–21288, doi:10.1073/pnas.0911197107.

Section II

Water Policy Issues Related to Climate Variability and Change

3

Key Legal Issues in Western Water Management and Climate Adaptation

Denise D. Fort

CONTENTS

ABSTRACT Western water law and the institutional framework for water management control the distribution and use of western water. After presenting an overview of the law and institutions, key topics are explored. Climate change is affecting every aspect of the hydrologic cycle, and the future holds more extremes of drought, flooding, and warmer temperatures, imposed on already strained water resources. The critical questions are how well the existing management regime serves the vast population, economy, and environment of the contemporary West and the choices that the nation and region face in water management as climate change takes effect. This chapter covers a lot of ground, so this discussion is intended to provide a taste of critical questions and an incentive to explore these questions more thoroughly in the wealth of water literature.

3.1 Introduction

What we see around us when we look at the engineering of water in our lives—the plumbing in our homes, dams, irrigated fields, power plants, and businesses—was created over generations and by a complex web of institutions. These institutions include not only agencies but also the legal system under which water distributions were established and the contemporary geography of water was created.

The story of water institutions in the West can be traced back to the Pueblo and Spanish canals used for irrigation in arid regions, to the canals built by the early Mormon settlers, and to disputes over water in the gold rush. Large-scale irrigation is often dated back to the federal government's support of agricultural projects for immigrants from the eastern United States, the so-called reclamation of the arid West, carried out by what would become the Bureau of Reclamation (BOR). The Corps of Engineers built reservoirs for power and flood storage. States created systems to protect existing uses of water and to provide for new uses. Much later, the nation became concerned about water pollution and the loss of habitat for species. Federal agencies played a prominent role in establishing and enforcing these laws, but often under programs that delegated authority to states. Tribes, long ignored by state and federal water development programs, asserted their water rights in litigation and created agencies to administer water on their lands. The legal and policy framework to manage groundwater has been slow to evolve, but it is receiving new attention because of the recognition of its connection to surface water, declining aquifer levels, and the intense use of groundwater during drought periods. It is now generally recognized as an essential element of state water management.

None of the institutional structure is settled, of course. Much of the West is in severe drought, and depleted reservoirs, aquifers, and rivers are pressed to meet the water expectations of growing populations, agricultural uses, species protection, and recreational uses. Climate projections indicate that the future will bring continued challenges in many regions of the West: reduced summer streamflows due to earlier snowmelt and increased potential evapotranspiration, changing precipitation patterns that may cause increased flooding, and greater demand for water for crops due to increased evapotranspiration. Thus, we need to understand the structures and practices that currently govern water, but we also need to think critically about what is needed to serve future generations. Water laws and institutions do change, albeit slowly, and informed citizens can help bring about needed reforms.

3.2 Federal, State, and Other Agencies

3.2.1 Federal Agencies

A complex web of players affects the management and control of water, including state and federal agencies, legislative bodies, the courts, water users, nongovernmental organizations (NGOs), and tribes. This brief overview begins with the major federal agencies, the Departments of the Interior and Agriculture, the Corps of Engineers, and the Environmental Protection Agency (EPA). The constitutional basis for the federal role is founded in the federal government's ownership of land (Forest Service and Bureau of Land Management lands, for example); the Commerce Clause, which permits congressional action on matters affecting commerce among the states (much regulation of pollution falls within this clause); and federal control over navigation (the original basis of Corps of Engineers' jurisdiction). Many constitutional questions are settled in this area, but as we discuss below, there are contentious exceptions, such as the definition of *navigable waters* and the question of when water users must be compensated by the government for environmental restrictions (the Takings Clause). But the burning question about the federal role is what it will be in the future. The federal investment in infrastructure has been large (dams, power plants, irrigation works) and has shaped water resources for much of the West. Will the federal government provide funding to rescue the West from its current crises? Will federal environmental regulations increase in scope or be drastically reduced?

The BOR is prominent in the western states and has expanded from its original mission of delivering irrigation water from reservoirs to an agency that also provides a loosely linked set of services to the region. Reclamation manages storage and water delivery for irrigators, generates power at its reservoirs, and is involved in other water supply projects, such as supporting water reuse projects.

The Corps of Engineers, perhaps surprisingly, is present on some of the West's largest rivers, with the Columbia River heading the list. The Corps' mission was originally established for the protection of commerce through the construction of canals in the East but expanded to flood control, building dams and levees, and hydropower facilities.

Both agencies were sharply criticized by environmentalists for their destructive effects on western waters. They have evolved to include environmental goals in their missions and consider a wider range of alternatives in project planning. The Corps has an established program under which it undertakes environmental restoration, with a favorable cost share for project proponents. The bureau is involved in restoration efforts on a number of western rivers, although there is no broad authorizing statute for restoration.

The EPA sets technology-based effluent limits for discharges of pollutants into water and regulates a few other types of water pollution, such as runoff from concentrated animal feeding operations. Water quality regulations are generally based on standards established by states, in conjunction with technology-based standards. Pollution regulation affects direct discharges of pollutants, but so-called nonpoint sources such as runoff from agriculture are exempted from regulation.

The Fish and Wildlife Service is an arm of the Department of the Interior. Its water-related responsibilities come with its management of wildlife habitat and fish hatcheries, but most attention recently has focused on its role in protecting threatened and endangered fishes under the federal Endangered Species Act (ESA). Some states have programs addressing aspects of biodiversity and endangered species, but the federal role predominates.

The Obama administration made its mark on water resources management through the release of Principles and Requirements for Federal Investments in Water Resources (CEQ 2013), which is supplemented by Interagency Guidelines (2014) for its implementation. These provide presidential direction to the federal agencies about how federal responsibilities should be approached. They require a rigorous evaluation of proposed water projects and other federal activities with the goal of reducing the costs and negative environmental effects of proposed actions. If followed by federal agencies, they would result in dramatic changes in standard assumptions within federal agencies, but their implementation is not a sure thing, as presidents change and agency cultures abide.

3.2.2 Other Water Management Institutions

If only it were sufficient to understand the roles of federal agencies on western rivers. In truth, there are multiple layers of state and local government agencies, water districts, tribes, and other water authorities. In some areas, there may be no visible federal presence (for example, no dams, restoration projects, or endangered fishes), and in others, the federal government may appear to be the most powerful actor. But regardless of appearances, one must look carefully to see what role might be played by any of these agencies.

State water resource and water quality agencies are on the front lines of water administration and policy, although in some states, water may be administered at a district level. As discussed below, state agencies determine water rights among water users. The Western States Water Council convenes state administrators and coordinates policy positions with the Western Governors' Association. Water quality agencies may administer delegated federal programs, as well as state initiatives, such as groundwater protection.

Irrigation districts typically are composed of member irrigators who share common storage and conveyance, often with elected leadership. City and regional governments manage water supply for their customers. Tribes and pueblos may manage water supply and water quality within their jurisdictions.

NGOs are key institutions in environmental disputes and other areas of public interest. For example, environmental groups are often the moving actors in bringing about protection for endangered fishes, overcoming the reluctance to act by federal agencies.

The governance of water is not limited to formal institutions. Across the West, new means of bringing stakeholders together over water issues, often involving environmental disputes, have been developed. These participatory mechanisms have been heralded as providing solutions that can include sustainability and other values that are not part of existing water law. Chapter 6 (Schlager, this volume) discusses this important development.

3.3 Western Water Law: Key Topics

3.3.1 Prior Appropriation

One summarizes the thorny field of water law at one's peril. Instead, we have highlighted several key principles that are generally applicable, with the important caveat that no generalization can be accurate. The states are responsible for administering the laws that control how a water right is created as well as other aspects of its use and protection. The legal system in use in most western states was founded on the concepts that a water right was established by putting water to use and that the earliest users should be protected against later users in times of shortfall. Administrative systems in which applications were required, reviewed by agencies, and perhaps subject to protest by other water users were created under the direction of *state engineers*. Indeed, the field of engineering has historically dominated water management.

The prior appropriation system made it possible for water to be acquired by those who were not directly situated on a river, and it protected investments made in reliance on a water right. But, it has been criticized for several reasons: (1) it does not provide for sharing in times of shortage, so *junior appropriators* may get no water in dry years; (2) it limited water rights to uses that took place out of a river, such as irrigation, so that the natural flows of rivers were not protected under the doctrine; and (3) it enshrined property rights in agricultural uses that are no longer critical to western economies.

Despite the prior appropriation doctrine's original requirement of putting water to use outside a stream, it has been modified to allow legal protections for environmental flows in most states. But the recognition of environmental values came long after most rivers were fully appropriated, so that its legacy leaves desiccated rivers in much of the West. We present a full discussion of this in Chapter 5 (Richter et al., this volume).

3.3.2 Water Transfers and Water Banks

Water transfers and water banking have addressed some of the economic dysfunctions attributed to the prior appropriation doctrine. Both of these mechanisms preserve the concept enshrined in prior appropriation: that water can be held as a property right. Water transfers are a means of allowing water rights to be moved from the land or for the use to which they were originally put and to be purchased (or leased) by a new user. While we are familiar with markets, a famous law professor wrote, "Water is not like a pocket watch or a piece of furniture, which an owner may destroy with impunity" (Sax 1988, p. 482). There is a difference because of the inherent public interest in how water is used. Water touches on a community's social fabric, its prospects for economic development, its ability to absorb new populations, as well as the recreation and fisheries that a river may support, and even a city's identity (San Antonio's River Walk, as an example). Downstream irrigators may be reliant on the return flow from agricultural lands and thus object to transfers when the return flow from a farm will be diminished. Conversely, the transfer of a water right upstream may harm aquatic ecosystems because of the earlier withdrawal of water. All of the intertwined interests in water may be invoked when an application to market water is made, although many of these interests are not protected under state laws governing transfers.

Water markets are not the only means by which water has been transferred from one use to another. Much of the West's irrigation water is controlled by irrigation districts, or the BOR, and delivered through infrastructure owned by these entities. Significantly, the BOR has permitted water provided under its contracts to move from agricultural to municipal uses.

Water banks have been established in some states to allow water rights holders to reduce their use (in an agricultural context, this might mean fallowing fields for a growing season) and to lease the foregone water through a water bank. Because most water in the West is used by agriculture, typically, these transfers would be agricultural to municipal, or to another agricultural user. Water banks vary in their structure; some are private arrangements between groups of farmers, for example. Others are operated by districts or other public entities, and may provide opportunities for protection of third-party interests, such as community employment or environmental concerns (Miller 2000). In a well-publicized transaction between the Palo Verde Irrigation District and the Metropolitan Water District in Southern California, individual farmers were paid for fallowing lands on a rotating basis, but the buyer also donated money to a community trust to benefit other stakeholders in the community (Waterworld 2002).

Because most water in the West is fully appropriated, water transfers have been a means of procuring water for environmental purposes. Many states have programs under which water rights can be purchased for these purposes, and occasionally, *environmental water* is a condition for large-scale water transfers.

The term *water banks* is also used to refer to the practice of storing surface water in groundwater basins, for withdrawal and possible transfer at a later time. When water is purchased and stored for later use, or stored for later sale, the value of water is increased by providing supplies in times of greater need. In effect, the groundwater aquifer functions as surface reservoirs have, but with more possibilities for commercial or private operations. Groundwater storage is also used regularly by municipal governments, such as those in Orange County, California.

Whether water transfers are temporary or permanent, the institutional framework is quite varied and depends on the state, whether federal actors are involved, whether environmental protections are available under state or federal law, the nature of the water being transferred (e.g., surface water or groundwater), the uses to which it is transferred, and a variety of other factors. Data regarding these transactions may be proprietary, and the evidence of prices paid is often anecdotal. The complexity of effecting water transfers cannot be overstated, and therefore, markets are not as widespread as proponents of markets would like (Bretsen and Hill 2009). Professor Mark Squillace (2013, p. 68) put it baldly: "The story of water transfers in the western United States is largely a story of market failure."

Nonetheless, most observers believe that significant quantities of water eventually will be moved from agriculture to cities, and from lower-valued agricultural uses to higher-value uses. A comprehensive study of California water transfers found that over 2 million acre-feet (MAF)* of water was committed for sale or lease in the early years of this century (Public Policy Institute of California 2012). The current drought in California is leading to greater interest in facilitating these transactions.

3.3.3 Tribal Water Rights

Tribal water rights may be established by treaty or, more commonly, through the judicially recognized doctrine of *reserved water rights*. The quantification of tribal water rights has proceeded slowly, but tribal water rights may displace existing uses, because they are older in time. Thus, western states are motivated to determine the extent of tribal water rights, and settlement of cases has been viewed as a desirable vehicle to accelerate the slow pace of litigation. In these agreements, federal funding often is used to ensure that all parties are put in a better position than before the settlement, even though no party prevails to the full extent of their claimed water. Thus, the western New Mexico town of Gallup receives water from a new pipeline that was built to deliver Navajo water, providing federally funded water delivery to two entities with very different legal rights. Settlements are preferred by the major parties because of the control that parties retain in negotiating, as compared to asking a court to make a determination. In specific instances, there are outside entities who object to the terms of a settlement.

New questions are arising around tribal water. Should it be used in agriculture or leased to higher-paying uses? Should tribes be allowed to lease across state lines? Is the use of agriculture as a standard for determining tribal water rights outdated for economic or climate reasons? How should groundwater be treated when a tribe did not historically use it?

* 1 million acre-feet (MAF) = 1.233 billion cubic meters (BCM).

3.3.4 Groundwater Law and Regulation

Groundwater law is not one of the law's shining achievements, despite the West's heavy reliance on groundwater for municipal, industrial, and agricultural purposes. Perhaps this is because groundwater is not visible, and it is difficult to know the size, quality, and movement of aquifers. Perhaps it is because it is in the interest of water users to keep their usage close to the chest, to avoid too much scrutiny from public stakeholders. Even municipalities may not want prospective industries to know how fragile their water supplies are—no chamber of commerce has ever advertised a precipitously declining water table. California assiduously avoided state regulation of groundwater until the current calamitous drought. The increasing use of groundwater proved the catalyst for state legislation, but the legislation might be characterized as a step toward groundwater control, rather than bringing the state into a sustainable relationship with its dependence on groundwater (Miller 2014; California State Assembly 2013–2014; California State Senate 2013–2014).

Groundwater allocation is almost entirely controlled by state law and, therefore, varies from state to state. In general, a water right must be established, but in some states, the right is a benefit of surface land ownership, and in others, it is dependent on the more familiar prior appropriation model. The goals of regulation may be to protect an adjacent river from depletion through groundwater pumping, to protect other groundwater users, or to protect the life of the aquifer.

The Ogallala aquifer illustrates the implicit policy questions in groundwater law. This enormous aquifer underlies the eight states of the High Plains aquifer. The rate of mining varies across the aquifer, but the United States Geological Service (USGS) estimates an overall decline of 14.2 ft. across the High Plains aquifer (McGuire 2012). The mining of the aquifer is a matter of significant concern to the federal government; it produces a large share of the wheat, corn, cotton, and cattle of the nation, and declining groundwater levels imperil that production (USDA NRCS 2015). The United States Department of Agriculture (USDA) has established programs to conserve water and retire agricultural lands in the region, and research is underway on new crops. At the same time, federal agricultural policies drive ethanol production, increasing the pressure to grow corn (Little 2009). The time remaining for this scale of agriculture is unknown because it is not clear what public policies will prevail.

Groundwater mining is a thorny problem. Most westerners would not regard it as realistic to prohibit groundwater mining, any more than it would be to prohibit the mining of other irreplaceable resources. But most would say that aquifers should be sustainably managed, as difficult as that might be to define. Should states cooperate across state lines in its management? Should state law protect the interests of future generations in this supply? No representatives of future generations are here to defend their interests, and the status quo has been to allow the mining of aquifers, with rare exceptions, such as Arizona's protection of key aquifers. Thus, as water levels drop, we will see irrigators abandon their farmland, to become a small data spot in the national database of water use.

The federal role in groundwater management has been very limited, in marked contrast to its role in surface waters. The USGS does provide studies and data concerning groundwater withdrawals. When groundwater levels are drawn down, states may successfully appeal for federal funds to rescue communities.

A project in New Mexico illustrates this. Eastern New Mexico is a sparsely populated region, anchored by a federal air force facility and a fast-growing dairy industry. The region relies on groundwater, primarily for alfalfa, in addition to municipal use. The prospect of a depleting aquifer led to a half-billion-dollar project in which water from the

Canadian River would be used to supplement groundwater, with the project to be largely funded by the federal government. No environmental impact statement was done on the project, and reducing groundwater mining did not receive serious consideration. While the project has been authorized by Congress, only a small amount of federal and state monies have been received (Fort and McKean 2011; see also *Village of Logan v. U.S. Dept. of Interior*, 577 Fed. Appx. 760, 2014).

Governance issues in groundwater management are very much in flux because the field is still evolving. Groundwater pumping has led to Supreme Court litigation when surface flows in a river are affected; conflicts between states over aquifer pumping have led to calls for compacts comparable to those governing surface waters, and some communities have organized among themselves for voluntary agreements on pumping rates. Scientists have shown the importance of protecting *groundwater-dependent ecosystems*, such as wetlands, from groundwater pumping. Australia has pioneered legal protection for these systems, but in the United States, there is only sporadic protection. Long ignored, groundwater law and policies are moving to the forefront of water discussions in some regions.

3.3.5 Environmental Law

Volumes of law and regulations affect the protection of water quality, ranging from regulation of point-source discharges to pollution from hazardous waste. As mentioned, the framework is largely one of federal regulation, but with a substantial role for states under key regulatory schemes, such as the Clean Water Act (CWA). Ever since the nation awoke to the seriousness of water pollution (roughly 1970), there have been calls to integrate water quantity and water quality law and management. Clearly, water pollution affects the quantity of water available for use. This threat is not hypothetical, as shown by communities losing water supplies due to contamination from farm runoff, the failure of dams at mining sites leading to interruptions in municipal water supplies, and other consequences of lax pollution regulation. Greater integration of quantity and quality management has been achieved within some states; it is still a distant goal for federal agencies.

Water quantity and quality also are related to land use. Activities on federal lands, such as grazing, logging, road construction, and fires, can have implications for water quality downstream. Construction practices are now recognized to affect runoff quality. Agricultural runoff, which may contain pesticides or fertilizers, is not regulated under federal law, and attention is focused on *dead zones* in estuaries and other locations far from the originating farms. *Wetlands* are the intersections between land and water, and activities affecting them may be regulated by the EPA and the Corps of Engineers. For several decades, the courts have struggled with the language of the CWA in determining what constitutes a wetland. The control of land is traditionally a matter of state control, but the protection of water quality has been a matter of federal control. This is a charged political issue that bounces between the courts, Congress, and the executive branch.

3.3.6 Interstate Compacts and Other Water Sharing Agreements

The division of rivers among states by means of compacts is another component of the legal setting for western water. The Colorado River and its tributaries provide water for seven western states and Mexico, with California the thirstiest of them. The compact was negotiated by the states and the federal government, approved by Congress, and has been supplemented by other laws that together form the Law of the River. But the physical development of water storage and distribution among the states by means of reservoirs and other infrastructure occurred through political logrolling, in which politics met hydrology on an uneven playing field.

As drought tightens around the Colorado River Basin, the nature of the law governing the river is revealed to be far more flexible than a first reading would indicate. High-stakes diplomacy was undertaken when the lower-basin states agreed on a shortage sharing agreement, under which some certainty was given to water-short states (USBR 2007; Grant 2008). But when Lake Mead, the source of water for Las Vegas, and other giant reservoirs are drawn down, it is not clear whether negotiations between the states will be sufficient.

The United States Supreme Court is the judicial body with jurisdiction over disputes between states, and compact cases regularly reach that body. Federal courts also can play a role where there is no compact governing a river.

There is always the possibility that Congress will reopen a compact, but the political odds of that occurring have always been dismissed. In our discussion of climate change that follows, we nonetheless entertain the possibility because nothing is sure under extreme conditions.

3.3.7 Water Storage and Pipelines

The western landscape is almost entirely plumbed, as discussed in Chapter 5. The laws governing federal facilities are complicated. They can include the authorizing language for a project; the accretions to that law, including the entire corpus of federal laws that relate to the environment; federal procurement; and other laws governing federal facilities. In the controversies over how dams affect the environment, particular attention is given to the Fish and Wildlife Coordination Act, which gives federal agencies the authority to operate facilities to protect fish and wildlife (Fish and Wildlife Coordination Act 2000).

Hydropower projects are regulated by the Federal Energy Regulatory Commission (FERC), in addition to other authorities. Because of the damage these dams cause to fishes and ecological functioning, FERC licensing has been a battleground between energy and environmental interests.

Pipelines and aqueducts are increasingly used to move water long distances from places of withdrawal to places of use. They are significant because their growing use indicates how many regions have exhausted nearby water supplies, such as groundwater aquifers; they have environmental impacts, including ecological effects associated with siting; and they often require large quantities of energy to move water uphill and over long distances. Southwestern water mavens voice dreams of pipelines from the Great Lakes and the Mississippi River. These projects raise questions about sustainability: How is the area of origin affected? Is a groundwater aquifer being mined, so that the project is limited by the size of the aquifer? How much energy is required to move the water? Citizens have successfully protested some of these projects, and one can safely predict that any project that moves water from one basin to another will lead to protests (Nelson 2012).

3.3.8 Water and Energy

Energy use is intimately connected to water development. A 2005 study done for the California Energy Commission found that "water-related energy use consumes 19 percent of the state's electricity, 30 percent of its natural gas, and 88 billion gallons of diesel fuel every year" (Klein 2005, p. 1). This is discussed in Chapter 13 (Wilkinson, this volume). Energy consumption is not accounted for in water projects, although environmental attorneys have begun raising greenhouse gas concerns over new federally funded projects.

But the more familiar linkage between water and energy is the substantial quantity of water required for the development and production of energy, from water used in fracking, or the produced water in oil and gas operations, to the cooling water required for coal-fired power plants. As oil and gas development increases dramatically in the West, water availability has become a point of contention with local people and of growing concern to energy developers. The legal framework is essentially one of state law, although there is limited federal regulation of water quality effects associated with mining, milling, power generation, and other aspects of the fuel cycle.

3.3.9 Water Reuse and Desalinization

As western cities encounter the conflict between supply and demand, due to declining groundwater, unreliable surface water, or greater use by growing populations, they have turned to technological approaches to water supplies. *Water reuse* can include the intensive treatment of wastewater to make it suitable for domestic uses, or it can refer to reusing less treated water for outdoor irrigation, such as golf courses. The legal regime protects both property rights in water and public health in the recycled water. Desalinization is a potentially helpful approach because of the ample supplies of seawater and saline groundwater. Both technologies are expensive and can be energy intensive, but the costs may be lower for cities than any of their alternatives and may provide more reliable water supplies.

3.4 The Evolving Institutional Framework and the Response to Climate Change

3.4.1 Trends in Water Resources

States and municipalities have become more prominent in western water management as the population grows, and the traditional BOR constituency of rural agricultural users shrinks in number. State and local governments have access to bonding revenues for water projects. These governments may practice *integrated water resources management* and take multiple perspectives into account in decision making. They also may be more cost conscious and choose lower-cost alternatives than governments do when federal funding is involved. But the federal government remains the de facto powerhouse, as it is in the Columbia River Basin and the Rio Grande. And large-scale projects are likely to involve a federal role in environmental review or federal funding.

It is more difficult to identify a trend in water quality and species protection in state–federal relations because the lack of congressional action, in these and many other areas, means that the status quo has been unchanged for a number of years. In contrast, within civil society academic disciplines developed an understanding of aquatic ecosystems and the ecosystem services that they provide. Clean water, for example, can be a benefit of undisturbed wetlands. Clean drinking water reduces public health risks. Polluting groundwater requires expensive remediation and imperils a resource for future generations. All of these propositions would be regarded as no-brainers by water professionals. But as discussed, the EPA is under attack for its attempts to protect wetlands and to set standards for toxins in drinking water. Agricultural and other *nonpoint* pollution is not directly regulated by the federal CWA, nor is it likely to be in the near future. While the electorate is generally supportive of environmental protection, an expanded CWA seems unlikely.

States, tribes, and many cities are moving to fill the gap and to take on regulation of polluters. The national debate over *fracking* illustrates this. Congress exempted the injection of fracking fluids from the Safe Drinking Water Act, but some jurisdictions have regulated or even banned fracking. To find the cutting edge and innovation in protecting clean water, or habitat, one would look to some of the western states.

This trend, however, is unequal, and that has implications for environmental justice, important ecosystems, and even national interests. Generally, affluent states are more likely to regulate their industries than poorer states. It is argued that people in poorer states choose employment over environmental regulation. Whether this is true in any meaningful way for most of the residents of poor states, the fact is that there is a higher rate of exposure to environmental contaminants for poor people. More affluent states also are more likely to provide protection for ecosystems above that provided by the federal ESA. Ecosystems are left unprotected in states that are hostile to species protection, illustrated by the western states that successfully removed protection for wolves. Federal agencies may step forward when there are conflicts between states over pollution, when species come so close to extinction that the federal government has to take measures under the ESA, and generally when a region needs to be rescued from the effects of poor environmental management. But the proper role of the federal government in environmental and ecological protection has been hotly disputed since the passage of the major federal environmental laws.

3.4.2 Climate Change and Water Management

Climate change is bearing down upon the nation with alarming intensity. While its causes are well understood (at least within the scientific community), the policy response has been anemic, at best. Our nation and the world have not risen to the challenge of reducing greenhouse gasses and mitigating the drivers of climate change. *Adaptation* to the changing climate has received more attention, but no jurisdiction has the answers.

The existing legal framework for natural resources and environmental management, including water rights, water quality, and endangered species protection, is challenged by climate change. One example is that of the Colorado River, where some 40 million people are affected by the drying and warming West. The physical and legal infrastructure of the river was created at a time when average flows were believed to be much higher than they are now, and far higher than they are expected to be in the future. Economic losses to California's agricultural economy from the current drought total in the billions; Las Vegas is scrambling for expensive new supplies, and experts scratch their heads over what a prolonged drought will do to the region.

When livelihoods are at stake, it is not surprising that litigation is following, along with an active interest from those congressional representatives whose districts are affected. In the sparsely populated state of New Mexico, empty reservoirs led farmers to increase groundwater pumping in the Rio Grande Basin, leading the well-populated state of Texas to march to the US Supreme Court in a case concerning the interpretation of the Rio Grande Compact. The case is unlikely to be decided soon, and both states are building new infrastructure for water storage and delivery (*Texas v. New Mexico and Colorado,* No. 220141, 2013, available at http://www.scotusblog.com/case-files/cases/texas-v-new-mexico-and-colorado/).

Earlier, we noted that the institutions surrounding western water are slow to change. The question we explore here is whether climate change is a disruptive force in water institutions, requiring a reorganization of the principles and institutions that have governed water thus far. Alternatively, some water managers view climate change as comparable to the familiar problems raised by droughts, repeating the mantra that the western hydrograph is notoriously unpredictable, and that droughts, flooding, and species extinction all predate climate change. Consider these perspectives as we examine the possible ways in which environmental and water law and institutions may respond to climate change.

3.4.2.1 Water Quality Regulation

Hotter temperatures, lower flows, and greater evaporation can exacerbate water quality problems. The nation has made progress in the years since 1972, when the CWA was passed, with its national goal of swimmable, fishable waters.

It will be difficult to maintain this progress. One course would be for the country to strengthen its commitment to clean water by addressing exemptions in the regulatory framework. The CWA does not address most pollution from nonpoint sources and exempts most agricultural activities from its scope. As a result, agricultural pollution is the single largest remaining source of impaired waters. Although academicians might agree about how the CWA should be improved, only a transformation in congressional attitudes toward environmental regulation will bring about national change.

Professor Robin Craig has explored these questions and points to an additional difficult problem that arises when water quality standards can no longer be met due to climate change, in some instances regardless of whether all sources are brought under controls. States set water quality standards, but it is currently difficult for states to relax these standards because of what is called the antidegradation clause. Should the CWA be revised to acknowledge that climate change will prevent maintaining water quality in some waters? This will be a difficult pill to swallow, but the physical reality may overwhelm the goals of the CWA (Craig 2013).

3.4.2.2 Water Quantity

Severe drought has brought crises to the West, affecting recreation, agriculture, and even tourism. Water shortages were addressed through the construction of water infrastructure through most of the twentieth century. Now that most dam sites have been used, western water managers are accelerating the importation of water through pipelines and aqueducts, moving it from rivers and groundwater aquifers, and some new water storage is also being built.

Most water resource managers now look for water supplies through water conservation, pricing, technologies (such as desalinization or water reuse), and water transfers. Because agriculture uses the lion's share of the water in the West, it is logical to look to it as a water supply source for other uses. Institutional issues are key to the relatively anemic volume of water transfers thus far. Major impediments include the control of much irrigation water by irrigation districts, and the restrictions that may have been imposed on federal water when the BOR built water projects. But markets are working in response to California's severe drought, illustrating how quickly institutional constraints can be worked through when significant amounts of money are at stake.

Australia transformed its water allocation system in response to a long-term drought in the Murray–Darling River Basin, which is a well-populated, agriculturally productive, and critical environmental resource for the country. The essence of the solution involved the protection of base flows in rivers and the creation of an active market for water leasing. Base flows for river functioning and environmental purposes were established, and further environmental water was acquired. For agricultural users, water rights were converted to shares of entitlements, which allowed use of yearly allocations. The yearly allocations reflect projected water availability. Entitlements and yearly allocations can be freely traded, with the result that a flourishing market was created. This system permits water users to respond to information about projected supply and to make sound economic decisions about agricultural practices for a given year (see McKenzie 2009; Pilz 2010; Sundareshan 2010).

Adopting the Australian model in the United States could address concerns about deficient environmental flows and the difficulty of effecting water transfers, but would be a costly and contentious undertaking. A considerable initial outlay of money would be required to acquire base flows (see Chapter 5), and questions about priority of use would need to be addressed in granting entitlements. The political ability to bring about such a fundamental transition in law is questionable, given the power of the agricultural senior water rights holders and the inherent conservatism of the water system. Incremental reforms that modify, rather than replace, prior appropriation are more likely.

3.4.2.3 Endangered Species and Protection of Biodiversity

The world is now experiencing a wave of extinctions that will increase under climate change. E.O. Wilson (2002) has warned that we face the loss of half of all species by the end of this century. There is no federal law that protects biodiversity as such, although the federal government does so in its administration of some of its lands and waters. Some states have programs to protect areas of high biodiversity and laws extending protection in some fashion, such as setting minimum environmental flows for rivers. The federal ESA protects threatened and endangered species, with mandates that are regarded as highly restrictive by some. But it is rapidly becoming clear that this act, passed in 1973, is not well suited for the extensive alterations that are caused by climate change. Professor J.B. Ruhl, a leading scholar of ESA, has said that "there soon may be no practical way to administer the ESA in its present form for [Endangered] species" and that "the pika is toast" (Ruhl 2008, pp. 2, 7). The nation has not put its weight behind species protection with the types of investments in staffing, science, habitat and water rights acquisitions, prohibitions of development in certain areas, and other measures that could be taken. A greater commitment to protecting biodiversity could protect species now at risk. But for some species, the writing is on the wall. Should the ESA be amended to provide for triage, or should a greater investment be made in water and land to protect fishes and other species that are dependent upon aquatic ecosystems? In both the context of water quality and species protection as well as many other areas of environmental protection, scholars are asking whether existing laws will be enforced or modified. Chapter 5 discusses biodiversity in greater detail.

3.4.2.4 Flooding

Most of the discussion about climate change in the West, which is characterized by aridity, concerns the lack of water, but new climate patterns point to an increase in flooding in many areas. The traditional approach was to channelize rivers and build levees and dams for flood control. Controlling land use is an alternative approach.

The federal Corps of Engineers is the federal agency that builds much of the flood-control infrastructure. The federal government plays an indirect role in land-use decisions by providing insurance subsidies for houses built in floodplains. The political cost of raising insurance costs on flood-prone neighborhoods proved too high for Congress. After years of policy analysis showing the negative consequences of encouraging growth in these areas, Congress toughened requirements and then shortly thereafter retracted provisions of the legislation (the Biggert-Waters Flood Insurance Reform Act of 2012, closely followed by the Homeowner Flood Insurance Affordability Act of 2014).

The economic consequences of subsidizing insurance for building in floodplains and of permitting construction in floodplains are becoming clearer, as climate modeling becomes more precise. Policy prescriptions are fairly straightforward, and there are other nations that have taken steps to prepare for more frequent flooding and rising sea levels.

3.4.2.5 Agricultural Policies

Water and agricultural policies bear directly on each other, but the linkages are rarely made explicit. The congressional committee structure, the disparate focus of NGOs and trade organizations on either water or agriculture, and the divided structure between agriculture and water in federal and state agencies do not make for a coherent dialogue on water and agriculture.

One example would be national agricultural policies that encourage corn production, which, in turn, encourages consumption of groundwater. We know that ethanol is a costly fuel in terms of water, land, and energy use. As corn production is encouraged, the Ogallala aquifer is drawn down, with no accounting in energy policy for the water that is used. Thus, federal policies are adding to a water crisis that will, in turn, be the subject of federal policy.

Irrigation is by far the largest use of water in the western United States accounting for 85% of all water withdrawn, and the West accounts for 74% of all irrigated acreage in the nation (Kenny et al. 2009). It was practiced by Native Americans and others using surface waters, and then increased dramatically with the use of groundwater and center-pivot sprinklers. Large farms account for 61% of the irrigated acreage in the West (Little 2009; McGuire 2012; USGS 2013; NRCS 2014). In fact, concentration of water use (15% of all irrigators use 66% of irrigation water) and of profits (larger farms account for 85% of sales from irrigated farms) is the rule in the West (USDA Economic Research Service 2013).

Sandra Postel (1999) sounded the alarm about the future of irrigated agriculture in her book *Pillar of Sand: Can the Irrigation Miracle Last?*. Schoolchildren know that salinity led to the end of great civilizations that were built on irrigation. To that, we must add declining aquifers, impaired water quality, and, of course, climate change, which can increase the requirement for water and bring about changes in precipitation patterns.

Policy discussions have focused on institutional changes, such as water pricing or subsidies for efficiency improvements, that could lead to better conservation practices by irrigators. Federal water policies led to the development of irrigated agriculture; some look to lands with rain-fed crops as the next focus of development. The search is underway for crops that can withstand higher temperatures and use less water.

The United States is a net exporter of agricultural products, adding to our trade balance sheet. A new generation of water scholars has suggested that regions also should consider the *water footprint* of food, in making agricultural and water decisions. An arid region would do well to import food, rather than paying exorbitant amounts to import water

(Postel 2010). A recent study looked at the water footprint of some of the agricultural products exported by state and concluded the following:

> The arid Southwest is a secondary recipient [of imported water through food products] at best, where New Mexico, Wyoming, Colorado, and Utah are net exporters, totaling 5.3 billion m^3, and Nevada, Arizona, and California are only modest net importers. As a whole, this seven-state arid region is a net exporter of over 1 billion m^3 of water per year. (Mubako and Lant 2013, p. 389)

Water footprint studies distinguish between rain-fed and irrigated crops (Hoekstra and Mekonnen 2012). Further studies are needed to distinguish between surface water and groundwater irrigation, and to determine what groundwater use is sustainable, that is, when the groundwater source is being mined.

The economic benefits of irrigation must also be taken into account. These can vary greatly depending on the crop. Because irrigation water often has a low cost (often the cost of energy for pumping, or a cost imposed by the irrigation district), low-value crops are grown with irrigation water, in addition to the fruits and nuts for which Arizona and California are well known.

As climate change intensifies and water supplies become more contested, we can expect more debate over where water is used in a state and the costs to the region of this water use. The question of why crops are grown in the desert will be revisited, more than a century after the country first committed to reclaiming the West.

3.4.2.6 Broad Shifts in Water Institutions

As the water crisis worsens in the West, Professor Robert Adler (2010) has questioned whether state sovereignty over water, which has never been the absolute rule, will continue as the national interest in water becomes more compelling. He argues that an explicit turn toward federal control will be necessary. It is clear that the nation would never tolerate state control over oil resources; we expect an open market, federal subsidies, and federal policies to control the movement and use of oil. Is it reasonable to think that Congress will step in to protect national interests that are affected by water practices? He offers the example that a future Congress might authorize national efficiency standards for agriculture.

Another example worth pondering is the persistent belief that some major body of water, for example, the Columbia or even the Great Lakes, should be piped to the arid West. The fundamental constitutional principle is that states cannot impede interstate commerce, and water has been held to be an article of interstate commerce. But Congress can control interstate commerce and currently has protected both the Columbia River and the Great Lakes against export schemes. Is it possible that Congress will permit these exports if the water crisis becomes sufficiently acute? While informed observers believe that none of these scenarios will come to fruition, the "pipe dream" of a grand water transfer periodically resurfaces.

The compacts between states governing water allocation on the great western rivers are often viewed as immutable. The arguments against controlling water use in the twenty-first century with instruments crafted 100 years earlier are becoming louder (Johnson 2014). Should extreme drought continue in western states, the unthinkable will become more of a possibility, as challenges to bans on interstate water transfers are raised (Matthews and Pease 2006) or movement grows to modify compacts (Schlager and Heikkila 2011).

The US Constitution has provided well for a booming capitalist economy and private entrepreneurial success. But this system also has created political barriers to the regulation of carbon pollution, such that the United States, with its excellent universities and scientific institutions, has yet to take action on the scientific understanding of the dangers of climate change. The Constitution has been held to create barriers to some forms of environmental regulation, and calls for its reform are growing.

For example, the Fifth Amendment requires compensation when the government "takes" private property, as when a house is condemned for a new highway. It is easy to conclude that the government should pay when it takes someone's house. The questions in environmental regulation arise from a different sort of governmental action, such as when a lot is declared to be a wetland and the owner is unable to fill it in for a housing development, or when a beach lot owner is unable to develop because of the effects of development on coastal preservation. Water is a peculiar type of property, and access to it has traditionally been subject to institutional constraints, as well as whether nature has provided enough precipitation.

Professor Dan Tarlock (2012, p. 738) has raised three questions regarding how climate change should affect the judicial understanding of property. "Climate change increases the risk of title and enjoyment disturbance, and thus three possible climate-change impact scenarios are possible. First, the existing property doctrines can adapt to climate change. Second, climate change will produce new doctrines that limit the exclusive enjoyment of rights and mandate greater resource sharing. Third, the change will come from legislatures and the role of courts will be to assess the constitutionality of the legislation."

He argues that there is inherent risk in property ownership, that climate change should make that risk implicit, and that there is a moral hazard associated with the improper or rash use of property. The public trust doctrine, founded on the concept that there is an abiding public trust in a shared natural heritage, is a jurisprudential articulation that may define the limits of private ownership of rivers.

Another perspective comes from Professor Robin Craig (2010), who argues that the common law doctrine of necessity may be used by the government to take action to provide water, including environmental water, without compensation to the water rights holder. Climate change is causing emergencies in water supply, including droughts, flooding, and loss of biodiversity. In general, we recognize that firefighters may destroy a structure in the course of fighting a fire; we do not expect governmental compensation for the building's owner. Her argument could stop the beating heart of a state water official:

> Within that authority, moreover, public necessity allows for prioritizing survival necessities at the expense of water luxuries—drinking water ahead of swimming pools, and water for climate-appropriate staple food crops and local food needs (including local fish) ahead of water for luxury, climate inappropriate, and/or export crops. As part of that prioritization, public necessity also allows governments, during times of shortage, to reallocate individual water rights to the aquatic ecosystems and ecosystem services that communities' larger survival and well-being depends upon, rather than forcing the death of streams and other water bodies that climate change renders effectively over-appropriated. (Craig 2010, p. 750)

These ideas are speculative at this time, and there is a hardening of the constitutional interpretations that privilege private property in the federal courts. Congress is not likely to increase environmental protections in the near future. There is limited predictability of the local-scale impacts of climate change, and it adds to other pressures on the existing water regime.

3.5 Conclusion

The West's response to climate change will determine whether people face unexpected restrictions on water use; whether rivers are dried up and ecosystems lost; whether irrigated agriculture dominates water and land use in the western states; and, ultimately, whether the region is resilient and sustainable. Water laws and institutions that have worked for decades will need to be reconsidered and adjusted to the exigencies of a different time. It is possible that the severity of the environmental crisis will lead to foundational changes in these familiar institutions. Many unresolved questions permeate this topic. Technology has transformed our lives in short order but, thus far, has played a smaller role in water. Are there breakthroughs that will allow greater efficiencies in water use? Government has been the driver for most water management decision making, but civil society has demanded more responsive and open decision making. Will citizen activism start to affect the big levers of water policies? Finally, food preferences and policies are drivers of water policy, but it is very difficult to predict where they will lead. Given the volume of water consumed by agriculture in the West, this sector is one to watch closely.

References

Adler, Robert W. 2010. Climate Change and the Hegemony of State Water Law. *Stanford Environmental Law Journal* 29:1–61.

Biggert-Waters Flood Insurance Reform Act of 2012, Public Law 112-141, Div. F, Title II, §§ 100201 to 100249, 126 Statute 916.

Bretsen, Stephen N. and Peter J. Hill. 2009. Water Markets as a Tragedy of the Anticommons. *William & Mary Environmental Law and Policy Review* 33:723–784.

California State Assembly. 2013–2014. Assembly Bill No. 1739 Groundwater management. Regular Session 2014, California. Available at http://leginfo.legislature.ca.gov/faces/billNavClient.xhtml?bill_id=201320140AB1739.

California State Senate. 2013–2014. Senate Bill No. 1168 Groundwater management. Regular Session 2014, California. Available at http://leginfo.legislature.ca.gov/faces/billNavClient.xhtml?bill_id=201320140SB1168.

California State Senate. 2013–2014. Senate Bill No. 1319 Groundwater. Regular Session 2014, California. Available at http://leginfo.legislature.ca.gov/faces/billNavClient.xhtml?bill_id=201320140SB1319.

CEQ (Council on Environmental Quality). 2013. Principles and Requirements for Federal Investments in Water Resources. Available at https://www.whitehouse.gov/sites/default/files/final_principles_and_requirements_march_2013.pdf.

CEQ (Council on Environmental Quality). 2014. Chapter III—Interagency Guidelines. Available at https://www.whitehouse.gov/sites/default/files/docs/prg_interagency_guidelines_12_2014.pdf.

Craig, Robin Kundis. 2010. Adapting Water Law to Public Necessity: Reframing Climate Change Adaptation as Emergency Response and Preparedness. *Vermont Journal of Environmental Law* 11:709–756.

—. 2013. Climate Change Adaptation, the Clean Water Act, and Energy: A Call for Principled Flexibility Regarding Existing Uses. *George Washington Journal of Energy & Environmental Law* 4(2):26–45.

Fish and Wildlife Coordination Act. 2000. 16 U.S.C. §§661-667.

Fort, Denise D. and Summer McKean. 2011. Groundwater Policy in the Western United States. *Idaho Law Review* 47:325–340.

Grant, Douglas L. 2008. Collaborative Solutions to Colorado River Water Shortages: The Basin States Proposal and Beyond. Symposium—Collaboration and the Colorado River. *Nevada Law Journal* 8:964–993.

Hoekstra, Arjen Y. and Mesfin M. Mekonnen. 2012. The Water Footprint of Humanity. *PNAS* 109 (9):3232–3237.

Homeowner Flood Insurance Affordability Act of 2014, Public Law 113-89, Mar. 21, 2014, 128 Statute 1020.

Johnson, Tarryn. 2014. What's Yours Is Mine and What's Mine Is Mine: Why Tarrant Regional Water District V. Herrmann Signals the Need for Texas to Initiate Interstate Water Compact Modifications. *Texas Tech Law Review* 46:1203–1228.

Kenny, Joan F., Nancy L. Barber, Susan S. Hutson, Kristin S. Linsey, John K. Lovelace, and Molly A. Maupin. 2009. Estimated Use of Water in the United States in 2005. *U.S. Geological Survey Circular* 1344.

Klein, G. 2005. *California's Water–Energy Relationship*. California Energy Commission. CEC-700-2005-011-SF. Sacramento, CA. Available at http://www.energy.ca.gov/2005publications/CEC-700-2005-011/CEC-700-2005-011-SF.PDF

Little, Jane Braxton. 2009. The Ogallala Aquifer: Saving a Vital U.S. Water Source. *Scientific American*, March 1. Available at http://www.scientificamerican.com/article/the-ogallala-aquifer/?page=4

Matthews, Olen Paul and Michael Pease. 2006. The Commerce Clause, Interstate Compacts, and Marketing Water Across State Boundaries. *Natural Resources Journal* 46:601–656.

McGuire, V.L. 2012. Water-Level and Storage Changes in the High Plains Aquifer, Predevelopment to 2011 and 2009–11. *U.S. Geological Survey Scientific Investigations Report* 2012-5291 15 pp.

McKenzie, Michael. 2009. Water Rights in NSW: Properly Property? *Sydney Law Review* 31:443–464.

Metropolitan, Palo Verde boards approve Colorado River water transfer. *WaterWorld*, 2002. Available at http://www.waterworld.com/articles/2002/10/metropolitan-palo-verde-boards-approve-colorado-river-water-transfer.html.

Miller, Jeremy. 2014. California's sweeping new groundwater regulations. *High Country News*, November 10. Available at https://www.hcn.org/issues/46.19/californias-sweeping-new-groundwater-regulations.

Miller, Kathleen A. 2000. Managing Supply Variability: The Use of Water Banks in the Western United States. In *Drought: A Global Assessment Volume II*, edited by D. A. Wilhite, 70–86. London: Routledge.

Mubako, Stanley T. and Christopher L. Lant. 2013. Agricultural Virtual Water Trade and Water Footprint of United States. *Annals of the Association of American Geographers* 103(2):385–396. doi: 10.1080/00045608.2013.756267.

Nelson, Barry and Denise Fort. 2012. Pipe Dreams: Water Supply and Pipeline Projects in the West. *Natural Resources Defense Council*, 2012.

NRCS (Natural Resources Conservation Service). 2014. *Ogallala Aquifer Initiative*. U.S.D.A. (United States Department of Agriculture). Accessed May 13, 2015. Available at http://www.nrcs.usda.gov/wps/portal/nrcs/detailfull/national/programs/initiatives/?cid=stelprdb1048809

Pilz, Robert David. 2010. Lessons in water policy innovation from the world's driest inhabited continent: Using water allocation plans and water markets to manage water scarcity. *University of Denver Water Law Review* 14:97–130.

Postel, Sandra. 1999. *Pillar of Sand: Can the Irrigation Miracle Last*? New York and London: Worldwatch W.W. Norton.

—. 2010. Water: Adapting to a New Normal. In *The Post Carbon Reader: Managing the 21st Century's Sustainability Crises*, edited by R. Heinberg and D. Lerch. Healdsburg, CA: Watershed Media.

Public Policy Institute of California. 2012. *California's Water Market, By the Numbers, Update 2012*, by E. Hanak and E. Stryjewski. Sacramento, CA.

Ruhl, J.B. 2008. Climate Change and the Endangered Species Act: Building Bridges to the No-Analog Future. *Boston University Law Review* 88:1–62.

Sax, Joseph. 1988. The Limits of Private Rights in Public Water. *Environmental Law* 19:473–483.

Schlager, Edella and Tanya Heikkila. 2011. Left High and Dry? Climate Change, Common Pool Resource Theory, and the Adaptability of Western Water Compacts. *Public Administration Review* 71(3):461–470.

Squillace, Mark. 2013. Water Transfers for a Changing Climate, *Natural Resources Journal* 53:55–116.

Sundareshan, Priyanka. 2010. Using the Transfer of Water Rights as a Climate Change Adaptation Strategy: Comparing the United States and Australia. *Arizona Journal of International and Comparative Law* 27:911–944.

Supreme Court of the United States. 2013. *Texas v. New Mexico and Colorado*. No. 22O141. Available at http://www.scotusblog.com/case-files/cases/texas-v-new-mexico-and-colorado/.

Tarlock, A. Dan. 2012. Takings, Water Rights, and Climate Change. *Vermont Law Review* 36:731–758.

USBR (United States Bureau of Reclamation). 2007. *Colorado River Interim Guidelines for Lower Basin Shortages and Coordinated Operations for Lake Powell and Lake Mead*. Available at http://www.usbr.gov/lc/region/programs/strategies.html.

USDA (United States Department of Agriculture) Economic Research Service. 2013. *Western Irrigated Agriculture: Production Value, Water Use, Costs, and Technology Vary by Farm Size*. Available at http://www.ers.usda.gov/amber-waves/2013-september/western-irrigated-agriculture-production-value,-water-use,-costs,-and-technology-vary-by-farm-size.aspx.

USGS (United States Geological Survey). 2013. *Water-Level and Storage Changes in the High Plains Aquifer, Predevelopment to 2011 and 2009–11*. Available at http://ne.water.usgs.gov/ogw/hpwlms/files/McGuire_gwlevels_2013_final.pdf.

Village of Logan v. U.S. Dept. of Interior, 577 Fed. Appx. 760 (10th Cir. 2014). Available at http://law.justia.com/cases/federal/appellate-courts/ca10/13-2082/13-2082-2014-08-25.html.

Wilson, Edward O. 2002. *The Future of Life*. New York: Knopf.

4

The West's Water—Multiple Uses, Conflicting Values, Interconnected Fates

David Lewis Feldman

CONTENTS

ABSTRACT This chapter considers the competing values that have animated western water policy. Diverse views toward exploitation and development, allocation, rights, environmental/in-stream needs, and the restoration of rivers have long characterized policy debates. We begin by examining the relative scarcity of water in the region; the drive to overcome shortages of surface water and groundwater through improvised and heavily engineered public works and law; and the encouragement—and consequences of—rapid population growth and urbanization. We then chronicle the role of irrigation boosters and their fervent belief in federally provided water as a means of encouraging self-reliant farmers. By the early twentieth century, Native American rights, environmental concerns, and fervent opposition to many interbasin diversions exacerbated issues over reserved rights and in-stream flow protection. By mid-century, these concerns became wedded to a strong preservationist ethic expressed by artists, writers, and resource protection advocates—while by the end of the twentieth century, changing public attitudes led to efforts to restore urban waterways—an issue that has become entangled in larger debates over gentrification, economic revitalization, open space, and community identity. Changing attitudes have also revived debates over the role of underrepresented groups in shaping water policy. We conclude by touching upon two looming challenges: growing controversies around innovations to relieve climate-related and urbanization-induced water stress, including conservation and supply innovations (e.g., wastewater reuse), as well as changes to the region's hydrology through reliance upon imported water.

4.1 Introduction—Conflicting Values over Water

The West's water is subject to multiple and competing uses as well as divergent values. A common critique of water management in the West is that it is biased toward exploitive, acquisitive goals. Moreover, the dominant ethos guiding water policy is said to embrace profligate use over conservation. Over time, this ethos has become problematic due to the onset of protracted drought and climate variability (Powell 2008).

As in much else claimed about western water policy, this critique is only partly valid. This is so for two reasons. First, there have long been confounding attitudes shaping the West's views toward water. These attitudes promote preservation of the region's natural amenities, fervent protection of freshwater, and equity in decision making.* While perhaps less influential than the mainstream views cited, these attitudes have influenced the West's values toward water and affected contemporary policies toward its management.

Second, westerners have not always acted alone in shaping policies toward water in the West, which historically reflected a ubiquitous and regionally transcendent ethos characterized by narrow, acquisitive views toward the management of natural resources in the United States (Feldman 1995, p. 3). As the historian Earl Pomeroy (1955) noted some 60 years ago, the West's approaches to resource management have been strongly influenced by external influences both private (e.g., land investors, settlers from the east, immigrants from overseas) and public (e.g., competing federal agencies with different missions). This has been true since various parts of the West became part of the United States in the early nineteenth century (Nugent 2001).

This chapter contends that the preeminent goals of western water policy have been the desire to foster settlement and sustain agriculture and cities. The highest values have been efficiency and the harnessing of scarce water resources. These aspirations—while dominant—have long been challenged by minority voices concerned with resource preservation and justice. Concerted efforts have been made to wed ethics and law to the dominant policy objective of development, with the cherished hope that, while support external to the region could be acquired to achieve the goal of efficient harnessing of water resources, water management would be locally determined. In essence, state governments would use federally provided tools and financing for ends desired by the region's population. As Norris Hundley Jr. (2009) put it, policymakers have sought "…to get the purse strings without the purse" (p. xi).

While policymakers have long focused on ways to use federal intervention to harness and allocate water, the future of water management in the West is likely to rely increasingly on local initiatives and local means to manage shortages exacerbated by growing demands and diminishing supplies. We begin with an analysis of water availability and use as policy drivers.

* See Kurt Repanshek, "Survey of Western Attitudes Shows Strong Support for National Parks, Clean Environment," *National Parks Traveler,* January 31, 2012, at http://www.nationalparkstraveler.com/2012/01/survey-western-attitudes-shows-strong-support-national-parks-clean-environment9390. See also the *2014 Conservation in the West Poll, Executive summary*, p. 35, Colorado College state of the Rockies project, at http://www.coloradocollege.edu/other/stateoftherockies/conservationinthewest/.

4.2 Water Availability as Policy Driver

An inexorable condition of the West's water is its relative scarcity. The West is the most arid region in the continental United States, a condition exacerbated by chronic scarcity relative to desired uses and high variability (Gleick and Heberger 2012). As Table 4.1 indicates, average rainfall in the 17 states west of the 100th meridian is well below the average of other states, while the 5 most arid US states, by average annual precipitation, are all in the West.

Scarcity and the quest for overcoming its constraints have long been drivers of the region's water practices, laws, and institutions. From the inception of European settlement, improvised public works were introduced to harness intermittent streamflow. Spanish settlers in the Southwest, for instance, utilized *zanjas*—literally, hand-tooled ditches—to divert water from streams and rivers to irrigate small farms and orchards, as well as large estates, pueblos, and missions. A long-standing practice in much of Spanish-settled western North America, this improvisation was continued by Anglo settlers who adopted their own variants, appending to these ditches water wheels, mills, and other engineered features (Hundley 2001). In other instances (e.g., Mormon settlers in the Salt Lake Valley), innovations consisted of cooperative efforts to dam local creeks and make ready "pools, vats, tubs, reservoirs, and ditches at the highest points of land" to be filled and drawn off as needed (Morgan 1947, pp. 198–207). Means to harvest groundwater, arising later, also used improvised advances (Bryner and Purcell 2003).

Ingenuity as a means of overcoming relative scarcity secured local supplies. It also had a deleterious effect: exacerbating pressures on these supplies through encouraging growth, sustained by the conviction that improvised public works could support larger settlements. Unlike other arid regions, the West has, in recent times, experienced dramatic population growth that has strained limited water availability (Hoerling and Eischeid 2007). And, while population growth has not been the sole or even dominant determinant of water use, it is an important driver of ongoing growth in demand.

In recent decades, some states—notably Arizona, Colorado, and Nevada—have experienced phenomenal rates of urban growth, and thus, huge increases in demands for outdoor landscaping and other uses associated with residential and commercial development (National Research Council 2007). In fact, the West has long been the most urbanized region in the United States in terms of the percentage of people living in cities. Prophetically, harnessing and diverting streams made possible the dramatic growth of cities as hydrologically diverse as Los Angeles, San Francisco, Denver, Albuquerque, Las Vegas, Phoenix, Seattle, and Salt Lake City—among others (Pomeroy 1965; Fradkin 1995; Phillips et al. 2011).

Since the 1970s, urban growth has prompted serious action—in some instances, unprecedented in scope—to adopt aggressive conservation measures, including pricing policies and metering, to encourage savings in per capita use (City of Los Angeles Department of Water and Power 2008, 2010; Hanak et al. 2011). Despite such measures, continued growth portends significant upward demands. Together with climate variability and drought, growth will make the task of providing additional water supply especially problematical as reliance upon traditional sources, such as large reservoirs, becomes more difficult and as *satisfying the unique demands* characteristic of the West—as discussed in Section 4.2.1—exacerbate these pressures.

4.2.1 An Interconnected Web—Water Use and Infrastructure

Water uses in the West are highly diverse. Hydrologists, water planners, and public officials identify six general categories of uses: agriculture; municipal and industrial use;

TABLE 4.1

Average Total Yearly Precipitation for Each State, 1971–2000

State	Inches	Millimeters	Rank
Alabama	58.3	1480	4
Alaska	22.5	572	39
Arizona	13.6	345	47
Arkansas	50.6	1284	8
California	22.2	563	40
Colorado	15.9	405	44
Connecticut	50.3	1279	9
Delaware	45.7	1160	16
Florida	54.5	1385	5
Georgia	50.7	1287	7
Hawaii	63.7	1618	1
Idaho	18.9	481	42
Illinois	39.2	996	27
Indiana	41.7	1060	26
Iowa	34.0	864	31
Kansas	28.9	733	35
Kentucky	48.9	1242	12
Louisiana	60.1	1528	2
Maine	42.2	1072	23
Maryland	44.5	1131	18
Massachusetts	47.7	1211	14
Michigan	32.8	833	32
Minnesota	27.3	693	37
Mississippi	59.0	1499	3
Missouri	42.2	1071	24
Montana	15.3	390	45
Nebraska	23.6	599	38
Nevada	9.5	241	50
New Hampshire	43.4	1103	20
New Jersey	47.1	1196	15
New Mexico	14.6	370	46
New York	41.8	1062	25
North Carolina	50.3	1279	9
North Dakota	17.8	452	43
Ohio	39.1	993	28
Oklahoma	36.5	927	30
Oregon	27.4	695	36
Pennsylvania	42.9	1089	21
Rhode Island	47.9	1218	13
South Carolina	49.8	1264	11
South Dakota	20.1	511	41
Tennessee	54.2	1376	6
Texas	28.9	734	34
Utah	12.2	310	49

(*Continued*)

TABLE 4.1 (CONTINUED)

Average Total Yearly Precipitation for Each State, 1971–2000

State	Inches	Millimeters	Rank
Vermont	42.7	1085	22
Virginia	44.3	1125	19
Washington	38.4	976	29
West Virginia	45.2	1147	17
Wisconsin	32.6	829	33
Wyoming	12.9	328	48

Source: NOAA National Climatic Data Center.

energy; minerals; fish, wildlife, and recreation; and tribal uses (US Bureau of Reclamation 2012).* While none of these is totally unique to the West, their *relative allocation* conforms to a distinct regional pattern (Figure 4.1).

Irrigation remains the dominant category of water use in the western states, despite recent urban population growth. Consumptive water demands for agriculture have long been a primary impetus for construction of large-scale, centralized provision systems. Water allocation to agricultural uses in the West is high, with some 80% of western supplies devoted to crop production. While modest shifts of agricultural water to municipal and industrial uses could meet growing urban demands, the losses that would be sustained by food production are thought by many to be a serious trade-off (US Department of the Interior 2012).

Per capita public water use is approximately 50% higher in the West than in the East, due mostly to landscape irrigation (US EPA 2004). Like agriculture, landscape uses are consumptive, placing extra pressure on the region's supplies, because once water is applied to crops, lawns, and golf courses, a large fraction of it evaporates and transpires—and thus, is not available for subsequent users. Some western cities, finding themselves short of water, have pursued two basic alternatives. The first is mandated water conservation measures that inhibit (or even prohibit) certain types of landscaping. These measures were introduced in places such as Las Vegas, Tucson, and California's Coachella Valley. Alternatively, cities have instituted strong economic incentives to induce switching to drought-tolerant landscaping and xeriscaped yards and gardens (Hanak et al. 2011). These measures often faced significant urban design and residential population hurdles before they became widely accepted. Compelling efforts making the case that they are not only environmentally sensible but also aesthetically attractive have helped to achieve public acceptance (e.g., Southern Nevada Water Authority, undated).

By the beginning of the twentieth century, water supply infrastructure, funded mostly by the federal government, included major dams and appurtenant projects (e.g., California's Colorado River Aqueduct, the All-American Canal, and a wide range of other distribution systems), which ensured that portions of the Upper and Lower Colorado Basins, and other watersheds, were provided with dependable supplies. These projects

* Agricultural water is used to meet irrigation requirements of crops, maintain stock ponds, and sustain livestock. Municipal/industrial (M&I) water is used to meet urban and rural population needs and industrial needs. Energy refers to water used for energy services and development as well as needs for energy generation. Minerals refer to water used for mineral extraction not related to energy. Fish, Wildlife, and Recreation means water used to meet National Wildlife Refuge, National Recreation Area, state park, and off-stream wetland habitat needs, as well as Endangered Species Act–listed species needs, and ecosystem and other instream needs. Finally, tribal uses include water to meet tribal needs and settlement of tribal water rights claims.

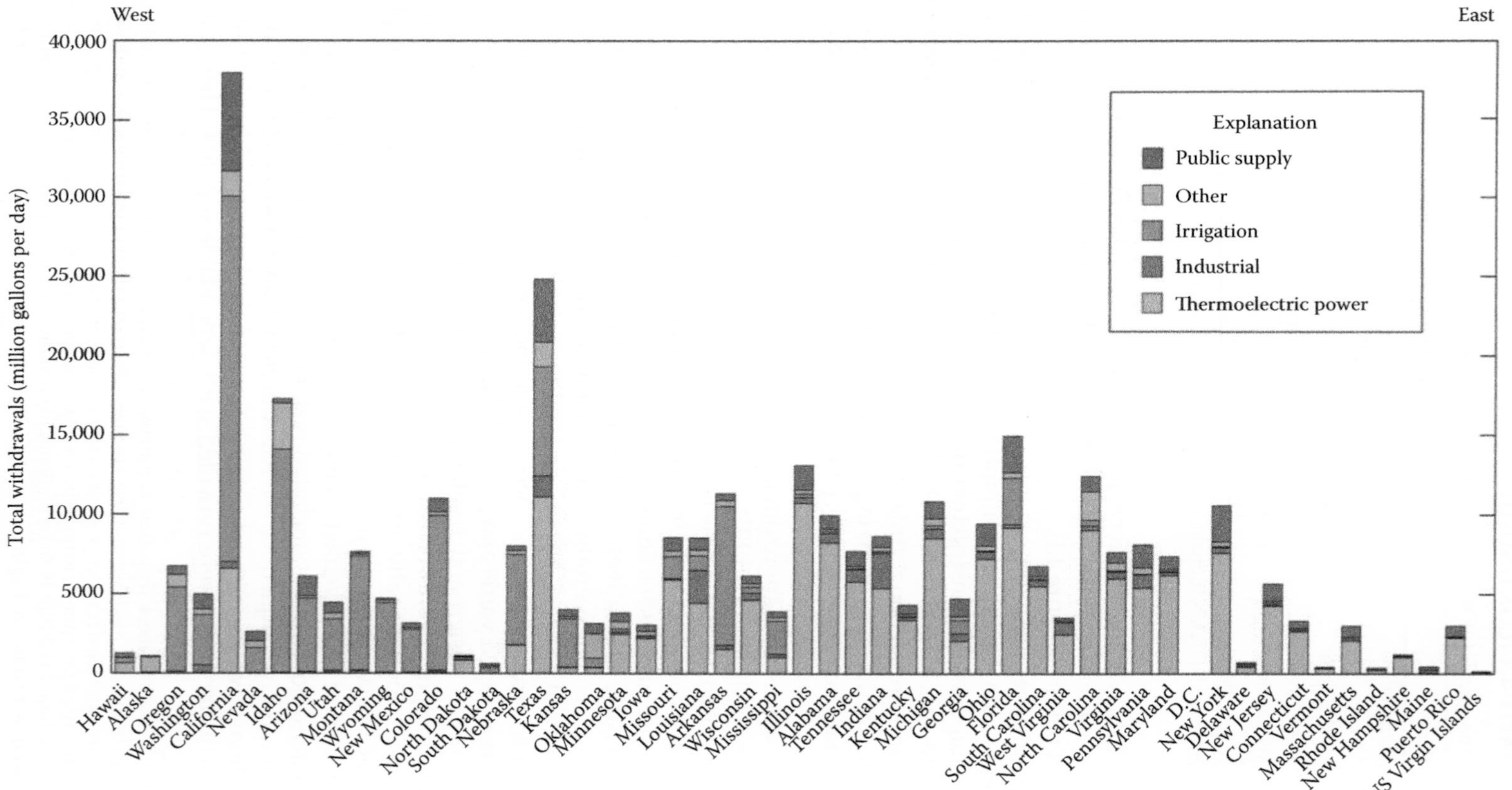

FIGURE 4.1
Total water withdrawals showing category of use by state from west to east, 2010. (From Maupin, M.A. et al. Estimated Use of Water in the United States in 2010, *U.S. Geological Survey Circular*, 1405, 56 pp., 2014.)

also provided flood control and hydropower for many of the region's cities as well as irrigation in California's Imperial and Central Valleys, Central Arizona, Utah, New Mexico, and the interior regions of Oregon, Washington, and Idaho. Cushioning the West against water shortages and at the same time ensuring dependable supplies were explicit objectives articulated in these projects' enabling statutes (e.g., Boulder Canyon Project Act 1928; Boulder Canyon Project Adjustment Act 1940; Colorado River Storage Project Act 1956; Colorado River Basin Project Act 1968; US Department of the Interior 2008). By the late twentieth and early twenty-first centuries, a gradual policy shift away from developing water resources, and toward improving efficiency of reservoir operations while optimally managing increasingly vulnerable supplies, took hold in the region. Two examples are *adaptive management* efforts on the lower Colorado River and changes to the operation of dams to mimic historical flows. Both of these have resulted in adoption of new reservoir operating regimes, despite growing challenges in meeting unabated demands in the face of overallocation of the Colorado River; see Figure 16.2 in this volume (US Department of the Interior, Bureau of Reclamation 1996, 2007, 2012). Despite changes in reservoir operations by federal agencies, one public expectation has largely remained unchanged: that the operation of this elaborate infrastructure should be determined by local needs, via local management. This desire to maximize regional benefit through water resource development provided by the national government on one hand, coupled with popular control over provision of water from within the region on the other, has been justified by a unique ethos. As we will see, ingenuity in law and policy advanced this ethos.

4.2.2 Another Interconnected Web—Ideology, Law, and Policy

Development of water infrastructure is only one means by which westerners overcame the constraints imposed by limited water. A more important vehicle has been the web of ideas, ideology, and—by derivation—the laws they begot. Throughout the nineteenth century, a fervent belief in the benefits of cooperative efforts to irrigate land, divert water to growing cities, and *equalize* the benefits of what water resource development could yield took root in the region. Thanks to figures such as Benjamin Cummings Truman, Charles Howard Shinn, and most notably, William Ellsworth Smythe, widespread public support for federal investments in water projects received an enormous boost (Starr 1990). Smythe was the most ardently evangelical of the proponents of irrigation as a means of developing a virtuous society of yeoman farmers. He contended that irrigation was a "miracle" whose introduction would further the cause of democracy by creating a West populated by self-reliant agrarians (Smythe 1900).

Aside from organizing *irrigation congresses* that had enormous influence upon the US Congress, these advocates linked the economic advantages of water resource development to deeper moral and even religious arguments. *Reclaiming* arid, but potentially fertile, land—it was argued—would produce a thriving middle class, a healthful rural lifestyle, and a new civilization that would unleash the talents and abilities of the common man. Such developments would further generate unprecedented economic opportunities through flood control, power generation, and industrialization benefitting the entire nation, not just the West. Irrigation enthusiasts helped accelerate the codification of laws and other policies favoring senior appropriators in the West—those who laid first claim to water rights within an area. Later, water markets, permitting water transfers to higher-valued uses, were established in part because of the advantages prior appropriation law afforded for the ability to move water *off site*, so long as the use was deemed beneficial.

In most instances, this meant support of growing agricultural and urban water demands (Gleick 2003; MacDonnell and Fort 2008).

While water law is discussed in Chapter 3 of this volume, its role in codifying user values must be briefly noted here. In general, western water law emphasizes beneficial use, not locking up water, and is based on the same utilitarian premises that gave rise to other water policies.* Like public works, legal institutional innovations focused largely on augmenting supply and only rarely on attenuating demand or encouraging efficiency of use.

A further example of utilitarian values in institutional innovation was the establishment of the Bureau of Reclamation. Its official rationale implied that the West was an underwatered wasteland: an ideal place for irrigation to create a productive agricultural paradise through using '…surplus fees from sales of land (to) be set aside for a *"reclamation fund"* for the development of water resources.' Only the federal government possessed the money, engineering skills, and bureaucratic means to establish a comprehensive plan for watering the region (National Reclamation Act of 1902 Section 1; Clayton 2009; Jones and Wills 2009).

Despite federal encouragement of irrigation, nonconsumptive, instream water uses have also been enshrined in western water law as a reflection, in part, of the value placed on *equity, as well as efficiency*. The federal government has long held so-called proprietary interests in water on or under federal lands. Once these lands were relinquished to states added to the union, the national government reserved water rights to ensure public supplies on military bases, and for fish and wildlife habitat, instream flows in national parks and forests, and—later—tribal reservations (*Winters v. United States 1908*; California-Oregon Power Company 1935).

An ongoing debate is the question of how much water can be *reserved* for federal purposes—in part, a matter of water rights seniority. In the face of increasing drought and diminishing flow, however, this issue may take on greater importance and lead to pronounced conflicts in the region (Gillilan and Brown 1997). Moreover, it reflects a larger value debate regarding the proper balance between *efficiency* (putting water to a beneficial use) and *equity* (ensuring that whatever use water is put to treats users fairly). This is a multifaceted issue.

4.3 Equity and Environmental Values versus Utilitarian Efficiency

Some scholars have long decried the West's water policies as epitomizing a *hydraulic society* based on centralized, elite control of water resources; bureaucratization of development decisions; and the marriage of capital accumulation and technology (Worster 1986). The dominant norms at the core of these policies overwhelmingly favor efficiency and all but ignore equity, it is claimed. These norms originated in the Progressive Era's doctrine of utilitarian conservation and the "gospel of efficiency" (Hays 1999); the exalting of agrarian-based democracy promoted by the frontier ingenuity of the region's early settlers (Webb 1931); or by the "industrial conquest of the West" fostered by a survival-of-the-fittest ethic, which embraced rapid growth regardless of ecological cost (Jones and Wills 2009). As noted previously, however, not everyone has embraced the value of extractive use and

* An 1853 California Supreme Court decision (*Irwin v. Philips*) ruled that senior appropriators could divert water to the detriment of downstream riparians, based on the rationale that prior appropriation (already in operation) led to installation of "costly artificial works" to sustain appropriators' rights. An 1882 Colorado Supreme Court case (*Coffin v. Left Hand Ditch Company*) also upheld the prior appropriation's seniority system (Clayton 2009, p. 21).

maximizing utilitarian efficiency through harnessing water. Equity and the social value of public access to environmental amenities have also been important concerns, even to irrigation boosters.

Nevertheless, defining equity is no easy matter. It is such a broad concept that debates over what is equitable in water policy span a wide range of issues. As an official of the Wilderness Society recently noted, an *equitable* framework for managing natural resources could include economic, racial, gender, religious, and international or political equity and embrace the rights and dignity of individuals and diverse cultures, as well as the moral standing of nonhuman species (Morgan 2009). Traditionally, discussions of water equity in the West have largely focused on *regional* equity. Examples include controversial inter-basin transfers ranging from diversion of water from the Owens Valley to Los Angeles in the early twentieth century to more recent plans by Las Vegas, thwarted as of this writing, to divert some 84,000 acre-feet*/year of groundwater from beneath portions of the Great Basin (EcoWatch 2013). In most cases, regional equity disputes revolve around the impacts to one region's historical water rights from efforts to enhance the economic benefits of another. These disputes inevitably generate feelings of oppression and exploitation (Walton 1993). While these conflicts remain important, they are increasingly subsumed under two larger sets of water equity debates. These are (1) the relative importance of development aspirations versus those accruing to environmental protection (or even restoration) and (2) who should participate in decisions.

4.3.1 Equity as the Public Interest in Environmental Preservation

A long-standing tradition of water equity in the West takes as its point of departure the importance of preserving and protecting water resources—and the species and habitat that depend on them—in as close to a natural state as possible. Increasingly, this argument is prompted by proposals to undertake water transfers or build new impoundments. There are roughly two facets to this argument. The first, articulated by western writers as diverse as John Muir, Mary Austin, and Wallace Stegner (and by numerous artists and photographers who have *visually* expressed these views, e.g., http://www.photographsofthewest.org/about/) posits that the intrinsic beauty of arid lands and their awe-inspiring richness, variety, and power shape the character of the region, while also serving to morally instruct its inhabitants in humble living. Muir's opposition to plans to dam the Hetch Hetchy River to provide water for San Francisco in the early twentieth century exemplified this argument. The second facet is represented by environmental activists such as David Brower and Edward Abbey, for whom writing about water was as much an act of political engagement as a literary endeavor.

Austin's work, like that of Wallace Stegner many years later (e.g., Stegner 1969), emphasized the frailty of nature and the vulnerability of people who choose to live in the West. A larger lesson of her work is that—ethically—survival in such an environment is an exercise in the art of harmonious living: but only if one seeks to live in harmony with the water limits imposed by the environment, and not by arrogantly surmounting them. Her arguments contained a semimystical quality that profoundly influenced later environmentalists. Particularly influential was the notion that prudent coexistence with the land, as opposed to its overdevelopment, leads to a fairer and more environmentally virtuous society (Austin 1903). By contrast, Brower and Abbey viewed their writing, and the arguments they espoused for direct action against the politics of dam building in the West, as

* 1 acre-foot (af) = 1233.48 cubic meters (m^3).

manifestos. To an underappreciated extent (given that neither prevented the construction of Glen Canyon Dam, for example), they achieved some measure of success.

Their writings helped articulate the critical stance that human *dominion* over magnificent, free-flowing rivers through the building of large impoundments achieves a kind of Pyrrhic victory. Such actions obliterate distinctive landscapes, transforming the West into a region overly dependent on a water-based, resource-extractive economy—anomic, alienating, socially vacuous. Moreover, Abbey, Brower, and their disciples gave voice to a political movement that, over time, placed advocates of large water projects on the defensive and thwarted other projects (e.g., damming Marble Canyon). As advocates for preservationist equity, Brower's Sierra Club and Abbey's literary disciples differed in strategy. Brower sought to fuse ethics and science, and to harness hard data and technical knowledge to justice-based arguments when battling water agencies. In later years, he sought to marshal science in efforts to decommission Glen Canyon Dam (Farmer 1999; Powell 2008). By contrast, Abbey was more poetic in his appeal, arguing that the damming of the Colorado River was a moral loss—not merely an economic or environmental one. Once dammed, free-flowing rivers were forever destroyed and forever beyond recovery, as was the life of tranquility they once promised (Abbey 1968; Abbey and Hyde 1971; Farmer 1999).

While their direct influence on later debates on water equity is difficult to measure precisely, in some respects, they served as apocryphal voices foreshadowing an attitudinal change in the West. A further impetus for change came from parallel efforts by environmental economists to measure the value of undisturbed natural environments and to call attention to the damage to public welfare that dam projects could entail (Pearce 2002). The power of the attitudinal shift is vividly apparent in the context of river restoration efforts in cities such as Los Angeles, Denver, Phoenix, and elsewhere. The possibility that urban waterways can be something other than navigation channels or floodways and that—restored to something resembling their natural condition—they can serve as both recreational amenities and focal points of community engagement has seized the imagination of environmental equity advocates. Moreover, local officials are embracing the view that restored rivers can be a source for urban health and economic revitalization. Rivers are increasingly being viewed as focal points for defining a sense of community and as reminders of the "primal force of clean, flowing water" (Gumprecht 2005).

There is another equity challenge entailed in these efforts: the balance between river restoration as a means of reinvigorating local economies on one hand and as means of providing open space and public park land on the other. In Los Angeles, a question that has arisen amidst the efforts of restoration groups such as the Friends of the Los Angeles River and their allies is the following: on whose behalf should restoration occur? Will the river become a civic amenity available to all residents or a gentrified landscape accessible to only a few (Kibel 2007)?

In sum, changing conditions influence how westerners value water. During initial settlement, rivers were harnessed and water withdrawn to support growth. As cities industrialized, streams were channelized and forcibly confined to their banks. The postindustrial West is preoccupied with restoring urban rivers and preserving wild and scenic rural streams (Orsi 2005). The fact that these values have changed underscores the increasingly cosmopolitan character of the West and the elusive waterscape of equity.

4.3.2 Equity as Decision Making

In recent years, this equity–efficiency debate has faced new challenges involving the protection of underrepresented groups. California's San Joaquin Valley affords an important case study of these challenges. In Kern and Tulare Counties, for instance, many rural, low-income communities—many unincorporated towns or mobile home parks—are served by small, private, not-for-profit water systems or by utilities that are entirely investor owned. These communities—and their water providers—arose during an era when farm workers were prohibited from living in larger towns, with the result that few water utilities provided service to them. According to the Center on Race, Poverty and the Environment in Delano, California, the result has been a chronic lack of safe drinking water as a result of contaminated farm runoff and small water systems being unable to ensure safe drinking water since they cannot rely on surrounding communities to finance improvements. In short, many communities have become "locked…into a cycle of environmental injustice" (Environmental Justice Coalition 2005).

This is not only a rural issue—nor one affecting only people of color. Many western states have launched a number of legislative efforts to ensure that equity in representation and decisional voice be incorporated into water-provider decision making in cities as well as rural areas. These include reforms that require that when redistricting water utilities, past actions that may have left some neighborhoods without adequate access to water supply or treatment facilities be considered. In addition, if low-income communities have been bypassed when past decisions to develop water and sewer infrastructure were made, or rate-payer decisions formulated, measures to assist low-income groups in retrofitting water appliances and repairing distribution systems should be provided (e.g., Burch 2005). From the standpoint of debates regarding equity and efficiency, water law has also affected representation in water decisions in other ways. In all western states, constitutions explicitly ensure that water is a "public resource" and that the state is responsible for managing water on behalf of all citizens (MacDonnell and Fort 2008). Even when private rights to water are established under prior appropriation systems (see Section 4.2.2), states reserve the power to ensure that water uses do not harm other water rights holders, or the environment. Only that amount of water reasonably necessary to accomplish the particular use is protected as a legally recognized *beneficial use* (MacDonnell and Fort 2008). Led by Wyoming, western states also developed distinct administrative and judicial systems to clarify rights, mediate conflicts, and enable consideration of broader interests—all to secure equitable water allocation and ensure that narrowly defined *efficient use* does not dominate decisions. As western states move to consider ways to manage extreme climate variability and competing demands, the use of these tools to address equity issues is likely to grow. A notable example is Arizona's Groundwater Management Act, instituted in 1980, which has made significant strides in helping reduce municipal reliance on groundwater, mandates an assured supply of freshwater for new development, encourages aquifer recharge and recovery in sensitive areas, issues permits for groundwater pumping, and monitors drawdowns. In trying to ensure *safe yields* for groundwater, the act seeks to secure sustained protection of the public interest as a consideration in its management (Megdal 2012).

4.4 Conclusions—Future Challenges

The West has long faced unique challenges generated by aridity, growth, and more recently, a changing climate. In forecasting further challenges to the West's water future, we first need to consider what makes the West's water problems truly distinctive. While water has long been in short supply, disputes over shortages are no longer unique to the region. Atlanta and much of the southeast, for instance, face conflicts over water allocation due to rapidly growing urban demands exacerbated by drought. Proposals for interbasin diversion in that region and others provoke strikingly similar debates to those long faced by western cities. Moreover, many eastern cities (e.g., Boston and New York) grew to economic prominence in part through the diversion of distant freshwater sources. Two challenges appear likely.

First, as growing cities and the continued needs of agriculture impose demands on diminishing supplies, a variety of innovations will receive increased attention. These range from transferring water rights from lower- to higher-valued uses through markets, to aggressive conservation measures such as tiered-pricing systems and potentially controversial supply innovations, including greater reliance on desalination and recycled wastewater. Throughout the West, debates are arising among stakeholders regarding the *equity* of these and other measures. Disputes regarding the impacts of supply options to environmental amenities (such as coastal zones in the case of desalination); acceptability to underrepresented groups suffering from environmental justice issues (in the case of plans for recycled wastewater plants); and fairness to middle-income farming communities (in the case of water transfers to protect endangered or threatened species) are growing in both frequency and intensity. These issues are less likely to generate intense, short-term social movements—as was true in, say, disputes over interbasin diversions—and more likely to result in long-term, low-intensity erosion of public trust, confidence, and civility, as well as political gridlock in decisions regarding the adoption of such innovations (Feldman 2011).

A second challenge is related to the first. Because of the dramatic modification of its water environment, the West may be transforming its hydrology. As in other arid regions, limited water resources have long been supplemented by reliance on imported water. In coastal southern California, for example, scarce local sources are enhanced by large volumes (~1.6 billion m^3 annually) of imported water from the San Joaquin–Sacramento River delta, Colorado River, and Owens Valley. Ironically, this has meant that rivers that used to flow seasonally with intermittent discharge are now experiencing increased summer discharge due to wastewater discharges and urban runoff, which depend on imported water. Moreover, an increased impervious surface exacerbates large nonpoint-source discharges, especially during storms that carry large amounts of contaminants: a trend that has been observed in San Diego, Los Angeles, Austin, Texas, Denver, and elsewhere (Townsend-Small et al. 2013). This change in hydrology undermines many of the strategies western water managers pursue to ensure future supplies, such as storm water harvesting. Storm water cannot be a source of supply unless contaminants can be safely removed, and it cannot contribute to groundwater recharge, since many shallow aquifers are already saturated.

The West's ability to meet both these challenges will require innovations in technology, law, and policy. Most of all, changes in attitudes and values will be needed. Reconfiguring hydrology can only be done by restoring rivers; adopting regionally sensitive integrated water planning to permit storm-water storage and reuse where appropriate; and decreasing reliance on imported water. All these changes will—as we have noted—require overcoming perceived inequities in the application of innovations. Given the West's emphasis on large-scale supply solutions, and recognizing that path-dependent policies are difficult to change, it is clear that reforms will not come easily.

References

Abbey, Edward. 1968. *Desert Solitaire: A Season in the Wilderness.* New York: McGraw-Hill.

Abbey, Edward, and Philip Hyde. 1971. *Slickrock: Endangered Canyons of the Southwest.* New York: Sierra Club/Scribner's.

Austin, Mary. (1903). 1997. *The Land of Little Rain.* Reprint. New York: Penguin Books.

Boulder Canyon Project Act, 45 Stat. 1057, 1928.

Boulder Canyon Project Adjustment Acts, 54 Stat. 774, 1940.

Bryner, Gary, and Elizabeth Purcell. 2003. *Groundwater Law Sourcebook of the Western United States.* Boulder, CO: Natural Resources Law Center, University of Colorado Law School.

Burch, Marsha A. 2005. Water Rights: Supply Issues for Local Agency Formation Commissions. Paper presented at California Local Agency Formation Commission Annual Conference, Monterey, CA.

California Oregon Power Company v. Beaver Portland Cement Company, 295 U.S. 142, 1935.

City of Los Angeles, Department of Water and Power. 2008. Securing L.A.'s Water Supply.

City of Los Angeles, Department of Water and Power. 2010. Urban Water Management Plan. http://www.ladwp.com

Clayton, Jordan. 2009. Market-Driven Solutions to Economic, Environmental, and Social Issues Related to Water Management in the Western USA. *Water* 1:19–31. doi:10.3390/w1010019.

Colorado College. 2014. *The 2014 Conservation in the West Poll-Executive summary.* State of the Rockies project. Available at http://www.coloradocollege.edu/other/stateoftherockies/conservationinthewest/

Colorado River Storage Project Act, 70 Stat. 101, 1956.

Colorado River Basin Project Act, 82 Stat. 885-6, 1968.

EcoWatch. 2013. Landmark Ruling Rejects Disastrous Water Diversion Project in Nevada and Utah. *EcoWatch—Transforming Green.* December 12. Available at http://ecowatch.com/2013/12/12/ruling-rejects-water-diversion-project-nevada-utah/.

Environmental Justice Coalition for Water. 2005. Thirsty for Justice: A People's Blueprint for California. Oakland, CA: Environmental Justice Coalition for Water, June.

Farmer, Jared. 1999. *Glen Canyon Dammed: Inventing Lake Powell and the Canyon Country.* Tucson: University of Arizona.

Feldman, David L. 1995. *Water Resources Management: in Search of an Environmental Ethic.* Baltimore: Johns Hopkins University.

Feldman, David. 2011. Integrated Water Management and Environmental Justice—Public Acceptability and Fairness in Adopting Water Innovations, *Water Science and Technology—Water Supply* 11(2):135–141. doi: 10.2166/ws2011.035.

Fradkin, Philip. 1995. A River No More: The Colorado River and the West. Berkeley: University of California.

Gillilan, David M., and Thomas C. Brown. 1997. *Instream Flow Protection: Seeking a Balance in Western Water Use.* Washington, DC: Island Press.

Gleick, Peter H. 2003. Water Use. *Annual Review of Environmental Resources* 28:275–314. doi: 10.1146/annurev.energy.28.040202.122849.

Gleick, Peter H., and Matthew G. Heberger 2012. The Coming Mega Drought. *Scientific American* 306:1–14.

Gumprecht, Blake. 2005. Who Killed the Los Angeles River? In *Land of Sunshine: an Environmental History of Metropolitan Los Angeles,* eds. William Deverell and Greg Hise, 115–134. Pittsburgh: University of Pittsburgh.

Hanak, Ellen, Jay Lund, Ariel Dinar, Brian Gray, Richard Howitt, Jeffrey Mount, Peter Moyle, and Barton "Buzz" Thompson. 2011. *Managing California's Water: From Conflict to Reconciliation.* San Francisco: Public Policy Institute of California.

Hays, Samuel P. 1999. *Conservation and the Gospel of Efficiency: The Progressive Conservation Movement, 1890–1920*. Pittsburgh: University of Pittsburgh.

Hoerling, Martin P., and Jon Eischeid. 2007. Past peak water in the Southwest. *Southwestern Hydrology*. January–February:18–35.

Hundley, Norris, Jr. 2001. *The Great Thirst—Californians and Water: A History*, rev. edition. Berkeley: University of California.

Hundley, Norris, Jr. 2009. *Water and the West—The Colorado River Compact and the Politics of Water in the American West*, 2nd edition. Berkeley: University of California.

Jones, Karen R., and John Wills. 2009. *The American West: Competing Visions*. Edinburgh: University of Edinburgh.

Kibel, Paul Stanton, ed. 2007. *Rivertown: Rethinking Urban Rivers*. Cambridge, MA: MIT Press.

MacDonnell, Lawrence J., and Denise D. Fort. 2008. *Policy Report—A New Western Water Agenda: Opportunities for Action in an Era of Growth and Climate Change*. Albuquerque, NM: University of New Mexico Law School, February.

Maupin, Molly A., Joan F. Kenny, Susan S. Hutson, John K. Lovelace, Nancy L. Barber, and Kristine S. Linsey. 2014. Estimated Use of Water in the United States in 2010: U.S. Geological Survey Circular 1405, 56 pp., http://dx.doi.org/10.3133/cir1405.

Megdal, Sharon B. 2012. Arizona Groundwater Management, in *The Water Report*, eds. David Moon and David Light, Issue no. 104, pp. 9–15. Eugene, Oregon: Envirotech Publications, October 15.

Morgan, Anne. 2009. Vice President of Public Lands, Wilderness Society, Remarks at the University of California, Berkeley, "As If Equity Mattered" Symposium, Panel on Land and Fairness, October 1.

Morgan, Dale L. 1947. *The Great Salt Lake*. Albuquerque: University of New Mexico Press.

National Reclamation Act of 1902. Public Law 57-161, as amended.

National Research Council. 2007. *Colorado River Basin Water Management: Evaluating and Adjusting to Hydroclimatic Variability*. Washington, DC: National Academy of Sciences. Available at http://www.nap.edu/catalog/11857.html.

Nugent, Walter. 2001. *Into the West—The Story of Its People*. New York: Vintage.

Orsi, Jared, 2005. Flood Control Engineering in the Urban Ecosystem. In *Land of Sunshine: An Environmental History of Metropolitan Los Angeles*, eds. William Deverell and Greg Hise, 135–151. Pittsburgh: University of Pittsburgh Press.

Pearce, David. 2002. An Intellectual History of Environmental Economics. *Annual Review of Energy and the Environment*, 27:57–81. doi: 10.1146/annurev.energy.27.122001.083429.

Phillips, Fred, G. Emlen Hall, and Mary E. Black. 2011. *Reining in the Rio Grande—People, Land, and Water*. Albuquerque: University of New Mexico.

Pomeroy, Earl. 1955. Toward a Reorientation of Western History: Continuity and Environment. *Mississippi Valley Historical Review* XLI:579–599.

Pomeroy, Earl. 1965. *The Pacific Slope: A History of California, Oregon, Washington, Idaho, Utah, and Nevada*. Lincoln: University of Nebraska.

Powell, James L. 2008. *Dead Pool—Lake Powell, Global Warming, and the Future of the American West*. Berkeley: University of California.

Repanshek, Kurt. 2012. Survey of Western Attitudes Shows Strong Support for National Parks, Clean Environment. *National Parks Traveler*, January 31. Available at http://www.nationalparkstraveler.com/2012/01/survey-western-attitudes-shows-strong-support-national-parks-clean-environment9390.

Smythe, William E. 1900. *The Conquest of Arid America*. New York and London: Harper and Brothers.

Southern Nevada Water Authority. Undated. *Simply Beautiful—Water Smart landscapes—Inspiration, Installation, and Maintenance*.

Starr, Kevin. 1990. *Material Dreams: Southern California through the 1920s*. New York: Oxford University.

Stegner, Wallace. 1969. *The Sound of Mountain Water: the Changing American West*. New York: Doubleday.

Townsend-Small, Amy, Diane E. Pataki, Hongxing Liu, Zhaofu Li, Qiusheng Wu, and Benjamin Thomas. 2013. Increasing Summer River Discharge in Southern California, USA, Linked to Urbanization. *Geophysical Research Letters* 40:4643–4647. doi:10.1002/grl.50921, 2013.

US Department of the Interior, Bureau of Reclamation. 1996. *Operation of Glen Canyon Dam Final Environmental Impact Statement*. Denver, Colorado: US Bureau of Reclamation.

US Department of the Interior, Bureau of Reclamation. 2007. *Colorado River Interim Guidelines for Lower Basin Shortages and Coordinated Operations of Lake Powell and Lake Mead Final Environmental Impact Statement*. Boulder City, Nevada: US Bureau of Reclamation.

US Department of the Interior, Bureau of Reclamation. 2008. Available at http://www.usbr.gov/uc/rm/crsp/lor.html (accessed July 20, 2015).

US Department of the Interior, Bureau of Reclamation. 2012. *Colorado River Basin Water Supply and Demand Study—Study Report*. Washington, DC: USBR.

US EPA (Environmental Protection Agency). 2004. How we use water in these United States. Washington, DC. Available at http://www.epa.gov/watrhome/you/chap1.html and http://esa21.kennesaw.edu/activities/water-use/water-use-overview-epa.pdf

Walton, John. 1993. *Western Times and Water Wars: State, Culture, and Rebellion in California*. Berkeley: University of California.

Webb, Walter Prescott. 1931. *The Great Plains*. Waltham, MA: Ginn and Company.

Winters v. United States, 207 U.S. 564, 1908.

Worster, Donald. 1986. *Rivers of Empire: Water, Aridity and the Growth of the American West*. New York: Pantheon.

5

Protection and Restoration of Freshwater Ecosystems

Brian D. Richter, Emily Maynard Powell, Tyler Lystash, and Michelle Faggert

CONTENTS

ABSTRACT During the past 150 years, the water flows of western rivers have been increasingly harnessed to supply water for farming, public water supplies, electricity generation, and other purposes, enabling rapid growth in cities and the regional economy, and in the production of agricultural goods distributed around the globe. However, this growth has come at great ecological cost. The excessive depletion of the region's rivers has driven many native species to the brink of extinction and severely compromised the health of freshwater ecosystems. With climate-change projections forecasting even greater imbalances between water supply and demand in coming decades, water managers are faced with an urgent need for water policy reform that can support a sustainable future for both people and nature.

5.1 Introduction

For thousands of years, a large population of a behemoth fish called the totoaba (*Totoaba macdonaldi*) emerged each spring from the depths of the Sea of Cortez in Mexico to gather at the mouth of the Colorado River, where the fish would await the annual flood of water arriving from snowmelt more than a thousand miles upstream in the Rocky Mountains. Those floodwaters would signal the fish to move upriver into the delta's deep green pools to spawn, and as the Colorado's nutrient-enriched water mixed with ocean water and warmed in the Mexican sun, it provided ideal conditions for the rapid growth of newborn fish. Because of its delicious taste and generous heft (a mature totoaba can weigh over 200 lb. at 6 ft. length), the fish was a highly sought-after meal for indigenous Cucapá villagers living in the delta.

By synchronizing its reproduction with reliable annual floods, water temperature, and salinity conditions in the delta, the totoaba optimized its reproductive success, creating thousands of offspring each year. By the late twentieth century, however, the totoaba had become an endangered species, on the brink of extinction. Ironically, when a species like the totoaba is really good at playing the game of evolution—tightly hitching its life cycle to nature's cycles—and humans then fundamentally change the environmental conditions essential to their existence, the special adaptations that served the species so well for so long suddenly become major liabilities.

The totoaba's undoing came in the form of big dams and large-capacity diversion canals that captured and sent the Colorado's annual floodwaters to distant cities and farms, robbing the delta of its annual replenishment (Figure 5.1). During the past half century, the Colorado River Delta has been transformed from a vast wetland wilderness into a dry, salt-caked, sunbaked wasteland, largely devoid of the fish, waterbirds, and jaguars that once frequented its freshwater habitats.

The totoaba's tragic story is, however, not unique. Human development of the western United States has drastically altered natural water flows by damming rivers and diverting their water, and in so doing has devastated freshwater ecosystems. This chapter tells the story of the West's development through the lens of river flow alteration, directly linking it to the ecologic damage wrought therein. It then explores what has been done to mitigate or stem this environmental destruction and shows that these actions have been woefully insufficient thus far. Considering these findings, we posit three steps for a more sustainable future in the American West.

5.2 Western Rivers: Dammed, Diverted, and Drained

Prior to 1850, ecosystems west of the Mississippi remained largely undisturbed by human influence, but this changed very rapidly as settlers of European descent migrated west en masse in the latter half of the nineteenth century. Federal governmental acquisition of territory stretching to the Pacific opened the doors to the West, and by the start of the twentieth century, millions of new settlers had moved in. They quickly harnessed the region's rivers with dams and diversion canals, leaving few rivers untouched and most of them heavily depleted.

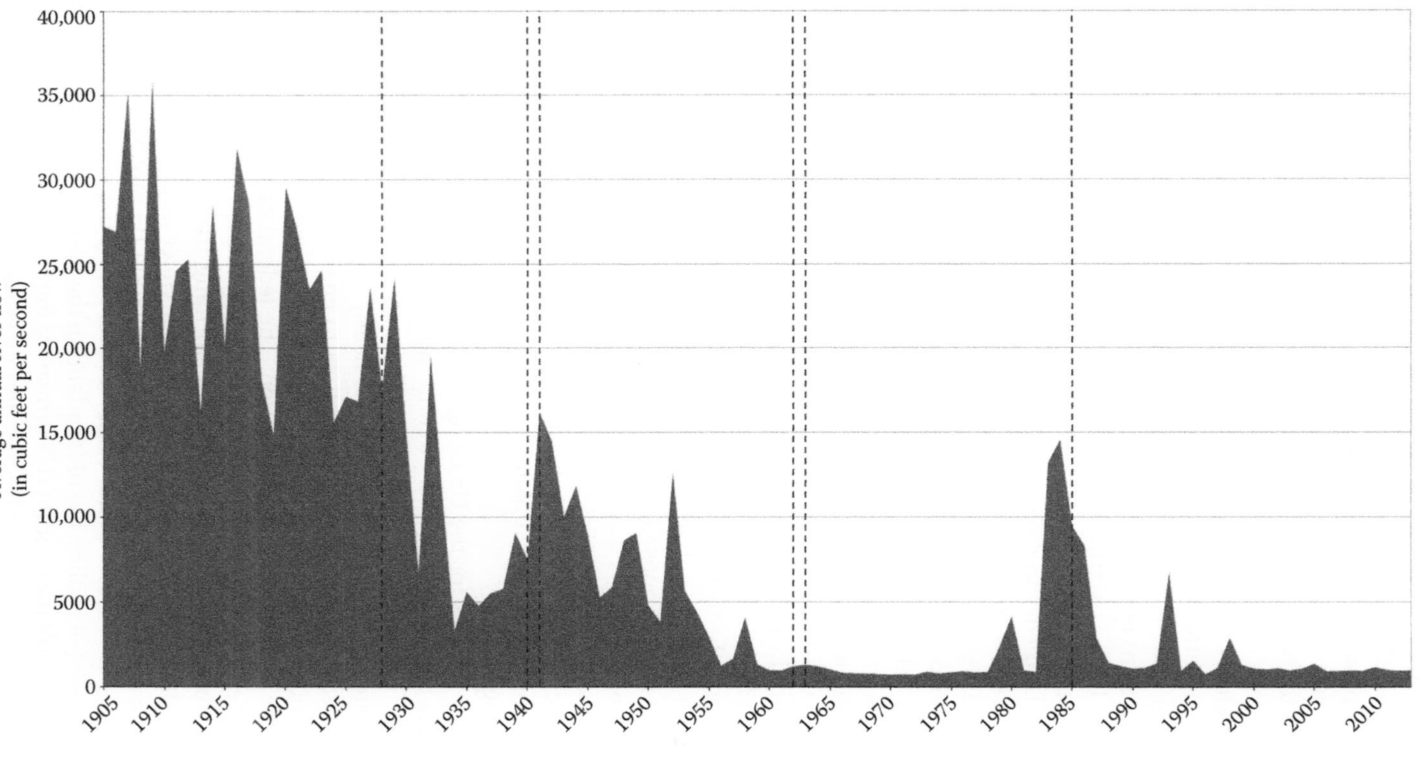

FIGURE 5.1
River flows in the lower Colorado River near Yuma, Arizona, have declined substantially since the completion of Lake Mead (Hoover Dam) in 1928, and the river now dries completely before reaching its delta in Mexico. Other large structures built on the river include Flaming Gorge Reservoir and Dam in 1962 and Lake Powell (Glen Canyon Dam) in 1963. Large diversion canals depleting the river's flow were completed in 1940 (All-American Canal), 1941 (Colorado River Aqueduct), and 1985 (Central Arizona Project). (Data from US Geological Survey, Colorado River near Yuma, 2014.)

5.2.1 Irrigating the West

The promise of free farmland; discoveries of gold in California, Montana, and the Black Hills of South Dakota and Wyoming; expansion of rail lines across the West; and passage of the Indian Appropriations Act of 1871—which stripped native tribes of their sovereignty and moved them onto reservations—stimulated a massive migration of Americans past the 100th Meridian to the Pacific Ocean during the latter half of the nineteenth century. The federal Homestead Acts, the first of which was passed in 1862, offered 160 acres (~65 ha) of land to individuals and in return asked only for a promise to farm and improve the land. Additionally, in its effort to stimulate western settlement, the federal government granted nearly 200 million acres (~81 million ha) of land along new rail routes to the railroad companies. These actions opened up huge swaths of land to a growing population of Americans who believed in their Manifest Destiny—a duty to conquer and repurpose the wild lands of the West to the agrarian, American model. This western migration also benefitted from a new populist theory in meteorology that explained why much of the plains states had become remarkably greener in the 1860s and 1870s: "rain follows the plow." Americans seeking new farmlands seized this theory as further proof of their Manifest Destiny and eagerly undertook to till as much land as they could (Reisner 1993).

Those unlucky enough to have followed this prophecy to the West, however, were soon confronted with scientific data showing that the "miraculous" greening was actually the result of a long humid-climate period that had brought above-average rainfall to the plains. This wetness would not last for long (Reisner 1993). Major John Wesley Powell, a Civil War veteran whose first descent in 1869 of the Colorado River through the Grand Canyon had boosted him to fame and the directorship of the US Geological Survey, cautioned that the future would not be so easy. In "A Report on the Lands of the Arid Region of the United States," Powell stated, "If it be true that increase of the water supply is due to increase in precipitation, as many have supposed, the fact is not cheering to the agriculturalist of the region.... We shall have to expect a speedy return to extreme aridity, in which case a large portion of the agricultural industries of these now growing up would be destroyed" (Powell 1879, p. 91). Pointedly and with icy reception in the US Congress, Powell asserted that the western two-fifths of the United States have a climate that generally cannot support farming without irrigation.

It took decades of farm failures for the US Congress to finally understand the wisdom in Powell's reasoning, including the need to federally subsidize irrigation development if farming was to continue. The Reclamation Act of 1902 (ironically, passed in the same year that Powell died) was intended to both stabilize the nearly 4 million acres (~1.6 million ha) of land already under private irrigation as well as to enable farmers to bring additional lands into production. Twenty-one new irrigation projects were initiated within the first 5 years of the Reclamation Act's implementation. The federal agency created by the act—originally coined the US Reclamation Service but later renamed as the US Bureau of Reclamation—soon became a powerful force for transforming the rivers of the West by building more than 600 dams that capture and divert river water in the service of irrigation, urban water supplies, and hydroelectric power generation (Figure 5.2). The agency is now the largest wholesaler of water in the United States, providing water to one in five farmers and 10 million acres (~4 million ha) of farmland in the West and public water supplies to one-third of the region's population (US Bureau of Reclamation 2014).

The US Army Corp of Engineers and the Bonneville Power Association joined the Bureau of Reclamation in a dam-building binge that reached its zenith in the 1960s but continued for decades beyond (Reisner 1993). By the time the National Park Service took inventory of

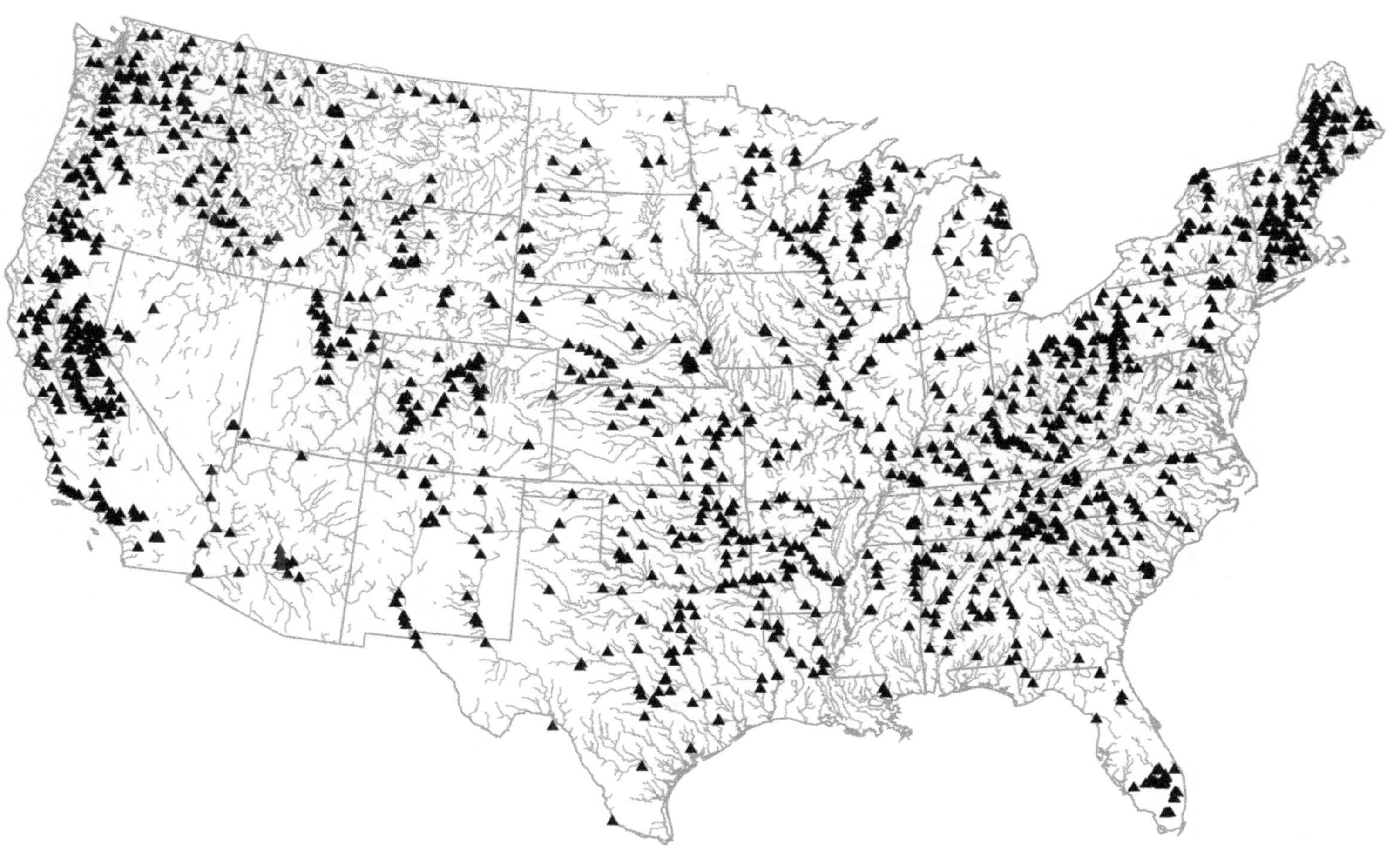

FIGURE 5.2
Locations of federal dams. (From Postel, S. and Richter, B.D., *Rivers for Life: Managing Water for People and Nature*, Island Press, Washington, DC, 2003.)

the country's rivers in 1980, more than 75,000 dams had been erected on the nation's 3000+ rivers, leaving only one major river—the Yellowstone—undammed (Fedarko 2013).

5.2.2 Failures of Water Governance

The rules governing water management in the West developed with little forethought as settlers began irrigating their lands, miners scoured streams for their gold, and cities rose up from the desert. Early western policy makers adapted and tailored legal systems borrowed from England and the eastern United States to fit the novel conditions posed by the West. Competition for water in the California gold fields during the mid–nineteenth century gave rise to a *prior appropriation doctrine* of water allocation that was eventually adopted by all western states. This established the principle of *first in time, first in right*, whereby the first person to put water to a beneficial use had priority over the second person to do so, and on down the line (Gillilan and Brown 1997). The prior appropriation system is administered by the state governments, who hold the water in trust and issue rights to use water to individuals, companies, water utilities, electricity producers, irrigation districts, and other water users (see Chapters 3 [Fort] and 4 [Feldman]).

The prior appropriation system worked well as long as the line of water users remained sufficiently short such that enough water was left over to sustain natural ecosystems. But as water demands approached and even exceeded the limits of water availability, five major shortcomings of this system became apparent, creating a precarious and highly undesirable water situation in the West for both people and nature.

First and most importantly, many states have issued a greater volume of rights to use water than can be satisfied, particularly during drier years or droughts. California, for example, has issued rights to use five times more water than its rivers can provide, on average (Grantham and Viers 2014). In theory, the states could have limited the total volume of water rights issued, such that all rights could be fully met in all years, and they could have left enough water in the rivers to support their ecological health and to provide a buffer reserve during unexpectedly dry years. Things did not work out that way in much of the West, however, due largely to the unplanned and rapid nature of western settlement and a lack of foresight that fish might someday run out of water.

The prior appropriation doctrine does allow for more recently issued water rights (*junior* rights) to be curtailed in part or in their entirety during periods of lower water availability, in order to allow older (*senior*) water rights to be fully realized. However, for this approach to work well, each water right and its associated *priority date*—the date at which the right was issued, and thus determining the right's place in line for fulfillment—must be clearly defined and recognized by the courts, and regulators must be able to monitor and act to curtail use by junior water-rights holders when there is not sufficient water available to serve all rights. This highlights a second major shortcoming of western water law: in practice, most states lack the high-quality records and administrative apparatus to do this efficiently, making it politically perilous and physically challenging to cut off junior water-rights holders. Overallocation, combined with lack of enforcement, leads to greater water use than should occur, often heavily depleting river flows and sometimes causing unwarranted impacts to higher-priority water-rights holders.

A third major shortcoming results from a perverse disincentive embedded in the prior appropriation doctrine that discourages water users from saving water when it is not needed. In order to prevent widespread speculation in water rights, the doctrine incorporated a principle known as *use it or lose it*: if a right holder does not use his or her water right to its full extent for its original purpose over the course of a number of years as defined by state law, the unused portion of the right becomes subject to forfeiture. The forfeited water then becomes available for other users.

A fourth major concern is that surface water regulation is, in most places, disconnected from groundwater regulation (if the latter exists at all), even though many rivers and aquifers are, in fact, strongly connected hydrologically. Many groundwater aquifers drain into rivers, and many rivers recharge aquifers, particularly during floods. Unfortunately, when surface water supplies are depleted or unavailable, many water users turn to groundwater pumping, often without restriction. This results in interception of groundwater flows that would have otherwise drained into rivers, thus reducing surface water supplies for the people and ecosystems that depend on these *base flows*. Legal reforms to address this problem have been imperfect, as discussed in Chapter 18 (Cech) of this volume.

A fifth shortcoming is that protection of fish, plants, invertebrates, and other organisms in western rivers was not originally recognized as a *beneficial use* of water under the prior appropriation system, preventing the issuance of water rights for environmental flow. That anachronism has been legally rectified by western states in recent decades, but because of the long delay in fixing this legal defect, the environment is now positioned at the back of the priority line with very junior water rights, thereby providing little protection for river flows in drier years when available water supplies are insufficient to satisfy all water rights. Moreover, legal and procedural hurdles greatly limit the applicability of this protection in some states (such as California, Texas, Utah, and New Mexico), rendering it nearly meaningless in practice. For example, in Texas, environmental flow standards have only recently been developed and adopted by the state's permitting agency, and those standards are applied only to water rights issued post-2011, which does virtually nothing to restore the state's heavily depleted streams (Sahs 2012).

5.2.3 Depleted Rivers in the Wake of Settlement

The West has thus ended up with heavily exploited rivers and a water governance system that has ossified into a complex hierarchy of water rights. Its landscape and river networks are pocked with huge water-storage reservoirs, and its natural water sources, pierced with thousands of pipe straws pulling water to supplement the rain that has not, alas, been bolstered by the plow. The heavy use of water in the farms, industries, and cities of the West removes half of the available river water in half of all rivers during the driest month of the year, and removes more than three-quarters of river water in one-quarter of all rivers* (Figure 5.3).

The existence of a half-full or even quarter-full river would seem to indicate that available water supplies have not bottomed out yet, suggesting that the region's water sources have more to give. But these statistics on river flow depletion—*based on long-term monthly averages*—obscure critical episodic risks to both people and nature that arise during drought. This heavy flow depletion across the West, coupled with a legal system that allows and even encourages maximum water use, has pushed rivers and aquifers to near exhaustion and made their dependent human populations and natural ecosystems increasingly vulnerable to water shortage.

* These statistics are based on an analysis conducted by Daren Carlisle of the US Geological Survey that assessed average monthly water flows at 3768 river monitoring stations across the western states.

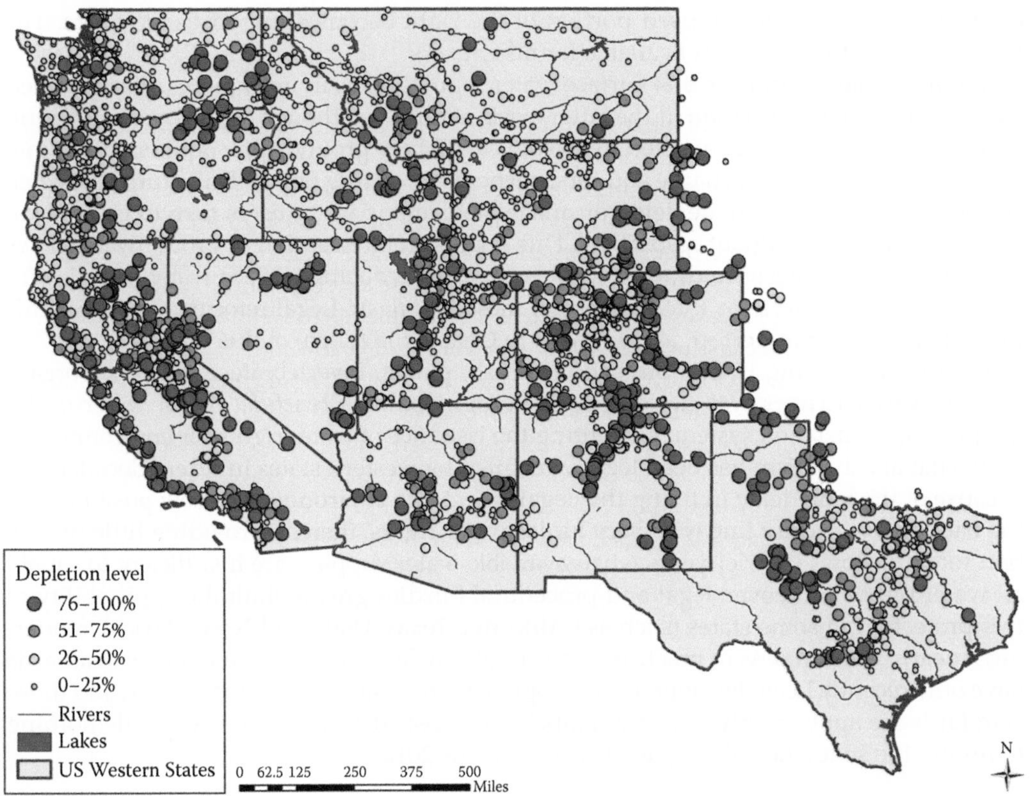

FIGURE 5.3
This map depicts the level of river flow depletion estimated at more than 3400 US Geological Survey gauging stations across the West, for the month of September. Flow depletion levels are estimated using a ratio of measured September flow averages to modeled estimates of natural September flows since 1950. (Data provided by Daren Carlisle, US Geological Survey.)

5.3 The West Today: A Precarious Situation

Episodic and recurring water shortages—resulting from the inability to manage the West's demands for water within the natural limits of availability—create a perilous situation for both people and nature. The complex systems of water-supply infrastructure and governance mechanisms—along with a very heavy dependency on available water sources—have produced a false sense of security that is regularly debunked with each new drought, often leading to severe economic consequences and ecosystem damage. The economic consequences are easily quantified, but what happens to creatures living underwater is more difficult to discern. This section will reveal some of the devastation to the West's biota wrought by human-induced flow alteration and explain how current climate science suggests these conditions will likely worsen in coming decades.

5.3.1 Economic Impacts of Water Shortages

Water shortages, like economic crises, result from an imbalance between supply and demand. During droughts such as the one experienced in California in 2014, the precipitation that naturally replenishes rivers and aquifers can decline by more than a third of the long-term average (California Department of Water Resources 2014; Howitt et al. 2014). In 2011, some parts of Texas received only one-quarter of their average rainfall (Combs 2012). If farmers and urban water users are unable to quickly pare down their water use to align with supply shortfalls, rivers and aquifers can be depleted to dangerously low levels, causing many water users to be cut off or endure reduced access to their water. The economic consequences of the resultant water shortages can be quite heavy, as evidenced by the projected loss of over $2 billion in agricultural revenues during the 2014 drought in California (Howitt et al. 2014) and the documented loss of more than $5 billion in Texas in 2011 (Combs 2012).

5.3.2 Ecological Vulnerability

People, however, are not the only casualties of drought and flow depletion. River species are vulnerable to a broad array of human impacts, ranging from pollution to habitat destruction to introduction of highly competitive nonnative species. Among these many influences, water flow is now widely recognized within the scientific community to be a "master variable" for river ecosystem health (Poff et al. 1997; Richter et al. 1997), meaning that alteration or depletion of natural water regimes has disproportionately large consequences for river habitats and species. The volume, rate, and timing of water flow strongly influence water temperature and chemistry; shape physical habitats through movement and rearrangement of sand, cobbles, and boulders; and can dictate whether or not competitive, nonnative species can persist. Plants and animals across the West, including the ancient totoaba, evolved to capitalize on the natural flow regimes of rivers and streams, but dams, diversions, and the ultimate draining of western rivers severely disrupted these natural regimes (Figures 5.1 and 5.3), with devastating ecologic consequences.

It is also widely understood that artificial *increases* in water flow—such as regularly occur when operating dams to generate electricity or to send stored irrigation water downstream to farmers during the dry season—can be just as damaging as flow depletion. For example, one of the causes of decline for the Colorado pikeminnow, a large fish endemic to the Colorado River, has been unnatural fluctuations in winter flows due to releases from large hydropower dams such as the Flaming Gorge Dam on the Green River (a major tributary to the Colorado). Higher-than-natural and erratic releases caused by the dam's generation of electricity during the winter cause fish to swim faster and move more often, burning up precious energy reserves in icy water (Stanford 1994).

The water diversions that decimated the totoaba population and flow regime alterations that have nearly wiped out the pikeminnow have similarly impacted two-thirds (33) of all native fish species in the Colorado River Basin. This story is playing out across the entire West (Figure 5.4), where humans, not nature, are now turning the knobs that govern the flows of water, nutrients, and sediments in rivers in ways that do not resemble the natural rhythms to which river species are adapted. Sixty-two percent of subbasins in the West have at least one imperiled species, and in total, 367 different plant and animal species dependent upon the riverine habitats of the American West have become imperiled by flow depletions or flow regime alterations.* Flow depletion has been singled out as the leading cause of fish endangerment in the United States as a whole, affecting nearly three-quarters of all fish species listed under the Endangered Species Act (ESA) (Reed and Czech 2005).

* This analysis has been based upon a review of the causes of decline for all species located in the western United States that are listed as threatened or endangered under the US Endangered Species Act or listed as globally at risk (G1–G3 rankings) by NatureServe.

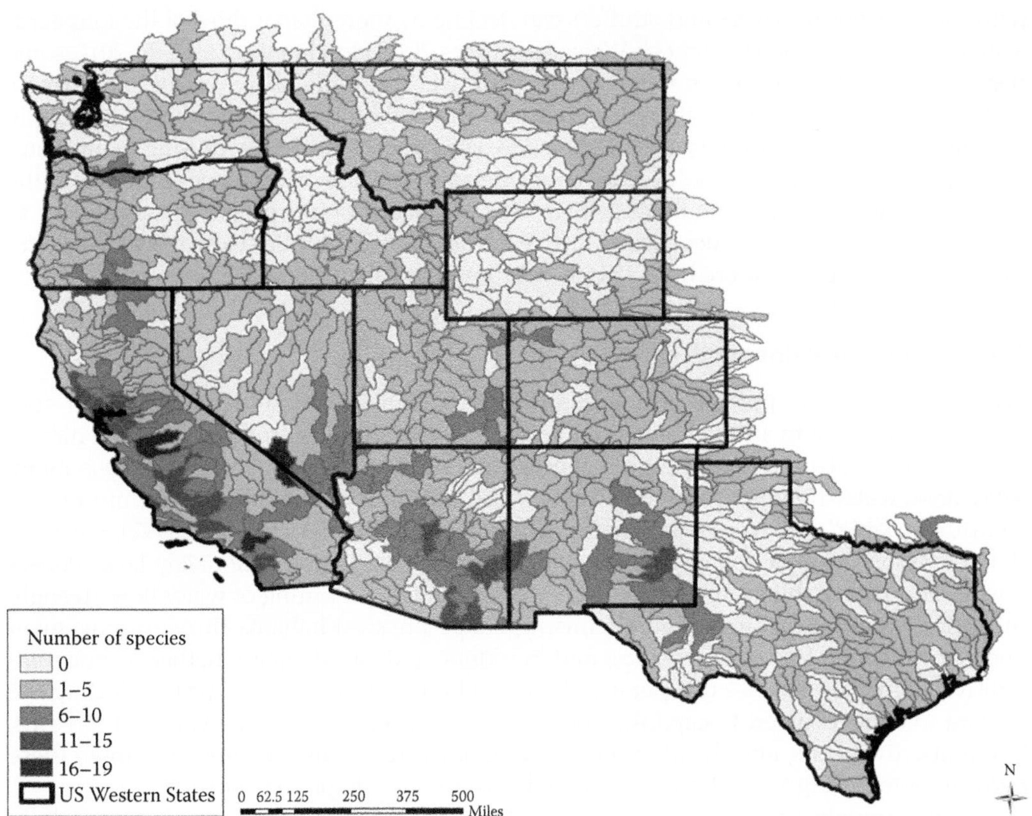

FIGURE 5.4
This map portrays the number of freshwater species that have become imperiled due to flow depletion in watersheds across the West, based on a review of the causes of decline for all species located in the western United States that are listed as threatened or endangered under the US ESA or listed as globally at risk (G1–G3 rankings) by NatureServe.

5.3.3 More Change on the Horizon

Recent assessments of likely changes in runoff associated with projected climate changes across the West suggest that balancing demands with water supplies may become even more difficult in coming decades. Summer-season runoff in the southwestern United States is projected to be most heavily affected (US Bureau of Reclamation 2011). Substantial decreases (i.e., minus 10–15+%) in summer flow are projected over a region spanning southern Oregon, the Southwest, and Southern Rockies, including the Klamath, Sacramento, San Joaquin, Truckee, Rio Grande, and Colorado river basins. Given the already heavy levels of flow depletion shown in Figure 5.3, the likelihood for even greater impacts on the West's ecosystems and species is of great concern.

Also of concern are shifts in the timing of runoff, particularly snowmelt flooding. The life cycles of many aquatic and riparian plants and animals are intimately tied to seasonal variability in both flow and water temperature. For example, the timing of seed dispersal in narrowleaf cottonwood trees (*Populus angustifolia*) found along Rocky Mountain–fed rivers from Canada to Arizona and New Mexico occurs during a short window of a few weeks in early spring. The seeds must fall on soils recently moistened by snowmelt floods to enable germination and early growth. With many rivers already experiencing floods that come weeks sooner due to earlier and more rapid melting of the mountain snowpack associated with climate warming, the timing of seed dispersal and floodplain wetting may become increasingly misaligned in coming decades.

Similarly, the Colorado pikeminnow discussed earlier is cued to begin its spring spawning migrations by the rise of snowmelt floodwaters. The pikeminnow must arrive on their spawning grounds at a time when water temperatures have warmed to 20–22°C. Shifts in the timing of snowmelt runoff and water warming are threatening to desynchronize the pikeminnow's spawning migrations and the availability of suitable water temperatures. In light of the intrinsic dependence on natural flow regimes by native species such as the narrowleaf cottonwood and the pikeminnow, human-induced flow depletion and alteration, and predictions of further climatic shifts, it is clear that the plants and animals of the West are facing significant and growing threats to their existence.

5.4 Restoring the Flow, One River at a Time

The American conservation movement developed in direct response to the loss of natural lands and wildlife habitats to unchecked agricultural and suburban development, particularly following World War II, when cities began sprawling rapidly across the landscape of the West. The early focus of the conservation movement was directed at preserving majestic landscapes and large animal populations—primarily through the establishment of nationally managed and state-managed parks and wildlife refuges—but by the 1960s, the conservation lens had widened to explicitly include rivers. This gravitation toward river conservation developed as conservationists and scientists realized there were severe biodiversity consequences to damming and diverting rivers on a broad scale (Figures 5.3 and 5.4).

For instance, Donald Tennant, a fisheries biologist working for the US Fish and Wildlife Service (USFWS) in the 1950s, became increasingly concerned about what was happening to the rivers and streams he was studying. Tennant recognized that dams and the increasing extraction of water posed a grave threat to aquatic life. In the journal *Fisheries*, he wrote, "Philosophically, it is a crime against nature to rob a stream of that last portion of water so vital to the life forms of the aquatic environment that developed there over eons of time" (Tennant 1976, p. 10). In the same article, Tennant outlined his proposed approach—which later became widely known as the *Tennant method*—for protecting the life-giving waters of a river or stream. To sustain "optimum" biological conditions, Tennant suggested that 60–100% of a river's average flow would need to be protected. To provide "excellent habitat," 30–50% of the flow would be needed, and 10–30% would provide "fair or degrading" conditions (Tennant 1976).

Tennant thus became one of the first scientists to attempt to gauge how much water a river needs to remain healthy. In the latter decades of the twentieth century, a plethora of methods and tools for assessing the water needs of fish or entire freshwater and estuarine ecosystems were developed (Postel and Richter 2003). These tools, including physical habitat simulation models, sediment transport models, and food-web dynamics models, can rapidly analyze large amounts of data to generate plots and other outputs representing the changing availability or condition of aquatic habitat in response to the amount and timing of water flows in a river.

Concurrently, the theoretical underpinnings of river science advanced markedly. A "natural flow paradigm," which highlights the importance of maintaining the full range of natural intra- and interannual variation of hydrologic regimes in sustaining the native biodiversity and integrity of aquatic ecosystems, has been broadly accepted in the scientific community (Poff et al. 1997; Richter et al. 1997). Based on review of many recent case studies and research papers, Richter et al. (2011) concluded that river ecosystems and species become increasingly vulnerable as flows are altered beyond 20%, corroborating Tennant's early findings that 60–100% of a river's flow must remain to support optimal biological health. Given the much higher levels of flow depletion in many watersheds portrayed in Figure 5.3, it is of little surprise that so many western river species have become imperiled due to flow depletion (Figure 5.4).

The scientists studying the rivers and aquatic species of the West are confident in their abilities to answer the question of how much water a river needs to ensure its ecological health. However, faced with ever-growing demands for increased water use in cities and industries, long uphill battles with senior water-rights holders—primarily irrigators—who defend fervently against any proposed adjustments in the status quo, and an increasingly unreliable water supply due to climate change, environmental advocates have been unable to implement sweeping flow restoration initiatives or laws. In absence of protective legislation, they have generally focused instead on restoring flow in one river at a time.

5.4.1 State-Based Flow Restoration Efforts

As state governments hold primary authority over water allocation, it would be logical to expect that leadership for water policy reforms to rectify the aforementioned shortcomings in water allocation policies and to meet the needs and values of the twenty-first century should come from the states. Due to the highly contentious nature of water governance and a passion for economic growth, however, states have generally been ineffective in proactively offering comprehensive policy reform to protect aquatic ecosystems and species. Taking the path of least resistance, states have undertaken comprehensive remedial action only when forced to do so by federal environmental laws or when the inadequacies in water governance lead to calamity. For instance, the state of Texas created a groundwater management authority for the Edwards Aquifer only when directed by a federal district court judge to do so in an ESA court case (Votteler 2004), and California passed groundwater management legislation in 2014 only when it became clear that the viability of the state's water-dependent economy was at risk (Moren et al. 2014).

That does not mean that western states have not taken any legislative steps to restore flow in recent decades, however. States have facilitated or enabled flow restoration in assorted ways, with some creating regulatory protections that specify certain flow regimes that are to be protected from other water allocations and others allowing a state agency, environmental organization, or individual to apply for a water right for environmental flow purposes (Table 5.1). Unfortunately, these approaches have yielded only very limited flow protection to date, simply because the protections afforded to environmental flows are superseded by other rights in most basins (Figure 5.5). In this void, private groups have begun creatively deploying existing state and federal legal tools to restore flow in individual rivers and streams.

Seeking a top-down but minimally invasive solution, some states, including Texas and Washington, have used legislation to enact regulatory protection that conditions new applications for water rights such that they do not deplete river flows below targeted environmental levels (Table 5.1). The caveat, but major appeal for senior water-rights holders, is that this type of regulation applies only to the issuance of new rights and, therefore, has no effect on rights issued prior to the time at which the regulatory protection was adopted. Texas, for example, passed state legislation in 2007 calling for scientific studies and a stakeholder review process to determine environmental flow needs for rivers throughout the state, with the intent to

TABLE 5.1

Summary of State Instream Flow Protections

	Allowable Holders of Permanent Instream Rights	State Actions to Benefit Instream Flow	Private Options to Benefit Instream Flow	Number of Water Rights for Instream Flow			Oldest Priority Date	Median Priority Date
				Water Rights	Regulatory	Voluntary Dedications		
Arizona	Anyone	Appropriation of new rights or conversion of existing rights	Conversion of existing rights but must transfer to state to retain priority date	38	0	0	1979	1990
California	Anyone	Closure of streams to new appropriations under Water Code sections 1205–1207	Nonbinding conversion of existing right through Water Code section 1707	0	0	52	1850	1926
Colorado	State only	Appropriate new rights or convert existing rights	Temporary or permanent donation of existing right to state	1653	0	0	1859	1982
Idaho	State only	Legislature creation of minimum flow water rights using unappropriated water only	Temporary or permanent donation of existing right to state water bank but cannot condition for instream use	288	0	0	1976	2005
Montana	State or federal entities only	New appropriation or conversion of existing rights, up to 50% of average annual flow	Temporary (30 years max) or permanent donation of existing right to state	422	0	0	1965	1985
Nevada	Anyone	New appropriation or conversion of existing rights	New appropriation, permanent or temporary conversion of existing rights	239	0	0	1859	1906
New Mexico	State only	Transfer of existing right to state Strategic Water Reserve	Temporary or permanent donation of existing right to state Strategic Water Reserve	21	0	0	1883	1935

(Continued)

TABLE 5.1 (CONTINUED)

Summary of State Instream Flow Protections

	Allowable Holders of Permanent Instream Rights	State Actions to Benefit Instream Flow	Private Options to Benefit Instream Flow	Number of Water Rights for Instream Flow			Oldest Priority Date	Median Priority Date
				Water Rights	Regulatory	Voluntary Dedications		
Oregon	State only	Appropriation of new rights, conversion of regulatory minimum flow standards and existing rights to instream rights	Donation of conserved water or new or existing right to state	1532	0	0	1857	1990
Texas	State only for full rights	Legislature creation of minimum flow standards that may condition new water rights only (Senate Bill 3)	Donation of full existing water right to state or nonbinding conversion of existing right by adding instream use as beneficial use type	5	103	0	1895	2011
Utah	State only	Temporary or permanent conversion of existing state rights	Permanent donation of existing right to state	1	0	0	2012	2012
Washington	State only	Legislature creation of minimum flow water rights	Temporary or permanent donation of existing water right to state	0	150	0	1976	1985
Wyoming	State only	New appropriation or conversion of existing rights	Temporary or permanent donation of existing water right to state	86	0	0	1949	1995

Source: Charney, S, An Analysis of Instream Flow Programs in Colorado and the Western United States, Colorado Water Conservation Board, available at http://cwcb.state.co.us/public-information/publications/documents/reportsstudies/isfcompstudyfinalrpt.pdf (accessed December 18, 2014), 2005; Gillilan, D.M., and Brown, T.C., *Instream Flow Protection: Seeking a Balance in Western Water Use*, Island Press, Washington, DC, 1997; Zellmer, S., Legal Tools for Instream Flow Protection, in *Integrated Approaches to Riverine Resource Stewardship: Case Studies, Science, Law, People, and Policy*, Locke, A. et al. (eds.), pp. 285–327, The Instream Flow Council, Cheyenne, WY, 2008.

Note: Each state has developed instream flow provisions, the basics of which are described in the qualitative columns here. The quantitative columns describe the degree to which each state's policies have been applied to create legal instream flow protection. Each state provided the authors with data on all instream flow water rights, regulations, or voluntary (optional, user-controlled instream use of their water right) dedications designated for instream flow. The results represent paper rights only and do not attempt to assess the degree to which each instream flow protection is fulfilled.

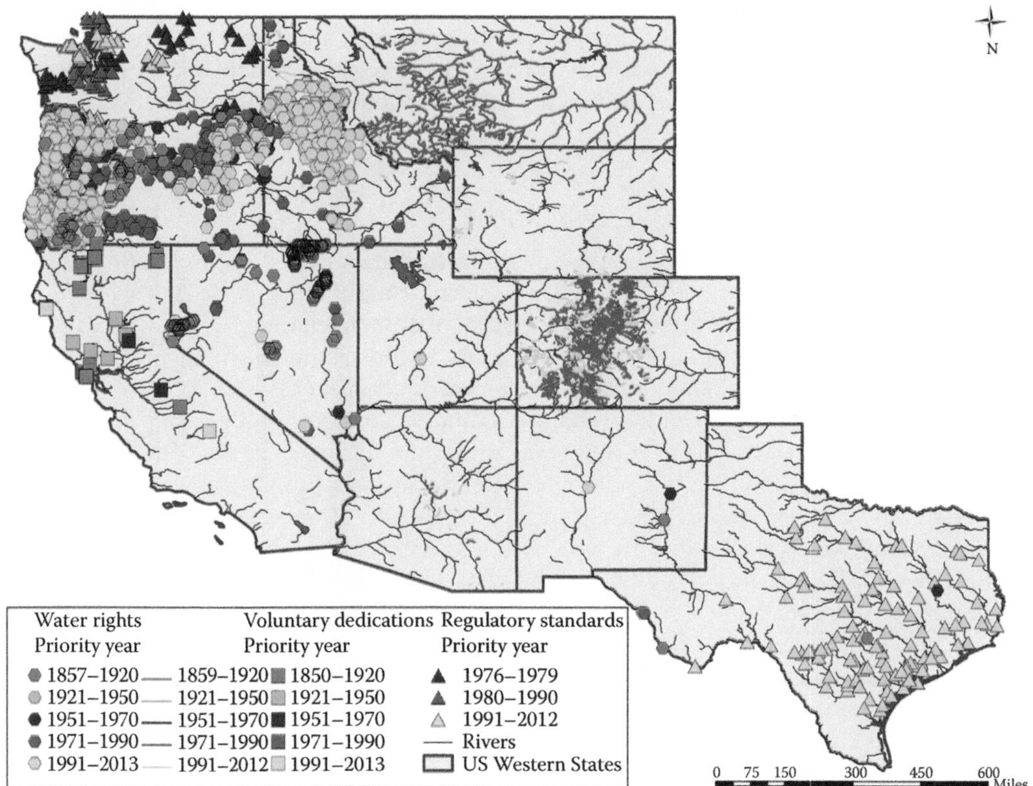

FIGURE 5.5
This map illustrates the location and priority dates associated with water rights or regulatory state designations dedicated to river flow protection in the West. Many of these flow protection provisions are tied to locations where river flow is monitored, such as at stream gauges, and are therefore portrayed as dots, but some are intended to apply to a stream reach and are shown here as lines. As illustrated here, most of the environmental flow protection that has been afforded is of recent (junior) priority.

condition any future water rights such that they do not impinge on the adopted flow standards. While the adopted standards (7 Tex. Admin. Code § 298.1-540) appear to reserve a substantial volume of water for the environment in rivers across the state, these protections will be largely ineffective in preventing preexisting water-rights holders from depleting river flow far below the environmental flow standards, particularly during drier years.

Oregon, on the other end of the spectrum, was the first state to explicitly recognize environmental flow as a beneficial use, doing so in 1955. The 1987 Oregon Instream Water Rights Act strengthened this recognition and opened the door for state agencies, environmental groups, and others to designate water rights for instream use through new appropriation or conversion of existing diversion rights. By this time, however, the flow of many of the state's rivers had already been overallocated, leaving little unappropriated water available for environmental flow (Neuman et al. 2006). Environmentalists thus quickly recognized the advantages of acquiring and converting existing (particularly senior) water rights as their best hope for restoring river flows in Oregon.

The Deschutes River, in Central Oregon, saw 90% of its water diverted for human use by the late twentieth century (Figure 5.6). The 1955 and 1987 acts provided little mitigation of the flow depletion in the Deschutes due to the junior status of these protections.

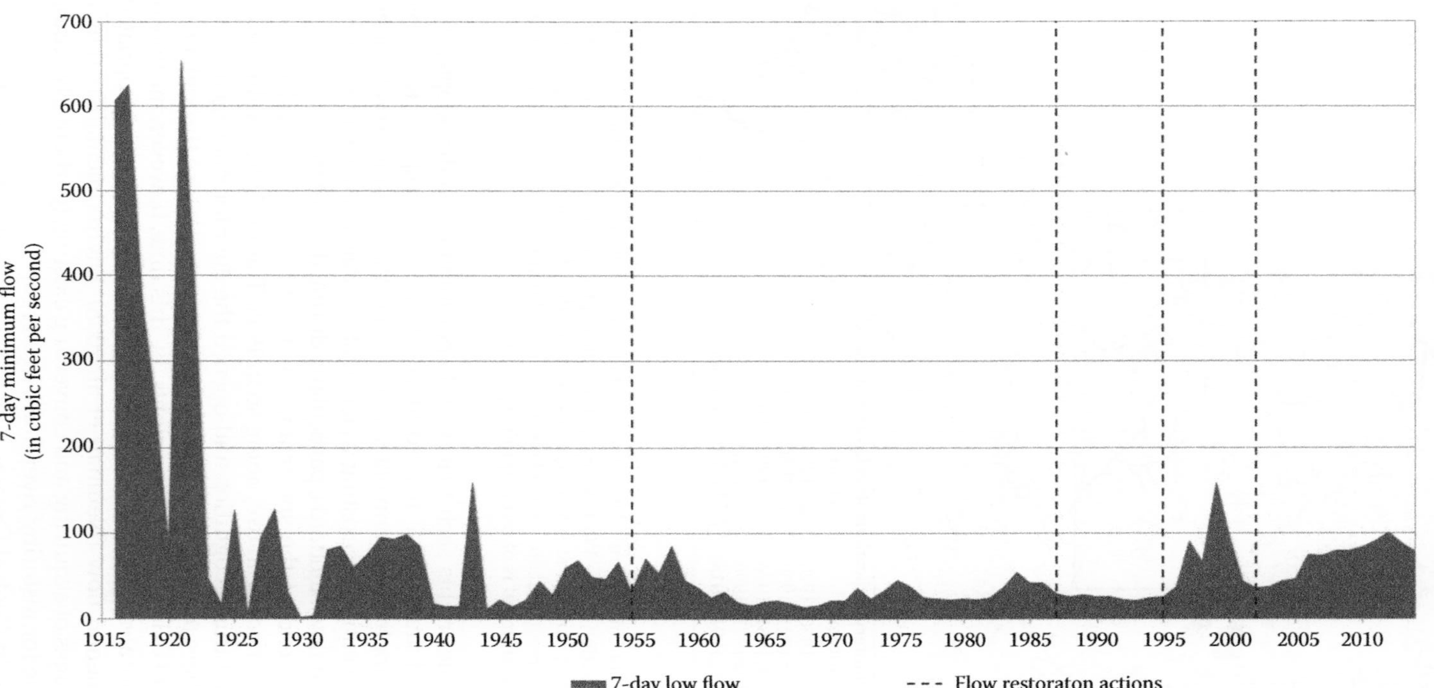

FIGURE 5.6
Seven-day minimum flows in the Deschutes River downstream of Bend, Oregon, were substantially reduced during the twentieth century, sometimes leaving less than 10 cfs in the river, and saw little improvement after the passage of the 1955 and 1987 state acts. Since the creation of the Deschutes River Conservancy in 1996 and Columbia Basin Water Transaction Program in 2002, the 7-day minimum flows have improved. (Data from US Bureau of Reclamation Pacific Northwest Regio, Deschutes River below Bend, Oregon, 2015.)

Local environmentalists formed the Deschutes River Conservancy in 1996 to seek market-based solutions. Through a combination of incentivizing water conservation in irrigation districts and acquisition of senior water rights through donation, lease, or purchase, this organization has protected over 200 cubic feet per second (cfs) (i.e., 5.66 m^3/s) for environmental flow (Deschutes River Conservancy 2015a), making a discernible improvement in the 7-day minimum low flows each year (Figure 5.6).

While this success is in many ways notable, it benefits the upper basin of just one river, and the majority of the water protection is through short-term leases, half of which are donated (Deschutes River Conservancy 2015b). Moreover, this success would never have been possible without two critical enabling conditions set forth in the 1987 act: recognition of instream flow as a beneficial use and a relatively streamlined transfer process to acquire and convert a right from a diversion to instream use. One other major factor limits the proliferation of market-based solutions: to have a market, one must have willing sellers.

Water rights are not readily relinquished by most water users in the West, and when they are, they can be quite expensive. Few state governments or environmental groups have the ability or willingness to pay large sums of money for their acquisition. With water rights typically selling at prices ranging from \$500 to more than \$5000 per acre-foot (1233 m^3), restoring ecologically meaningful volumes of water to any sizeable, heavily depleted river can be enormously expensive, requiring thousands or hundreds of thousands of acre-feet of water. For example, scientists have estimated that 100,000 acre-feet (~123 million m^3) of water rights are needed in the Guadalupe River of Texas to provide adequate flows from the river into San Antonio Bay to maintain habitat for highly endangered whooping cranes (Trungale 2011). Reduced inflows of freshwater into the bay have, in recent years, caused a highly undesirable rise in the estuary's salinity levels, impacting the crane's food resources and leading to the death of 23 cranes in 2008. Even if water rights could be acquired as cheaply as \$500 per acre-foot, the price tag for restoring adequate inflows into just this one estuary would require \$50 million—a number many times larger than the operating budgets of many environmental organizations. These numbers lay bare a rational explanation for why flow restoration has been successful only in smaller rivers or tiny headwater streams where measurable ecological benefits can be gained with modest volumes of water.

5.4.2 Role of the Federal Government

The work of the Deschutes River Conservancy has also been enabled by the federal government in recent years. Much of the success realized in western flow restoration efforts to date, including on the Deschutes, would not have been possible without one key federal statute: the ESA of 1973. The express purpose of this act is to protect and recover imperiled species throughout the country, and it is one of the most powerful legal tools for restoring flows (Gillilan and Brown 1997). It has been used in a variety of ways by environmental interests, including accessing governmental funds set aside for purchasing water rights, renegotiating operations of existing dams, and capping water withdrawals to protect river and aquifer levels. The major flaw in leveraging the ESA for water management changes, however, is that this law often kicks in too late, long after river flows have been depleted and species populations have declined to perilous levels.

5.4.2.1 Leveraging the ESA to Acquire Water for Flow Restoration

One of the large-scale applications of the ESA in river restoration can be found in the Columbia River Basin, which has seen nearly all of its rivers and tributary streams dammed and diverted over the past 150 years. This massive flow alteration and the barriers to fish migration posed by dams have brought numerous salmon populations to the brink of extinction. Salmon migrate upstream to spawn every year to small headwater streams, where excessive water extraction has left insufficient flow to facilitate migration or egg laying and fertilization. In 1991, there was only one spawning ground still accessible to Snake River sockeye salmon, and only four fish returned to spawn. The same year, just 1% of the Snake River summer chinook salmon run returned to its spawning grounds; historically, this run included 1.5 million fish. As a result, the Snake River sockeye and two populations of Snake River chinook were listed as endangered species under the ESA in 1991 and 1992, respectively (Blumm 2002). Following these initial listings, 11 more distinct populations of the Salmonidae family in the Columbia River Basin have been listed as endangered or threatened, for a total of 14 (US Fish and Wildlife Service 2015).

The Bonneville Power Administration (BPA) owns and operates dams throughout the Columbia River Basin, but BPA-led efforts during the 1990s to restore endangered salmon populations were largely unsuccessful. The National Marine Fisheries Service, as part of its responsibilities under the ESA, wrote a Biological Opinion in 2000 mandating BPA to take more aggressive action to restore flows in historic spawning grounds in order to bring salmon populations back to life. In response, BPA and the Northwest Power Planning Council initiated the Columbia Basin Water Transactions Program (CBWTP) in 2002. This program is operated by the National Fish and Wildlife Foundation (NFWF) and is tasked with acquiring water rights for flow restoration in the Columbia Basin, with funding provided by BPA. Through this program, NFWF distributes BPA funds to small, locally managed water trusts, such as the Deschutes River Conservancy, that acquire water rights for instream use, focusing on flow restoration in salmon spawning grounds. As of 2013, this program has executed 340 water transactions to improve over 1500 miles of tributary streams. While salmon recovery is dependent on many factors, putting water back into streams that previously ran dry because of diversions is an obvious prerequisite for salmon spawning and recovery.

While these results are inspiring, the CBWTP program is limited in scope to headwater streams in the Pacific Northwest. The program's long-term benefits are also somewhat uncertain, as many of the transactions undertaken to date have been temporary (short-term) leases of water rather than permanent acquisitions of water rights. Nonetheless, the program illustrates that demonstrable ecological benefits can be gained with relatively small volumes of flow restoration in headwater streams, and that given sufficient investment, the problem of overallocation of water rights can be effectively addressed through purchases of water rights for the environment.

5.4.2.2 Reoperating Dams to Restore Natural Flow Patterns

The ESA has been invoked to restore stream flows in other areas of the West as well, with notable success in altering dam operations to increase flows in specific basins or aquifers. Dams in the West were originally constructed and managed to optimize water storage for supply, flood control, and power generation. In light of scientific evidence demonstrating the ecological impacts of flow alteration and the listing of numerous affected species under the ESA, federal dam management agencies, such as the Bureau of Reclamation, BPA, and the Army Corps of Engineers, have been directed to redesign the management plans of specific dams to improve downstream flows. Before a dam's release schedule can be reformed, however, lengthy studies and flow assessments must be undertaken by the operating federal agency to determine the volume of water needed to achieve the desired environmental benefits, how the various alternate flow regimes would affect the dam's ability to meet its economic purposes, and the associated costs to the operating agency through lost revenues. These studies are both costly and time consuming, sometimes taking years to complete (Gillilan and Brown 1997).

The endangered totoaba is a beneficiary of a recent prescription for dam-released environmental flows that has been decades in the making. The United States and Mexico have signed numerous agreements governing Colorado River water, beginning with the Mexico–US Water Treaty of 1944. This treaty required the United States to deliver 1.5 million acre-feet (1.85 billion m^3) of water to Mexico annually, but as the Colorado was nearly fully allocated in the United States, no provisions were made to provide environmental flows downstream into the delta (Stanger 2013).

In light of the ESA listing of multiple delta species as endangered, including the totoaba and a small porpoise known as the vaquita, binational agreements began formally investigating possible mechanisms for flow restoration measures with Minute 306 in 2000 (an amendment to the 1944 treaty). These discussions and associated environmental assessments culminated with the signing of Minute 319 in 2012, which created formal mechanisms for 158,088 acre-feet (195 million m^3) of water to be released into the delta as base and pulse flows over 5 years from Morelos Dam located just north of the US–Mexico border, which is managed by the International Boundary and Water Commission in partnership with the Bureau of Reclamation (Stanger 2013). This treaty marks a momentous escalation in flow restoration efforts, one to which western US states can look as a model for future efforts. That said, this success must again be caveated: just one dam is being reoperated, and while optimists anticipate continuation and expansion of this pilot program, it is only currently guaranteed to run from 2013 to 2017 (International Boundary and Water Commission 2012).

Longer-term dam reoperations for environmental flows have been sought and granted, on nonfederal hydropower dams. The Federal Energy Relicensing Commission (FERC) issues operating licenses to privately owned hydropower dams, with license terms generally running for 30–50 years, at which time they must be renewed. Before a license is issued, FERC undertakes an in-depth review of the dam and its operations, in which FERC is required to give environmental consequences "equal consideration" in weighing the costs and benefits of the dam (Gillilan and Brown 1997). More importantly, if one or more listed species is present in the stretch of river below the dam, FERC is required to comply with the ESA in the requirements set forth in the new license. This opens the door for environmental advocates to press for license conditions that provide for environmental flows, fish passage provisions, and other habitat restoration measures. Of the 320 current FERC licenses for hydropower projects (FERC 2015), at least 156 have included provisions for environmental flow augmentation. FERC licensing has been one of the more successful processes for restoring flows across the West because licenses can dictate the volume and timing of required water releases, but it is limited as a regional strategy for a number of reasons. First and foremost, it is only applicable on river stretches below hydropower dams that are regulated by FERC, which include only private hydropower dams. The licensing process for a single dam can take years, and as it usually involves multiple stakeholders, environmental flow needs are balanced with the water needs of other stakeholders, including for power generation and flood control. Additionally, any water released for environmental flow as per a FERC license agreement is just that: environmental water released from the dam but not protected from diversion for human use downstream of the dam (Gillilan and Brown 1997), which can significantly reduce the spatial extent of downstream benefits. Lastly, the long-term nature of FERC licenses can be problematic, making it difficult to change dam-release requirements when new or better information becomes available. Of particular note here is the expectation that our understanding of likely future climate changes will continue to improve in coming decades, but incorporating that new knowledge into dam operations will be difficult if not impossible under FERC licensing conditions.

5.4.2.3 Using the Courts to Restore Flows

A third primary application of the ESA for restoring environmental flows is through litigation. Any individual or group can use the ESA in litigation to force changes in water management if the current practices can be shown to be damaging one or more listed species. These lawsuits are extremely expensive for all parties and can take years to move through the court system, and success for the environment is not guaranteed. Given the myriad of factors that can affect the health of an ecosystem or species, proving that water management practices are directly causing a taking is no small feat. The results of recent court decisions in Texas shed light on the challenges of gaining flow protection from ESA litigation.

The Edwards Aquifer is a shallow aquifer in Texas that feeds multiple natural springs that provide habitat for multiple ESA-listed species that inhabit both the springs and the bowels of the aquifer itself, including the fountain darter (a fish), the Texas blind salamander, and Texas wild rice. The Edwards Aquifer is also a major water source for more than 2 million people, providing water for irrigation and municipal use, particularly for the city of San Antonio (Debaere et al. 2014). Use of the aquifer grew rapidly following World War II, and in the absence of any groundwater regulation to restrict pumping, aquifer levels and springflows began plummeting repeatedly during dry periods when the aquifer was not being recharged quickly enough to offset pumping.

Responding to the threat posed to listed species from overpumping of the aquifer, the Sierra Club sued the USFWS under the ESA, contending that the listed species dependent upon springflow from the Edwards Aquifer were not being adequately protected (Votteler 2004). The US district court ruled in favor of the Sierra Club, ordering two major actions to bring the Edwards Aquifer management policies into ESA compliance. The court required that the USFWS identify the springflows necessary to protect the listed species and commanded the state of Texas to manage pumping from the Edwards Aquifer to maintain the aquifer levels needed to provide the aforementioned flows. Both parties eventually complied, and Texas created the Edwards Aquifer Authority (EAA) to regulate pumping from the aquifer.

The EAA imposed a cap on the amount of water that could be pumped and issued use permits, without which one could not legally pump water from the aquifer (Gulley and Cantwell 2013). Thus far, this has represented a successful case in which the ESA was used in court to alter water management practices to restore flows for the environment: populations of listed species dependent upon the Edwards have increased or remained steady since 2000 (Debaere et al. 2014). Subsequent legislation, litigation, and management decisions have deviated from the initial court ruling, both increasing the cap and allowing permits to be issued in excess of the cap, while shifting to an alternate strategy of reducing allowable pumping levels as aquifer levels decline (Debaere et al. 2014). It remains to be seen whether this new approach of reducing pumping—as opposed to enforcing a maximum cap on permits—will achieve the desired ecological outcomes. Hence, while litigation can help force a move toward more sustainable water management, success in perpetuity is far from guaranteed.

This point is well illustrated with another ESA-related court case in Texas. The Guadalupe River is joined by the San Antonio River, both of which are fed in part by springflows from the Edwards Aquifer, and then flows into San Antonio Bay on the Gulf Coast. The marshes of San Antonio Bay depend on a delicate balance of salinity levels to support the species found there, and freshwater inflows from the rivers are a critical input to this balance. The Texas Commission for Environmental Quality (TCEQ) issues water rights in the Guadalupe and San Antonio river basins and, as of September 2014, has issued rights to withdraw more than three times the mean annual flow available in these basins.* This overallocation results in severe flow depletion throughout the river system and substantially reduces freshwater inflows to San Antonio Bay during drier periods.

San Antonio Bay provides wintering grounds for the endangered whooping crane, and during a dry period in 2008, 23 of the birds perished. The Aransas Project, an environmental nongovernmental organization, sued TCEQ under the ESA, claiming that the allocation of water-use rights in the Guadalupe Basin did not ensure sufficient inflows to San Antonio Bay, and thus directly caused the deaths of the whooping cranes. The US District Court for the Southern District of Texas ruled in favor of the Aransas Project in 2013, but this decision was overturned in 2014 by the US Court of Appeals, Fifth Circuit. The appellate court did not challenge the findings that 23 whooping cranes died and that their deaths were the result of reduced food supplies, which in turn was caused by the heightened salinity levels due to reduced inflows. However, the court overturned the initial ruling on the grounds that these deaths were not "foreseeable" by TCEQ in their water-rights permitting process, and that even if they were, there were too many other variables outside of their control, such as drought, that could also lead to reduced freshwater inflows and increased salinity levels (*The Aransas Project v. Shaw et al.*, No. 13-40317, U.S. 5th Cir. 2014). This appellate court decision, if not eventually overturned itself in the US Supreme Court, could substantially weaken the potency of the ESA in restricting water use to protect species.

5.5 Principles for an Ecologically Sustainable Water Future

The undeniably dire water situation in the West calls for major water policy reforms not just to protect the freshwater ecosystems of the region but to also enhance overall water security for all water users. In this closing section of our chapter, we highlight three actions necessary for a sustainable water future, drawing from principles articulated by Richter (2014).

5.5.1 Set Limits on Water Consumption in Each Watershed or Aquifer

As illustrated in Figures 5.3 and 5.4, many rivers of the West still retain half or more of their natural flow, even during drier months such as September, and many of these rivers harbor imperiled or declining species. Western states can follow Washington's example,

* Based on an analysis comparing (1) the mean annual naturalized flow at the most downstream control points on the Guadalupe and San Antonio Rivers in the Texas Commission on Environmental Quality Water Availability Models (https://www.tceq.texas.gov/permitting/water_rights/wam.html) and (2) the volume of active water rights allocated the San Antonio and Guadalupe River Basins, as listed in the Texas Commission on Environmental Quality Water Rights Database (http://www.tceq.state.tx.us/permitting/water_rights/wr_databases.html).

as the state has closed many of its watersheds to further issuance of water rights, thereby "stopping the bleeding." The capping of water withdrawals might also be accompanied by a state-level commitment to cap water withdrawals, such as the Oklahoma Water for 2060 Act, which states that the following:

> The Legislature hereby declares that, in order to protect Oklahoma citizens from increased water supply shortages and groundwater depletions by the year 2060 in most of the eighty-two watershed planning basins in the state..., the public policy of this state is to establish and work toward a goal of consuming no more fresh water in the year 2060 than is consumed statewide in the year 2012, while continuing to grow the population and economy of the state and to achieve this goal through utilizing existing water supplies more efficiently and expanding the use of alternatives such as wastewater, brackish water, and other non-potable supplies.

In both the Washington and Oklahoma examples, these states have taken assertive action to prevent further damage to their freshwater ecosystems and to enhance water security for their citizens.

However, as discussed earlier, many of the West's river basins are grossly overallocated already. In those river basins, any attempts to cap and reduce water use by retiring existing water rights are certain to ignite a political and legal firestorm. A more politically palatable solution would be acquisition of senior water rights from willing sellers, and dedication of those rights to environmental flow purposes. These water rights must be of sufficient seniority to ensure their effectiveness even during droughts. The aforementioned BPA-funded program in the Pacific Northwest is an example of an public effort to acquire water rights; another extraordinary example of public funding being allocated for environmental water rights purchases can be found in the Murray–Darling Basin of Australia, where the federal government has dedicated more than $14 billion to date for the purpose of retiring water rights (Richter 2014).

5.5.2 Facilitate Private Water Trading

Buying and selling of water rights is taking place across the West, but many water experts have pointed to the need for a much greater volume of water trading to help reduce water shortages and restore environmental flows (Squillace 2012; Culp et al. 2014; Debaere et al. 2014). One of the major hurdles to water trade in the West is transaction cost, in terms of both money and time. The review and public vetting of proposed transactions can take years to complete, greatly discouraging trade. Some of the more complicated aspects of these reviews involve a determination of the volume of water that has been historically consumed by the seller (as contrasted with the volume of water withdrawal, which is the basis of water rights), which sets the volume that can be traded, and an assessment of potential third-party or environmental impacts resulting from a change in the place of use. Government agencies must find ways to use decision-support systems in their evaluations, or adopt simple rules of thumb for estimating the volumes of water consumed by sellers.

5.5.3 Invest in Water Conservation to Its Maximum Potential

There is no more cost-effective or environmentally friendly way of restoring river flows or avoiding water shortages than to reduce water use by investing in water conservation or improvements in water-use efficiency (Richter et al. 2013; Richter 2014), and dedicating the conserved water to the environment. Given that more than two-thirds of all consumptive water use in the West goes to irrigated agriculture, the first place to look for potentially large water savings should always be on farms, but opportunities to reduce watering of urban lawns and other landscape areas should not be overlooked (Richter et al. 2013; Richter 2014).

A number of cities have invested in water conservation programs that subsidize the costs of installing more efficient plumbing infrastructure such as low-flow toilets or pay homeowners to remove water-guzzling lawns. However, much greater state and federal investment, on the magnitude of the Australian government's actions, is needed. In comparison to the more than $14 billion invested by the Australian government in just one river basin (Murray–Darling), the US federal government's WaterSMART grant program for funding water conservation in the West has spent only a paltry $134 million to date (Bureau of Reclamation 2014).

5.6 Conclusion

The rivers of the West have withered over more than a century of settlement and development, and the changes in climate and population growth forecast for coming decades suggest that attaining true sustainability in water management will become ever more challenging. But we also heed Wallace Stegner's urging, "One cannot be pessimistic about the West," a region he called the "native home of hope" (Stegner 1969, p. 313). In this chapter, we have chronicled the checkered history of water development and ecosystem damage of the past century and a half, but we have also attempted to point a way forward through key water policy reforms. The adoption of such reforms will require, as Stegner asserted, that the West abandon its image of rugged individualism and move toward a more cooperative way of living. Only then, said Stegner, will it have "a chance to create a society to match its scenery."

References

Blumm, M.C. 2002. *Sacrificing the Salmon: A Legal and Policy History of the Decline of the Columbia Basin Salmon.* Den Bosch, The Netherlands: BookWorld Publications.

Bonneville Power Association and National Fish and Wildlife Foundation. 2004. Columbia Basin Water Transactions Program: Program History. Available at http://www.cbwtp.org/jsp/cbwtp/program/history.jsp (accessed December 12, 2014).

California Department of Water Resources. 2014. Water Year 2014 Ends as 3rd Driest in Precipitation. State of California. Available at http://www.water.ca.gov/waterconditions/ (accessed October 14, 2014).

Charney, S. 2005. An Analysis of Instream Flow Programs in Colorado and the Western United States. Colorado Water Conservation Board. Available at http://cwcb.state.co.us/public-information/publications/documents/reportsstudies/isfcompstudyfinalrpt.pdf (accessed December 18, 2014).

Combs, S. 2012. *The Impact of the 2011 Drought and Beyond.* Austin: Texas Comptroller of Public Accounts.

Culp, P.W., R. Glennon, and G. Libecap. 2014. Shopping for water: how the market can mitigate water shortages in the American West. Stanford Woods Institute for the Environment, Discussion Paper 2015-05, 40 pp.

Debaere, P., B.D. Richter, K.F. Davis, M.S. Duvall, J.A. Gephart, C.E. O'Bannon, C. Pelnik, E.M. Powell, and T.W. Smith. 2014. Water Markets as a Response to Scarcity. *Water Policy* 16:625–649.

Deschutes River Conservancy. 2015a. Accomplishments. Available at http://www.deschutesriver.org/about-us/accomplishments/ (accessed January 9, 2015).

Deschutes River Conservancy. 2015b. Water Right Leasing. Available at http://www.deschutesriver.org/what-we-do/streamflow-restoration-programs/water-rights-leasing/ (accessed January 9, 2015).

Fedarko, K. 2013. *The Emerald Mile.* New York: Scribner.

FERC (Federal Energy Regulatory Commission). 2015. Issued Licenses. Available at http://search.usa.gov/search?utf8 =%E2%9C%93&affiliate = ferc&query = issued+licenses (accessed January 16, 2015).

Gillilan, D.M. and T.C. Brown. 1997. *Instream Flow Protection: Seeking a Balance in Western Water Use.* Washington, DC: Island Press.

Grantham, T.E. and J.H. Viers. 2014. 100 years of California's water rights system: Patterns, trends, and uncertainty. *Environmental Research Letters* 9:1–10.

Gulley, R.L. and J.B. Cantwell. 2013. The Edwards Aquifer Water Wars: The Final Chapter? *Texas Water Journal* 4(1):1–21.

Howitt, R., J. Medellín-Azuara, D. MacEwan, J. Lund, and D. Sumner. 2014. Economic Analysis of the 2014 Drought for California Agriculture. Center for Watershed Sciences at University of California Davis, UC Agricultural Issues Center, and ERA Economics, Davis, CA. Available at https://watershed.ucdavis.edu/files/biblio/DroughtReport_23July2014_0.pdf (accessed December 18, 2014).

International Boundary and Water Commission. 2012. Minute No. 319. Available at http://www.ibwc.gov/Files/Minutes/Minute_319.pdf (accessed January 11, 2015).

Luckingham, B. 1989. *Phoenix: The History of a Southwestern Metropolis.* Tucson: The University of Arizona Press.

Moren, T., J. Choy, and C. Sanchez. 2014. The Hidden Costs of Groundwater Overdraft. Available at http://waterinthewest.stanford.edu/groundwater/overdraft/ (accessed December 30, 2014).

NatureServe. 2014, March. Listed and Imperiled Species by Watershed Database. Arlington, VA, U.S.A.

Neuman, J., A. Squier, and G. Achterman. 2006. Sometimes a Great Notion: Oregon's Instream Flow Experiments. *Environmental Law* 36:1125–1155.

Poff, N.L., J.D. Allan, M.B. Bain, J.R. Karr, K.L. Prestegaard, B. Richter, R. Sparks, and J. Stromberg. 1997. The Natural Flow Regime: A New Paradigm for Riverine Conservation and Restoration. *BioScience* 47:769–784.

Postel, S. and B.D. Richter. 2003. *Rivers for Life: Managing Water for People and Nature.* Washington, DC: Island Press.

Powell, J.W. 1879. *Report on the Lands of the Arid Region of the United States.* Washington: Government Printing Office. Available at http://pubs.usgs.gov/unnumbered/70039240/report.pdf (accessed December 18, 2014).

Reed, K.M. and B. Czech. 2005. Causes of Fish Endangerment in the United States, or the Structure of the American Economy. *Fisheries* 30(7):36–38.

Reisner, M. 1993. *Cadillac Desert: The American West and Its Disappearing Water, Revised Edition.* New York: Penguin Books.

Richter, B. 2014. *Chasing Water: A Guide for Moving from Scarcity to Sustainability.* Washington, DC: Island Press.

Richter, B.D., J.V. Baumgartner, R. Wigington, and D.P. Braun. 1997. How Much Water Does a River Need? *Freshwater Biology* 37:231–249.

Richter, B.D., M.M. Davis, C. Apse, and C. Konrad. 2011. A Presumptive Standard for Environmental Flow Protection. *River Research and Applications* 28(8):1312–1321.

Richter, B.D., D. Abell, E. Bacha, K. Brauman, S. Calos, A. Cohn, C. Disla et al. 2013. Tapped Out: How Can Cities Secure Their Water Future? *Water Policy* 15:335–363.

Sahs, M.K., ed. 2012. *Essentials of Texas Water Resources*. Austin: Texas Bar Books.

Stanford, J.A. 1994. Instream flows to Assist the Recovery of Endangered Fishes of the Upper Colorado River Basin. Biological Report 24. Washington, DC: U.S. Department of the Interior, National Biological Survey.

Stanger, W.F. 2013. The Colorado River Delta and Minute 319: A Transboundary Water Law Analysis. *Environs* 37(1):73–104.

Stegner, W. 1969. *The Sound of Mountain Water*. New York: Penguin Books.

Squillace, M. 2012. *The Water Marketing Solution*. Washington, DC: Environmental Law Institute.

Tennant, D.L. 1976. Instream Flow Regimens for Fish, Wildlife, Recreation, and Related Environmental Resources. *Fisheries* 1(4):6–10.

Trungale, J. 2011. Effect of diversions from Guadalupe San Antonio Rivers on San Antonio Bay. Prepared for The Aransas Project as expert testimony in *The Aransas Project v. Shaw et al.* (No. 13-40317, US 5th Cir. 2014).

US Bureau of Reclamation. 2011. SECURE Water Act Section 9503(c)—Reclamation Climate Change and Water, Report to Congress. Denver, CO.

US Bureau of Reclamation. 2014. WaterSMART Grants Available from Reclamation to Conserve Water and Improve Energy Efficiency. Available at http://www.usbr.gov/newsroom/newsrelease/detail.cfm?RecordID=48072 (accessed December 31, 2014).

US Bureau of Reclamation. 2014. Bureau of Reclamation—About Us. US Department of the Interior. Available at http://www.usbr.gov/main/about/ (accessed December 18, 2014).

US Bureau of Reclamation Pacific Northwest Region. 2015. Northwest Region Hydromet Historical Data Access. Stream Daily Average Discharge at DEBO (Deschutes River below Bend, OR) from January 1, 1916 through December 31, 2014. Available at http://www.usbr.gov/pn/hydromet/arcread.html (accessed January 9, 2015).

US Fish and Wildlife Service. 2015. Environmental Conservation Online System. Species Profiles for Chum Salmon, Coho Salmon, Sockeye Salmon, Chinook Salmon, Bull Trout, and Steelhead. Available at http://www.fws.gov/endangered/ (accessed January 13, 2015).

US Geological Survey. 2014. National Water Information Services. Annual statistics taken for gages 09521000 and 09521100 for full approved period. Available at http://waterdata.usgs.gov/nwis/sw (accessed December 17, 2014).

Votteler, T. H. 2004. Raiders of the Lost Aquifer? Or, the Beginning to the End of 50 Years of Conflict over the Texas Edwards Aquifer. *Tulane Environmental Law Journal* 15(2):257–335.

Zellmer, S. 2008. Legal Tools for Instream Flow Protection. In *Integrated Approaches to Riverine Resource Stewardship: Case Studies, Science, Law, People, and Policy*, A. Locke, C. Stalnaker, S. Zellmer et al. (eds.), pp. 285–327. Cheyenne: The Instream Flow Council.

6

Climate Variability and Adaptive Capacities of Intergovernmental Arrangements: Encouraging Problem Solving and Managing Conflict

Edella Schlager

CONTENTS

ABSTRACT Residents and policy makers of western river basins face three substantial challenges. First, hydrologic cycles are changing, but how rapidly and toward what new and different patterns is unknown. Thus, uncertainty is substantial, complicating adaptation efforts. Second, reorienting infrastructure and governing arrangements that were matched to particular hydrologic regimes and that historically have focused on water and hydropower production and flood control will be a slow and difficult process. Path dependencies are difficult to overcome. Third, incorporating a wider range of ecosystem services in management efforts is becoming increasingly more critical not just to recover and protect threatened and endangered species but also to bolster resilience to climate change impacts. This chapter explores how multiscale governing arrangements are better suited for addressing these challenges than are more centralized river basin organizations, and suggests ways in which to bolster the adaptive capacities of intergovernmental arrangements.

6.1 Introduction

One of the most compelling titles of a scientific article in the last decade is "Stationarity is Dead: Whither Water Management?" in which the authors argue that water managers can no longer assume that key dimensions of rivers, such as streamflows or floods, "fluctuate in an unchanging envelop of variability" (Milly et al. 2008, p. 573). Rather, human disturbances, especially anthropogenically caused climate change, have disrupted stationarity assumptions such that the past is not a good predictor of future hydrologic regimes. Water management and managers need to adapt to these changing patterns so that the many services that enhance human welfare—flood control, power generation, recreation, municipal and agricultural supplies—may continue with minimal interruption.

The article was soon followed by social science and law review articles examining the implications of changing hydrologic cycles for the governance of river basins and watersheds and the associated ecosystem services they provide (Craig 2010; May and Plummer 2011). Craig (2010) argued that the foundational assumptions of most US environmental laws—restoration and preservation—are no longer appropriate, because they are oriented to the past. Grounded in stationarity, the goals of restoration and preservation assume that policy actions may be taken that will result in achieving a state of the world that existed at some prior time, or that it is possible to keep a resource within "an unchanging envelop of variability" (Craig 2010, p.15 quoting from Milly et al. 2008, p. 573). Craig proposes grounding environmental laws in adaptation, or what she calls "principled flexibility" (Craig 2010, p. 17). May and Plummer (2011) note that management approaches advocated by international organizations as models for adaptation, such as risk management, are themselves grounded in assumptions of stationarity. They propose explicitly grounding risk management in adaptation by emphasizing learning and encouraging widespread participation in risk management decisions. Craig (2010) and May and Plummer (2011) are drawing attention to how our environmental practices and policies are predicated on control through stability and predictability, which will not serve us well in a changing environment. Instead, environmental policies and governance need to be grounded in adaptation—embracing change in a predictable fashion.

Figuring out how to get from preservation, restoration, and stability, to principled flexibility is no easy matter as three significant challenges must be confronted. First, hydrologic cycles are changing, but how rapidly and toward what new and different patterns are unknown. Thus, uncertainty is substantial, complicating adaptation efforts. Second, reorienting infrastructure and governing arrangements that were matched to particular hydrologic regimes and were also historically focused on water and hydropower production and flood control will be a slow and difficult process. Path dependencies are difficult to overcome. Third, incorporating a wider range of ecosystem services in management efforts is becoming increasingly more critical, not just to recover and protect threatened and endangered species but to bolster resilience to climate change impacts.

These grand challenges form the context of this chapter, which examines the challenges of adapting to a nonstationary future in a federal system of governance. Elazar (1987) famously described federalism as a system of governance characterized by self-rule and shared rule. Self-rule is meant in the sense that governments (local, special districts, counties, states, and a national government) are independent entities that may take actions and make decisions based on their own authority and in response to the demands of their citizens. Shared rule is the concept whereby governments exercise concurrent and overlapping powers and must take one another into account in devising policies and providing goods and services. In any given state, municipal water utilities, local flood control districts, irrigation districts, counties, regional water authorities, state agencies, and federal agencies all have authority to make decisions about water distribution and water quality. Consequently, adaptation in a "federal" river basin means adaptation of relationships among citizens and governments in relation to diverse ecosystems and services, as will be spelled out in Section 6.2. Such institutional and organizational messiness is often met with efforts to tidy things up and impose order, as discussed in Section 6.3. However, given the grand challenges posed by climate change as well as practical experience, efforts to impose order are likely to fail, and there are good reasons to believe that complex governing arrangements may have substantial adaptive capacity, as discussed in Section 6.4. In Section 6.5, how the adaptive capacities of federal river basins may be enhanced is spelled out.

6.2 Ecological and Social Complexities of River Basins

River basins are complex assemblages of ecological systems situated in regions characterized by particular hydrologic patterns or cycles. Even among the many western river basins that are *snowmelt dependent,* hydrologic patterns and associated ecological systems are different in the Pacific Northwest than in the Southwest, and different still from the Intermountain West. For instance, the Klamath River is fed by snowmelt from the mountains in the upper part of the basin in southern Oregon. River flows are highest in the spring as the accumulated snowpack melts, feeding the river, and taper off in the summer and fall as precipitation occurs in the form of thunderstorms (National Research Council 2008). However, the lower reaches of the river in northwestern California, lower in elevation and closer to the coast, experience much higher levels of precipitation in the form of rain that primarily occurs in the winter months, averaging over 50 in. per year (US Climate Data 2014). Likewise, in the Southwest, the Rio Grande River Basin experiences distinct hydrologic patterns throughout its course. From its headwaters in the San Luis Valley of south central Colorado to the Elephant Butte reservoir in southern New Mexico, it is snowmelt dominated, with the highest flows occurring between April and June (Woodhouse et al. 2012). The portion of the lower Rio Grande that forms the international boundary with Mexico relies heavily on a major tributary, the Rio Conchos, for its flows. The Rio Conchos, which originates in the Sierra Madre Occidental in northern Mexico, is dominated by the North American monsoon and receives just over 50% of its flow during the summer months (Woodhouse et al. 2012).

Both river basins are ecologically diverse as well. For instance, the upper Klamath is characterized by shallow lakes and wetlands that are home to a number of fishes, such as suckers and trout. In addition, waterfowl rely heavily on lakes and wetlands for feeding and resting grounds as they migrate in the spring and fall through the North American flyway. During winter, the region is home to the largest concentration of bald eagles in the lower 48 US states (Klamath Basin Audubon Society 2014). The lower Klamath is characterized by anadromous fishes, such as steelhead trout and salmon, and the river and its tributaries are surrounded by heavily forested areas (National Research Council 2008). Diverse ecosystems provide a variety of services. For instance, the remaining floodplains of the Rio Grande River moderate flood flows as bosques absorb flood pulses. In the uplands areas of the watershed, juniper forests limit silt runoff, contributing to maintaining water quality. River basins, situated in regions experiencing particular hydrologic cycles, are collections of ecosystems at diverse spatial and temporal scales, providing ecosystem services.

The hydrologic and ecologic diversity of river basins is matched by the diversity of their associated social systems. Social systems are collections of communities, economies, and governing arrangements organized at diverse spatial and temporal scales. If one were to peer beneath the surface of any western river basin, one would find a breathtaking array of governments, intergovernmental agreements, and numerous private and nonprofit organizations and associations representing diverse interests and values in the decision-making processes of these governing arrangements. The usual suspects would be present, such as the general-purpose governments who are organized with little regard for river basin or subbasin boundaries. The state and tribal governments and counties and municipalities would fall in this category, with each type of government exercising general powers across a variety of sectors, not just water, and providing numerous goods and services, again, not just water. The functionally specialized federal and state agencies are often organized with river basin and watershed boundaries in mind. Examples include Nebraska's natural resource districts, organized by river basin and placed in charge of groundwater management; and the divisions of the Colorado State Engineer's Office and the water courts organized by basin. Special districts, such as irrigation or flood control or wastewater treatment, devoted to a single activity, join these other governments. Given the many governments, with different relationships with one another, intergovernmental agreements are common, from interstate river compacts in which states agree to water allocation rules for dividing the waters of shared rivers, to agreements between federal agencies and special districts for managing water infrastructure. It does not matter how small, e.g., the Costilla Creek, or how large, e.g., the Rio Grande River—relatively rich collections of governments and governing arrangements will be found. Thus, river basins are overlaid with complex and diverse social systems consisting of societies, economies, and governments organized and structured by diverse governing arrangements.

These complex and diverse river basin governing arrangements were created and exploited to pursue economic development goals. Economic development rested on a few highly valued or highly desired uses, such as timber harvesting; cattle grazing; water storage and delivery for irrigation, municipal, and industrial uses; hydropower production; and flood control. Furthermore, to optimize these human uses, the variability in the ecological sources of this wealth was diminished, and certainty, predictability, and stability of flows, whether of trees, grass, or water, was pursued and successfully realized for decades. In short, the ecological systems of river basins were transformed into human-engineered systems to serve human purposes (Schlager and Blomquist 2008).

The goal of economic development was more than realized, but it came at the cost of degraded river basin ecosystems. Virtually every major river basin in the western United States is home to endangered or threatened species because of the numerous human disturbances that have not only eliminated riparian and aquatic habitat but have also undermined processes key for a variety of ecosystem services. In addition, people's values and goals have changed. Besides stable supplies of water, flood control, and electricity generation, people desire recreational opportunities, scenery and wildlife viewing, and the protection of iconic species, among other, more esoteric values. Unprecedented challenges will confront efforts to incorporate a wider variety of ecosystem services in management and governing arrangements, while simultaneously responding to climate change impacts and reorienting governing arrangements and infrastructure to nonstationary hydrologic cycles.

6.3 River Basin Governance in a Federal System

Given the grand challenges facing the many users, managers, and decision makers within river basins, the temptation may exist to advocate for a more centralized, integrative governance approach, whereby the diverse resources and ecosystem services may be considered in a more holistic fashion, under the umbrella of a single governing body that has regulatory, planning, and management powers. Popularly known as *integrated water resources management*, it is a more centralized, comprehensive approach to governing river basins that, at least on paper, appears to hold great promise for addressing multiple, overlapping, and interacting problems. If everything is related to everything else in a river basin, then should a governing body, or a river basin organization, not be granted the authority to attend to all parts of the system? Water quantity and quality are closely related; both are affected by land-use practices, not to mention infrastructure development; and all four affect the health of riparian and aquatic habitats and their associated species. Attending to the interactions and spillover effects of different uses seems to best be handled at the river basin level. Efforts to restore critical habitat, coordinating infrastructure development and management, integrating water conservation practices across sectors, such as agriculture, industrial, and municipal uses, seems to best be dealt with by a river basin organization.

6.3.1 Efforts toward River Basin Government

In practice, however, there are very few examples in the United States of a centrally controlled river basin government with the authority to regulate diverse uses and coordinate various activities. This is not for a lack of effort, with much of this effort focused on coordinating the infrastructure development within basins as well as attending to water quantity and quality issues (Schlager and Blomquist 2008). In the first half of the twentieth century, as the federal government became increasingly involved in building dams, reservoirs, and water distribution systems, more centralized coordination occurred at the national level through a planning board, which, over time and across presidential administrations, was known by various names (Schlager and Blomquist 2008). It was meant to advise the president on a host of issues, and its water committee developed river basin plans and water project proposals to be forwarded to Congress for approval. This national planning effort was dogged by interagency conflict and dissatisfaction by the states, as they had only an indirect say in the decision-making process through their state's congressional representatives. Furthermore, water development, hydropower, and flood control were the only goals pursued, ignoring increasing demands to include a wider array of values such as water quality and fish and wildlife protection. By the early 1960s, convinced that the United States was facing an impending water crisis that required stronger national leadership, the 1965 Water Resources Act was adopted and replaced the national planning board. Among other things, it allowed the states to create commissions, which were forums for coordination and planning. Commissions, composed of representatives from each state in a basin, were charged with developing comprehensive plans. States were only required to implement the plans if state legislatures adopted them into law. No plan was ever adopted by a legislature. In fact, many commissions never developed comprehensive plans, and as Martha Derthick (1974) concluded, having reviewed completed plans, they were mostly wish lists of projects with no effort to prioritize or coordinate projects. By 1980, the newly elected Reagan administration eliminated the commissions as part of a government downsizing effort (Schlager and Blomquist 2008).* No further efforts have been made at the national level to pursue integrated water management.

* As Schlager and Blomquist (2008, p. 50) conclude, "Historically, efforts at comprehensive integrated watershed management have failed. Not only has there been no one best way, but whatever way was chosen proved neither comprehensive nor integrated."

6.3.2 River Basin Government Failure

Examining some of the reasons these efforts have failed is instructive for understanding the forms of decentralized, polycentric river basin governance found throughout the United States and, consequently, what is possible in revising them and bolstering their adaptive capacity to respond to a nonstationary future.

The larger institutional context provides one reason why centralized governance is not widely practiced. Martha Derthick, a highly respected and long-time student of intergovernmental relations in the US federal system of government, asserts, "There has never been a sustained movement for regional organization that left its impress *across* the United States" (Derthick 1974, p. 3; emphasis added). Why? According to Derthick, the answer lies in the design of the US federal system. For a regional organization, such as a river basin agency, to thrive, state and federal governments and agencies would have to cede authority and autonomy to it, allowing it to make decisions and take actions that otherwise would be reserved for them. Given such unwanted impingement on the authority of constitutionally defined governments, robust river basin organizations—and there are very few—come about as political accidents, "the product of ad hoc coalitions whose success was fortuitous in important respects" (Derthick 1974, p. 192).*

Derthick (1974) is careful to note that addressing regional-scale problems is important in a federal system. Regional coordination and cooperation are necessary to address impacts and problems that spill over state boundaries, guide and direct the activities of agencies whose functional activities overlap, and address problems and issues unique to a region. However, in a federal system of government, regional coordination is likely to take the form of voluntary cooperation among governments, and if governments do create enforceable, governing arrangements, it is because they address a pressing need. Derthick (1974) counsels pragmatism as the best policy, enabling, but not commanding, regional responses. Centrally planned regional organizations will be weak and ineffectual (Derthick 1974). Regional organizations voluntarily entered into will be more effective as they are more likely to have the support of their creators, who want them to work.

In addition to being sensitive to what is possible, attention must also be paid to what is desirable. Extensive and well-developed literature on the limitations and shortcomings of centralized, hierarchical organizations and governments exists and is widely known. Rather than detailing the numerous limitations of centralized control, for the purposes of this chapter, two issues will be critically explored: information and values.

The complexity of river basin social–ecological systems combined with the different types and sources of information about processes and activities occurring at different scales requires the development, monitoring, and dissemination of information to citizens and policy makers alike. For instance, efforts are underway in a number of western rivers to reincorporate important dimensions of the rivers' hydrographs, such as peak floods, in order to maintain floodplains or assist in species recovery. This often requires that attention be centered on information at the scale of the basin. In contrast, restoring riparian habitat along streams and tributaries necessitates a careful focus on smaller-scale dynamics and activities, such as local land-use and agricultural practices, not to mention the location and sources of feeding, resting, and nesting for different species of wildlife. Developing and attending to information at different spatial and temporal scales, as well

* Using colorful language to drive home her argument, Derthick (1974, p. 4) asserts, "Regional organizations are excrescences on the constitutional system, unusual things that must be superimposed on the universe of functionally specialized federal and state agencies. The odds are against their being formed and, if formed, against their flourishing."

as to the appropriate level of detail, have long bedeviled centralized arrangements. More localized and fine-grained levels of information are typically neglected as the organization's decision-making bodies attend to basin-wide issues (Lee 1993; Dietz et al. 2003). For instance, Lee (1993), in an early classic on adaptive management, pointed out the limitations of experimenting with main-stem Columbia River flows and wondered how many years it would take to learn about floods and species and habitat recovery if only one experiment occurred per year.

Information shortcomings also relate to the diversity of types and sources of information. Where attention is paid is often dictated by the professional expertise and backgrounds of managers and decision makers as well as the organization's legal mandates (Knott and Miller 1987; Wilson 1993; Brunner and Steelman 2005). Often neglected are the demands of the problems at hand or the perspectives and experiences of the different users and uses of the river basin. Brunner and Steelman (2005) contrast the types of information used by more centralized organizations with adaptive management processes. Centralized organizations tend to pay attention to a few types of information, particularly technical sources of information, whereas adaptive management processes often encompass both technical and experiential sources. Centralized organizations tend to have informational blind spots, both in terms of scale and diversity. Rather than providing the desired comprehensive and integrated approach to governing a river basin, a centralized organization is more likely to simplify, neglecting important sources of information at different scales.

Finally, it is not just a matter of information scale and diversity limitations, but it is also a matter of processing information. Centralized organizations are designed to be serial information processors, not parallel information processors (Jones and Baumgartner 2005; May et al. 2008). That is, decision making is concentrated at the higher levels of the organization in a single decision maker or decision-making venue. In contrast, parallel information processing occurs among multiple, semiautonomous organizations. A single decision maker, or decision-making body, possesses limited capacity to process information and attend to numerous issues, in contrast to multiple decision-making venues. As a result, fewer opportunities to experiment and to learn exist. Instead, some problems receive the bulk of the organization's relatively limited attention, and many issues receive little or no attention, until some threshold of urgency is crossed. Then attention is diverted to the urgent problem, pushing other issues off of the organization's agenda (May et al. 2008). Given the constraints of serial processing, a river basin organization does not have the capacity to attend to the multiple interactions in a river basin in a comprehensive fashion as is hoped for by advocates of integrated management.

Information and values are tightly linked. People pursue a variety of values, both instrumental, such as water supply reliability and flood protection, and noninstrumental, like cultural identity or physical and spiritual renewal, within a river basin. Desired values are realized by directing information and action toward them. Thus, the types and forms of information sought after and attended to are those that relate to values. Ingram (1990) argues that the values and water users left behind are those not recognized during the creation of large water projects and that such initial decisions are not easily corrected. In Ingram's (1990) study of the Colorado River Basin Act, initially, environmental values, future generations, and poor rural communities were neglected. And while most of the projects authorized by the act were built over time, the efforts to address neglected values made little headway, prompting Ingram (1990) to argue that only substantial hydrologic changes will open opportunities for a wider array of values to be considered in large-scale water projects.

Whose values should matter often raises issues of scale and incommensurability. An expansive notion of whose values should matter, such as anyone who has an interest in endangered species, or anyone who receives water from a watershed, even if they are not watershed residents, means that local interests are likely to be overlooked or neglected. For instance, Denver does not reside in the Colorado River Basin, but the Denver Water Authority relies on the Colorado for a significant portion of its water. How much of a say should Denver have in the governance of the basin? The Center for Biological Diversity, a strong advocate for endangered species, is located in Tucson, Arizona, but was a central player in a settlement with the US Fish and Wildlife Service to protect hundreds of species, from the Oregon spotted frog to the Miami blue butterfly, that will affect watersheds across the country (Center for Biological Diversity 2011). Conversely, local interests and values may not adequately capture the numerous values related to a watershed (Margarum 2007). Oftentimes, local values mean maintaining livelihoods or investing in economic development, such as developing water for local irrigation and local industries. Glennon (2002, p. 57), examining the conflicts surrounding multiple efforts to protect streamflows in the San Pedro River located in southeastern Arizona, quotes a local elected official who opposed regulating groundwater to protect riparian habitat as saying, "What benefit do these animals have for humans? We are the ones who rule supreme, and if a plant or animal cannot adapt to our needs, then too bad." This quote captures the fears of many people who suspect that local resource users have little interest in habitat protection. The quote also captures the fact that some values are simply incommensurable. They are not readily traded off against one another, and one has to be chosen over another.

Addressing issues of scale and incommensurability around values is difficult, no matter the governing arrangement; however, centralized organizations are particularly clumsy in this area. Values emerge at different scales, and it is difficult for a centralized organization to be sensitive to scale. In addition, demanding that a single decision venue make tradeoffs among incommensurable values is a recipe for conflict and gridlock. Alternatively, it could be a recipe for ramrodding through decisions that trample on minority interests.*

On paper, at least, integrated watershed management pursued through a river basin organization appears attractive. The complexity of social–ecological systems encompassed in a river basin and the incredible diversity and, at times, intractability of water problems seem more manageable if they can be brought together under a single umbrella and addressed. But decades of experience and research suggest otherwise.

6.4 Opportunities and Limitations of Complex Governing Arrangements

In western US river basins, what is possible and what is desirable substantially overlap, with some notable caveats. As Schlager and Blomquist (2008, p. 20) suggest, "In the uncertain world of complex social and ecological systems, institutional richness may be preferable to institutional neatness." In the US federal system of governance, what is possible is institutional and organizational diversity. If one were to take a census of organizations in a river basin, one would find a combination of general-purpose and special-purpose governments, a diverse array of associations, and a variety of private and nonprofit organizations actively engaged in different aspects of river basin governance, management, and use. Furthermore,

* Thank you to the editor for this insight.

there would be discernable patterns in the river basin governance mosaic. There would be limited duplication of services but noticeable levels of redundancy and overlap in the provision of those services. For instance, multiple irrigation systems may populate a watershed, or multiple water utilities may serve different towns and neighborhoods in a metropolitan area, each with their own portfolios of water that they draw upon to service their own areas. Thus, there is no duplication of services. But redundancy is present. If one water delivery system experiences a major failure in equipment or infrastructure, it is highly likely that it can tie into the infrastructure of its neighbor to ensure that its members continue to access water. Then, special districts or authorities organized at different scales, depending on the good or service being provided, would populate the river basin. Consider the Northern Colorado Water Conservancy District, the operator of the Colorado–Big Thompson Project, which delivers Colorado River water to a variety of irrigation districts and organizations and municipalities in the South Platte River Basin (http://www.northernwater.org/). Furthermore, the river basin may be part of an intergovernmental agreement in which the states that encompass the basin coordinate endangered species recovery efforts, such as the Platte River Recovery Implementation Program entered into by Colorado, Wyoming, Nebraska, and the Department of the Interior (https://www.platteriverprogram.org/Pages/Default.aspx). Such a diverse array of governments, government agencies, and governing arrangements often emerges over time in response to problems, conflicts, and opportunities.

Why is such institutional diversity desirable? Multiscale institutional arrangements, including small and local organizations linked horizontally with each other and vertically with larger-scale organizations, which are, in turn linked horizontally and vertically, may be able to achieve monitoring of subsystem conditions at the appropriate scale. In addition, the array of organizations, each with its own members and constituencies, is likely to represent diverse interests and values associated with different physical components of the system as a whole. Also, the multiple organizations, if appropriately linked, provide "opportunities to communicate and exchange information across subsystem elements and to discuss subsystem interactions and systemwide conditions without necessarily trying to manage all parts of the system with a comprehensive organization" (Schlager and Blomquist 2008, p. 20). Finally, if a policy pursued by a government performs poorly or creates negative effects for citizens of another government, error correction is more likely to happen in a timely fashion. Depending on the source and magnitude of the error, correction may be as simple as adjusting a water accounting system or as complex, time-consuming, and costly as filing a US Supreme Court case. Take the Rio Grande Compact Commission as an example. It boasts a history of good working relations among the three states and federal water agencies punctuated by the occasional serious conflict that requires the mediation of a water master appointed by the US Supreme Court. The key to good relations among the parties is mutual monitoring supported by transparency. Thus, monitoring, learning, representation of values, policy experimentation, and error correction are built into relatively well-designed multiscale governing arrangements.

But, complex governing arrangements are always a work in progress. Citizens, businesses, interest groups, and public officials face difficult choices around whether to create another government, add an activity to or extend the authority of an existing government, or simply support the creation of a collaborative venue in an effort to bolster better coordination (Feiock 2013). In addition, tensions, irritations, and conflicts will emerge repeatedly due to a variety of spillover effects and incommensurable values. The boundaries of a jurisdiction cannot be perfectly aligned with the geographic scope of the problems it seeks to address or the ecosystems it impacts. In addition, participants in governing arrangements may act strategically, even opportunistically, to benefit themselves at the expense of others (Bednar 2009). Ask any state official who actively participates in interstate river compacts, for instance. States jockey for better outcomes for their water users. There is no once-and-for-all governing arrangement; rather, these sources of tension and irritation will be visited and revisited, and organizations will be created, dissolved, restructured, and revised over time as new problems and issues arise.

This constant tinkering may be thought of as a form of principled flexibility, as discussed in the introduction of this chapter (Craig 2010). Stability of institutional arrangements allows for predictability, goal setting, and goal achievement. Individuals and governments may be held accountable for actions that violate agreed-upon rules and regulations, while at the same time encouraging long-term collective action to achieved shared goals. However, too much stability leads to institutional arrangements incapable of responding to changing environmental pressures and societal values. Flexibility allows for governing arrangements to be adapted to changing circumstances; however, too much flexibility weakens collective action efforts and maintaining consensus on shared goals. Given the significant challenges of reorienting watershed governance to address the uncertainties and impacts of climate change while also managing a broad array of values that include numerous ecosystem services in addition to reliable, long-term storage, flood control, and hydropower, greater flexibility will be necessary, but it must rest on a stable foundation. The next section outlines a series of recommendations designed to rebalance the stability–flexibility trade-off so as to better cope with climate change.

6.5 Bolstering the Adaptive Capacity of Intergovernmental Arrangements to Climate Change

Efforts to enhance the capacity of multiscale governing arrangements must take advantage of the strengths of such arrangements, such as addressing multiple issues in parallel, generating information at appropriate scales, and attending to a range of values. Such efforts must also emphasize the relations among governments, in particular, ensuring that spillovers and conflicts are addressed. The following policy recommendations seek to take advantage of intergovernmental arrangements and strengthen their adaptive capacity. The recommendations do so by attempting to rebalance the stability–flexibility trade-offs in existing forms of river basin governance. Recommendations 1 and 2 support greater flexibility by leaving open a variety of options for human uses of river basins. Diverse ecosystem services will better buffer river basin communities from climate change impacts. Recommendation 3 builds a bridge between flexibility and stability. Intentionally investing in information gathering and monitoring allows people to better anticipate problems (flexibility) but also provides greater transparency so that rule-following behavior may be better tracked (stability). Recommendations 4 and 5 support stability. Equity and conflict management work hand in hand to support compliance with governing arrangements. Finally, Recommendation 6 provides a supportive context in which citizens and government officials may experiment with different ways of trading off stability and flexibility as they wrestle with climate change impacts.

1. Reorient state and federal natural resources and environmental quality laws from an emphasis on how much can be removed to an emphasis on how many uses can be allowed.

The central focus of most natural resource laws and environmental quality and protection laws is that of *how much*. How much salmon can be harvested without decimating the population? How much water can be diverted from a river without harming the diversions of historic users? How many dams can be built on a river without their interfering with one another's operations? How much of a pollutant can be deposited in a river without contaminating drinking-water supplies? Or, conversely, in response to demands for environmental protection and restoration, how much salmon must remain unharvested in order to maintain a viable population? Or, how much water must be returned to or remain in the river to prevent the extirpation of the silvery minnow? Or, how many dams may a river contain and still have viable riparian habitat and water quality for multiple endangered species? Basing natural resources and environmental quality laws and policies on how much can be taken or used has three corrosive effects. First, it narrows the vision of an ecosystem to a single stream of resources, erasing the context the stream of resources is connected to and interacts with. Second, it sets up zero-sum trade-offs among users and uses of ecosystems, and sparks conflict. Water used to protect aquatic habitat is not available for irrigation; or, filling in wetlands for an industrial park eliminates the ecosystem services such wetlands provided. Third, it undermines the credibility and value of science and scientists. Scientists are asked to do the impossible: answer *how much* questions with a high level of certainty. When they fail to do so, they are discredited, even though sound scientific practices were used to develop the forecasts (Rose and Cowan 2003). Failure is inevitable because forecasts are typically based on questionable assumptions, such as a stationary hydrologic system and constant water use, and inadequate data.* Thus, forecasts are associated with substantial levels of uncertainty, which, when used for management decisions, result in constant questioning and challenges from managers, resource users, and other stakeholders who are jockeying for their preferred *how much* outcome. Wilson (2002) argues that not only are scientists discredited, but policies and policymakers are discredited as well when policies do not have their desired effects. A prime example is the battle over science that erupted in the Klamath River Basin in the early 2000s in the midst of a severe drought and desperate water shortages (Levy 2003). Scientists attempting to answer the *how much* question around the water needs of endangered and threatened species came to different conclusions, prompting charges by resource users and their elected representatives of *junk science*.

Natural resources and environmental quality laws and policies need to be reoriented to *how many*—how many uses and users river basin ecosystems may support. River basins are home to many, diverse ecosystem services. All contribute to the well-being of the many inhabitants of a river basin, and those contributions need to be recognized and valued. As Frischmann (2012) argues, the value of ecological systems that generate a variety of services can best be realized by recognizing the demand for such services and allowing that demand to be expressed and satisfied through policies oriented toward allowing for and protecting many different uses.

This recommendation is not as radical as it may appear at first glance. Diversity of uses is embedded within different property rights and regulatory systems.

* Rose and Cowan (2003, p. 142) call this "forecasting with disbelief."

Many western states recognize a variety of beneficial uses of water, for instance. Furthermore, as Henry Smith (2002), a prominent property scholar, has argued, the notion of private property includes within it the authority of owners to recognize and pursue a variety of uses as opposed to a single use. That is, it allows owners to pursue many uses, a theme reflected in Wilson's (2002) argument that single-species regulation be replaced with planning and managing for many uses across regions. Anderies and Janssen (2013) concur, suggesting that policies are less costly to monitor and generate fewer conflicts if they avoid a single-species, *how much* approach.

2. One of the key steps in achieving a diversity of uses is by protecting the irreplaceable or the irrecoverable. This may mean mitigating effects of different forms of human activities and development to prevent key ecological processes and linkages from being extinguished or severed. For instance, in many river basins, restoring, maintaining, and protecting floodplains and other links between land and river are key for maintaining important ecosystem services. In other instances, it may require moratoria on certain activities. For instance, the Nebraska Department of Natural Resources has the authority to declare river basins overappropriated. Such a designation comes with a moratorium on new wells and surface water diversions. This should not be interpreted as a backdoor way of reintroducing restoration goals. Rather, this should be interpreted as leaving many future options open for a variety of uses and for exploration and experimentation.
3. Create information-rich river basin social–ecological systems. As Craig (2010) recommends, study everything all of the time; while, of course, that is impossible, the point is well taken. Oftentimes, information generation unfolds in a haphazard manner, typically when organizations and government have a pressing need to invest in particular forms of information. For instance, Colorado has invested in an extensive stream gauge monitoring program as well as a modeling program that allows water administrators to more effectively administer water allocations under the prior appropriation doctrine. Likewise, the Nebraska Department of Natural Resources has invested in developing forecasting models of the Republican River that allow administrators to anticipate the amount of water likely to be available in the next water year and to regulate water users accordingly so that use complies with the Republican interstate river compact. But this process of developing information on an as-needed basis, while important, needs to be supplemented with more systematic forms of information gathering.

   Being more intentional about information generation requires attention be paid to the appropriate scales at which information should be gathered. "This information must be congruent in scale with environmental events and decisions" (Dietz et al. 2003, p. 1908). For instance, monitoring nonpoint-source pollution and its sources is likely to be more localized and fed into collaborative management processes involving actors, whereas monitoring for the effects of wetland disturbances on spawning habitat may be more regional and technical in nature.

   In addition to monitoring human–environment interactions, data must also be collected on values and goals of different resource users, organizations, and decision makers and how values are affected by human–environment interactions.

   This recommendation is not just another call for more data that can be used to engage in better forecasting or more sophisticated modeling of ecosystem processes. In fact, it is much broader than that. Generating information on values, goals, and human–environment interaction that is congruent with the scale of events and decisions supports adaptation.

4. Pay attention to risk sharing in governing the many uses of a river basin. One of the consistent findings in studies of the performance of intergovernmental river basin agreements is that those based on proportionate water-sharing rules experience less conflict than those based on other types of water allocation rules (Yoffe et al. 2003; Schlager and Heikkila 2009; Garrick et al. 2013). In contrast, fixed water allocation rules tend to place shortages on a single party. For instance, several interstate river compacts in the West require the upstream state to maintain minimum streamflows across state lines, regardless of drought, placing the burden of water shortages on the upstream state (Schlager and Heikkila 2009). Risk sharing, or more generally, the equitable distribution of benefits and burdens, must occur across a variety of dimensions. E. Ostrom (2005, p. 262) proposes "proportional equivalence between benefits and costs," where the benefits one party receives are proportionate to the costs or burdens borne by that party. For instance, the proportion of costs an organization pays for the operation and maintenance of a water project is equivalent to the amount of water it receives. E. Ostrom (2005) argues that most actors view such an approach as fair and thus will be more likely to cooperate over the long term and less likely to violate agreed-upon rules.
5. Invest in the capacity to better manage conflicts. Richer information environments and the more equitable sharing of benefits and burdens likely reduce the incidence and intensity of conflicts. If resource use is transparent so that actors may readily determine if others are following the rules, and if the rules are considered fair, then conflict is less likely. In addition, multiple venues for information sharing, negotiation, and bargaining that provide actors with opportunities to address shared problems and resolve their conflicts are also needed, and such venues must be backstopped with more conventional conflict resolution mechanisms, particularly courts. Admittedly, different forms of conflict resolution often only work to the extent that they build trust among the parties so that they make good-faith efforts at working together to resolve their differences.
6. Enable multiple and overlapping forms of coordination among multiple, diverse, overlapping, limited governments so as to encourage experimentation and problem solving. The key word in this recommendation is *enable* in contrast to *impose*. Adaptive multiorganizational governance arrangements must have a problem-solving orientation. That means as problems, issues, and conflicts arise, people should have access to policy tools that allow them to experiment with solutions, rather than having solutions imposed. Policy tools represent an array of methods, from market-based mechanisms, such as allowance or credit trading, to planning activities, to rules and regulations, to achieve environmental goals. For instance, California recently adopted a groundwater governance law that provides local governments and groundwater users with a variety of tools to sustainably manage groundwater basins (Weiser 2014). While all priority basins must eventually have a management plan, how that plan is devised and which local authorities devise it largely rest in the hands of citizens and local public officials. Furthermore, given the spatial extent of some basins, multiple authorities and plans may be created for a single basin (Weiser 2014). California has a long history of local and regional governance of water, with the state providing planning and information resources

to local governments and water users (Blomquist 1992). The latest groundwater legislation builds on that local capacity to realize the statewide value of sustainable groundwater management.

6.6 Conclusion: Responding to Climate Variability

Winston Churchill is credited for saying, "It has been said that democracy is the worst form of government except all the others that have been tried." Much the same could readily be said about multiscale river basin governing arrangements. Such governing arrangements are complex, messy, and sometimes difficult to navigate. In addition, governments may refuse to cooperate, or problems may be neglected, or policy experiments may fail. Yet, in spite of these challenges, intergovernmental arrangements are more likely to match governing arrangements to the proper scale of environmental events, generate and pay attention to different types of information, and be responsive to multiple interests and values, while also allowing for the management of conflict. Moving forward, the demands placed on intergovernmental arrangements will be even greater as climate change impacts become more widespread and severe. In order to adapt to such impacts, substantial investments will be required to reorient water governance to changing hydrologic cycles, while also recognizing and protecting diverse demands and uses of a wide range of ecosystem services. Reorienting multiscale governing arrangements requires that attention be paid to encouraging coordination, managing conflict, and generating a variety of sources of information at appropriate spatial and temporal scales to be useful to resource users and policy makers alike.

References

Anderies, J., M. Janssen. 2013. Robustness of ecological systems: Implications for public policy. *Policy Studies Journal* 41:513–536.

Bednar, J. 2009. *The Robust Federation: Principles of Design*. Cambridge: Cambridge University Press.

Blomquist, W. 1992. *Dividing the Waters: Governing Groundwater in Southern California*. San Francisco: ICS Press.

Brunner, R., T. Steelman. 2005. *Beyond scientific management. Adaptive Governance: Integrating Science, Policy, and Decision Making*. Brunner, R., T. Steelman, L. Coe-Juell, C. Cromley, C. Edwards, D. Tucker. New York: Columbia University Press, pp. 1–46.

Center for Biological Diversity. 2011. Landmark agreement moves 757 species toward federal protection. Available at http://www.biologicaldiversity.org/programs/biodiversity/species_agreement/index.html.

Craig, R.K. 2010. Stationarity is dead – long live transformation: Five principles for climate change adaptation law. *Harvard Environmental Law Review* 34:10–70.

Derthick, M. 1974. *Between State and Nation: Regional Organizations of the United States*. Washington, DC: Brookings Institute.

Dietz, T., E. Ostrom, P. Stern. 2003. The struggle to govern the commons. *Science* 302:1907–1912.

Elazar, D. 1987. *Exploring Federalism*. University, AL: University of Alabama Press.

Feiock, R. 2013. The institutional collective action framework. *Policy Studies Journal* 4:397–425.

Frischmann, B. 2012. *Infrastructure*. Oxford: Oxford University Press.

Garrick, D., L. De Stefano, F. Fung, J. Pittock, E. Schlager, M. New, D. Connell. 2013. Managing hydro-climatic risks in federal rivers: A diagnostic assessment. *Philosophical Transactions of the Royal Society* A371:20120415.

Glennon, R. 2002. *Water Follies*. Washington, DC: Island Press.

Ingram, H. 1990. *Water Politics: Continuity and Change*. Albuquerque, NM: University of New Mexico Press.

Jones, B., F. Baumgartner. 2005. *The Politics of Attention: How Government Prioritizes Problems*. Chicago, IL: University of Chicago Press.

Klamath Basin Audubon Society. 2014. "What Makes the Klamath Unique?" http://eaglecon.org/birds/index.shtml. Accessed September 24, 2014.

Knott, J., G. Miller. 1987. *Reforming Bureaucracy: The Politics of Institutional Choice*. Englewood Cliffs, NJ: Prentice-Hall, Inc.

Lee, K. 1993. *Compass and Gyroscope: Integrating Science and Politics for the Environment*. Washington, DC: Island Press.

Levy, S. 2003. Turbulence in the Klamath River Basin. *BioScience* 53(4):315–320.

Margarum, R. 2007. Overcoming locally based collaboration constraints. *Society & Natural Resources* 20:135–152.

May, B., R. Plummer. 2011. Accommodating the challenges of climate change adaptation and governance in conventional risk management: Adaptive collaborative risk management (ACRM) *Ecology and Society* 16(1): 47. [online] URL: http://www.ecologyandsociety.org/vol16/iss1/art47/.

May, P., S. Workman, B. Jones. 2008. Organizing attention: Responses of the bureaucracy to agenda disruption. *Journal of Public Administration Research and Theory* 18:517–541.

National Research Council. 2008. *Hydrology, Ecology, and Fishes of the Klamath*. Washington, DC: National Academies Press. Available at http://www.nap.edu/catalog.php?record_id=12072&utm_expid=4418042-5.krRTDpXJQISoXLpdo-1Ynw.0&utm_referrer=http%3A%2F%2Fwww.nap.edu%2Fopenbook.php%3Frecord_id%3D12072%26page%3DR1.

Ostrom, E. 2005. *Understanding Institutional Diversity*. Princeton: Princeton University Press.

Rose, J., K. Cowan. 2003. Data, models, and decisions in U.S. marine fisheries management: Lessons for ecologists. *Annual Review of Ecology and Evolutionary Systems* 34:127–151.

Schlager, E., W. Blomquist. 2008. *Embracing Watershed Politics*. Boulder, CO: University Press of Colorado.

Schlager, E., T. Heikkila. 2009. Resolving water conflicts: A comparative analysis of interstate river compacts. *Policy Studies Journal* 37:367–392.

Smith, H.E. 2002. Exclusion versus governance: Two strategies for delineating property rights. *Journal of Legal Studies* 31:453–487.

U.S. Climate Data, Orleans, CA. http://www.usclimatedata.com/climate/orleans/california/united-states/usca0815. Accessed September 24, 2014.

Weiser, M. 2014. California posed to restrict groundwater pumping. Sacramento Bee. Available at http://www.sacbee.com/2014/09/15/6706392/california-poised-to-restrict.html.

Wilson, J.M. 2002. *Scientific uncertainty, Complex Systems and the Design of Common Pool Institutions. Drama of the Commons*. E. Ostrom et al., eds. Washington, DC: National Academy Press, pp. 327–359.

Wilson, J. Q. 1989. *Bureaucracy: What Government Agencies Do and Why They Do It*. New York: Harper-Collins.

Woodhouse, C.A., D.W. Stahle, J. Villanueva-Díaz. 2012. Rio Grande and Rio Conchos water supply variability from instrumental and paleoclimatic records. *Climate Research* 51:147–158.

Yoffe, S., A. Wolf, M. Giordano. 2003. Conflict and cooperation over freshwater resources. *Journal of the American Water Resources Association* 39:1109–1126.

7

Support for Drought Response and Community Preparedness: Filling the Gaps between Plans and Action

Kelly Helm Smith, Crystal J. Stiles, Michael J. Hayes, and Christopher J. Carparelli

CONTENTS

ABSTRACT This chapter examines which levels of government handle various aspects of drought, as well as interactions between levels of government, providing examples from states across the western United States. It also takes a look at aspects of drought that fall outside traditional lines of authority and disciplinary boundaries. As part of a discussion on how states support local drought response, the chapter details and contrasts how California and Colorado track public water supply restrictions, and describes Colorado's process for incorporating input from river basins across the state into its water plan. Case studies focus on drought planning in the Klamath River and Upper Colorado River basins through the lens of collaborative environmental planning. The chapter concludes that drought planning will be more effective as more states coordinate and align goals and policies at multiple levels of government.

7.1 Introduction

Flying over a landscape reveals patterns of land use that are not visible from ground level. You may notice a patchwork of crops, with the rows and leaf textures of a field showing up as distinct shades of green, or circles of crops under center-pivot irrigation, indicating availability of groundwater. Farther west, bare, arid ground gives way to riparian strips of irrigated, cultivated land. Rivers and mountains form natural boundaries, but state borders are indistinguishable. The difference in perspective between 30,000 ft. and ground level is also true for socioeconomic systems. The pressures and opportunities for conservation, development, and sustainable resource use look very different, depending on whether you're sitting in a small-town mayor's office, a regional planning office, a tribal council, or the state capital. Although the most local levels of government may have the fewest resources and the least ability to gain a big-picture perspective higher than the town water tower, traditionally municipal matters such as zoning and water supply decisions may have some of the greatest effects on patterns of water use and drought resilience. Fortunately, just as technology now makes it possible for people to view the planet as if from space in one instant and to zoom down to their own backyard in the next, an increasing array of collaborative methods is evolving to support working across and between traditional lines of authority and levels of jurisdiction. This chapter will provide an overview of recent developments in drought planning that ensure that decision makers at state, tribal, municipal, and other levels of authority are working within a consistent understanding of opportunities and constraints. Failing to share data and create a shared perspective may result in decision making based on short-term economic interests that diminish drought resilience for an entire watershed.

Droughts are a normal part of the climate across the western United States and thus have had a tremendous influence on both the cultures and the environments across the region, most recently in 2011–2015. In 2011, a very intense drought combined with record heat struck the southern Plains. The 2012 drought was more widespread and affected large parts of the United States, including regions outside the West. Drought conditions and the associated dust storms in early 2014 revived images of the Dust Bowl in the southern High Plains. Meanwhile, very intense drought was entrenched in California and adjacent states in 2014–2015. The resulting impacts have occurred within agriculture (both to crops and livestock), affected drinking water supplies in both rural and urban areas, enhanced wildfire potentials, and created challenging wildlife management issues. According to the National Climatic Data Center's list of billion-dollar weather disasters, drought cost the United States $4 billion in 2014, $10 billion in 2013, $30 billion in 2012, and $12 billion in 2011 (NCDC 2015). The University of California–Davis estimated that drought in 2014 would cost the state $1 billion in agricultural revenue and $0.5 billion in additional pumping costs, with a total statewide economic cost of $2.2 billion, and a loss of 17,100 seasonal and part-time agricultural jobs (Howitt et al. 2014).

These recent drought examples in the western United States occurred amid growing concerns from scientists and officials about food security, water shortages, energy supplies, climate change, and the complex interactions of these issues. The recent National Climate Assessment report highlights the southwestern United States as a region facing future increases in both drought frequency and intensity (Garfin et al. 2014). Trying to anticipate future drought impacts in the western United States requires an understanding of the past, as well as an understanding that the past may not represent the best analog for the future given the changing climate and shifting vulnerabilities across the region (Milly et al. 2008). New challenges loom—growing populations place increasing demands on limited water resources, and elevated temperatures compound the effects of low precipitation, a phenomenon that Overpeck (2013) calls "hot droughts."

The issue of drought in the West fits well into a larger context pointed out within a 2013 United Nations report. In that report, it was estimated that the direct losses from natural disasters globally since 2000 are potentially in the US $2.5 trillion range (UNISDR 2013). Natural disasters such as drought will be significant issues for all societies in the future. In a press release accompanying the report, UN Secretary-General Ban Ki-moon argued that the "economic losses from disasters are out of control" and that these losses will continue to escalate unless actions are taken to reduce disaster risks in the future. In the western United States and other regions with a projected increase in drought under a changing climate, it is critical that we understand how to reduce vulnerability and that we act on that knowledge.

Social planners Rittel and Webber in 1973 coined the term "wicked problems" to refer to complex problems that are difficult to define; involve complex sets of actors, issues, and trade-offs; and are impossible to isolate in laboratory conditions. Botterill and Cockfield (2013) observe that planning for drought is wicked in the following ways:

- There is no definitive formulation of the problem.
- The problems have no stopping rule.
- Solutions are not true or false, but rather, bad or good.
- Every solution is a "one-shot" operation; because there is no opportunity to learn by trial and error, every attempt counts significantly (Botterill and Cockfield 2013, pp. 9–10).

Adding to the complexity of drought planning is that no single discipline, profession, or sector has a monopoly on defining, experiencing, and managing drought. Agricultural policy and practices, laws governing water management and delivery, and urban and environmental land-use policies each affect our collective vulnerability to drought yet may often evolve in parallel, nonintersecting contexts. Much agricultural policy comes from the federal government. States are the primary level of authority for enacting laws governing water. Many land-use decisions that affect water consumption, runoff, or absorption happen in local offices that are typically separate from the utilities that actually deliver water to households.

One more good reason to plan for drought is that it is a good starting point and capacity-building exercise in planning for a changing climate. Both involve a slow-moving phenomenon that is difficult to detect until it is already well underway. Planning for both drought and climate change requires processes that involve many sets of stakeholders with different interests who are interested in and respond to different kinds of information and data. Both also involve a shift of perspective, from planning what to do in case of emergency (response) to planning to avoid an emergency (hazard mitigation, climate adaptation).

7.2 Institutional Approaches to Drought Risk Management

Traditionally, most of the efforts focusing on drought impacts have dealt with responding to these impacts after an event. Beginning in the 1970s and 1980s, scientific and policy organizations described ad hoc responses to drought as uncoordinated and untimely (GAO 1979; Wilhite and Pulwarty 2005; GSA 2007) and began calling for better coordination of government responses to drought (Wilhite 1991). The alternative to treating every drought as a separate, unforeseeable emergency is to emphasize improving drought monitoring, planning, and mitigation strategies to reduce impacts from future droughts (Wilhite et al. 2005). This approach is in accord with the shift in focus from disaster response to disaster risk reduction that was officially recognized when the United Nationals General Assembly declared the 1990s the International Decade for Disaster Risk Reduction (Hellmuth et al. 2011). This approach requires identifying who and what are at risk, why they are at risk, how individuals or organizations respond to events, and what steps can be taken ahead of time to reduce risk. Underscoring a shift toward anticipating droughts as a recurrent feature of climate, the National Drought Policy Commission (NDPC) in 2000 recommended favoring "preparedness over insurance, insurance over relief, and incentives over regulation" (National Drought Policy Commission 2000, p. 35). The NDPC also recommended passing a national drought preparedness act and creating a national drought council to coordinate national drought policy in the United States (National Drought Policy Commission, cited in Whitney 2013, pp. 73–74). As of 2014, the broader recommendations for a comprehensive US drought policy had not been enacted, but monitoring and early warning provisions had been implemented.

In 2006, Congress established the National Integrated Drought Information System (NIDIS), to consolidate the nation's drought early warning and monitoring capabilities, with the National Oceanic and Atmospheric Administration (NOAA) as the lead agency. Through its website (http://www.drought.gov), NIDIS provides one-stop access to drought monitoring products from many federal agencies, including NOAA, the US Department of Agriculture (USDA), the National Aeronautics and Space Administration (NASA), the US Geological Survey, and more. Most recently, the National Drought Resilience Partnership, introduced in 2013 as part of President Barack Obama's Climate Action Plan, calls for the development of long-term planning and resilience strategies to improve the nation's drought preparedness (Bergman 2014; NIDIS 2014), with the USDA as the lead agency. The USDA is also the agency most active in providing drought relief to agricultural producers, particularly through crop insurance.

As a slow-moving hazard, drought falls outside the traditional emergency management responses to natural disasters, as set forth in the Stafford Act of 1988 and amended by the Disaster Mitigation Act of 2000. The purpose of the Stafford Act is "to reduce the loss of life and property, human suffering, economic disruption, and disaster assistance costs resulting from natural disasters" (FEMA 2013, pp. 1–2), and it specifically includes drought in a list of major disasters. However, the Federal Emergency Management Agency (FEMA) has not typically been involved in official federal responses to drought, nor has the Stafford Act been invoked specifically for drought within the continental United States. Although drought may be as cumulatively disruptive as faster-moving disasters, legal and environmental scholar Jeremy Brown describes it as lacking charisma or screen presence. He notes that if FEMA or others responded to drought under the Stafford Act, it would permit additional benefits including unemployment, supplemental nutrition assistance (formerly food stamps), and crisis counseling (Brown 2014).

The Disaster Mitigation Act of 2000 emphasizes state and local mitigation planning, noting that "high priority should be given to mitigation of hazards at the local level" (Section 101[4]). Multihazard mitigation plans are a prerequisite for local governments to access various FEMA mitigation grants. Although drought is not one of the handful of disasters specifically named in the Disaster Mitigation Act of 2000, emergency and hazard mitigation planners can include drought mitigation in multihazard mitigation plans. Many of the strategies that reduce vulnerability to drought also reduce vulnerability to other hazards.

States, which have the legal authority to regulate water, have been more active in implementing drought plans. In the aftermath of the late 1970s drought, Don Wilhite, a professor at the University of Nebraska-Lincoln, began investigating what drought risk management would look like in context of state drought planning efforts. Wilhite's first efforts were to work with states on the concept of drought planning, and he first codified states' approaches to drought planning in 1990 as the 10-Step Drought Planning Process. He has been helping disseminate the fundamentals of drought planning since then. The 10 steps that Wilhite identified, and that have been adapted by states, tribes, and countries around the world, are as follows:

1. Appoint a drought task force.
2. State the purpose and objectives of the drought plan.
3. Seek stakeholder participation and resolve conflict.
4. Inventory resources and identify groups at risk.
5. Establish and write the drought plan.
6. Identify research needs and fill institutional gaps.
7. Integrate science and policy.
8. Publicize the drought plan.
9. Develop educational programs.
10. Evaluate and revise the drought plan (Wilhite et al. 2005, pp. 93–94).

Wilhite worked with the USDA and NOAA to establish the National Drought Mitigation Center (NDMC) in 1995 at the University of Nebraska–Lincoln. The center's mission is to reduce societal vulnerability to drought. One of the NDMC's main emphases has been fostering drought planning, and the center has done so at scales from the individual farm or ranch to communities, tribes, river basins, states, and countries around the world. In 1995, 27 states had drought plans. As of 2014, 45 states had drought plans, according to the NDMC's ongoing collection and catalog of plans on its website (http://drought.unl.edu/Planning/PlanningInfobyState.aspx).

7.2.1 State Drought Plans

A recent assessment of drought plans from all 19 Western Governors' Association states identified several common themes. Typical state drought *response* strategies include increasing communication, issuing water restrictions, facilitating and/or expediting water transfers or temporary permits, purchasing water rights to keep water in streams, financial support for public water suppliers, recommending measures such as permitting roadside haying, and activating state assistance and technical support (Fontaine et al. 2014). State officials identified the following as mitigation strategies, which are implemented proactively to reduce drought vulnerability: increasing water conservation, especially for development and growth; enhancing water supplies; improving delivery infrastructure and intersystem connections; increasing availability of monitoring data; promoting rangeland fire insurance; and requiring public water systems (PWSs) to address drought in their planning documents. Many states' drought communication processes include opportunities for input from different localities and sectors.

> Some states have established local groups that provide information to the state on drought conditions and impacts, enabling the states to focus response efforts. Other state monitoring groups use field agents to report on local impacts. Many state drought committees have individual state agencies that report on drought-impact information from specific sectors, and then provide assistance as needed.... (Fontaine et al. 2014, p. 97)

Asked to identify factors contributing to a successful drought program, state drought coordinators highlighted the need for communication and coordination with local entities, including encouraging local governments to develop their own drought plans (Fontaine et al. 2014). States take a variety of approaches to fostering drought planning by local entities, including delegating authority to regional or municipal entities and providing technical assistance, data, and model plans.

7.2.2 State Support for Local Drought Mitigation and Response

Some states work with municipalities or water suppliers to provide resources that improve response to and mitigation of drought. For example, the Colorado Water Conservation Board (CWCB) has developed the Drought Tool Box (http://cwcb.state.co.us/technical-resources/drought-planning-toolbox/Pages/main.aspx), a resource that includes discussion of and links to drought monitoring resources, a granting program to help municipal water providers develop drought management plans, drought planning guidance for municipalities, and background information on climate change and drought. The town of Firestone, Colorado, used the Tool Box and worked with a consultant to complete a drought management plan in 2012 (http://www.firestoneco.gov/DocumentCenter/View/72). (Also see the discussion in this chapter on Colorado's basin-level planning.)

In California, water suppliers over a certain size must file an Urban Water Management Plan every 5 years detailing how they will maintain reliable water supplies under different conditions, which is part of the state's Integrated Regional Water Management planning. Water suppliers must also file water shortage contingency plans. Passage of the Urban Water Management Planning Act of 2001 linked water-use and land-use planning, making approval of new developments contingent on adequate water supplies (Brislawn et al. 2013). The Department of Water Resources worked with the California Urban Water Coalition and others to produce the *Urban Drought Guidebook* in 2008.

A succession of dry years in California beginning in 2012 and continuing as of this writing in 2015 led to heightened awareness of the drought risk that California faces. Drought reduced agricultural production, led to mandatory reductions in urban water uses, made it harder for wildlife to find food and water, and revealed the vulnerability of some rural residents, who were confronting dry wells. In 2013, California began holding workshops for rural water providers to increase awareness of the issues likely to arise and actions they could take if drought continued (Weiser 2013). When Governor Jerry Brown proclaimed a drought State of Emergency in January 2014, he was flanked by several department heads, including the leaders of Cal Fire, the Department of Food and Agriculture, the Department of Water Resources, the Water Resources Control Board, and the Governor's Office of Emergency Services (California 2014). The Office of Emergency Services played a key role, with its Incident Command System providing a coordinating structure for the interagency effort, and allowing the state to tap into funds under the California Disaster Assistance

Act to help address problems such as dry domestic wells (Davis-Franco 2014). In 2014, the California Governor's Office of Planning and Research created the Local Government Drought Toolkit and Local Drought Clearinghouse (http://www.opr.ca.gov/s_drought resources.php) to support local agencies in coping with drought. As part of the January 2014 drought emergency proclamation, Governor Brown called for Californians to reduce water consumption by 20%. Some communities responded much more actively than others, but a state survey found that in May, usage increased by 1% statewide (Lovett 2014). In midsummer, the State Water Resources Control Board imposed emergency conservation regulations, banning outdoor uses of potable water, such as car washing and nonrecirculating fountains, and requiring water suppliers to impose restrictions on outdoor irrigation and to report on water use each month. In November 2014, voters approved Proposition 1, allocating $7.5 billion for infrastructure and environmental projects. In early 2015, the drought had further intensified, prompting the governor to issue mandatory water conservation targets for urban water utilities (State of California, Executive Department, 2015).

7.2.3 Drought Planning for Communities

In 2010, the NDMC and partners published the *Guide to Community Drought Preparedness*, which expresses the core elements of drought planning in a way that may make sense for smaller communities. The guide includes work sheets and many ideas that may contribute to a community's drought planning process, and communities are encouraged to select the pieces that seem most appropriate for the issues that they are facing. The guide deliberately refers to *community*, a looser term than *municipality*, anticipating the need for intergovernmental, transboundary processes. A logical drought planning entity in many towns and cities is the public water utility, but water utilities generally have a fairly specific mission related to delivering a continuous supply of drinkable water, and drought may have broader impacts than that. It may make sense to address drought planning at the watershed level, uniting agricultural areas and small towns. The guide simplifies the 10-step process, focusing on establishing the planning team and connecting with stakeholders, establishing monitoring, understanding the community's drought history and vulnerability, establishing a public education and outreach program, and identifying and implementing steps to reduce vulnerability to drought (Svoboda et al. 2011).

7.2.4 Integrated Planning for Communities

Many states, tribes, and cities have the resources to conduct stand-alone drought planning processes. With an understanding of policies and processes that contribute to drought resilience, it is also possible to incorporate elements of drought planning into other planning processes. More entities are beginning to include drought in multihazard planning processes. Integrating mitigation measures for drought and other hazards into other kinds of plans increases the likelihood of reducing drought vulnerability.

The American Planning Association published *Planning & Drought* in 2013, a Planning Advisory Service report that introduced concepts of drought planning to urban planners, developed in collaboration with the NDMC and NIDIS. Authors of the chapter on how planners can address drought emphasize that "establishing a fully integrated framework merging land-use and water resource management planning at a regional level might be considered the 'gold standard' in terms of facing the challenges of drought and climate change" (Brislawn et al. 2013, p. 40).

Opportunities to mainstream drought planning include wrapping it into multihazard planning, zoning, comprehensive planning, infrastructure planning, water supply management, storm water and water treatment planning, climate adaptation planning, capital improvement planning, riparian and floodplain planning, conservation planning, watershed protection planning, and more. In short, there is no end of opportunities to incorporate elements that increase resilience to drought into other kinds of planning and practice. The challenge is creating awareness so that planners and decision makers can recognize and act on opportunities as they arise.

7.2.5 Technical Assistance and Organizational Capacity

Inventorying data and involving data providers are main steps in drought planning, and one of the key ways that state and federal agencies can support community drought planning is by providing data on water, climate, agriculture, and the environment. NIDIS' 2006 congressional mandate includes providing data and drought early warning information to decision makers at all levels, including local government (NIDIS 2007). NIDIS has identified several pilot areas as Regional Drought Early Warning Systems (RDEWS), which aim to develop partnerships between agencies at multiple levels to identify drought risk reduction strategies through monitoring and prediction. These pilot projects also focus on delivering timely, spatially relevant information to test regions across the country.

Agencies can contribute to planning efforts by providing technical services such as geographic information systems (GIS) or by sharing communication capacity to coordinate activities; organize meetings; and distribute agendas, findings, and other information. Research by Floress et al. (2011) found that watershed management groups that included more agency personnel participating as part of their job responsibilities were more effective and better networked than all-volunteer groups.

7.2.6 Tracking Public Water Supply Restrictions

California and Colorado are among the handful of states that have online systems for tracking local PWSs that have imposed either voluntary or mandatory water-use restrictions on customers, although Colorado's system is maintained by the state and California's system is maintained by an industry association. These systems may serve several purposes, including increasing awareness of water supply conditions; helping people figure out what, if any, restrictions apply in their area; and helping government agencies target assistance more effectively. Examples from Colorado and California, described here, illustrate differences in how much effort statewide agencies or organizations have chosen to invest in verifying information from PWSs. Colorado's system, which is a permanent fixture, links users to their PWS for the most recent information. California's system, operational only under serious drought conditions, relies on organization staff to find and verify information from PWSs.

The CWCB launched its website http://www.coh2o.co in the spring of 2013 in response to severe drought conditions that had been affecting large portions of the state since 2012. The website offers a search feature that allows users to search water restrictions by entering a city, county, or zip code. Users are then redirected to the web page of the PWS that serves the area they entered. The CWCB saw a need to create the website because there was confusion among the public as to what the specific water-use restrictions were for each area. The Denver metropolitan area is home to several different water providers with service area boundaries that are not common knowledge to the public. Furthermore, the mainstream media coverage in the state is largely focused on the Denver metro area, so Denver-based television news stations or newspapers often do not report drought-related PWS water-use restriction information for other communities throughout the state. The CWCB wanted to ensure that the state would not infringe upon local control over water-use restriction information and messaging, so it chose to redirect users to PWS websites rather than collect the information from PWSs for dissemination by the state. The CWCB contacted most of the PWSs in the state and offered them the opportunity to opt into voluntary participation with the website (http://www. coh2o.co). Not all PWSs opted in, and as a result, some website users are affected by these information gaps (Taryn Finnessey, drought and climate change technical specialist, CWCB, Colorado Department of Natural Resources, in discussion with Chris Carparelli, November 12, 2013).

The Association of California Water Agencies (ACWA) is an industry group that represents the interests of 440 PWSs that deliver water to roughly 90% of California's population. In January 2014, ACWA began posting a map of PWS water-use restrictions on its website (http://www.acwa.com/content/drought-map) in response to escalating drought conditions that began in 2013. The map uses a Google Maps interface with color-coded dots that, when clicked upon, convey local water-use restriction information at that location. ACWA only posts the map during serious drought conditions. Information for the map is collected in a variety of ways. ACWA staff monitor the news, contact member agencies by phone and e-mail, and then update the map based on the information gathered. Some member PWSs will also proactively contact ACWA to submit information. The frequency of information collection and map updates is variable, and it is voluntary for PWSs to provide information. The primary purpose of the map is to publicize the impacts of the drought to create public awareness. Member agencies use it to communicate with their customers, and other entities, including the California Department of Water Resources, have embedded the map or provided links to it on their websites. ACWA chose not to design an information collection system that would put the onus on PWSs to provide and maintain data, because it felt that the information would be more accurate if ACWA was the keeper of the information (Matt Williams, communications specialist, ACWA, in conversation with Chris Carparelli, February 12, 2014).

7.3 Barriers to Drought Risk Management

As planners and other resource managers look for opportunities to increase resilience to drought, they should be aware of some pitfalls. As a slow-moving natural hazard, drought tends to be off the radar screen of many people, especially those living in cities, where water and food supplies are more mediated by technology that buffers the population from natural variation. The separation of systems that have evolved to govern and manage water supplies, land use, and food needs to be addressed directly in dealing with drought.

7.3.1 Disciplinary Silos: Land and Water Use

A major challenge for drought resilience is that planning for water supply and for urban land use is traditionally handled by different departments and different professions, and much agricultural decision making happens separately from both of these. Within the boundaries of an incorporated municipality, land-use practices can have a big effect on the demand for water, but water suppliers and city planning and zoning departments are traditionally separate decision-making entities.

In workshops held to update California's *Urban Drought Guidebook* in 2008, participants identified the following issue:

> The disconnect between planning departments and water suppliers, even within the same city government, continues to stymie efforts to develop new projects with built-in water efficiency and to enforce landscape ordinances. Furthermore, conflicting state and local regulations and policies—especially those concerning state housing mandates and the ability to serve water, and especially during times of water shortages—need to be addressed. (Schwab 2013, p. 37)

Vivek Shandas, an urban ecologist at Portland State University, and G. Hossein Parandvash, an economist at the Portland Water Bureau, have teamed up to research the gap in land and water planning, and what might be gained by closing it. They note that "the provision of sufficient quantities of water for all forms of development while ensuring adequate supplies for agricultural and nonhuman use is arguably the most significant challenge faced by urban planning agencies" (Shandas and Hossein 2010, p. 112) and that urban planners observed the need to incorporate water into their work as early as 1978, but the administration of water-use and land-use planning have gone the opposite direction since then, becoming increasingly separate. Shandas and Parandvash compared water use at the level of individual parcels of land and found that smaller development size and higher-density single-family residential land use was correlated with lower water use, and they found zoning and development regulations to be key factors in predicting water demand.

7.3.2 Limits of Local Government

The water supply for a farm or for a community may originate some distance away (in some cases, hundreds of miles), which limits the scope of what a single planning entity can accomplish. Many planners' preference is to situate water and drought planning within a regional or landscape-scale effort. Carleton Montgomery, editor of *Regional Planning for a Sustainable America*, observes, "Regional resources, such as large intact forests and aquifers, are far beyond the power of most municipal governments to conserve, yet development decisions made by local governments represent the greatest threat to these resources" (Montgomery 2011, p. 3).

Hazard planner Raymond J. Burby also criticized local governments for weighing development interests too favorably:

> One of the most serious limitations of the land use approach [to hazard planning] is that without strong mandates from higher-level governments, few local governments are willing to protect against natural hazards by managing development. (Burby 1998, p. 14)

Recognizing the need for a different level of authority, some states have created new entities to govern land and/or water use. In 1980, Arizona's Groundwater Management Act established active management areas (AMAs) to limit groundwater overdraft in and around urban areas. The Assured and Adequate Water Supply program is a key element of the AMAs, ensuring that any new development has enough water and will not exacerbate groundwater depletion. In a review of Arizona's groundwater management policy, Jacobs and Holway (2004, p. 64) stated the following:

> In the case of the assured water supply program a strong state-level regulatory approach was essential. The standards for establishing a program like assured supply must be set at a level of government higher than the local governments that have the responsibility to approve or disapprove individual zoning and subdivision proposals.

7.3.3 Short-Time Horizons

Gene Whitney, who worked at the science–policy interface in Washington, DC, for many years, collaborating with both the executive and legislative branches of government, observed the following:

> Developing public policy to address the long-term preparedness, mitigation and impacts of drought is difficult in a political culture that increasingly operates with a short-term perspective. In the United States, politicians tend to focus on short-term problems because policies are promulgated by elected officials who operate on...reelection cycles.... [E]ven though recovery may be much more expensive than prevention, it is often politically simpler to address an incident after it occurs than it is to prevent an incident or reduce its impact. (Whitney 2013, pp. 72–73)

7.4 Governance, Government, and New Ways of Working

Much of the literature on governing complex human and environmental systems explicitly recognizes that problems will never align neatly with preexisting boundaries, and that working across organizations, both vertically and horizontally, will be necessary. Drought crosses jurisdictional boundaries, and responses are most effective when local, state, and federal plans and policies align. Although this adds complexity, separate entities working together may voluntarily achieve better results than the old command-and-control regulatory approach. In fact, it is this recognition of the opportunity to have more control over the outcome and the opportunity to achieve a more appealing outcome that provides incentives for parties to participate in collaborative processes. Particularly when working outside traditional lines of authority, creating common understanding through dialog among diverse stakeholders is crucial. One of the most important elements in this new way of working is the recognition that good information alone is not enough. It has to be embedded in human experience, mediated by dialog and collaborative learning, to be effective. Although scientists, planners, and many other professions subscribe to the implicit idea that good data and accurate information can resolve issues, research has shown that it takes more than good information to penetrate to the level of changing behavior or policy, and that decision makers are more likely to use knowledge cocreated through dialog (Innes 2010). (See Edella Schlager's excellent discussion in Chapter 6 of the pros and cons of decision making across watersheds, scales, and jurisdictions.)

Scholars studying governance of combined social and environmental systems use various composite terms such as *adaptive comanagement* or *collaborative adaptive management* to describe an approach that combines adaptive management, an ecological concept, with collaborative governance of water systems. They say that adaptive comanagement should incorporate (1) polycentric governance, with overlapping functions providing increased resilience; (2) public participation, because buy-in comes through understanding created in dialog, and many types of knowledge must be incorporated; (3) willingness to learn from experience or to build from pilot projects; and (4) a bioregional, watershed perspective, either empowered from above or composed of overlapping jurisdictions (Huitema et al. 2009). Or, as summarized by McNutt et al. (2013, p. 152),

> Numerous theoretical and empirical studies have shown that effective management of common-pool resources is easier to realize when communities develop social capital through a distributed and dense social network that develops trust and common understanding and stimulates learning and formulation of alternative response options.

One of the first steps in initiating a polycentric, collaborative governance process is to define the boundaries of the problem, if possible. In the case of water management issues, river basins are frequently a logical area. Recognizing that neither the state nor the local level is ideal for many water management tasks, states have created river basin organizations to provide advice and guidance. In 2005, Colorado passed the Colorado Water for the 21st Century Act, creating roundtables for each of eight river basins and the Denver metropolitan area. Their charge was to incorporate state data as well as input from local governments and water suppliers in a process of issue identification and assessment, and make recommendations back to the state. The state released a draft water plan in December 2014 incorporating recommendations from each basin (online: http://www.coloradowaterplan.org). The plan draws from basin recommendations to address major challenges such as the need to move water from the western to the eastern side of the Continental Divide; diverting water from agricultural to urban use; the need to preserve the environment for fish, wildlife, and recreation; a changing climate; and the need to fund increased water security (Colorado Water Conservation Board 2014).

Water suppliers in many parts of the country are also involved in protecting upstream watersheds. Land management practices that contribute to good water quality, such as maintaining healthy forests, also work to slow runoff and increase infiltration, which increase drought resilience.

7.4.1 Drought Planning by Transboundary River Basins

One planning method that can be used to address drought planning for river basins is collaborative environmental planning (CEP). CEP emerged as a subdiscipline of planning in the 1990s to address complex environmental issues. Perhaps the most important element of CEP is stakeholder involvement. Stakeholders are people who effect change and also those who are affected by it. CEP is most effective when stakeholders are identified and engaged early in the planning process, they are given full participation and allowed to take ownership of the process, they establish and build trust among one another, and they recognize collaborative learning as their primary goal (Randolph and Bauer 1999). Stakeholders who collaborate would not only formulate a plan to solve an environmental issue, but they would implement the plan and continue to update it, as CEP is an iterative process. CEP can be applied to a wide range of environmental issues, including the management of water resources. CEP is particularly useful for river basin planning because many rivers are transboundary in nature, meaning they cross more than one geopolitical jurisdiction, and that may require coordination of planning activities among several groups.

A recent study looked at how CEP is used for drought planning at the river basin level in the United States (Bergman 2014). The study found 12 basins that are engaged in drought planning to some extent. Interviews that were conducted as part of the study revealed six critical areas in which CEP is beneficial for basin-level drought planning: (1) identifying and engaging key stakeholders; (2) increasing collaboration and coordination among stakeholders; (3) enhancing the quality and quantity of information and data upon which decisions are based; (4) increasing communication among stakeholder groups; (5) developing the planning process into one that ensures implementation and continuous updating of the plan; and (6) enhancing awareness of government and legal matters, such as litigation, that can undermine the planning process. Interview participants were also asked if they thought drought planning was best implemented at the scale of a river basin. Some participants stated that planning for drought at the basin scale was best, especially if it is a transboundary river basin, while other participants said that integrated planning, such as integrating river basin planning with state planning, was ideal.

Two of the basins found by Bergman (2014) to be engaged in drought planning are in the West: the Klamath River Basin and the Upper Colorado River Basin (UCRB). The Klamath River Basin is shared by Oregon and California and supports several uses, including irrigation, hydroelectric generation, tribal water rights, and habitat for wildlife and endangered species. Issues resulting from these competing uses were amplified by drought in the early 2000s, prompting the development of the Klamath Basin Restoration Agreement (KBRA) in 2010. The goals of the KBRA are to restore habitat for fish species in the basin, ensure a sustainable water supply for the various uses of the river, and resolve disputes that arise between the competing users. One section is dedicated to addressing drought planning and providing an assessment of climate change in the basin. The KBRA recognizes the importance of including stakeholders from both Oregon and California to increase cooperation throughout the basin. According to Bergman (2014), federal authorization of the KBRA has been delayed because of resistance at the congressional level, and as of 2012, the drought-plan portion of the agreement had not been implemented, because basin conditions had not been dry enough to warrant further action. Drought in 2014 led to renewed legislative efforts. Senator Rob Wyden introduced the Klamath Basin Water Recovery and Economic Restoration Act in May, and as of November 2014, the bill was making its way through congressional committees (Klamath Restoration Agreements 2014; Clevenger 2014).

Water resource planning officials are also engaged in drought planning for the UCRB. As defined by the Colorado River Compact of 1922, the UCRB is the portion of the Colorado River Basin that drains above Lees Ferry and includes part of the states of Arizona, Colorado, New Mexico, Utah, and Wyoming. The UCRB has experienced multiple drought episodes and has generally been in drought since the early 2000s. The UCRB was identified by NIDIS as a pilot area where an RDEWS would be implemented. The Upper Colorado RDEWS has focused on providing local input to the US Drought Monitor, coordinated through the Colorado Climate Center. Weekly drought assessment webinars started in early 2010 and have brought together representatives of federal and state agencies, water conservation districts, and recreation and tourism to discuss current conditions, water supplies, and outlooks. The Colorado Water Availability Task Force, which has overlapping membership, can make use of the information from the drought webinars to alert relevant decision makers to emerging drought conditions. The Upper Colorado RDEWS has served as a mechanism for strengthening local input into the US Drought Monitor process, which enhances credibility and legitimacy at federal, state, and local levels (McNutt et al. 2013).

The Colorado River has been highly modified for human use, including the creation of the large reservoirs of Lake Powell and Lake Mead. The combination of ongoing drought, the uncertainty of climate change impacts on the basin, and a rapidly growing population in the Lower Basin has caused concern over the future of water supply of the Colorado River. The US Bureau of Reclamation attempted to address these concerns through the conduct of a water supply and demand study for the entire Colorado River Basin that was completed in 2012 (USBR 2012). The study determined that imbalances between the supply and demand of water in the basin are expected in the future, and a collaborative approach to planning will be needed to address the issue. Collaboration between multiple agencies from each of the basin states will be especially important because of the vastness of the basin.

Drought planning for river basins in the West is becoming more important now than ever because of the rapidly growing population of the region and the uncertainty of how climate change will impact precipitation and temperature patterns in the future. International river basins in the West, such as the Rio Grande and the Columbia River basins, must contend with the additional complexity of coordinating water resource management activities with Mexico and Canada, respectively. Collaboration between agencies and coordination of water resource policies is crucial for managing drought effectively in western river basins.

7.5 Conclusion

Daniel Connell of the Australian National University recently described drought as "a force for truth" for Australia in how they, as a nation, must look at climate change (Connell 2010, slide 10). In other words, a thorough examination of drought could reveal important information on how to address climate change. Because of similarities between Australia and the West, Connell's three main points may provide some insights into the issues identified within this chapter. First, Connell emphasized that an analysis of drought risk management is the starting point for a comprehensive institutional analysis, which is necessary in understanding how to deal with climate change. It is a call to action for the West to investigate drought risk management approaches. Second, the stress from droughts highlights both the strengths and weaknesses in how a society deals with long-term threats like climate change. The current and recent droughts in the West provide windows of opportunity to identify these strengths and weaknesses, and to understand the political priorities and underlying cultural values revealed by difficult choices in these situations. Third, for better or worse, societies are likely to manage climate change in the same way they manage drought. Therefore, the intentional drought risk management approaches that are taking place and that need to take place may help us in coming years as we face the larger array of risks posed by climate change.

References

Bergman, C.J. 2014. Improving Drought Management for Transboundary River Basins in the United States through Collaborative Environmental Planning. PhD diss., University of Nebraska–Lincoln.

Botterill, L.C., and G. Cockfield. 2013. Science, policy and wicked problems. In *Drought, Risk Management and Policy*, edited by Linda Courtenay Botterill and Geoff Cockfield, 1–14. Boca Raton, Florida: CRC Press.

Brislawn, J., M. Prillwitz, and J.C. Schwab. 2013. Drought: How planners can address the issue. In *Planning and Drought*, edited by James C. Schwab. Planning Advisory Service Report No. 574. Chicago: American Planning Association.

Brown, J. 2014. Always a Disaster, Never a Presidentially Declared One. Blog post, The Center for Global Energy, International Arbitration and Environmental Law, University of Texas at Austin. July 11. http://www.utexas.edu/law/centers/energy/blog/2014/07/drought-always-a-disaster-never-a-presidentially-declared-one/.

Burby, R.J., ed. 1998. Introduction to *Cooperating With Nature: Confronting Natural Hazards With Land-Use Planning for Sustainable Communities*. Washington, D.C.: Joseph Henry Press.

California Department of Water Resources. 2008. Urban Drought Guidebook. http://www.water.ca.gov/pubs/planning/urban_drought_guidebook/urban_drought_guidebook_2008.pdf.

California, State of. 2014. Governor Brown Declares Drought State of Emergency. Press release and proclamation. Jan. 17, 2014. http://ca.gov/Drought/news/story-27.html.

California, State Water Resources Control Board. 2014. Emergency Conservation Regulations. http://www.swrcb.ca.gov/waterrights/water_issues/programs/drought/emergency_regulations_waterconservation.shtml.

Clevenger, A. 2014. Senate Bills on C&C lands, Klamath Basin Advance. *The (Bend, Oregon) Bulletin*. November 14. http://www.bendbulletin.com/localstate/2583079-151/senate-bills-on-oc-lands-klamath-basin-advance?entryType = 0.

Colorado Water Conservation Board, Department of Natural Resources. 2005. Colorado Water for the 21st Century Act http://cwcbweblink.state.co.us/weblink/0/doc/105662/Electronic.aspx?searchid = f7f87ad7-7a52-45c7-8b7f-2469076e69c8.

Colorado Water Conservation Board, Department of Natural Resources. Drought Planning Toolbox. http://cwcb.state.co.us/technical-resources/drought-planning-toolbox/Pages/main.aspx.

Colorado Water Conservation Board. 2014. Colorado Water Plan. http://www.coloradowaterplan.org.

Connell, D. 2010. Drought Past and Future: Developing an Auditing Framework for the Governance of Rivers in Federal Systems. Presentation at the International Drought Symposium, Water Science and Policy Center, University of California–Riverside, March 24. http://cnas.ucr.edu/drought-symposium/presentations/Daniel%20Connell.pdf.

Davis-Franco, D. 2014. Drought Planning Toolbox: State Strategies for Mitigation and Adaptation. Webcast. American Planning Association. https://www.youtube.com/watch?v=WU0HLo-lIOc&list=UUvqWCr2888S3boRqcOCc0HA.

Disaster Mitigation Act of 2000. 2000. Public Law 106-390, an amendment to U.S. Code Chapter 68, Title 42.

Federal Emergency Management Agency. 2013. Local Mitigation Planning Handbook.

Firestone (Colorado), Town of. 2012. 2012 Drought Management Plan. http://www.firestoneco.gov/DocumentCenter/View/72.

Floress, K., L.S. Prokopy, and S. Broussard Allred. 2011. It's who you know: Social capital, social networks, and watershed groups. *Society & Natural Resources: An International Journal* 24:9, 871–886. doi: 10.1080/08941920903493926.

Garfin, G., G. Franco, H. Blanco, A. Comrie, P. Gonzalez, T. Piechota, R. Smyth, and R. Waskom. 2014. Chapter 20: Southwest climate change impacts in the United States. In *The Third National Climate Assessment*, J. M. Melillo, Terese (T.C.) Richmond, and G. W. Yohe, Eds., U.S. Global Change Research Program, 462–486. doi: 10.7930/J08G8HMN.

General Accounting Office (GAO). 1979. Federal Response to the 1976–77 Drought: What Should Be Done Next? CED-79-26. http://babel.hathitrust.org/cgi/pt?id=uiug.30112033947851;view=1up;seq = 1.

Geological Society of America (GSA). 2007. Managing drought: a roadmap for change in the United States. *A Conference Report From Managing Drought and Water Scarcity in Vulnerable Environments*—Creating a Roadmap for Change in the United States, held in Longmont, Colorado, September 18–20, 2006. http://www.geosociety.org/meetings/06drought/roadmapHi.pdf.

Fontaine, M.M., A.C. Steinemann, and M.J. Hayes. 2014. State drought programs and plans: Survey of the western United States. *Natural Hazards Review* 15: 95–99.

Haigh, T., L. Darby, N. Wall, and D. Bathke. 2014. Regional Drought Early Warning System Pilot in the Apalachicola–Chattahoochee–Flint Basin: Evaluation of Activities and Outcomes. National Integrated Drought Information System: Boulder, Colorado. Publication pending on drought.gov.

Hellmuth, M.E., S.J. Mason, C. Vaughan, M.K. van Aalst, and R. Choularton (eds) 2011. *A Better Climate for Disaster Risk Management*. New York: International Research Institute for Climate and Society (IRI), Columbia University.

Howitt, R.E., J. Medellin Azuara, D. MacEwan, J.R. Lund, and D.A. Sumner. 2014. *Economic Analysis of the 2014 Drought for California Agriculture*. Report, Center for Watershed Sciences, University of California, Davis, California. https://watershed.ucdavis.edu/files/biblio/DroughtReport_23July2014_0.pdf.

Huitema, D., E. Mostert, W. Egas, S. Moellenkamp, C. Pahl-Wostl, and R. Yalcin. 2009. Adaptive Water Governance: Assessing the Institutional Prescriptions of Adaptive (Co-) Management from a Governance Perspective and Defining a Research Agenda. *Ecology and Society* 14(1):26. http://www.ecologyandsociety.org/vol14/iss1/art26/.

Innes, J.E., and D.E. Booher. 2010. *Planning With Complexity: An Introduction to Collaborative Rationality for Public Policy*. London and New York: Routledge, Taylor and Francis Group.

Jacobs, K.L. and J.M. Holway. 2004. Managing for sustainability in an arid climate: lessons learned from 20 years of groundwater management in Arizona, USA. *Hydrogeology Journal* 12:52–65. doi: 10.1007/s10040-003-0308-y.

Klamath Restoration Agreements. 2014. Wyden Introduces Bill to Implement Klamath Agreements. http://www.klamathrestoration.org/.

Lovett, I. 2014. California Approves Forceful Steps Amid Drought. *The New York Times*, New York.

McNutt, C.A., M.J. Hayes, L.S. Darby, J.P. Verdin, and R.S. Pulwarty. 2013. Developing early warning and drought risk reduction strategies. In *Drought, Risk Management and Policy*, edited by Linda Courtenay Botterill and Geoff Cockfield, 151–170. Boca Raton, Florida: CRC Press.

Milly, P.C.D., J. Betancourt, M. Falkenmark, R.M. Hirsch, Z.W. Kundzewicz, D.P. Lettenmaier, and R.J. Stouffer. 2008. Stationarity is dead: Whither water management? *Science* 319(5863):573–574. doi: 10.1126/science.1151915.

Montgomery, C.K., ed. 2011. *Introduction to Regional Planning for a Sustainable America: How Creative Programs Are Promoting Prosperity and Saving the Environment*. New Brunswick, New Jersey: Rutgers University Press.

National Climatic Data Center (NCDC). 2015. Billion-Dollar Weather/Climate Disasters: Table of Events. National Oceanic and Atmospheric Administration. Accessed June 11, 2015. http://www.ncdc.noaa.gov/billions/events.

National Drought Policy Commission. 2000. Report: Preparing for Drought in the 21st Century. http://govinfo.library.unt.edu/drought/finalreport/fullreport/pdf/reportfull.pdf.

National Integrated Drought Information System (NIDIS). 2007. The National Integrated Drought Information System Implementation Plan. June. http://drought.gov/media/imageserver/NIDIS/content/whatisnidis/NIDIS-IPFinal-June07.pdf.

NIDIS. 2009. The National Integrated Drought Information System (NIDIS) Apalachicola–Chattahoochee–Flint River Basin Drought Early Warning Pilot. http://www.drought.gov/drought/regional-programs/acfrb/goals.

NIDIS. 2014. National Drought Forum Summary Report and Priority Actions: Drought and U.S. Preparedness in 2013 and Beyond. http://drought.gov/drought/content/national-drought-forum-summary-report-and-priority-actions.

Overpeck, J.T. 2013. The challenge of hot droughts. *Nature* 503: 350–351. doi: 10.1038/503350a.

Randolph, J. and M. Bauer. 1999. Improving environmental decision-making through collaborative methods. *Policy Studies Review* 16:168–191.

Rittel, H.W.J. and M.M. Webber. 1973. Dilemmas in a general theory of planning. *Policy Sciences* 4: 155–169.

Schwab, J.C., ed. 2013. *Planning and Drought*. Planning Advisory Service Report No. 574. Chicago: American Planning Association.

Shandas, V. and G. Hossein Parandvash. 2010. Integrating urban form and demographics in water-demand management: an empirical case study of Portland, Oregon. *Environment and Planning B: Planning and Design* 37(1):112–128. doi:10.1068/b35036.

State of California, Executive Department. 2015. Executive Order B-29-15. http://www.waterboards.ca.gov/waterrights/water_issues/programs/drought/docs/040115_executive_order.pdf

Svoboda, M., K.H. Smith, M. Widhalm, M. Shafer, C. Knutson, M. Spinar, D. Woudenberg, M. Sittler, R. McPherson and H. Lazrus. 2011. Drought Ready Communities: A Guide to Community Drought Preparedness. http://drought.unl.edu/portals/0/docs/DRC_Guide.pdf.

United Nations International Strategy for Disaster Reduction (UNISDR). 2013. From Shared Risk to Shared Value—The Business Case for Disaster Risk Reduction. Global Assessment Report on Disaster Risk Reduction. Geneva, Switzerland: United Nations Office for Disaster Risk Reduction. http://www.preventionweb.net/english/hyogo/gar/2013/en/gar-pdf/GAR13_infographic_min.pdf.

U.S. Bureau of Reclamation (USBR). 2012. Colorado River Basin Water Supply and Demand Study: Executive Summary. http://www.usbr.gov/lc/region/programs/crbstudy/finalreport/Executive%20Summary/CRBS_Executive_Summary_FINAL.pdf.

Weiser, M. 2013. State Urges Steps to Prepare for Drought in 2014. *Sacramento Bee*, Sept. 6.

Whitney, Gene. 2013. Scientists and drought policy: A US insider's perspective. In *Drought, Risk Management and Policy*, edited by Linda Courtenay Botterill and Geoff Cockfield, 71–85. Boca Raton, Florida: CRC Press.

Wilhite, D.A. 1990. Planning for Drought: A Process for State Government. IDIC Technical Report Series 90-1. International Drought Information Center, University of Nebraska–Lincoln.

Wilhite, D.A. 1991. Drought planning and state government: Current status. *Bulletin of the American Meteorological Society* 72(10):1531–1536.

Wilhite, D.A., M.J. Hayes, and C.L. Knutson. 2005. Drought preparedness planning: Building institutional capacity. In *Drought and Water Crises: Science, Technology and Management Issues*, edited by Donald A. Wilhite, 93–135. Boca Raton, Florida: CRC Press.

Wilhite, D.A. and R.S. Pulwarty. 2005. Drought and water crises: Lessons learned and the road ahead. In *Drought and Water Crises: Science, Technology, and Management Issues*, edited by Donald A. Wilhite, 389–398. Boca Raton, Florida: CRC Press.

8

Providing Climate Science to Real-World Policy Decisions: A Scientist's View from the Trenches

Andrea J. Ray

CONTENTS

ABSTRACT This chapter presents three cases of providing climate science for water policy decisions and the tangible challenges and opportunities encountered by scientists. These challenges include (1) different time frames typical to funding and implementing research projects versus the often short time frames of policy and legal processes; (2) the implications of divergent incentive structures for academic research compared to applied science and services; (3) the need to translate, synthesize, and interpret scientific results to serve the context of the decision or policy at hand; and (4) meeting the *best available science* criteria for many federal policy decisions. An additional critical function of scientists working in boundary organizations is identified—the framing of relevant climate questions. Finally, the efforts to participate in coproduction of policy-relevant science have value beyond the report or synthesis product itself. The outcomes of these stakeholder engagement efforts are described, and the benefits reaped by the science organization from the resources invested beyond the usual research models. Finally, it is argued that meeting the demands for usable science for policy requires ongoing support for the participation of research scientists and institutions in service functions.

8.1 Introduction

This chapter describes some of the challenges in connecting climate science into policy and management decisions from the standpoint of scientists working in and with boundary organizations. To produce usable science for decision making, scientists are often required to stretch beyond their usual reward structures, out of typical professional motivations and timetables, and to learn new professional languages. These activities typically take time and other resources that might have been used in more traditional research. Three cases of incorporating climate science into policy making are reviewed, and reflections are provided on the practical experience they offer for creating usable science, in particular though boundary organizations. The cases illustrate tangible examples of several challenges for scientists and their organizations that are not well described in the literature, including the different timetables and reward structures for research versus science for policy; significant time and resource demands of synthesizing, interpreting, and translating research results; and crafting the best available science for federal decisions. However, we find that this investment yields tangible benefits and opportunities over time for scientists and their organizations.

This chapter describes three efforts to provide usable climate science under policy process deadlines in the form of science synthesis products. The efforts connected research to specific policy contexts and two specific federal decisions: (1) a climate analysis for an environmental impact statement (EIS) for Colorado River operations (US Bureau of Reclamation 2007a); (2) the Climate Change in Colorado (CCC) report for state policy makers and planners (Ray et al. 2008); and (3) a climate study proposal for a Federal Energy Regulatory Commission (FERC) hydropower licensing decision (NMFS 2012, 2013; Ray et al. 2014). These efforts were carried out with significant involvement by the National Oceanic and Atmospheric Administration-funded (NOAA) Western Water Assessment (WWA) at the University of Colorado, a Regional Integrated Sciences and Assessments (RISA) Program, and its local federal partner, the NOAA Physical Sciences Division (PSD). For over 15 years, they have experimented with strategies to provide more useful climate information and other tasks described by Pulwarty et al. (2009), beginning primarily with serving stakeholders in water resource management. This chapter focuses on my experience in these cases through my affiliation with these organizations.

The WWA self-identifies as a boundary organization (Gordon et al. 2015; McNie 2007), with RISA goals to produce regionally specific knowledge assessments (Pulwarty et al. 2009); support informed policy development; serve as expert interpreters of climate science; and identify entry points for climate information in operations and planning (Binder and Simpson 2009). NOAA/PSD is primarily a research and applied research organization, but with long-term collaborations with boundary organizations including RISAs, Department of Interior (DOI) Climate Science Centers (CSCs), and operational service organizations including the DOI Bureau of Reclamation and NOAA river forecast centers, the Climate Prediction Center, and Regional Climate Centers. Together, WWA and PSD have accumulated insight into user needs and the context for information provision in a variety of decision processes and have developed trusted relationships with stakeholders including local, state, and federal policy makers (Gordon et al. 2015; Ray et al. 2015; Lowrey et al. 2009) and the organizations of these stakeholders, called action networks by Pulwarty et al. (2009).

These efforts illustrate the challenges in producing usable science articulated by McNie (2007) and Dilling and Lemos (2011) and others, and additional challenges and opportunities that are not well described in the current literature. These challenges include the (1) different time frames typical to funding and implementing research projects versus the often short time frames of policy and legal processes; (2) the divergent incentive structures for academic research compared to service and applied work useful to decision makers; (3) the challenge of interpreting and synthesizing scientific results that may be relevant to the decision or policy at hand but are not generated specifically for that issue; and (4) the *best available science* criteria for many federal policy decisions. Furthermore, we identify an additional critical function of scientists working in boundary organizations beyond the four described by Cash et al. (2006)—*framing the relevant climate question.* Finally, the value of this assessment process is more than the report or synthesis product generation: I describe the downstream opportunities and outcomes of these stakeholder engagement efforts, and the benefits to the science organization in working in this coproductive mode, which result from time and other resources invested beyond the usual research models.

8.2 On Supplying Usable Climate Science

A growing literature focuses on how scientists may participate in the production of usable science. Collaborative processes in which scientists participate with stakeholders to coproduce knowledge are seen as a key strategy to reconcile the supply of science with demands and to produce usable science (Dilling and Lemos 2011; McNie 2007; Sarewitz and Pielke 2007). Coproduction requires establishing trust between organizations with very different cultures and goals; joint learning of science, its limitations, and the overall decision environment including policy constraints; and developing and maintaining boundary organizations that span the gap from science to decision making. When successful, the coproduction process is iterative and time-consuming but results in insights not achievable without the shared knowledge and understanding. As individuals, scientists often participate through boundary organizations, and often take on nontraditional roles, such as convening, translating, and mediating science (Cash et al. 2006; Gordon et al. 2015;

Guston 2001). Other roles include serving as an embedded expert (Dilling and Lemos 2011); a boundary-spanning individual (McNie 2007); or a knowledge broker employing strategies including informing, consulting, matchmaking, engaging, collaborating, and building capacity (Michaels 2009).

However, producing usable science also has institutional challenges for science organizations and for the scientists whose careers are within those organizations. These include cultural differences such as the costs and time of participatory research processes; incentives and reward structures, including tenure and promotion; policies that favor basic research over *use-inspired* research; and differences between the status quo for production of science (including disciplinary scholarship) and producing usable science (McNie 2007). McNie (2013) suggests that end-to-end research agendas and organizational commitment to support applied work, including resources and time of scientists, would enhance institutions' ability to produce usable science. Dilling and Lemos (2011) also point out the need for funding flexibility and metrics that include and thus incentivize usable science. The synthesis and interpretation strategies that are described in this chapter can be thought of as a specific form of the translation function of Cash et al. (2006).

The literature also makes recommendations regarding the determinants of the value of the scientific information itself (McNie 2007), e.g., its credibility with, and legitimacy and salience for, users (Cash 2006; Gordon et al. 2015). Lemos et al. (2012) suggest that usability can be improved by adding value, retailing, wholesaling, or customization of the data into contextualized knowledge for particular user groups. For many federal decisions, a particular legitimacy of the science to be supplied is that it is the best available science, and the choice of this may be challenged in court. The Council on Environmental Quality regulations implementing the National Environmental Protection Act and its associated EIS process demand information of *high quality* and professional integrity (NEPA 1969). The Magnuson Act of 1996, which guides many NOAA fisheries management activities, requires that "the national fishery conservation and management program utilizes, and is based upon, the best scientific information available;" (Public Law 94-265, p. 3) and the Endangered Species Act of 1973 (ESA) requires the Secretary of the Interior to use the "best scientific and commercial data available" as the basis for decisions (ESA, Section 4, subsection [a][1]). The FERC includes the standard that "the proposed methodology is consistent with generally accepted practice in the scientific community" (Federal Power Act of 1995, Section 5.11 [b][5]). Creating the best available science can be considered a specific form of legitimacy, which I later define as including an honest assessment about the state of the science, the assumptions, and the unresolved issues.

8.3 Cases: Three Science Synthesis Products

The following sections describe the cases in chronological order, including the motivation for the interaction with the target stakeholders, the collaborative process used, and the challenges encountered. The first two efforts preceded or were in parallel with the literature described above on collaborative processes and usable science; however, cross-fertilization through RISAs meant that these efforts and the literature informed each other; thus, their terminology is used to describe the efforts. The outcomes beyond the synthesis product itself are described, termed *downstream results*, which contribute to the benefits of the resources invested in the interactions. Table 8.1 provides a summary of the cases.

TABLE 8.1

Cases and Their Policy Focus, Funding, Time Frame, Boundary Functions, and Strategies Used

Case/Policy Focus	Funding	Time Frame	Strategies and Boundary Functions Used	Outcome and Downstream Impacts
Appendix U/ Reclamation NEPA EIS process[a]	PSD base; CIRES and RISA discretionary funding; National Integrated Drought Information System (NIDIS)	May 2005–December 2007; much of the analysis done in 2006	Engaging; convening; boundary organizations; boundary-spanning individuals	Groundbreaking study incorporating climate change in a Reclamation EIS; fostered innovation within the project as well as downstream
Climate Change in Colorado/State water policy and planning[b]	CWCB contract; PSD base; WWA discretionary funding	May–October 2008	Boundary organization; convening; information brokers	Raised climate interest and literacy of managers and policy makers; fostered innovation leading to more detailed climate and water supply studies aimed at water policy decisions; CWCB commissioned an update released 2014 (Lukas et al. 2014)
Susitna Study Request/FERC ILP[c]	PSD base funding	October 2011–May 2012 for study proposal; 12 months of appeals	Engaging; embedded capacity; boundary-spanning individuals	Precedent-setting decision for FERC to include a limited climate study in licensing; NMFS developing process to consistently submit climate studies for future ILPs

Note: All projects used boundary-spanning individuals and strategies including some form of translation, collaborative processes, and framing of climate questions. References for the synthesis document or report are footnoted.

[a] Bureau of Reclamation 2007a.

[b] Ray et al. 2008.

[c] NMFS 2012.

8.3.1 Climate Analysis for the EIS on Interim Guidelines for the Colorado River and Lakes Mead and Powell (Appendix U)

In May 2005, Secretary of Interior Gayle Norton called on the Colorado River Basin states to develop an agreement on managing the Colorado River under "shortage," referring to the ongoing drought conditions unprecedented since the completion of Lake Powell. The US Bureau of Reclamation (hereafter, *Reclamation*) led a process to develop an EIS, which resulted in a secretarial decision on interim guidelines for river operations, signed in late 2007 (US Bureau of Reclamation 2007b). The WWA and NOAA/PSD had a working relationship with Reclamation's Colorado River regional offices since the mid-1990s, based on interest in reservoir management and seasonal climate variability (Gordon et al. 2015; Ray and Webb 2015). WWA and PSD were part of a larger team that developed a climate analysis that became widely known by its place as one of many appendices to the EIS, *Appendix U*.

8.3.1.1 Collaborative Process and Timeline

Reclamation convened a climate technical work group (CTWG), including scientists from NOAA/PSD, WWA, and other western RISAs; I participated in meetings of the CTWG and contributed to written comments from WWA and NOAA. The RISAs and NOAA/PSD had significant input in recommending participants and bringing in expertise as needed. This project was completed over a 2-year timeline, beginning in summer 2005 with Reclamation soliciting comments on developing the EIS. Reclamation published proposed operating alternatives in May 2006, a draft EIS for public review in December 2006, and the final EIS in September 2007, and issued the Record of Decision in December 2007. Some relevant studies of climate change impacts on hydrology and river flows already were underway, funded by the NOAA Climate Program Office (e.g., Global Energy and Water Cycle Experiment, RISA), the National Science Foundation, the National Aeronautics and Space Administration, and other sources, and in support of the Intergovernmental Panel on Climate Change (IPCC) Working Group II on adaptation. These projects were often inspired by previous work with water management stakeholders. However, additional analyses specific for the EIS were needed, with initial results required in about a year, much sooner than a typical grant cycle would provide results. To provide the needed analyses, the RISAs involved proposed and received special funding from the NOAA Climate Program Office, individual RISAs provided discretionary funds, and federal scientists participated.

8.3.1.2 Framing the Climate Science

WWA comments on the EIS scoping included noting that climate change impacts should be considered as well as climate variability (WWA 2005). Reclamation's charge to the CTWG, in fall 2005, asked for information on the state of knowledge regarding climatic processes and the numerical simulation of long-term future conditions and guidance on what methods would be appropriate given the uncertainty arising from imperfections in those numerical simulations (US Bureau of Reclamation 2006). Studies such as that of Christensen et al. (2004) using climate projections to drive hydrology models had indicated the risks of climate change to reservoir inflows. Thus, a key question for the CTWG was whether there was enough confidence in the results of the latest climate models to take action in policy and management (US Bureau of Reclamation 2006). Recall that phrases such as *stationarity is dead* had not yet reached the lexicon: Milly et al. (2008) had not been published, nor had the 2007 IPCC report with its higher confidence in projected patterns of warming and other regional-scale features than reported by the IPCC in 2001 (IPCC 2007). Reclamation did recognize the weakness of conventional assumptions of stationarity, and the severe drought of 2002 and ongoing conditions accelerated efforts of the agency to investigate alternatives (US Bureau of Reclamation 2007a). Ultimately, the CTWG recommended a qualitative discussion of climate change and variability accompanied by a quantitative sensitivity analysis using paleoclimate evidence, which became Appendix N of the final EIS. The CTWG also recognized the potential to use existing reconstructions of streamflows as proxies for climate change impacts. Paleoclimate research had shown that droughts of the twentieth century had been of moderate severity and short duration relative to the paleorecord (Woodhouse 2003).

Working with Reclamation, the TWG framed the key climate issues to include the risks for reservoir inflows of longer droughts, which were thought to be representative of climate change as it was understood in 2005, and the risks of climate change for inflows in the latest climate model runs being performed for the 2007 IPCC. This framing of the key climate issues required two new threads of science to be developed in short order analyzing reservoir operations and a major need for planning reservoir operations. First was the nontrivial task of developing methods for generating streamflows from the combination of observational and paleoreconstructed data (Prairie et al. 2008; Appendix N in US Bureau of Reclamation 2007b) and understanding the range of runoff projections in studies of climate and long-term drought impacts on river flows.

8.3.1.3 Synthesis Issues

Intercomparing the streamflow estimation studies that had been conducted with different assumptions and methods was recognized as a major scientific challenge but was not resolved during the EIS. The team identified a range of estimated streamflow reductions by 2050 from +6% to −45%, and a more complete analysis was conceived by team members that became known as the *reconciling flows* project. The TWG concluded that the uncertainty around future runoff projections would never be reduced to zero despite the best efforts of scientists; they urged water managers to incorporate the full range of futures into their planning efforts.

8.3.1.4 Downstream Results of This Project

The climate analysis was published as a formal part of the EIS, as a substantial, 110-page appendix (Reclamation 2007a). The flow reconciliation issues were documented by Hoerling (2009), and findings were published several years later (Vano et al. 2013). The effort also resulted in significant cross-fertilization among RISAs and participating scientists, for example, on climate model downscaling techniques and hydrologic models. Work developed by one RISA, such as the gridded meteorological data sets developed by the Climate Impacts Group RISA, was used by others. Furthermore, this effort resulted in the RISAs and scientists involved improving their skills for communicating climate science effectively, in particular with Reclamation and its water management stakeholders. The questions raised in this effort influenced climate research priorities in RISAs and in funding agencies such as the NOAA Climate Program Office. The product also came at a critical point in the interpretation of climate change risks: between the EIS scoping and signing of the Record of Decision, the 2007 IPCC report was released with higher confidence in the science of climate change, and there was increasing interest in impacts in the West (e.g., Seager et al. 2007). The Appendix U analysis prompted increased interest in climate change and its impacts on western rivers and water supply, and stimulated funding and attention for these topics from agencies. Arguably, the Colorado Climate Report (discussed next) is a result of Appendix U, because of the interest it generated in the impacts of climate on water supply across the state.

8.3.2 Climate Change in Colorado 2008 Report

In 2007, his first year in office, Colorado Governor Bill Ritter issued the Colorado Climate Action Plan (CCAP) (Ritter 2007). The major western drought of the early 2000s, as well as Appendix U, had elevated interest about the risks of climate change (Lowrey et al. 2009). The plan set out a goal to prepare the state to adapt to those climate changes "that cannot be avoided" (Ritter 2007, p. 3) and included recommendations to assess the vulnerability of Colorado's water resources to climate change, to analyze impacts on interstate water compacts, and to plan for extreme events such as drought and flooding. The Colorado Water Conservation Board (CWCB), the state water policy and planning agency, subsequently funded WWA to do a synthesis of the science on climate change aimed at state water planners, decision makers, and policy makers, which became the 2008 CCC report (Ray et al. 2008). The CWCB was among the water managers that WWA and NOAA/PSD had been working with since the late 1990s, and the idea of an assessment of climate science relevant to Colorado water managers emerged from those ongoing interactions. CWCB's commission of the report is evidence of the trust CWCB placed in WWA and PSD as sources of credible and legitimate information.

8.3.2.1 Timeline and Collaborative Process

This project was conceived in the late spring of 2008 to provide information and a document for the governor's drought conference planned for the fall of 2008. Thus, the timeline of the Appendix U project seemed generous, compared to the 5-month timeline for this effort. Accomplishing this project involved rearranging work plans by a number of scientists for the summer, and others revised/adjusted ongoing research to do analyses needed for the report. Water managers joined the planning meetings and provided comments on short notice. While CWCB provided funding, it did not completely cover the time of those involved; base funding to PSD and Cooperative Institute for Research in Environmental Sciences (CIRES)/WWA discretionary funding also covered time scientists spent on the project.

The report was informed by ongoing interactions with the water management community but also through a bottom-up process of engaging water managers specifically for this report. Activities included convening a meeting of scientists, water managers, state officials, and other agencies to discuss the outline of the report, and a second meeting to review the draft. Scientists were recruited to contribute their expertise in global, regional, and Colorado climate and hydrology. The outline and draft were reviewed by a wide range of scientists and stakeholders, including water managers, state officials, and other agencies. This bottom-up process also built credibility and legitimacy for the report by including scientists from around the state.

8.3.2.2 Framing Climate Questions

At the initial scoping meeting, water managers described topics of greatest interest to them, and climate scientists and hydrologists discussed the level of scientific understanding of different topics. For example, managers were interested in better understanding how climate models work, so a primer on climate models became a focus with a separate chapter. Other key topics included understanding causes of drought, the 2000s drought in particular, and uncertainty in precipitation projections. Although the hydrologic effects of increased temperatures were key topics, the science of hydrologic projections was evolving quickly, and efforts such as the flow reconciliation work inspired by Appendix U were not yet available. The report instead discussed major hydrologic uncertainties as among the unresolved issues.

The CWCB and managers who participated in scoping were also concerned that the report be framed so that it would be accessible to a wide range of water managers and suited for presentation to their upper management and boards. Thus, a goal for the product was that it be accessible to technically savvy professionals who were not trained in climate science but were willing to learn more about climate science for their jobs. The complexity of the climate science and presentation of research knowledge in the peer-reviewed literature was often in language not appropriate to communicate to this audience and was a challenge to translate. The authors took advantage of insights on communicating about climate, including that within the water resource engineering community, the stationarity assumption is a fundamental element of professional training; that the century timescales of climate change exceed typical planning and infrastructure design horizons and are remote from human experience; and that even individuals trying to stay up-to-date can face confusion in conceptually melding or synthesizing the diverse and expanding climate change impact literature—and in knowing what is the best available science at any point in time (Hartmann 2008; Moser and Dilling 2007).

The report authors anticipated that some readers would be familiar with the recent IPCC Fourth Assessment (IPCC 2007), and intentionally adopted a chapter structure that was largely based on the IPCC Working Group I report on physical climate: observed climate, a primer on climate models, climate attribution, and projections. We also chose the model of lead and contributing authors. Working Group II (impacts) issues were beyond the scope of the report. However, a final chapter provided a high-level overview of some implications of climate changes for water resources. One of the authors (Averyt) was a member of the Technical Support Unit for the IPCC Fourth Assessment; her participation provided experience and cross-fertilization from that effort.

The report authors were intentionally explicit in framing some issues as unresolved; admitting this was part of credibility. Unresolved issues included the confidence in climate modeling and downscaling and that regional and local processes and their role in Colorado's climate must be better modeled; that precipitation projections and related phenomena are key uncertainties; and that finer spatial resolution is needed in models to better represent Colorado's mountainous terrain and precipitation processes.

8.3.2.3 Synthesis Issues

We found that there was little science that could be used off the shelf, i.e., from the peer-reviewed published papers, with fairly simple translation or value-adding (Lemos et al. 2012). The authors summarized Colorado-related findings from peer-reviewed studies, conducted new analyses based on published methodologies with existing data sets and projections, and took advantage of recent syntheses in the new IPCC report (IPCC 2007) and the US Climate Change Science Program's Synthesis and Assessment Products (e.g., Hartmann 2008). As with Appendix U, the project faced significant intercomparison issues to synthesize results from different projects, including different time periods and spatial scales or domains used in analysis. For example, the months included in "spring" and definitions of other general terms were inconsistent, e.g., whether a daily value or monthly average was represented by "April 1st." Details were often buried or not described in publications, which required contacting authors for clarification. Furthermore, few Colorado-specific studies existed, and our work to translate or customize often involved reanalyzing or replotting the Colorado-relevant data from West-wide studies that provided little specific discussion of Colorado. The Arkansas, Rio Grande, and Platte River basins were not well covered in published studies. Furthermore, graphics from the research literature were often inappropriate for audiences not accustomed to certain conventions, so we redesigned several key graphics to improve understandability. Thus, even given research products with relevant information, the translation and interpretation of the climate knowledge for the decision-making context was a time-consuming endeavor.

8.3.2.4 Downstream Results

The report and the process to create it were intended to support a dialogue about climate risks (cf. Ray and Webb 2015) and to raise the level of interest in climate and climate literacy among Colorado water managers. Although the other cases address a specific federal policy decision, this report was intended to support Colorado state agencies and the water management community in order to better understand the risks of climate change and inform later policies and decisions. Based on the attention it garnered and demand for briefings and additional information, this report was well received (see Gordon et al. 2015 for more on this project). It was valued enough that several years later, CWCB commissioned an update, completed in 2014 (Lukas et al. 2014). The updated report was adopted as the science basis for the public review draft of the Colorado Water Plan (CWCB 2015). The report authors considered this a successful synthesis product based on the saliency, legitimacy, and credibility criteria (Gordon et al. 2015).

The report also was intended to inform and motivate later assessments of climate risks that were beyond its scope, including specific sensitivities and vulnerabilities of water supply and ecosystems. It was also intended to inform integrated resource planning and adaptation. This report, Appendix U, and ongoing flow reconciliation studies also generated interest in other studies to better understand risks to water supplies from Colorado's basins and to the Colorado Basin as a whole (see Table 5.1 in Ray et al. 2008). It built interest and literacy among water managers and policy makers at many levels in the state, spurring demand for additional work and more technical reports and leading to the use of climate variability and change information in drought mitigation and adaptation planning (Gordon et al. 2015).

8.3.3 Susitna Climate Study Request

The US FERC Integrated Licensing Process (ILP) is the process for licensing nonfederal hydroelectric projects. The ILP gives affected parties, including the NOAA National Marine Fisheries Service (NMFS), the opportunity to request studies needed to understand baseline and preproject conditions, effects of the project, and ultimately to develop license terms and conditions to mitigate unavoidable environmental impacts. Hydropower project construction and operations in conjunction with the impacts of climate change may result in direct, indirect, and cumulative effects on important habitat components such as water temperatures; riparian vegetation; and timing of flows, fish migration, side-channel habitat for spawning and rearing, and food availability. Under authorities including the National Environmental Policy Act (NEPA) and Fish and Wildlife Coordination Act, NMFS is mandated to assess the effects of a proposed project in order to adequately prepare and support appropriate license terms and conditions, and to develop effective measures to protect, mitigate, and possibly enhance resources for which NOAA is the *trustee* (in this section, words in italic indicate formal terms in the process), including endangered species and anadromous fish. Under the Federal Power Act of 1995, NMFS has mandatory authority to prescribe the passage of anadromous fish.

In the past several years, NMFS has requested a number of studies to evaluate the potential combined effects of a project and climate change on anadromous or marine species. FERC has consistently rejected these climate study requests with variations of the following: "[FERC is] not aware of any climate change models that are known to have the accuracy that would be needed to predict the degree of specific resource impacts and serve as the basis for informing license conditions. FERC typically relies on historical data and project-specific studies to evaluate project effects and inform license conditions" (FERC 2009, p. 25). Beginning in 2011, NOAA/PSD worked with NMFS Alaska Region to develop a climate study request for a large hydropower dam in south-central Alaska on the Susitna River, proposed by the Alaska Energy Authority; a study is *requested* because the needed analyses or synthesis products do not exist.

8.3.3.1 Collaborative Process

To develop the study request, NMFS brought me in as a climate expert for the team from NOAA/PSD. I served as a boundary-spanning individual between NMFS and PSD (as in McNie 2007) and in a role of embedded capacity (as in Dilling and Lemos 2011). I communicated regularly with the NMFS hydropower coordinator, often several times a week, and with others on the NMFS team. The PSD role was to provide the expertise to help NMFS make the argument that climate studies were necessary. Developing the proposal involved assessing the state of the climate science for the region; determining what knowledge was needed to explain the climate nexus with the project and NOAA *trust* resources; climate risks to the project and *trust* resources; and synthesizing the knowledge for this process.

8.3.3.2 Timetable and Professional Challenges

Alaska NMFS initially contacted PSD in October 2011, only 4 months before the original study request due date in early 2012 (later extended to May). Although scientists affiliated with the Alaska RISA did participate in discussions of climate science to inform the proposal, the short timetable limited participation due to academic commitments, including teaching courses. Scientists on tenure track also were concerned about delaying manuscripts for peer review in order to participate. Ultimately, the study request effort extended for over a year of appeals beyond the request submission. In fall 2012, FERC *rejected* the study request (NMFS 2012). NMFS and PSD then worked together to develop *comments* in November, and when those were not accepted, NMFS filed a *study dispute*. The study proposal team participated in a full-day dispute resolution hearing in April 2013. Each step required preparation and occurred with a short timeline for response—at most several weeks.

8.3.3.3 Framing the Climate Science

A major challenge was the need to counter the FERC concerns that the climate models were not accurate enough and that the license conditions could be opened later if "unanticipated changes" in climate occurred, referring to changes from assumed stationarity. FERC's professional standards relate to long-standing hydrologic engineering analysis that we argued had not been updated for the latest climate science and was based on ideas of stationarity of climate that were becoming less prevalent among many hydrologic engineers in the western United States Therefore, an important part of arguing for the use of the latest climate science was making the case for a different framing of perception of risk and the need to shift from a "predict then act" framework described by Weaver et al. (2013) and prevalent in FERC. We advocated for using climate knowledge as part of a shift to a risk framework (paradigm 2 in Weaver et al. 2013). The proposal reframed the need for climate studies as a need for information to support risk assessment. We argued that the licensing process would benefit from a better understanding of the state of climate science and the current use of that science in order to make an informed decision using the best available science. The science syntheses in the public review drafts of the IPCC Fifth Assessment (IPCC 2013) and the National Climate Assessment (NCA) Alaska chapter and climate technical reports (Chapin et al. 2014; Markon et al. 2012), which synthesized published results, helped make our case about the current understanding of the skill of the climate models. To address FERC's standard that methodologies be consistent with generally accepted scientific practice (Federal Power Act of 1995), we described the use of climate models by other water management agencies (e.g., Brekke et al. 2009; US Bureau of Reclamation 2011).

The proposal then framed the climate science issues for FERC, that climate change has the potential to affect hydrology across the basin—by affecting not just glacial melting but also altering snow distribution and timing, evaporation/evapotranspiration, and soil moisture—so that climate impacts across the catchment basin needed to be considered to understand changes in thermal regime and flows over much of the year. Thus, the proposal (and subsequent documents in appeals) explained how analysis of climate change was necessary for NMFS to fully evaluate potential project effects on anadromous or endangered species and to inform NMFS's authority to prescribe fish passage as part of a hydropower license. We argued that considering climate-related changes that are already happening is critical to establishing a baseline against which the impacts of project effects on flows could be assessed.

8.3.3.4 Synthesis Issues

The task was to document the need for a study to inform license requirements (Federal Power Act 1995), not to document impacts. The Alaska-related chapters for the 2014 NCA, available initially as public review drafts, provided important background and science synthesis for the team (Chapin et al. 2014; Markon 2012). Responding to FERC's concerns also required a synthesis of the state of understanding about the state of climate models as input into a risk assessment; we had to make the case for using the best available science. However, since 2008, when the primer on climate models was written as part of the CCC, more information on the state of climate models has become available (e.g., Hawkins and Sutton 2010), some of it aimed at water managers (e.g., Barsugli et al. 2012). These analyses made the synthesis task easier than in 2008 but still needed interpretation and translation for the FERC audience.

8.3.3.5 Downstream Impacts

In mid-2013, the study dispute resolution panel found the climate change study to be of crucial importance to confirming the viability of the project, in terms of sufficiency of runoff, over the 100-year expected lifetime of the project and agreed that it would be useful to assist NMFS in the exercise of its NEPA authority to prescribe fish passage. The FERC then approved a climate study that was limited in scope to a literature review relevant to glacial retreat and a summary of the understanding of potential future changes in runoff associated with glacier wastage and retreat. Other changes to the hydrology were not included in the study. However, NMFS considered this to be a precedent-setting decision, to have justified for the first time ever that a climate study was needed to support understanding of the risks of hydropower projects to other affected resources, and to make informed decisions on licensing. The determination also gives NMFS an opening to request a *study modification* in the near future that will be based on the rapidly accruing new research on climate change effects on hydrology in the project area.

8.4 Discussion: Challenges and Opportunities

This section discusses the responses to cope with the challenges described and the opportunities that have resulted from the investment in interactions with decision makers. The cases are all efforts in which scientists participated in boundary organizations or served as boundary-spanning individuals, and all used synthesis and translation functions to provide usable science for decision making (Table 8.1).

8.4.1 Decision Time Frames

Projects to meet the science needs of short-turnaround policy decisions face challenges of both inadequate funding and limited time to understand the context. Scientists may need to invest time learning about the stakeholder issues before moving forward. In the Appendix U and CCC cases, previous investment in collaborations with those stakeholders paid off in that climate scientists already had a significant level of understanding of water management issues, so they were able to quickly begin framing the issues for the effort. However, in the Susitna study request, scientists negotiated learning curves on the NMFS context and the ILP policy. Funding is a second major challenge. Although participating scientists often had existing grants or federal lab funding that was relevant to the project, in all cases, the analyses required were beyond their planned and funded work. Although CCC was partly funded by CWCB, considerable base and discretionary funds were used to create the report. Thus, in all cases, base or discretionary funds were used in order to provide science to meet short deadlines. However, this funding is intended for ongoing science, assessment, and research, not service activities; this is not a sustainable funding mode as the demand increases for science services for particular policy decisions.

8.4.2 Divergent Incentive Structures and Work Expectations

Traditional research expectations and incentive structures do not necessarily support applied work and services useful to decision makers. As described by Dilling and Lemos (2011), the standard metric for most scientists is number of peer-reviewed publications, especially in highly regarded journals, and citations to that work; these metrics are typical for federal research labs, university research centers, and certainly for academic tenure. Often, science synthesis products, or other products supporting policy processes, are in the so-called grey literature and count for less in reward structures. Reports and other documents that synthesize and interpret existing published work may not be considered original enough for many journals; these grey literature products may require significant rewriting and potentially additional work to convert to a manuscript for a peer-reviewed journal. Thus, scientists producing synthesis products for policy may face an extra burden to meet metrics.

Furthermore, there are often work commitments and expectations that limit the flexibility of scientists to participate in these service activities. For the CCC and the Susitna study request, scientists delayed activities; some academic scientists were unable to actively participate because of teaching commitments and tenure concerns. RISAs often step into a service role but can only do so if they have flexibility and funding for personnel and related needs such as project-related travel. However, policy decisions needing science are often not always known at the beginning of a fiscal or planning year. Providing science synthesis products on the timetables demanded by decisions requires flexibility in funding, personnel time, work expectations, and reward structures.

8.4.3 Translating, Synthesizing, and Interpreting

Translating and synthesizing results from multiple sources is crucial to making science credible and salient, and in all cases was a significant challenge requiring time and professional judgment. Relevant results not generated for the decision or policy at hand often required translation and interpretation to be usable, including putting the science into the context of related and sometimes conflicting studies. Often, the original authors had made analysis choices that were relevant to their own scientific, local, and regional interests but were often not easily comparable with other studies. Intercomparison issues included variables analyzed, time periods, model choices, and assumptions. Generally, it was not feasible to redo the analyses, and we needed to interpret the results in the context of other analyses. This challenge is illustrated by the effort to reconcile Colorado River flow estimates but also to simply understand the implications of different climatological reference periods used for studies that were inputs to the CCC.

Translation may be needed to adjust complex scientific language for a new audience. The CCC, for example, was intended to be more technical than a guide for the general public but less technical than the IPCC reports. Translation involved minimizing jargon; converting many figures to Fahrenheit units, with which state water managers were more familiar (customization); and including a glossary. Reworking figures from the research literature into a more accessible form added value, but the makeover took considerable creative effort. For example, CTWG member Brad Udall called a figure in Appendix U (Figure U.22 in Reclamation 2007a) a "horrendogram," and it was redone after several iterations as a set of box-and-whisker plots, a more familiar format for water resource engineers (Figure 5.9 in Ray et al. 2008). Other published figures were replotted for simplicity and to focus on Colorado. While usability can be improved by translation (Cash 2006; Michaels 2009), value-adding and customization into contextualized knowledge for particular user groups (Lemos et al. 2012) all required significant effort and nontrivial investment in time. Furthermore, these actions—even if envisioned while the research is being done—primarily take place after the information is generated as research products, requiring additional resources and attention after the typical products of research.

8.4.4 Determining the Best Available Science

Many federal policy decisions require the use of the best available science, for legitimacy, with little guidance on criteria to establish this. We have observed that some policy makers define *best available* as the most recently published; however, this is rarely a good definition, because of the evolving nature of science and the process of published findings inspiring additional analysis. Understanding what was best available in 2006–2007 for Appendix U also involved recognizing what science was *not* available, and would not be, in the time frame of the decision process. Both Appendix U and the CCC faced the challenge of synthesizing multiple papers with different results regarding future Colorado River flows; participating scientists recognized the need to resolve the differences, resulting in downstream publications assessing the estimates and risks to water supply on the Colorado River. For example, in 2008, Barnett and Pierce (2008) published a paper about the risk of "Lake Mead going dry," based on an analysis of the whole Colorado River Basin. The assumptions of that analysis were rapidly questioned (e.g., Barsugli et al. 2009) and resulted in additional analysis. Barnett and Pierce (2009) stood by their results using an improved model focused on the reservoirs and more targeted metrics. But others concluded that the situation, though presenting significant risk, was not as dire as assessed in the 2008 paper and was amenable to risk mitigation strategies (Rajagopalan et al. 2009). Even with the subsequent efforts at reconciling Colorado River flows (Vano et al. 2013), a large range of flow projections, and uncertainty, remains.

Another example of determining the best available science is the potentially greater rate of temperature change at higher elevations in response to rising global temperatures, an emerging topic important in the CCC. Diaz and Eischeid (2007) reported a faster rate of warming at higher elevations, but this estimate was contradicted by alpine observations by two Colorado scientists consulted by the CCC report team (Baron and Williams, cited as personal communication, in Ray et al. 2008). The report acknowledged the inconsistency and described it as an unresolved issue. The 2014 update of the CCC revisited the elevation issue. Lukas et al. (2014) described the results of a paper on the physical basis for greater warming at high elevations (Rangwala and Miller 2012) and concluded that the earlier finding of Diaz and Eischeid (2007) "is no longer believed to be robust…there is no consistent and robust observational evidence that higher-elevation regions in Colorado are warming at a different rate than lower-elevation regions" (Lukas et al. 2014, p. 24).

The *best available science*, I argue, is more than simply accepting the most recently published work. To be credible in determining the best available science, scientists must consider the implications of the latest studies in the larger context of that work and must be able to consider others' work as well as their own. In these cases, PSD and WWA scientists drew on work they were aware of through knowledge networks, for example, presentations at scientific meetings or work of collaborators that was not yet published, including other synthesis products in review (e.g., public review drafts of NCA for the Susitna project, and IPCC for the CCC). These project teams determined the *best available* from not only the most recent published literature but also a careful triangulation of the state of understanding across multiple sources, which typically included discussions with authors and other scientists, a synthesis of related information, and a close look at the available results and observations. Ultimately, determining the best available science included an honest assessment about the state of the science, the assumptions, and the unresolved issues.

8.4.5 Framing the Climate Question, as a Boundary Function

In all of these cases, the scientists involved worked with the decision-making organization to decide together what the key science questions were. The climate scientists on their own might pursue analyses on questions that are viable as climate analysis and potentially relevant to, but not usable for, the decision (e.g., the multiple different sets of choices made for Colorado River flows, climate projections before interpreted and adjusted for elevation). On the decision-maker side, stakeholders may not have enough specialized knowledge about climate science to know what to ask for (especially stakeholders new to climate issues) or do not have a sense of what is available or scientifically feasible. In the framing process, initial ideas on both sides were revised, and the project teams arrived at new ideas of what science questions were most relevant and critical to the decision. In the case of Appendix U, the iteration about long-term drought as a proxy for climate change led the scientists to propose methods to use the paleoclimate record to generate hydrologic scenarios. In the Susitna case, the framing led to justifying the use of the current climate models to provide a range of climate futures in a risk assessment framework.

This framing of climate questions must be iterative and cannot be done by either side alone; it moves beyond a consulting role or strategy (Michaels 2009) in which one side or the other owns the issue. Nor is it either top-down or bottom-up in which either the climate science or the stakeholder interests define an issue (Miller and Yates 2006). Each side learns enough about the other side to develop more nuanced and second-order questions for the assessment (Ray and Webb 2015). Scientists involved in this boundary framing function participate in this negotiation. The combination of iterative framing of climate questions and determining the best available science is critical to creating science for policy from the state of the science at any given time.

8.4.6 Benefits of Stakeholder Interactions for Science Organizations

It is well documented that working with decision makers and other stakeholders in climate information requires additional time and resources beyond the usual research modes (e.g., Dilling and Lemos 2011; McNie 2013). Morss et al. (2005) and McNie (2007) suggest that these interactions can not only inform use-inspired science but also describe engagement with stakeholders as a procedural dimension in which producers and users of information engage in dialogue aimed at shaping research agendas based on the context and needs of decision makers.

From these cases, I argue that the value of this engagement with decision makers is more than the report or synthesis product: there are significant downstream opportunities for the science organizations. These relationships have led to codevelopment, with participation and, often, funding from the stakeholder groups, of science synthesis products aimed at providing climate information usable for real-world policy and planning issues. In addition to the products discussed in this chapter, WWA and PSD have led or been major collaborators on numerous synthesis and assessment reports that summarize the state of climate knowledge with respect to a particular issue and, in some cases, assess vulnerability and impacts, e.g., the Colorado Climate Preparedness Project (Averyt et al. 2011) and the Gunnison Basin Climate Change Vulnerability Assessment (Neely et al. 2011). For PSD, a federal lab, and WWA, a federally funded RISA, these engagements help ensure that their science, assessment, and research are mission relevant and of value to the nation.

Stakeholder interactions foster innovation (Kirchoff 2013) by scientists and are a source of inspiration for research. The interactions and the efforts to develop science synthesis products generate new ideas and hypotheses and identify research gaps. That research—and reports in the grey literature—often ultimately lead to papers published in peer-reviewed journals, helping to establish the legitimacy, credibility, and salience of the work. Furthermore, there is a snowball effect of engagement with stakeholders. Over the last 20 years, we have observed both a capacity to anticipate needs of stakeholders and a faster "spin-up" with new user groups. After several years of interactions with water managers on seasonal-to-interannual climate variability (not discussed here; see Ray and Webb 2015), WWA and PSD were able to engage quickly with water managers on climate change issues, given trust relationships and the managers' familiarity with related climate science. Since 2010, PSD has sought out opportunities to work with ecologists and has found that spin-up time with ecologists has been faster than it was initially with the water managers. Scientists knew to begin asking questions about the policy or user context and knew that our default way of seeing the world would be different from theirs. PSD and WWA scientists now actively work with the DOI CSCs, and several DOI bureaus on climate and public lands and ecosystem management. Indeed, iterative modes have become second nature for many PSD and WWA scientists.

8.5 Conclusions

Providing information for federal and state policy decisions is a particular demand for usable science: these cases are services in that they responded to the decision-support needs of federal and state agencies for climate information. They illustrate the critical role of research scientists in the iterative framing of policy-relevant climate questions and determining the best available science for policy. These interactions have benefits both for the science organization as well as to the policy process. However, the service of synthesizing science for particular agency decision needs is not funded as an anticipated ongoing annual activity, and this is not a sustainable funding mode. PSD and WWA have now stepped into this science services role many times. However, given current research funding and personnel constraints, this approach risks delaying the science advances of the future. Meeting the demands of policy processes and decisions requires ongoing support for the participation of research institutions and scientists in service functions.

Acknowledgments

The author is grateful to Joseph J. Barsugli and Heather Yocum (University of Colorado/CIRES), Kathleen Miller (National Center for Atmospheric Research), Alan Hamlet (University of Notre Dame), and Lisa Darby (NOAA), for thoughtful and constructive comments, and to the following people for discussions contributing to the cases described: Robert S. Webb, Roger S. Pulwarty, and Susan Walker (NOAA); Kelly Redmond (Western Regional Climate Center); Joseph J. Barsugli and Kristen Averyt (University of Colorado/CIRES); Bradley Udall (Colorado State University); Terrence Fulp (US Bureau of Reclamation); and Veva DeHeza (then CWCB, now NOAA). Funding to support this chapter and the cases described came from the NOAA Office of Oceanic and Atmospheric Research, the NOAA Climate Program Office to the WWA, and the CWCB.

References

Averyt, K., K. Cody, E. Gordon et al. 2011. *Colorado Climate Preparedness Project Final Report*. Western Water Assessment for the State of Colorado, 108 pp. http://wwa.colorado.edu/CCPP_report.pdf.

Barnett, T.P., and D.W. Pierce. 2008. When will Lake Mead go dry? *Water Resour. Res.* 44: W03201, doi:10.1029/2007WR006704.

Barnett, T.P., and D.W. Pierce. 2009. Sustainable water deliveries from the Colorado River in a changing climate. *Proc. Natl. Acad. Sci. USA* 106: 7334–7338, doi:10.1073/pnas.0812762106.

Barsugli, J.J., J. Vogel, L. Kaatz, J. Smith, M. Waage, and C.A. Anderson. 2012. Two faces of uncertainty: Climate science and water utility planning methods. *J. Water Resour. Plann. Manage.* 138: 389–395.

Barsugli, J.J., K. Nowak, B. Rajagopalan, J.R. Prairie, and B. Harding. 2009. Comment on "When will Lake Mead go dry?" by T. P. Barnett and D. W. Pierce. *Water Resour. Res.* 45: W09601, doi:10.1029/2008WR007627.

Binder, L.W., and C. Simpson. 2009. RISA 2020 Vision Document: RISA community vision for future Regional Integrated Sciences and Assessments (RISA) efforts to match advances in climate impacts science with the needs of resource managers and planners. White Paper, NOAA Climate Program Office. http://cpo.noaa.gov/ClimatePrograms/ClimateandSocietalInteractions/RISA Program/RISAHistory.aspx.

Brekke, L., J.E. Kiang, J.R. Olsen, R.S. Pulwarty, D.A. Raff, D.P. Turnipseed, R.S. Webb, and K.D. White. 2009. Climate change and water resources management—A federal perspective: U.S. Geological Survey Circular 1331, 65 p. http://pubs.usgs.gov/circ/1331.

Cash D.W., J. Borck, and A. Patt. 2006. Countering the loading-dock approach to linking science and decision making. *Science, Technology and Human Values* 31: 465–494.

Chapin, F.S., III, S.F. Trainor, P. Cochran et al. 2014. Ch. 22: Alaska. Climate change impacts in the United States: The Third National Climate Assessment, J.M. Melillo, T.C. Richmond, and G.W. Yohe, Eds., *U.S. Global Change Research Program*, 514–536. doi:10.7930/J00Z7150.

Christensen, N.S., A.W. Wood, N. Voisin, D.P. Lettenmaier, and R.N. Palmer. 2004. The effects of climate change on the hydrology and water resources of the Colorado River Basin. *Climatic Change* 62: 337–363.

Colorado Water Conservation Board. 2015. Colorado Water Plan, public review draft, July 2015, https://www.colorado.gov/pacific/cowaterplan/july-2015-second-draft-colorados-water-plan (accessed 15 July 2015).

Council on Environmental Quality. Regulations implementing NEPA, 40 CFR 1500-1508.

Diaz, H.F., and J.K. Eischeid. 2007. Disappearing "alpine tundra" Köppen climatic type in the western United States. *Geophys. Res. Lett.* 34, L18707, doi:10.1029/2007GL031253.

Dilling, L., and M.C. Lemos. 2011. Creating useable science: Opportunities and constraints for climate knowledge use and their implications for science policy. *Glob. Env. Ch* 21: 680–689.

Endangered Species Act of 1973. Public Law 93-205. 16 U.S.C. § 1531 et seq.

FERC. 2009. Study Plan Determination for the Yuba-Bear, Drum-Spaulding, and Rollins Projects. FERC Office of Energy Projects, Washington, DC, 32 pp.

Federal Power Act 1995. Public Law 114-38. Federal Regulation and Development of Power. 18 CFR Section 5.11(b)(5).

Fish and Wildlife Coordination Act of 1965. 16 USC §661-667e.

Gordon, E.S., L. Dilling, E. McNie. 2015. Navigating scales of knowledge and decision making in the Intermountain West: Implications for science policy. In: *Climate in Context: Science and Society Partnering for Adaptation*, A. Parris et al., eds. Wiley Interscience, New York.

Guston, D.H. 2001. Boundary organizations in environmental policy and science: An introduction. *Science, Technology, and Human Values* 26: 399–408.

Hartmann, H. 2008. Decision support for water resources management. In: *Uses and Limitations of Observations, Data, Forecasts, and Other Projections in Decision Support for Selected Sectors and Regions*. U.S. Climate Change Science Program Synthesis and Assessment Product 5.1. pp. 45–54.

Hawkins, E., and R.T. Sutton, 2011. The potential to narrow uncertainty in projections of regional precipitation change. *Climate Dynamics* 37: 407–418.

Hoerling, M., D. Lettenmaier, D. Cayan, and B. Udall. 2009. Reconciling future Colorado River flows. *Southwest Hydrology* 8: 3.

IPCC. 2013. *Climate Change 2013: The Physical Science Basis. Contribution of Working Group I to the Fifth Assessment Report of the Intergovernmental Panel on Climate Change.* Stocker, T. F., D. Qin, G.-K. Plattner et al., eds. Cambridge: Cambridge University Press.

IPCC. 2007. *Climate Change 2007: The Physical Science Basis. Contribution of Working Group I to the Fourth Assessment Report of the Intergovernmental Panel on Climate Change.* Solomon, S., D. Qin, M. Manning et al., eds. Cambridge: Cambridge University Press.

Kirchoff, C.J. 2013. Understanding and enhancing climate information use in water management. *Climatic Change* 119: 495–509, doi:10.1007/s10584-013-0703-x.

Lemos, M.C., C.J. Kirchoff, and V. Ramprasad. 2012. Narrowing the climate information usability gap. *Nature Climate Change* 2: 789–794. doi:10.1038/nclimate1614.

Lowrey, J.L., A.J. Ray, and R.S. Webb. 2009. Factors influencing the use of climate information by Colorado municipal water managers. *Climate Research* 40: 103–119.

Lukas, J., J.J. Barsugli, N. Doesken, I. Rangwala, and K.E. Wolter. 2014. Climate change in Colorado—A synthesis to support water resources management and adaptation. Report by the Western Water Assessment for the Colorado Water Conservation Board. 108 pp. http://cwcb.state.co.us/environment/climate-change/Pages/main.aspx.

Magnuson–Stevens Fishery Conservation and Management Act, 1996. Public Law 94-265, as amended through October 11, 1996 (P.L. 109-479).

Markon, C.J., S.F. Trainor, and F.S. Chapin III, eds., 2012. *The United States National Climate Assessment—Alaska Technical Regional Report.* U.S. Geological Survey Circular 1379, 148 pp.

McNie, E.C. 2013. Delivering climate services: Organizational strategies and approaches for producing useful climate-science information. *Wea. Climate Soc.* 5: 14–26. doi: http://dx.doi.org/10.1175/WCAS-D-11-00034.1.

McNie, E.C. 2007. Reconciling the supply of scientific information with user demands: An analysis of the problem and review of the literature. *Environ. Sci. Pol.* 10: 17–38.

Michaels, S. 2009. Matching knowledge brokering strategies to environmental policy problems and settings. *Environ. Sci. Pol.* doi:10.1016/j.envsci.2009.05.002.

Miller K., and D. Yates. 2006. Climate Change and Water Resources: A Primer for Municipal Water Providers, National Center for Atmospheric Research (NCAR), AWWA Research Foundation Report #91120. http://nldr.library.ucar.edu/repository/assets/osgc/OSGC-000-000-021-450.pdf.

Milly, P.C.D., J. Betancourt, M. Falkenmark et al. 2008. Stationarity is dead: Whither water management? *Science* 319: 573–574. doi: 10.1126/science.1151915.

Morss, R.E., J.K. Lazo, and J.L. Demuth. 2010. Examining the use of weather forecasts in decision scenarios: Results from a US survey with implications for uncertainty communication. *Met. Apps.* 17: 149–162. doi: 10.1002/met.196.

Moser, S.C., and L. Dilling, eds., 2007. *Creating a Climate for Change: Communicating Climate Change and Facilitating Social Change,* Cambridge: Cambridge University Press.

National Environmental Policy Act of 1969. Public Law 91-190. 42 U.S.C. §§4321-4370h.

National Marine Fisheries Service. 2012, 2013. Susitna Study Request and related documents—https://alaskafisheries.noaa.gov/habitat/letters/2012/May/fercsusitnamasterstudy.pdf.

Neely, B., R. Rondeau, J. Sanderson et al., eds. 2011. *Gunnison Basin: Climate Change Vulnerability Assessment for the Gunnison Climate Working Group.* Prepared by The Nature Conservancy, Colorado Natural Heritage Program, Western Water Assessment, and University of Alaska–Fairbanks. Project of the Southwest Climate Change Initiative. https://www.conservationgateway.org/ConservationByGeography/NorthAmerica/UnitedStates/Colorado/science/Pages/gunnison-basin-climate-ch.aspx.

Prairie, J., K. Nowak, B. Rajagopalan et al. 2008. A stochastic nonparametric approach for streamflow generation combining observational and paleoreconstructed data. *Water Resour. Res.* 44: W06423, doi:10.1029/2007WR006684.

Pulwarty, R.S., C. Simpson, and C. Nierenberg. 2009. The Regional Integrated Sciences and Assessments (RISA) program: Crafting effective assessments for the long haul. In *Integrated Assessment of Global Climate Change,* C.G. Knight and J. Jaeger, eds. Cambridge, UK: Cambridge University Press.

Rajagopalan, B., K. Nowak, J. Prairie et al. 2009. Water supply risk on the Colorado River: Can management mitigate? *Water Resour. Res.* 45: W08201, http://doi:10.1029/2008WR007652.

Rangwala, I., and J.R. Miller. 2012. Climate change in mountains: A review of elevation-dependent warming and its possible causes. *Clim. Change* 114: 527–547.

Ray, A.J., and Webb, R.S. 2015. Understanding the user context: Decision calendars as frameworks for linking climate to policy, planning and decision-making. In: *Climate in Context: Science and Society Partnering for Adaptation*. A. Parris et al., eds. Wiley Interscience, New York, in press.

Ray, A.J., and S. Walker. 2014. Hydropower licensing and evolving climate: Climate knowledge to support risk assessment for long-term infrastructure decisions. AGU Abstract GC52A-0497.

Ray, A.J., J.J. Barsugli, K.B. Averyt, K. Wolter, M. Hoerling. 2008. Colorado Climate Change: A Synthesis to Support Water Resource Management and Adaptation, a report for the Colorado Water Conservation Board by the NOAA-CU Western Water Assessment. http://www.cwcb.state.co.us.

Ritter, Bill, Jr. 2007. Colorado Climate Action Plan: A Strategy to Address Global Warming. http://www.colorado. http://cnee.colostate.edu/graphics/uploads/ColoradoClimateActionPlan.pdf.

Sarewitz, D., and R. Pielke Jr. 2007. The neglected heart of science policy: Reconciling supply and demand for science. *Environ. Sci. Pol.* 10: 5–16.

Seager, R., M. Ting, I. Held et al. 2007. Model projections of an imminent transition to a more arid climate in southwestern North America. *Science* 316: 1181–1184.

US Bureau of Reclamation. 2011. SECURE Water Act Section 9503(c)—Reclamation Climate Change and Water Report to Congress. http://www.usbr.gov/climate.

US Bureau of Reclamation. 2007a. Appendix U—Review of Science and Methods for Incorporating Climate Change Information into Reclamation's Colorado River Basin Planning Studies. In: *Colorado River Interim Guidelines for Lower Basin Shortages and Coordinated Operations for Lakes Powell and Mead, Final Environmental Impact Statement*, 110 pp. http://www.usbr.gov/lc/region/programs/strategies/FEIS/AppU.pdf.

US Bureau of Reclamation. 2007b. Final environmental impact statement, Colorado River interim guidelines for lower basin shortages and coordinated operations for lakes Powell and Mead. Boulder City, Nev. http://www.usbr.gov/lc/region/programs/strategies/FEIS/.

US Bureau of Reclamation. 2006. Technical Work Group Meeting Minutes, November 8, 2006.

Vano J.A., B. Udall, D.R. Cayan et al. 2013. Understanding uncertainties in future Colorado River streamflow. *Bull. Am. Met. Soc.* 95: 59–78. doi:10.1175/BAMS-D-12-00228.

Weaver, C.P., R.J. Lempert, C. Brown et al. 2013. Improving the contribution of climate model information to decision making: The value and demands of robust decision frameworks. *WIREs Clim. Change* 4: 39–60. doi:10.1002/wcc.202.

Western Water Assessment RISA. 2005. Memo, comments on the Reclamation EIS on operations of Lake Powell and Mead, dated August 31, 2005.

Woodhouse, C.A. 2003. A 431-year reconstruction of western Colorado snowpack from tree rings. *J. Clim.* 16: 1551–1561.

Woolsey, P. 2012. *White Paper on the Consideration of Climate Change in Federal EISs, 2009–2011*. Sabin Center for Climate Change, Columbia Law School. http://web.law.columbia.edu/climate-change/resources/nepa-and-state-nepa-eis-resource-center.

9

Using the Past to Plan for the Future—The Value of Paleoclimate Reconstructions for Water Resource Planning

Connie A. Woodhouse, Jeffrey J. Lukas, Kiyomi Morino, David M. Meko, and Katherine K. Hirschboeck

CONTENTS

ABSTRACT Drought, growing demand on limited water supplies, and the impacts of climate change are all challenging the management of water resources in the western United States. While there is an increasing focus on climate change, information from the past in the form of tree-ring reconstructions of hydrology can provide information critical to water resource management. These reconstructions document the hydrologic variability that has transpired over past centuries, including events more extreme than those experienced in the modern period. Long-term natural variability, including droughts, will underlie the effects of anthropogenic climate change in the future. Reconstructions of past flows have been generated by researchers since the 1970s but have only recently been incorporated into resource planning and management. A number of challenges exist for the incorporation of this information into management, but a variety of motivations have prompted collaboration between researchers and water resource practitioners to develop ways this information can be applied. Examples from Denver Water, the Salt River Project, and the Bureau of Reclamation illustrate both the challenges and innovative uses of the reconstructions to address management questions. Looking into the future, tree-ring records supply important information about past climate, which, when combined with projections for future climate change, can provide a basis for robust water resource planning.

9.1 The Role of the Past in Planning for the Future

Since 2000, widespread drought conditions in the western United States have coincided with the emergence of climate change on the agenda for water resource management and planning in the region. While the causal linkage of the drought conditions with climate change is still debated (e.g., Cayan et al. 2010; Trenberth et al. 2013), it is clear that hydrologic conditions in the West are increasingly perceived through the lens of climate change. And as the impacts of climate change have become more widely anticipated, if not experienced, the past as a paradigm for future climate and hydrology has been called into question. The recent proclamation that "stationarity is dead" grabbed the attention of water managers (Milly et al. 2008, p. 573), and indeed, human activities (greenhouse gas emissions, land use, and land cover changes) are now a major influence on climate, and thus on the hydrologic cycle, unlike in the past.

Paleoclimatology, however, has long demonstrated the inherent nonstationary nature of the hydroclimate, clearly showing that the statistical characteristics of hydrology over the twentieth century do not represent the full range of variability that has occurred over longer timescales. Proxies of climate extending centuries into the past have provided evidence of climate variations on timescales of decades to centuries that have not been detectable in the shorter modern records (e.g., Gray et al. 2003). These records, by virtue of their length, almost invariably show extremes in droughts (and floods) more severe than those documented in instrumental records, reinforcing the need to look beyond the gage record to understand the range of conditions that occur under natural climate variability alone. This range of variability will continue in the future, underlying anthropogenic climate change (Meko and Woodhouse 2011). While projections of future climate from global climate models (GCMs) are critically important for anticipating the impacts of humans on future hydrology, even the most recent models appear unable to capture the range of decadal to multidecadal variability documented in paleoclimatic records (Ault et al. 2014). Because recent and ongoing droughts are likely due to both natural variability and the effects of climate change (e.g., Wang et al. 2014), understanding the extents of drought severity and duration that the natural system can generate is as important as understanding how humans may be exacerbating drought conditions through anthropogenic climate change.

In addition to the longer-term challenges of climate change, many water resource systems in the western United States are currently at, or approaching, capacity. For example, decadally averaged demands have recently exceeded supply in the Colorado River Basin (USBR 2012). Consequently, the recurrence of previously experienced variability in supply (i.e., droughts) over the next few decades could have major impacts on water management, even before projected anthropogenic changes are felt significantly. The range of streamflow variability possible in this time frame is arguably better captured by tree-ring data than either observed records or climate projections. One of the earliest studies that elucidated the nature of decadal-scale hydrologic variability and its implications for water resource management was the first tree-ring reconstruction of Colorado River annual streamflow (Stockton and Jacoby 1976). This reconstruction extended the gage record back to the mid-sixteenth century and clearly showed that the Colorado River Compact negotiations were based on an anomalously high period of flow, but also that low flows much more severe than those of the gaged record had occurred in the past. Updated reconstructions of the Colorado River identified the 1100s as the period with the lowest 25-year average flow in 12 centuries (Meko et al. 2007). Statistical analysis further indicated that such a drought might be expected about once every five centuries from the natural variability of the observed flow itself, and application of a long-term planning model showed that the recurrence of such a persistent period of low flow could reduce Lake Mead levels to "dead pool" (i.e., no usable storage) in two decades, given current allocations (Meko et al. 2012).

Tree-ring data provide a way to bridge traditional resource management with planning that considers long-term future climate and a plausible range of hydrologic variability. There are a number of ways the tree-ring information is being used by water resource managers (Woodhouse and Lukas 2006a; Rice et al. 2009). In many cases, water providers and agencies seek a context for assessing recent or current drought events. Sometimes, a qualitative assessment is most applicable, and a visual comparison of the gaged record with a centuries-long reconstruction of past flow provides the awareness needed to understand the limitations of the gage record (Woodhouse and Lukas 2006a). In other cases, a probabilistic assessment of critical features of the gaged record is most useful, such as for the severe drought years in the Colorado River Basin, 2000–2004. Tree-ring records classify that drought as a severe event in the long-term context but also indicate that at least one other drought of similar magnitude likely occurred over the past 500 years (Woodhouse et al. 2006; Meko and Woodhouse 2011). Tree-ring reconstructions have also been used to generate streamflow scenarios to test current management practices under highly stressful but plausible conditions, providing awareness of possible challenges in the future and an impetus for change. In an example of this, the Salt River Project (SRP), the Phoenix metropolitan area's largest water provider, assessed the impact of severe droughts over past centuries on reservoir depletions in the context of newly created operating guidelines (Meko and Hirscboeck 2008; Phillips et al. 2009). Another type of application examines reconstructions of past streamflow in the context of projected anthropogenic climate changes, anticipating a future with hydrology and water system outcomes well beyond the bounds of twentieth-century variability. In California, new and updated reconstructions of Sacramento River and San Joaquin River flows are being used to evaluate possible future hydrologic droughts as generated by models incorporating effects of changes in greenhouse gases, relative to droughts occurring under natural variability over the past thousand years (CADWR 2014).

9.2 Tree Rings as Records of Past Hydrologic Variability

Tree-ring reconstructions of past hydrology are based on the fundamental principles of *dendrochronology*, the science and dating of annual growth rings in trees (Fritts 1976).

These principles dictate, first, that trees growing in sites that are climatically stressful will record variations in the climate factor most limiting to growth. In much of the arid and semiarid western United States, the factor most limiting to growth is moisture; thus, wide rings reflect wet years, and narrow rings reflect dry years. Sampling trees at lower elevations and at sites with thin, well-drained soils, south-facing aspects, and little competition from other trees helps enhance the moisture signal in the ring widths (Fritts 1976). While the relationship between tree growth and precipitation is obvious, the link between tree growth and hydrology is less direct. The same climate factors, precipitation and evapotranspiration, mediated by the soil, are integrated in a similar manner into both annual streamflow and annual ring widths of trees (Meko et al. 1995; St. George et al. 2010). In the semiarid West, it is primarily the cool-season precipitation that links tree growth and hydrology; in their annual increments (i.e., ring width and water year flow), both reflect the cumulative precipitation that builds the snowpack and/or replenishes soil moisture during the period of low evaporative demand.

Trees from climatically stressed sites and species known for longevity and clearly discernable annual rings are targeted for collection. Living trees are sampled with increment borers, while cross sections from dead trees may be cut with a chainsaw. The samples from dead trees are intended to extend the tree-ring record and can often add centuries to the record available from living trees (e.g., Meko et al. 2007). Replication, another principle of dendrochronology (Fritts 1976), serves to enhance the climate information common to all trees, while minimizing the tree-to-tree variability in the ring widths. In arid and semiarid locations, about 20 trees (and two cores per tree) are typically sampled at each collection site.

In the laboratory, cores and cross sections are prepared for analysis by sanding each sample until the cells that make up each ring are clearly visible under a microscope. Rings in each sample are dated using a method called *cross dating*, in which the patterns of wide and narrow rings are matched between trees (Stokes and Smiley 1968). This process is undertaken instead of merely counting rings to ensure each and every ring is correctly dated to the exact calendar year. This process also allows the dating and incorporation of nonliving tree samples. Dated rings are measured to the nearest 0.001 mm using a sliding-stage micrometer. The ring-width series for each sample is then detrended to remove the effects of tree geometry (wider rings in the center of the tree typically grade to narrower rings on the outside) and age-related features unrelated to climate. Detrended sequences of ring width for each sample are averaged together to create a site chronology. The site chronology is the fundamental unit used for reconstructing past climate and hydrology.

A reconstruction of streamflow is an estimate of past conditions based on a statistical model of the relationship between a selected set of tree-ring chronologies and gaged flows (Figure 9.1). A reconstruction model is typically calibrated using tree-ring chronologies (or aggregates of chronologies) as predictors and a gage record as the predictand. Some version of multiple linear regression has been the most common approach for model calibration, but a number of other statistical approaches have been explored in recent years (e.g., Saito et al. 2008; Gangopadhyay et al. 2009). Reconstruction model calibration requires observed records that reflect natural flow conditions, either by virtue of the hydrology being relatively unaffected by diversions or depletions or through explicit estimation of natural flows. Tree-ring chronologies selected as candidate predictors should be significantly correlated with the gage record of interest. The period of time common to both gage record and tree-ring data is ideally 30 years or more.

The accuracy of a streamflow reconstruction model is usually measured by the proportion of variance of the gage record explained by the model and by the size of the departures (errors) between the estimated and observed flows. Differences in the statistical properties

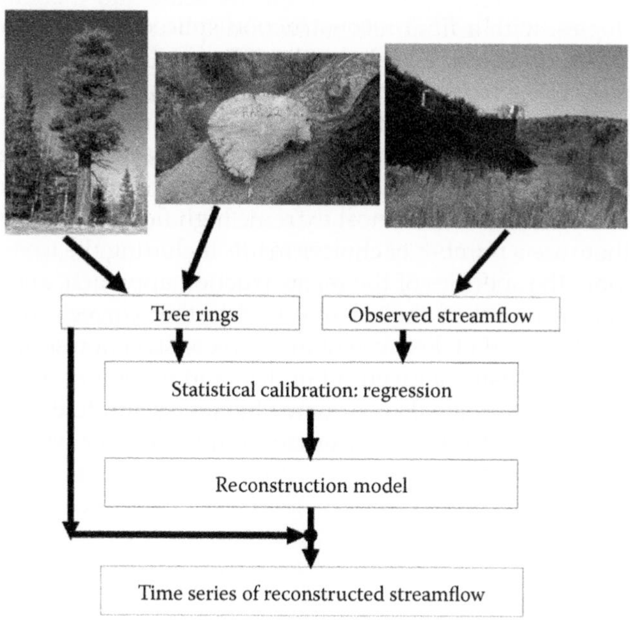

FIGURE 9.1
Flowchart showing reconstruction steps.

(e.g., mean, variance, range, autocorrelation) of the observed and reconstructed records are also compared.

The model is validated with data not used in fitting the model and estimating the regression coefficients. A visual comparison is very useful in showing how well the reconstruction replicates particular years of interest (Figure 9.2). Once the reconstruction model is tested and evaluated, the tree-ring data for the full period are applied to the model to generate the full-length reconstruction. If the time coverage of individual chronologies varies greatly, the entire procedure of calibration and validation may be repeated for different

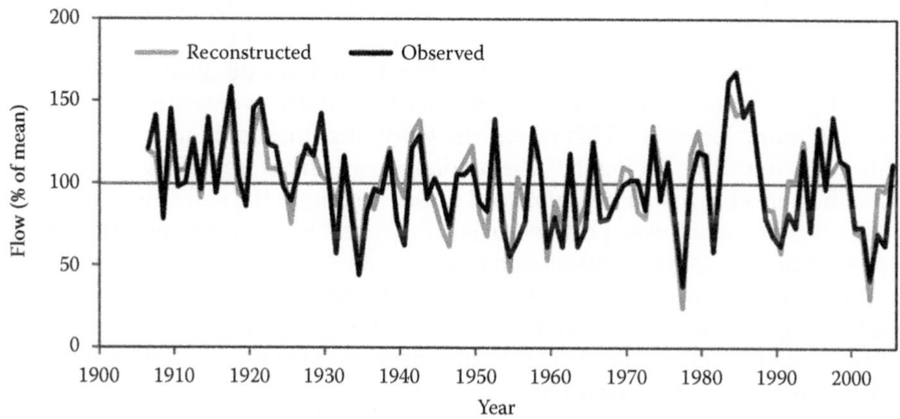

FIGURE 9.2
Time plots illustrating agreement of observed and reconstructed annual flows of Colorado River at Lees Ferry, Arizona. Example shown is for the time-nested regression model used to generate 1365–2002 CE segment of Meko et al. (2007) reconstruction. Flow is plotted as percentage of the 1906–2005 observed mean flow.

subsets of chronologies, with a final reconstruction spliced together from the segments provided by the subset models (e.g., Meko et al. 2007).

One important aspect of the reconstruction that is not always well communicated to water managers is the uncertainty in the reconstruction. Even the most skillful tree-ring reconstructions do not explain more than 70–80% of the variance in a gage record. Most often, the unexplained variance is in the failure of the reconstruction to replicate the most extreme values, and in particular, the most extreme high flows (Figure 9.2). In developing a reconstruction, there are a number of choices made, including the tree-ring data selected for the reconstruction, the specifics of the reconstruction approach, and the years used in the calibration. There is no set of choices that lead to the "correct" reconstruction; each solution will be slightly different. Because of this, a reconstruction should be considered a plausible estimate of flows using a given set of data and modeling decisions. The earliest years of flow reconstructions may also be somewhat less accurate than implied by model statistics because of lower replication (fewer trees) in the earlier parts of tree-ring chronologies (Meko et al. 2012).

9.3 Why Reconstructions Have Gained Acceptance by Water Managers in Recent Years

While streamflow reconstructions from tree rings have been available since the 1970s and 1980s (e.g., Colorado River flow, Stockton and Jacoby 1976; Sacramento River flow, Earle and Fritts 1986), the water resource management community in the western United States has not incorporated this information into their planning and management activities until recently. A convergence of multiple factors over the last 10–15 years likely motivated this, including severe and prolonged drought, water demands outpacing supplies, the growing reality of climate change, a set of outside-the-box-thinking resource professionals, and a shift in the way that researchers view the utility of their science and their interactions with stakeholders.

One of the first motivations for using tree-ring data was to assess the 2000s drought, which strongly affected water resources in the Colorado River Basin and elsewhere. The iconic droughts of the twentieth century, the 1930s Dust Bowl, the 1950s, and the late 1980s to early 1990s, were matched or exceeded in severity by early twenty-first-century drought conditions in many locations. This raised the following question: Have there been worse droughts in the more distant past? The extended records from tree rings almost invariably show that the period of instrumental record does not contain the full range of natural variability that has occurred over past centuries. In some cases, worse conditions may have occurred within the past few centuries (e.g., Rice et al. 2009); in other cases, the twentieth-century drought extremes have not been exceeded since the 1100s (e.g., Meko et al. 2014). Although droughts may be measured in a variety of ways, these assessments indicate that at least by some measures, the droughts that have occurred since 2000 are not unprecedented, in the context of the past 1000–1200 years.

Another motivation for the use of extended records of flow from tree rings is the increase in society's vulnerability to drought conditions in recent decades. The publication of the first reconstruction of Colorado River streamflow (Stockton and Jacoby 1976) coincided with the start of the driest water year (1977) in the twentieth century, but the two decades that followed saw mostly above-average streamflows in the Colorado River Basin. This was also a period of rapid growth and development in the urban areas—Los Angeles, Las Vegas, Phoenix, Tucson, Salt Lake City, Denver—that at least partly depend on Colorado River water supplies. When persistent drought conditions commenced in 2000, the impacts were strongly felt, but by some measures, conditions were no worse than the 1950s drought (e.g., Weiss et al. 2009). The vulnerability to drought had increased due to greater demands on limited water resources, and in the case of the Colorado River Basin, the demand has exceeded the overallocated water supply on a decadal basis since 2003 (Kenney et al. 2010; USBR 2012).

Confronting the reality of climate change and its potential impacts on water resources has been challenging for many water resource managers. Planning has traditionally been based on the period of instrumental record, and in many parts of the western United States, political sensitivities have inhibited a discussion of the impacts of climate change on water resources. Tree-ring reconstructions of streamflow have become an acceptable alternative, opening the door to considerations of time frames beyond the twentieth century, in the absence of open discussions of climate change (Woodhouse and Lukas 2006b; Rice et al. 2009). Management decisions based on the longer flow records can result in more resilient water supply systems for dealing with natural climate variability or anthropogenic climate change. The tree-ring reconstructions have become a stepping-stone, complement, or alternative to explicit projections of future hydrology, depending on needs and sensitivities of institutions.

The increasing use of tree-ring data in management has also come about in large part due to leadership from water resource professionals who are willing and able to grapple with new types of information for decision making, even if they are not required by the existing planning frameworks. This may reflect a more diverse training of water resource professionals. Key personnel at the Bureau of Reclamation (e.g., Prairie et al. 2008; Gangopadhyay et al. 2009), California Department of Water Resources, SRP (Phillips et al. 2009), Denver Water (Rice et al. 2009), and consulting firms (e.g., Harding 2005; Kenney et al. 2010) recognized the usefulness of tree-ring data and have been instrumental in the development of innovative techniques to apply these data to resource management. On the part of researchers, new sources of funding for decision-support science such as the National Oceanic and Atmospheric Administration (NOAA) Regional Integrated Sciences and Assessments (RISA) program and Reclamation's WaterSmart program have motivated collaborations between researchers and practitioners. On both sides, the emergence of researchers and practitioners serving as *information brokers*—people who understand both the scientific and management perspectives and challenges—has facilitated the use of scientific information in resource applications (Ferguson et al. 2014).

As new reconstructions have been generated for specific agencies and basins in the western United States, they have caught the attention of other water resource professionals, leading to requests for additional reconstructions. Since 2009, the ever-expanding West-wide network of over 60 gage reconstructions has been made accessible in a central location, along with supporting information, on the TreeFlow web resource (http://www.treeflow.info/). A series of over 15 workshops has facilitated the education of water managers and peer-to-peer interactions among practitioners who are interested in using the reconstructions (http://www.treeflow.info/workshops.html). With easier access to both the data and information about applications, over the past decade, a core group of paleo-savvy practitioners has evolved who are using tree-ring reconstructions in a variety of novel ways. This has provided case studies and motivation to other practitioners to consider the benefits of streamflow reconstructions in their planning process.

9.4 Challenges and Applications of Streamflow Reconstructions to Water Resource Management

In spite of the motivating circumstances cited previously, significant challenges exist to the application of streamflow reconstructions to water resource management. Fundamental mismatches between what is being produced by the scientific community and what is useful to the management community have hampered applications. For example, gages that could be most feasibly reconstructed by researchers were not necessarily those of interest to specific water agencies. Spatial coverage and timescales of the reconstructions (constrained to be annual, correlating to annual growth rings) have often not been compatible with an agency's water system's modeling or planning parameters (e.g., Woodhouse and Lukas 2006a). In addition, the capacity to make use of new types of information varies among water agencies, with some needs best met by relatively straightforward basic awareness raising, while others require specific model input to test system sensitivity over a range of conditions. Yet another challenge has been finding ways to translate the uncertainty in the reconstruction into the planning methodology. In recent years, many of these issues have been addressed in collaborations between researchers and water resource practitioners, leading to greater use of the streamflow reconstructions in water resource management. Several case studies that demonstrate the outcomes of these collaborations, in the South Platte River Basin and the Upper and Lower Colorado River Basins (Figure 9.3), are described next.

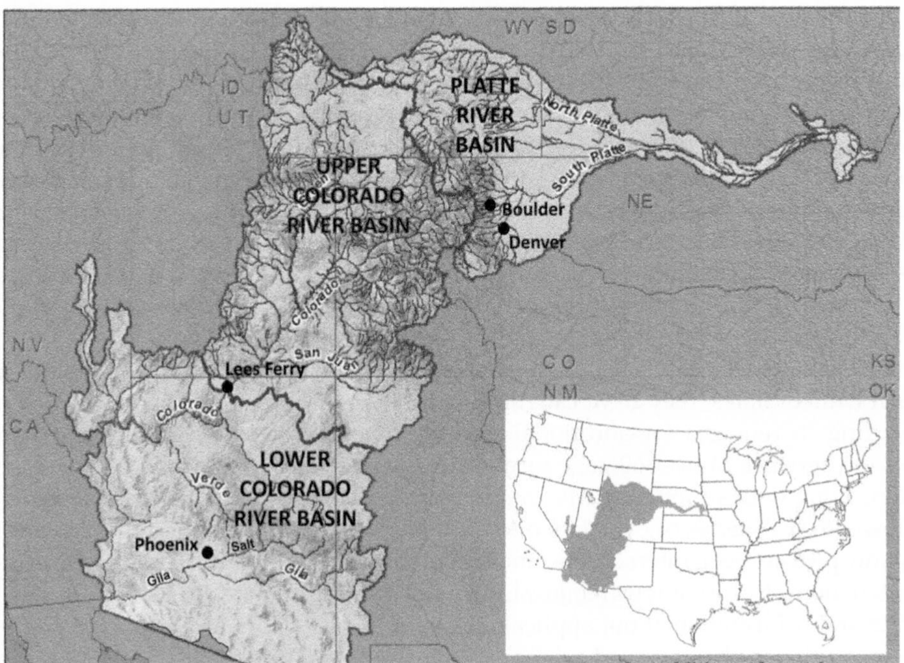

FIGURE 9.3
Rivers, basins, and cities mentioned in the case studies in this chapter.

9.4.1 Colorado Front Range Water Utilities: Stress-Testing Systems for Severe Drought and Climate Change

Colorado's Front Range region is home to over 80% of the state's 5.2 million residents. The several dozen municipal water utilities on the Front Range vary widely in their size, their technical capacity, and the makeup of their water supply portfolio. Starting around 1998, sustained engagement by tree-ring scientists based at NOAA and the University of Colorado (CU) in Boulder led to the use of tree-ring data in planning by several of these water providers. The work with two of these providers, Denver Water and the city of Boulder's Water Utilities Division, highlights the benefits and challenges inherent in the application of tree-ring data.

9.4.1.1 Denver Water

Denver Water is the oldest and largest urban water provider in Colorado, serving over 1.3 million people in Denver and the surrounding suburbs. About half of their water supply comes from the South Platte River Basin, in which Denver is located, and the other half from the Colorado River Basin (Figure 9.3), with two tunnels transporting water from the latter to the former. The impacts of droughts are buffered by 10 major reservoirs, located in both basins, which have a combined storage capacity of 670,000 acre-feet (830 million cubic meters [MCM]), equivalent to over 2 years of the current water demand.

Denver Water has developed a sophisticated model of its water system: the Platte and Colorado Simulation Model (PACSM), which simulates streamflows, reservoir operations, and water supplies in the South Platte and Colorado River basins. The model runs on input from 450 locations on a daily time step, using data from 1947 to 1991. The 1950s drought (1953–1956) had been considered the worst-case scenario for planning and still remains the most severe multiyear drought on record for Denver Water. In 2002, however, the third and most severe year of drought conditions led to record-low annual flows at several of Denver Water's gages, and raised concerns about the adequacy of the 1947–1991 model period as the baseline for planning.

After the Boulder-based tree-ring scientists developed preliminary reconstructions of their gages of interest, Denver Water funded the re-collection of about 20 tree-ring sites in both the South Platte and Colorado River basins to update the chronologies through 2002. The new streamflow reconstructions, generated in 2004 with the updated chronologies, very closely captured the drought conditions of 2002 (Woodhouse and Lukas 2006a). The fidelity of the trees in representing this recent extreme event enhanced the perceived credibility of the reconstructions by Denver Water's managers and board.

Using the new reconstructions as inputs to the PACSM system model faced major technical challenges—spatially and temporally disaggregating the tree-ring reconstructed values (annual, at two locations) to the daily time steps at 450 locations required by the model. The Denver Water engineers arrived at an "analog" method, in which the daily model inputs for each paleo year were derived from the year in their 1947–1991 model period that most closely matched the annual flow of that paleo year. The model inputs representing paleo years with higher or lower annual flows than any in this reference period were scaled up or down accordingly. Running the full tree-ring record (1634–2002) through PACSM, they could determine the impacts of the broader range of droughts seen in the paleorecord on their system under different assumptions of demand level and policy interventions during droughts. They found that a 4-year drought in the 1840s (1845–1848) and another one in the 1680s had greater modeled impacts than the 1950s drought but that their system would be able to meet water demands through the paleodroughts with progressive restrictions on outdoor water use.

While this was the most significant application of the tree-ring data by Denver Water, discussions between this engineering group and the tree-ring scientists over a period of 6 years led to other useful developments. For example, Denver Water was interested in reconstructing climate-driven variations in their water demand (i.e., outdoor watering) to complement the supply-side streamflow reconstructions. Variability in demand is influenced mainly by late spring and summer precipitation and temperature over the urban service area, while annual streamflow is driven by winter–spring precipitation over high-elevation catchments. These quantities are often related but in some years are well correlated. The reconstruction of water demand in the Denver Water service area allowed Denver Water to assess the frequency of periods with particularly high demand and low supply.

9.4.1.2 City of Boulder

The city of Boulder's Water Utility Division serves about one-tenth the number of customers as Denver Water. About 70% of its water supply comes from the Boulder Creek drainage in the South Platte River Basin, with the remainder coming from the Colorado River Basin via the Colorado–Big Thompson (C-BT) and Windy Gap projects (Smith et al. 2009). Boulder has ample annual average supply relative to the current demand of 25,000 acre-feet (31 MCM) but relatively less storage than Denver Water, equal to about 1 year of current demand, so their supply is vulnerable to severe multiyear drought. In addition to assessing this drought risk, Boulder has been ahead of the curve in considering how anthropogenic climate change might impact its water supply.

Collaboration between consultants for the city of Boulder and the NOAA and CU tree-ring scientists first led to a reconstruction of annual streamflows for Middle Boulder Creek from 1703 to 1987. As with Denver Water, the reconstructed flows were run through a system model to examine impacts on water deliveries and other metrics. Again, the multiyear drought from 1845 to 1848 was identified, along with another drought in the 1880s, as leading to greater stress on the system in terms of expected delivery reductions than any droughts in the twentieth century. These results were included in Boulder's comprehensive Drought Plan in 2003. At about this time, a separate simulated stress test of the city's system was performed, assuming a flat 15% reduction in annual streamflows due to climate change. The obvious next step was to assess the joint risks of climate variability—as more fully seen in the tree-ring record—and climate change, by examining what would happen if the droughts of the past occurred again under a warmer and possibly drier future climate.

Starting in 2006, Boulder conducted a multiyear study to assess the vulnerability of its water supply to climate change, with support from NOAA, CU, the National Center for Atmospheric Research (NCAR), and Stratus Consulting (Smith et al. 2009). The study was one of the very first to explicitly combine projected future climate from GCMs with paleoclimate reconstructions from tree rings. First, a new tree-ring reconstruction of annual streamflow for Boulder Creek, from 1566 to 2002, was generated using some of the same chronologies developed for the Denver Water work. A nonparametric *k*-nearest-neighbor technique was used to conditionally resample the 1953–2002 historic period climate, based on matching each paleo year with the five historic years that are nearest in terms of annual flow. The effect is similar to the Denver Water analog method in that it disaggregates the annual reconstructed streamflows, in this case, into monthly climate variables

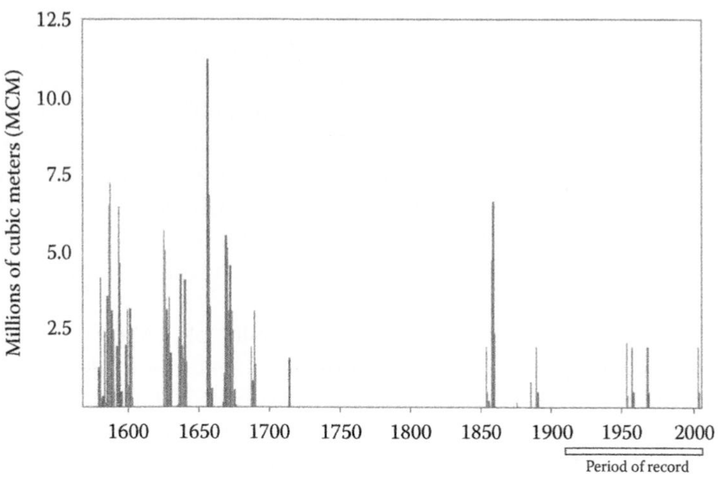

FIGURE 9.4
Modeled shortages in water deliveries by the city of Boulder, from simulation run that used temperature and precipitation changes from a global climate model (GCM) under the A2 (high) emissions scenario to adjust a 437-year (1566–2002) trace conditionally resampled from a tree-ring reconstruction of Boulder Creek. Using only the period of record for risk assessment (1907–2002) would dramatically underestimate the risk of shortage seen over the longer period. (From Smith, J.B. et al., The potential consequences of climate change for Boulder Colorado's water supplies, Stratus Consulting, for the NOAA Climate Programs Office. Boulder, CO, 2009.)

(precipitation and temperature). That allowed those variables to be separately adjusted according to the projected changes from nine different GCMs, and then input into snowmelt-runoff and water-balance models to produce multiple simulations conditioned on the paleodata and future climate conditions. The simulated Boulder Creek hydrology was then run through Boulder's water system model.

All of the system simulations showed that undesirable outcomes such as shortages in water deliveries tended to be concentrated in the 1600s and 1700s, when there were more frequent paleodroughts than in later centuries (Smith et al. 2009) (Figure 9.4). The frequency and size of those shortages also depended on the future changes seen in the particular GCM projection driving the simulation; the high-emissions (i.e., warmer) projections that also showed reduced precipitation had the most frequent and largest shortages, as would be expected. Overall, the combination of the tree-ring record and GCM output was more stressful to the system than either one alone.

9.4.2 Upper Colorado River Basin: Supply Scenarios Using Tree Rings

The Colorado River Basin covers over 240,000 mi.^{2} (620,000 km^2), including portions of Arizona, California, Nevada, Colorado, New Mexico, Utah, and Wyoming, collectively referred to as the *Basin States*, as well as northwestern Mexico (Figure 9.3). In the western United States, the Colorado River supplies water to about 40 million people, supports 3 million acres (1.2 million ha) of irrigated agriculture, and produces vital hydropower for the region. Two large reservoirs, Lakes Mead and Powell, account for the bulk of storage in the Colorado River system, which, combined, are able to store up to 60 million acre-feet (MAF), or 4 years of the twentieth-century average annual flow.

Two recent Bureau of Reclamation (hereafter *Reclamation*) planning efforts highlight the use of tree-ring data in Colorado River management. The first is an environmental impact statement (EIS) completed in 2007 in response to rapidly declining water levels during the early 2000s in Lakes Powell and Mead (USBR 2007, hereafter *FEIS*). This document provides important background and justification for the new operating policies developed to manage shortage and low reservoir conditions on the Colorado River. The second is the Colorado River Basin Water Supply and Demand Study, completed in 2012, arising from concerns regarding the reliability of Colorado River water supplies given the supply and demand imbalances that had persisted since the early 2000s (USBR 2012, Technical Report A; hereafter *Basin Study*). In contrast to the FEIS, the Basin Study did not inform any changes in policy but will undoubtedly serve as a rich source of information for future policy decisions.

Reclamation lists its collaboration with the University of Arizona, including tree-ring scientists, as one of the five key components in the research and development program that contributed to the analysis and content of FEIS (USBR 2007). Since 2004, Reclamation had been working with University of Arizona researchers in a project aimed at enhancing water supply reliability through improved understanding of reservoir system responses to natural and anthropogenic drivers. Identifying how tree-ring data could be used in Colorado River management was one of the primary objectives of this project. This objective was realized during the EIS process: tree-ring data were used to "analyze the sensitivity of the hydrologic resources to alternative future hydrologic scenarios" (USBR 2007, pp. 4–11).

9.4.2.1 Challenges of Integrating Tree-Ring Data into Reclamation's Management Model

Prior to using a streamflow reconstruction to conduct a sensitivity analysis in the FEIS, it was first necessary to address two issues: (1) reconstruction uncertainties and (2) the incongruity between spatial and temporal scales of the reconstruction and input requirements for Reclamation's management model. Several reconstructions have been developed for the Colorado River at Lees Ferry (e.g., Stockton and Jacoby 1976; Hidalgo et al. 2000; Woodhouse et al. 2006). A comparison of these reconstructions revealed that for any given year, there was a range of flow magnitudes. This uncertainty, however, was offset by the consistency among reconstructions of the hydrologic state information—whether it was a wet or dry period (Woodhouse et al. 2006). To address the uncertainty in the annual flow magnitudes, Reclamation developed a methodology that utilized state information from the reconstruction and magnitude information from the instrumental data (Prairie 2006). For each year in the reconstruction, conditional upon the state (wet or dry) of the current and previous year, a set of candidate years was identified in the instrumental record. From this set, one year was randomly selected and used to represent flow for that year. Blending the tree-ring and instrumental data in this way, it was possible to generate many different reconstruction-length sequences, each with novel sequences of instrumental data. Reclamation used both the blended (*paleo-conditioned*) and unblended tree-ring data to evaluate the Colorado River system.

The tree-ring reconstructions also needed to be disaggregated before they could be used for system analyses. Reclamation's management model required hydrological input at a monthly time step for 29 input locations within the basin, whereas Colorado River streamflow was reconstructed at an annual resolution for a single location in the basin, Lees Ferry. A nonparametric method was used to identify analog years in the instrumental record, enabling the disaggregation of the streamflow reconstruction into monthly data for multiple input locations. This approach was developed for the FEIS (Prairie 2006; Prairie et al. 2007). A similar methodology was used in the Basin Study (Nowak et al. 2010).

9.4.2.2 Role and Use of Tree-Ring Data in Reclamation Studies

The main findings in the FEIS were based on instrumental data used as input to Reclamation's system model. Tree-ring data were used to conduct a sensitivity analysis of the effect of hydrological inputs on system behavior because Reclamation considered it important to "...understand the potential effects of future inflow sequences outside the range of historic flows..." (USBR 2007, pp. 4–14). Thus, although results from the sensitivity analysis based, in part, on tree-ring data were not used in the primary evaluation of proposed actions, they did inform the Secretary of Interior's decision to designate the guidelines as *interim*, extending them only through 2026 (USBR 2007).

In contrast, for the Basin Study, tree-ring data played a central role in the primary analysis of evaluating system vulnerability and reliability. Here, Reclamation used four, equally weighted hydrological data sets to drive their system model: instrumental, unblended tree-ring, blended tree-ring, and downscaled projection. Thus, for any given evaluation of the Colorado River system, half of the supply scenarios were based either wholly or partly on tree-ring data.

While the role of tree-ring data in each planning effort was quite different, the use of tree-ring data was very similar. In both cases, tree-ring data provided multiple sequences of hydrological data for Reclamation to use as input for its management model. This generated an ensemble of outputs for any given system indicator, which was analyzed and evaluated in a variety of ways. By and large, the blended tree-ring data set tended to simulate more extreme outcomes for system indicators when compared to instrumental and unblended tree-ring data. For example, an FEIS analysis showed that by 2060, the tenth percentile of Lake Powell elevations simulated by blended tree-ring data was about 150 ft. lower than the tenth percentile of elevations simulated by instrumental and unblended tree-ring data sets (USBR 2007).

In the Basin Study, Reclamation examined how, in conjunction with multiple scenarios of demand and two management alternatives, the different supply scenarios affected system vulnerability over time (USBR 2012). In this analysis, each supply scenario contributed an equal number of input sequences, or traces. Vulnerability here was defined as the percentage of traces in which Lake Mead elevation was below 1000 ft. for any given month. In the short term, supply scenarios accounted for the greatest amount of variability in vulnerability, with instrumental (called *observed resampled* in the Basin Study) and unblended (called *paleo resampled* in the Basin Study) tree-ring data associated with the lowest vulnerabilities and downscaled projections with the highest vulnerabilities. By mid-century, however, vulnerability is largely governed by a combination of demand scenarios and management policy. For example, vulnerabilities based on instrumental data range from 0% to over 50% for all combinations of demand scenario and management alternative. A similar spread is indicated by the two tree-ring-based supply scenarios. However, within each supply scenario, there are differences in the relative vulnerability of each combination of demand scenario and management alternative. These results suggest that changes in demand scenario and operations guidelines alone, as illustrated by the blended tree-ring data, can increase the likelihood of Lake Mead levels dipping below 1000 ft., even without considering climate change projections (Figure 9.5).

9.4.3 Salt River Project—A Problem of Extremes: Integrating Tree-Ring Information into Water Resource Planning and Management

The SRP manages a 13,000 mi.^{2} (33,670 km^2) watershed area that includes dams, reservoirs, wells, canals, and irrigation infrastructure to supply water to its service area in the

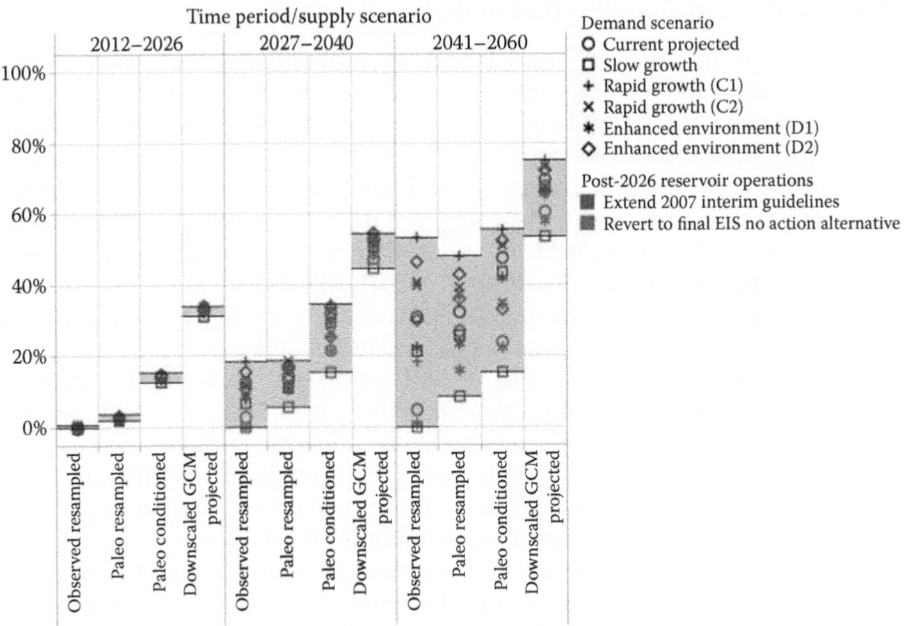

FIGURE 9.5
Percent of traces vulnerable without options and strategies by scenario and time period, Lake Mead elevation indicator metric (below 1000 ft. in any 1 month). (Figure from US Bureau of Reclamation's Colorado River Basin Water Supply and Demand Study, Final Report, December 2012.)

Phoenix Metropolitan Area (SRP 2014). Compared to Lake Mead, the amount of storage in the SRP reservoirs is small, resulting in a greater sensitivity to extreme high or low annual streamflow contributions originating in the Salt and Verde watersheds (Figure 9.3). A water exchange agreement with the Central Arizona Project (CAP), which delivers Upper Colorado River Basin (UCRB) water to Arizona, gives SRP an additional source of surface water, if needed. Such a need occurred in response to a multiyear drought that began in 1996 and by the end of 2003 had exceeded the historic 7-year drought of record for the SRP reservoir system. In fact, SRP's Roosevelt Reservoir would have been "close to empty" in 2002 without the CAP contributions of Colorado River water (Philips et al. 2009, p. 117). The severity of this event prompted a concern that extreme regional drought conditions might affect both Upper and Lower Colorado Basin water supply areas at the same time, presenting a major management challenge. To assess the probability of such an occurrence, SRP initiated a collaboration with the University of Arizona Laboratory of Tree-Ring Research to estimate the probability of synchronous extremes in the Upper Colorado and Salt–Verde River Basins using existing streamflow reconstructions from tree rings. The following section describes how the integration of paleoinformation into SRP's operational guidelines emerged from two studies, each focusing on a key planning and management question.

9.4.3.1 *SRP I: Extreme Streamflow Episodes in the Upper Colorado and Salt–Verde River Basins*

The SRP I study addressed the long-term synchrony of droughts and wet periods in the UCRB and Salt–Verde Basin, Arizona (Figure 9.3). The conventional view at the time of the study was that droughts in the Salt–Verde Basin would be buffered or offset by normal or above-normal supplies from the UCRB. Existing tree-ring chronologies were assembled and applied in multiple linear regression (Hirschboeck and Meko 2005) to generate reconstructions of total annual flow of the Colorado River at Lees Ferry and flow summed over the Salt, Verde, and Tonto Rivers (SVT). The former represents the Colorado River outflow from the UCRB, and the latter, the "local" river flow into SRP reservoirs in Arizona. Time-nested regression models (Meko 1997) explained up to 70% and 76% of annual flow for the SVT and the Colorado, respectively, and indicated that the tree-ring network was sufficiently dense for reliable inference of joint occurrences of high or low extreme flows over 1521–1964.

Dry years and wet years, or lows and highs, in the reconstructions were defined by the 25th and 75th percentiles of reconstructed flow for the common period. A count of events revealed that same-sign flow anomalies in the two basins occurred much more often than opposite-sign anomalies. Large opposite-sign events were found to be especially rare: in only 2 of the 444 reconstruction years was one basin below its 25th percentile, while the other was above its 75th percentile. The clear message from the reconstructions was that large flow deficits on the Salt–Verde are unlikely to be offset by high flows on the Colorado. Rather, extreme dry years or wet years were often shared by both basins. Synchronous lows, especially, tended to cluster in time, reflecting the occurrence of multiyear widespread drought conditions. Of interest from a management perspective was the observation that these extended low-flow periods could be interrupted by years of near-normal flow: the longest stretch of synchronous extreme dry years was three. Figure 9.6 depicts one of the longest clusters of synchronous low-flow years as revealed by counts in a 5-year moving time window.

9.4.3.2 SRP II: The Current Drought in Context

The main topic of the follow-up study was an assessment of the severity of the "current" drought (as of 2004) on the Salt–Verde as revealed by the long-term tree-ring record. This drought was judged at the time to have begun as early as 1996 and to have intensified during 1999–2004. Because this period postdated most of the previous collections of drought-sensitive chronologies in the Salt–Verde Basin, new tree-ring collections were needed. Accordingly, 14 tree-ring sites were collected in 2005, and the resulting chronologies were incorporated with existing data in a multivariate reconstruction model to generate a reconstruction of SVT annual flow, 1330–2005 (Meko and Hirschboeck 2008).

The results indicated that two of the current drought years—1996 and 2002—were unsurpassed in single-year severity of reconstructed drought since the fourteenth century. The long-term context of these exceptionally dry years is illustrated in a time-series plot of reconstructed flow (Figure 9.7). Despite these single-year anomalies, however, the multiyear severity of extended low streamflow during the current drought was not unprecedented. Running means of various lengths failed to identify this drought as exceptional. For example, the reconstructed running mean for 1999–2004 ranked 14th driest of all 6-year running means in the reconstruction. The two most severe multiyear events in the reconstruction were characterized by 5 consecutive years of flow below the long-term median, and both of these occurred in the distant past (Figure 9.7).

Throughout SRP I and SRP II, feedback from SRP scientific staff was important in guiding the research and often posed new questions. One question that arose concerned the length of the maximum interval of time between *drought-relieving* wet years in the reconstructed low flows. This information was deemed critically important by

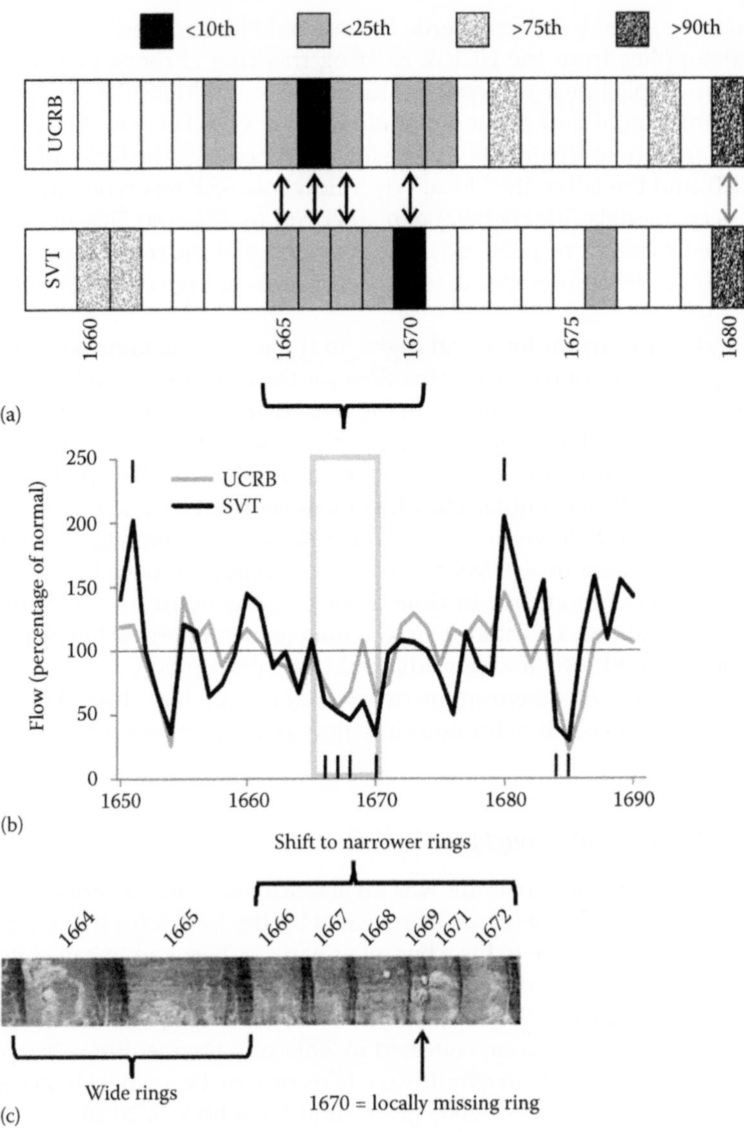

FIGURE 9.6
Synchrony of extreme low-flow years in the watersheds of the Upper Colorado River Basin (UCRB) and the Salt–Verde–Tonto River Basins (SVT) in the late 1600s. (a) Demarcation of high- and low-flow years in the UCRB and SVT tree-ring reconstructions based on percentiles of annual flow, 1660–1680. Synchrony between the basins in extreme low-flow years (<25th percentile) and extreme high-flow years (>75th percentile) is indicated by double-headed arrow. (b) Reconstructed flows for the period 1650–1690 are plotted as percentage of the 1521–1964 long-term mean. The moving window highlights four common dry years (ticks at bottom) during the 5-year period 1666–1670. High-flow years in common in both basins are marked by ticks at top. See text for description of reconstructions and definition of dry and wet years. (c) Tree-ring core segment from a site near Show Low, Arizona, in the Salt River watershed showing the narrow ring sequence during the 5-year period highlighted in (b). Growth during 1670 was so stressed that a complete ring did not form on this tree at the point sampled with the core.

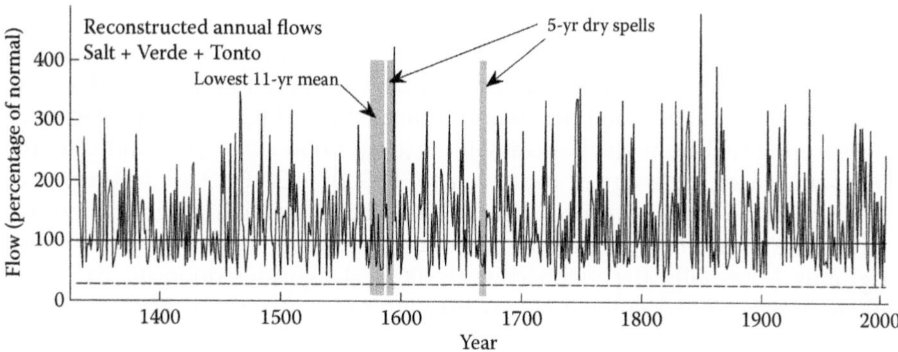

FIGURE 9.7
Time plot illustrating the long-term context of the 1996–2004 drought period near the start of the twenty-first century. Depicted are reconstructed annual flows summed over the Salt, Verde, and Tonto Rivers for the period 1330–2005. The dashed line shows an unprecedented single-year tree-ring-based low-flow response to the driest (2002) and next-driest (1996) years of recent drought. Shading marks the only two periods with flow below the long-term median for 5 consecutive years. The 5-year dry spell in the late 1600s is also shown in Figure 9.5. Normal (solid horizontal line) defined as median of 1914–2006 observed flows.

THE PALEORECORD AND FLOODS

The use of the paleorecord to transfer climate change information into operational flood hazard management presents substantially different challenges than those faced in employing the paleorecord in long-term planning for future low flow or drought extremes. Unlike droughts, which evolve from cumulative climatic conditions with an impact on tree growth that can be resolved in annual and seasonal tree-ring records, floods arise from distinct, watershed-specific precipitation events that do not appear as discrete features in the annual rings of tree growth.

Can Flood Events Be "Seen" in Streamflow Reconstructions? In watersheds where the maximum annual flood peak is well correlated with mean annual streamflow, inferences about the magnitude and frequency of past high flows (and presumed floods) can be made based on streamflow reconstructions, but the timing, frequency, and magnitude of the flooding that contributed to the high flow will be unknown, especially because reconstructions typically capture the magnitude of dry years better than wet years (see Figure 9.2 discussion). Floods produced by storms that contribute substantially to water supply are more likely to be correlated with high-flow years in a streamflow reconstruction than short-lived flash floods. However, years with high mean annual flow in either gaged or reconstructed time series do not always indicate major flooding. Reconstructed high-flow years may arise from sustained moderate flow, multiple floods of varying size, one or more long-lasting extreme events, snowmelt contributions unrelated to flooding, or other combinations of events.

"Paleo Flood Stage" Indicators. Floods in the pregaged record are best recorded by physical features in the environment that reveal past streamflow stages or the influence of high flows at or near the stream channel, such as botanical and geologic paleostage indicators or field evidence of exceedance or nonexceedance of a given

flood stage in the landscape (Baker 1987, 2008). Tree rings can serve as paleostage indicators when there is evidence of cell damage or alteration due to a flood event. Scars in rings caused by flood debris can indicate both stage and year of event, and changes in ring widths and cell or vessel size (e.g., "flood rings") can indicate timing and, in some cases, the stage of past events (St. George 2010). Such features are site specific, and flood scars are more likely to be found in small, high-gradient streams, but given the right kind of channel conditions, systematic field studies of alluvial stratigraphy (especially the detailed analysis of slack-water deposits) can provide excellent evidence of paleoflood stages throughout a watershed. Although the dating resolution varies when using non-tree-ring-based paleostage indicators to assess past flooding, with dating control and other supporting evidence, systematic field studies of paleoflood evidence can be combined with gaged data to augment records of naturally occurring extreme floods and serve as benchmarks for future flooding scenarios, including defining the upper bounds of physically plausible flood occurrence in a region (Enzel et al. 1993). Comprehensive overviews of various paleoflood approaches can be found in House et al. (2002) and Ballesteros-Cánovas et al. (2015).

Paleofloods and Flood Hazard Management. Global recognition of the importance of paleoflood information for flood risk assessment is increasing (Benito et al. 2004), although the integration of paleoflood methods into management practices has been hindered by policies that rely on standardized risk-assessment approaches (e.g., use of the *100-year flood*). To date, in the United States, the operational use of paleoflood information has generally been limited to dam safety applications that address flood hazards of extreme magnitude (US Bureau of Reclamation 1999; Levish 2002; Swain et al. 2006; Raff 2013). The need to provide flood hazard estimates for extremely large events has spawned innovative approaches that combine hydrometeorology, flood hydrology, and paleoflood hydrology to integrate physically based runoff modeling, stochastic and observed storm information, gaged streamflow, and paleoflood data (England et al. 2014). Other studies have used well-established hydraulic modeling and innovative flood-frequency analysis techniques applied to paleofloods to extend the gaged record, compute recurrence intervals, and estimate the magnitude and frequency of floods. An example of this latter approach for the western United States can be found in the work of Greenbaum et al. (2014), who used a 2000-year record of paleofloods to augment the systematic gaged flood record of the Upper Colorado River near Moab, Utah, and found that the gaged record "greatly underestimates the frequencies of extreme floods on a river that is critical to the water security of the nation."

SRP in view of their reservoirs' responsiveness to high flows; a single extreme wet year during a prolonged drought could refill a reservoir. The 676-year tree-ring record indicated that 3 years was the median interval between one or more drought-relieving years (defined as flow above the 75th percentile). The longest interval without relief was 22 years (1382–1403). In contrast, during the more recent 1913–2005 reconstructed record, the longest interval was 12 years. This length of interval occurred twice: 1953–1964 and 1993–2004. In sum, the message for SRP from these findings was that the drought they were currently managing was by no means unique in the context of the long-term tree-ring record.

Since the completion of the projects just described, SRP has made direct use of the SVT reconstruction from SRP II to quantitatively assess the robustness of the water supply system to droughts more severe than those of the instrumental period. The analysis tool for the assessment is SRP's storage planning diagram (SPD), a graph showing the relation between total reservoir storage, inflow, groundwater pumping production as a percentage of demand, and water allocation (Phillips et al. 2009). Applying this graph, SRP concluded that the unusually persistent 11-year tree-ring drought of 1575–1585, with an average flow of just 70% of the historical median, would severely stress the system under current operating rules but that the drought would be manageable with feasible changes to the allocation and pumping guidelines. Accordingly, in 2006, based on the tree-ring record, SRP's operational planning horizon was increased from 7 to 11 years. This appreciation for the long-term variability in the past has set the stage for additional adjustments in operating guidelines that may be necessary due to future climate changes.

9.5 Conclusions

Paleoclimate information from tree rings can assist the water management and policy community in planning for future risk—including that due to anthropogenic climate change—by providing a better understanding of natural climate variability on long time-scales. Knowledge of the range of natural variability possible is a useful baseline for planning, which also should include projected impacts of climate change. While the potential benefits of using tree-ring data in water resource management are widely appreciated, the actual application of these data to management has required considerable efforts on the parts of both researchers and water resource practitioners. In the case studies described in this chapter, the application of tree-ring data by agencies and institutions was facilitated by a number of factors. Water management organizations had sufficient technical capacity to engage with the new information, institutional prerogatives encouraged that engagement, and individuals within these organizations were willing and able to lead the effort to incorporate a new kind of data into their planning and management toolboxes. Researchers were interested in working outside the regular channels of academia, had the flexibility to do so, and were motivated by the potential application of research results into decision making. On both sides, and in all cases, the commitment of adequate time was also a factor, as the projects evolved in an iterative manner, built on multiple meetings over a period of several years.

Over time, the use of tree-ring data has made a difference in the way water resource planning in the western United States is framed. A paradigm shift has occurred, from relying on the gage record for planning and worst-case scenarios to considering the deeper past as documented in the tree-ring reconstructions. Understanding that natural climate variability includes a broader range of variability than contained in the twentieth- and twenty-first-century gage records has been an important step in beginning to plan more comprehensively for the future. The use of tree-ring data may also have opened the door to considering other types of data, including projections of future climate from GCMs.

Looking ahead, there is great potential to expand the use of tree-ring reconstructions to plan for and anticipate hydroclimatic events. New research methods combine tree-ring data and climate projections to assess future risks of extreme events (e.g., Ault et al. 2014). Other measures of climate important to water resource management such as temperatures, soil moisture, snowpack, and seasonal precipitation have been reconstructed or have the potential to be reconstructed, and work is underway to incorporate this information into management questions. Continued collaborations between researchers and water resource practitioners will be necessary to find ways to integrate relevant and useful results into water resource management, and to explore new uses, roles, and applications of these data to address important water resource issues.

References

Ault, T., J. Cole, J.T. Overpeck, G. Pederson, and D.M. Meko. 2014. Assessing the risk of persistent drought using climate model simulations and paleoclimate data. *Journal of Climate*, 27:7529–7549, doi:10.1175/JCLI-D-12-00282.1.

Baker, V.R. 1987. Paleoflood hydrology and extraordinary flood events. *Journal of Hydrology* 96:79–99.

Baker, V.R. 2008. Paleoflood hydrology: Origin, progress, prospects. *Geomorphology* 101:1–13.

Ballesteros-Cánovas, J.A., M. Stoffel, S. St George, and K. Hirschboeck. 2015. A review of flood records from tree rings. *Progress in Physical Geography* 31:794–816.

Benito, G.B., M. Lang, M. Barriendos, M.C. Llasat, F. Francés, T. Ouarda, V.R. Thorndycraft, Y. Enzel, A. Bardossy, D. Coeur, and B. Bobée. 2004. Use of systematic, paleoflood and historical data for the improvement of flood risk estimation. *Review of Scientific Methods: Natural Hazards* 31:523–643.

California Department of Water Resources (CADWR). 2014. Severity of Past Droughts Quantified by New Streamflow Reconstructions, March 10, 2014. http://www.water.ca.gov/news/newsreleases/2014/031014.pdf (accessed September 15, 2014).

Cayan, D.R., T. Das, D.W. Pierce, T.P. Barnett, M. Tyree, and A. Gershunov. 2010. Future dryness in the southwest U.S. and the hydrology of the early 21st century drought. *Proceedings of the National Academies of Science* USA. 107: 21271–21276, www.pnas.org/cgi/doi/10.1073/pnas.0912391107.

Earle, C.J. and H.C. Fritts. 1986. Reconstructed riverflow in the Sacramento River basin since 1560. Report to California Department of Water Resources, Agreement No. DWR B55398, Laboratory of Tree-Ring Research, University of Arizona, Tucson, Arizona.

England, Jr., J.F., P.Y. Julien, and M.L. Velleux. 2014. Physically-based extreme flood frequency with stochastic storm transposition and paleoflood data on large watersheds. *Journal of Hydrology* 510:228–245.

Enzel, Y., L.L. Ely, P.K. House, V.R. Baker, and R.H. Webb. 1993. Paleoflood evidence for a natural upper bound to flood magnitudes in the Colorado River Basin. *Water Resources Research* 29:287–229.

Ferguson, D.B., J. Rice, and C.A. Woodhouse. 2014. Linking Environmental Research and Practice: Lessons from the Integration of Climate Science and Water Management in the Western United States. Climate Assessment for the Southwest, University of Arizona. http://www.climas.arizona.edu/sites/default/files/pdflink-res-prac-2014-final.pdf (accessed September 7, 2014).

Fritts, H.C. 1976. *Tree Rings and Climate*. London: Academic Press.

Gangopadhyay, S., B.L. Harding, B. Rajagopalan, J.J. Lukas, and T.J. Fulp. 2009. A non-parametric approach for paleo reconstruction of annual streamflow ensembles. *Water Resources Research* 45: W06417, doi:10.1029/2008WR00720.

Gray, S.T., J.L. Betancourt, C.L. Fastie, and S.T. Jackson. 2003. Patterns and sources of multidecadal oscillations in drought-sensitive tree-ring records from the central and southern Rocky Mountains. *Geophysical Research Letters* 30:1316, doi:10.1029/2002GL016154.

Greenbaum, N., T. Harden, V.R. Baker, J. Weisheit, M.L. Cline, N. Porat, R. Halevi, and J. Dohrenwend. 2014. A 2000 year natural record of magnitudes and frequencies for the largest Upper Colorado River floods near Moab, Utah. *Water Resources Research* 50:5249–5269.

Harding, B. 2005. SSD redux? Comparison to a historic drought. *Southwest Hydrology* 4:24–25.

Hidalgo, H.G., T.C. Piechota, and J.A. Dracup. 2000. Alternative principal components regression procedures for dendrohydrologic reconstructions. *Water Resources Research* 36(11):3241–3249.

Hirschboeck, K.K. and D.M. Meko. 2005. A tree-ring based assessment of synchronous extreme streamflow episodes in the Upper Colorado & Salt–Verde–Tonto River Basins. http://srp.ltrr.arizona.edu/SRP/SRP-I-Final-Report-2005.pdf (accessed July 30, 2014).

House, P.K., R.H. Webb, V.R. Baker, and D. Levish (eds.). 2002. Ancient floods, modern hazards: Principles and applications of paleoflood hydrology. *Water Science and Application*, Vol. 5, American Geophysical Union.

Kenney, D., A. Ray, B. Harding, R. Pulwarty, and B. Udall. 2010. Rethinking vulnerability on the Colorado River. *Journal of Contemporary Water Research and Education* 144:5–10.

Levish, D.R. 2002. Paleohydrologic bounds—Non-exceedance information for flood hazard assessment. In *Ancient Floods, Modern Hazards: Principles and Applications of Paleoflood Hydrology*, Water Sci. and Appl. Ser., Vol. 5, P.K. House et al. (eds.), pp. 175–190, Washington, D.C.: AGU.

Meko, D.M. 1997. Dendroclimatic reconstruction with time varying subsets of tree indices. *Journal of Climate* 10:687–696.

Meko, D.M. and K.K. Hirschboeck. 2008. The current drought in context: A tree-ring based evaluation of water supply variability for the Salt–Verde River Basin. http://srp.ltrr.arizona.edu/SRP/SRP-II-Final-Report-2008.pdf (accessed July 30, 2014).

Meko, D.M. and C.A. Woodhouse. 2011. Dendroclimatology, dendrohydrology, and water resources management. In *Tree Rings and Climate: Progress and Prospects*, M.K. Hughes, T.W. Swetnam, and H.F. Diaz (eds.), pp. 231–261. New York: Springer.

Meko, D.M., C.W. Stockton, and W.R. Boggess. 1995. The tree-ring record of severe sustained drought. *Water Resources Bulletin* 31:789–801.

Meko, D.M., C.A. Woodhouse, C.H. Baisan, T. Knight, J.J. Lukas, M.K. Hughes, and M.W. Salzer. 2007. Medieval drought in the upper Colorado River basin. *Geophysical Research Letters* 34m L10705, doi:10.1029/2007GL029988.

Meko, D.M., C.A. Woodhouse, and K. Morino. 2012. Dendrochronology and links to streamflow. *Journal of Hydrology* 412–413:200–2009, doi:10.1016/j.jhydrol.2010.11.041.

Meko, D.M., C.A. Woodhouse, and R. Touchan. 2014. Klamath/San Joaquin/Sacramento Hydroclimatic Reconstructions from Tree Rings. Final Report to California Department of Water Resources Agreement 4600008850. February 2014, http://www.water.ca.gov/waterconditions/docs/tree_ring_report_for_web.pdf (accessed July 30, 2014).

Milly, P.C.D., J. Betancourt, M. Falkenmark, R.M. Hirsch, Z.W. Kundzewicz, D.P. Lettenmaier, and R.J. Stouffer. 2008. Stationarity is dead: Whither water management? *Science* 319:573–574, doi:10.1126/science.1151915.

Nowak, K., J. Prairie, B. Rajagopalan, and U. Lall. 2010. A nonparametric stochastic approach for multisite disaggregation of annual to daily streamflow. *Water Resources Research* 46, doi: 10.1029/2009WR008530.

Phillips, D.H., Y. Reinink, T.E. Skarupa, C.E. Ester, III, and J.A. Skindlov. 2009. Water resources planning and management at the Salt River Project, Arizona, USA. *Irrigation and Drainage Systems* doi:10.1007/s10795-009-9063-0.

Prairie, J.R. 2006. Stochastic nonparametric framework for basin wide streamflow and salinity modeling: Application for the Colorado River Basin. Civil Environmental and Architectural Engineering Ph.D. Dissertation. University of Colorado. Boulder, Colorado.

Prairie, J., B. Rajagopalan, U. Lall, and T. Fulp. 2007. A stochastic nonparametric technique for space–time disaggregation of streamflows. *Water Resources Research* 43(3), doi: 10.1029/2005WR004721.

Prairie, J., K. Nowak, B. Rajagopalan, U. Lall, and T. Fulp. 2008. A stochastic nonparametric approach for streamflow generation combining observational and paleo reconstructed data. *Water Resources Research* 44:W06423, doi:10.1029/2007WR006684.

Rice, J.L., C.A. Woodhouse, and J.J. Lukas. 2009. Science and decision-making: Water management and tree-ring data in the western United States. *Journal of the American Water Resources Association* 45:1248–1259.

Raff, D. 2013. Appropriate application of paleoflood information for the hydrology and hydraulics decisions of the U.S. Army Corps of Engineers, Report CWTS 2013-2, 45 p., US Army Corp. of Eng.

Saito, L., F. Biondi, J.D. Salas, A.K. Panorska, and T.J. Kozubowski. 2008. A watershed modeling approach to streamflow reconstruction from tree-ring records. *Environmental Research Letters* 3, doi:10.1088/1748-9326/3/2/024006.

Smith, J.B., K. Strzepek, L. Rozaklis, C. Ellinghouse, and K.C. Hallett. 2009. The potential consequences of climate change for Boulder Colorado's water supplies. Stratus Consulting, for the NOAA Climate Program Office, Boulder, CO.

Salt River Project (SRP). 2014. Facts about SRP. http://www.srpnet.com/about/facts.aspx (accessed July 23, 2014).

St. George, S. 2010. Tree rings as paleoflood and palseostage indicators. In *Tree Rings and Natural Hazards: A State-of-the-Art, Advances in Global Change Research* 41, M. Stoffel et al. (eds.), pp. 233–239. New York: Springer.

St. George, S., D.M. Meko, and E.R. Cook. 2010. The seasonality of precipitation signal embedded within the North American Drought Atlas. *The Holocene* 20:983–988, doi:10.1177/0959683610365937.

Stockton, C.W. and G.C. Jacoby. 1976. Long-Term Surface-Water Supply and Streamflow Trends in the Upper Colorado River Basin. Lake Powell Research Project Bulletin No. 18, Institute of Geophysics and Planetary Physics, University of California at Los Angeles.

Stokes, M.A. and Smiley, T.L. 1968. *An Introduction to Tree-Ring Dating*. Chicago: University of Chicago Press.

Swain, R.E., J.F. England Jr., K.L. Bullard, and D.A. Raff. 2006. *Guidelines for Evaluating Hydrologic Hazards*. Bureau of Reclamation, Denver, CO, 83 pp.

Trenberth, K.E., A. Dai, G. van der Schrier, P.D. Jones, J. Barichivich, K.R. Briffa, and J. Sheffield. 2013. Global warming and changes in drought. *Nature Climate Change* 4:17–22, doi:10.1038/nclimate2067.

US Bureau of Reclamation. 1999. A framework for characterizing extreme floods for dam safety risk assessment. Prepared by Utah State University and Bureau of Reclamation, Denver, CO, 67 pp.

US Bureau of Reclamation. 2007. Colorado River Interim Guidelines for Lower Basin Shortages and Coordinated Operations for Lake Powell and Lake Mead, http://www.usbr.gov/lc/region/programs/strategies/FEIS/index.html (accessed July 30, 2014).

US Bureau of Reclamation. 2012. Colorado River Basin Water Supply and Demand Study. Final Report, December 2012. Executive Summary: http://www.usbr.gov/lc/region/programs/crbstudy/finalreport/Executive%20Summary/CRBS_Executive_Summary_FINAL.pdf (accessed July 30, 2014).

Wang, S.-Y., L. Hipps, R.G. Gillies, and J.-H. Yoon. 2014. Probable causes of the abnormal ridge accompanying the 2013–2014 California drought: ENSO precursor and anthropogenic warming footprint. *Geophysical Research Letters* 41:3220–3226, doi:10.1002/2014GL059748.

Weiss, J.L., C.L. Castro, and J.T. Overpeck. 2009. Distinguishing pronounced droughts in the southwestern United States: Seasonality and effects of warmer temperatures. *Journal of Climate* 22: 5918–5932.

Woodhouse, C.A. and J.J. Lukas. 2006a. Drought, tree rings, and water resource management. *Canadian Water Resources Journal* 31:297–310.

Woodhouse, C.A. and J.J. Lukas. 2006b. Multi-century tree-ring reconstructions of Colorado streamflow for water resource planning. *Climatic Change* 78:293–315, doi:10.1007/s10584-006-9055-0.

Woodhouse, C.A., S.T. Gray, and D.M. Meko. 2006. Updated streamflow reconstructions for the upper Colorado River basin. *Water Resources Research* 42:W05415, doi:10.1029/2005WR004455.

Section III

Case Studies: Regional Issues and Insights on Adaptation Pathways

10

The Columbia River Treaty and the Dynamics of Transboundary Water Negotiations in a Changing Environment: How Might Climate Change Alter the Game?

Barbara Cosens, Alexander Fremier, Nigel Bankes, and John Abatzoglou

CONTENTS

ABSTRACT Review of the Columbia River Treaty between the United States and Canada presents an opportunity to consider how governance might be made more adaptive in the face of climate change while maintaining an appropriate level of stability. Climate change scenarios applied to the Columbia River Basin predict an increasing water deficit due primarily to change in timing of runoff and increased vegetative demand as the result of warming. Current users that will suffer the most from change in timing of runoff are those dependent on late summer flow—fish and farmers. The intersection of climate change with normal climate variability also suggests greater extremes that will require planning for both drought and flood beyond the historic recurrence and magnitude. A problem-solving approach to bridge the gap between the status quo and a modernized system must be combined with new approaches to governance that are both more flexible and more responsive to change if the basin is to navigate the future. This challenges not only the conventional wisdom that regulatory stability is essential to economic stability and achievement of societal goals, but also the existing distribution of benefits among powerful players in the basin. Yet the goals of even the current recipients of benefits will not be met if a rigid approach is maintained as climate change unfolds. Adaptive water governance requires attention to institutional structure, introduction of adaptive authority and local participatory capacity including knowledge building, and process design that balances stability with flexibility. Models for implementation of these factors are found in international law and provide a pathway to adaptation to climate change through a modernized transboundary water agreement on the Columbia River.

10.1 Introduction

This chapter uses the review of the 1964 Columbia River Treaty (CRT) between the United States and Canada to analyze how process and the use of mechanisms for flexibility in water management may aid in bridging gaps between divergent opinions on transboundary water management as well as in preparing the basin to adapt to an uncertain future. The CRT was negotiated to achieve shared benefits from hydropower production and flood control with tightly bounded means to adjust to change in water supply through technical operational changes. It has been held throughout the world as the pinnacle of international cooperation on freshwater sources because of its approach to shared benefits and cooperative management of certain river infrastructure located in both countries (Barton and Ketchum 2012). However, the assured flood control provisions of the treaty expire in 2024. This change, combined with the right of either country to terminate the power provisions of the treaty by giving 10 years' notice any time after September 2014, has created an opportunity to review the treaty. While ostensibly triggered by these concerns, changes in the basin's social–ecological system since 1964 and increasing recognition of the impact of climate change have substantially broadened the scope of review.

Negotiators of transboundary agreements governing water have historically sought clear lines between the rights and obligations of nation-states on shared watercourses to assure limited interference with control inside sovereign borders and to provide stability for water-based economic growth. Climate change is not only affecting the water balance by altering vegetative demand, supply, and precipitation patterns including increasing frequency, duration, and degree of extreme conditions of flood and drought, but also, substantial uncertainty exists in how it will play out. The need to adapt to a nonstationary future has led to a search for an approach to water governance that replaces the policy of reliance on technical means to constrain water supply variability with one that accounts for uncertainty while balancing the need for economic stability—i.e., adaptive water governance. The high level of cooperation already practiced by the United States and Canada in transboundary water management has raised hopes that the window of opportunity provided by the CRT review may allow the basin to take a leading role in the negotiation and implementation of a modern and flexible form of water governance.

By the time of this publication, the United States and Canada may have chosen their course on the Columbia River. Regardless of their choice, this case study is intended to shed light more broadly on the law and policy aspects of a major transboundary water negotiation and the potential for adaptive governance mechanisms to provide management flexibility in the face of climate change to bridge the gap between a stationary past and a nonstationary future. We believe these lessons will be useful whether or not these issues are addressed in the current round of review of transboundary management of the Columbia River.

This chapter begins with an introduction to the river and the context and substance of the CRT. It then examines the changes affecting the basin since 1964 and the predicted impacts of climate change. Finally it introduces the concept of adaptive water governance and describes models for flexibility in international water law in the context of the current review of the CRT. We begin with the river.

10.2 The Columbia River and the United States–Canada Treaty

To understand the river and the role of governance, the following sections provide a view of both the physical system and those aspects of historic development that have legacy effects today.

10.2.1 The River

The Columbia River Basin covers 259,500 mi.^{2} (672,100 km^2),* with 15% in Canada and the remainder in the United States (Barton and Ketchum 2012) (Figure 10.1), but due to the high snowpack, Canada contributes 38% of the average annual flow and 50% of the peak flow measured at The Dalles (located on the mains tem between Oregon and Washington) (Shurts 2012). The river's headwaters are in the Rocky Mountains, with high snowpack serving as natural storage and its rate of runoff playing a key role in flood risk. Although the average annual flow is 200 million acre-feet (247 BCM)† at the mouth, the basin experiences considerable seasonal variability, with the ratio between unregulated low and high season flow as much as 1:34 (Barton and Ketchum 2012; Hamlet 2003).

Human contact with the Columbia River Basin occurred at least 9000 years ago [*Bonnichsen v. U.S.*, 367 F.3d 864 ([9th Cir. 2004]). Indigenous communities remained the dominant society until the mid-1800s. These communities relied on salmon runs as a primary protein source, and salmon were a critical aspect of cultural life (Josephy 1997; Deur 1999; Barber 2006; Landeen and Pinkham 2008; Pearson 2012; Cosens and Fremier 2014).

The transition to a dominant Euro-American society spans the mid- to late 1800s. European settlement and development in the region transformed the upland arable land to monoculture and exploited the salmon fishery for commercial canneries. Indigenous communities experienced a dramatic decline in health and were moved onto reservations much smaller than their aboriginal territory (*United States v. Washington*, 384 F.Supp. 312 [D.Wash., 1974], aff'd by 520 F.2d 676 [9th Cir. 1975], cert denied *Washington v. U.S.*, 423 U.S. 1086 [1976])

* 1 square mile (mi.^{2}) = 2.59 square kilometers (km^2).

† 1 million acre-feet (MAF) = 1.233 billion cubic meters (BCM).

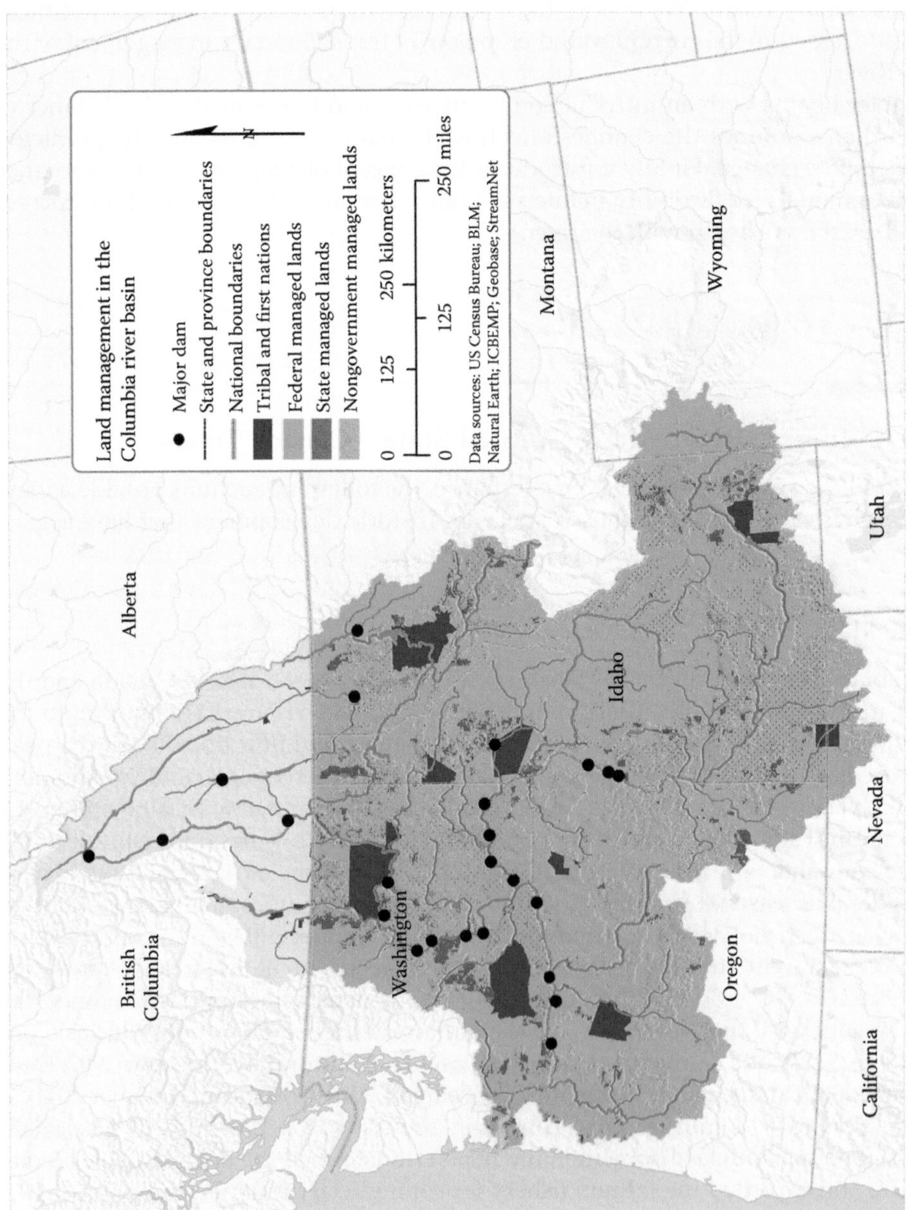

FIGURE 10.1
Columbia River Basin showing land management in the United States portion of the basin. BLM, U.S. Bureau of Land Management; ICBEMP, Interior Columbia Basin Ecosystem Management Project.

(White 1995; Josephy 1997; Goble 1999; Fiege 1999; Harrison 2012; Cosens and Fremier 2014). The treaties negotiated during this period play a major role in the renewed voice of indigenous governments in the US review of the CRT. On the Canadian side of the boundary, the province of British Columbia historically resisted the negotiation of treaties with the First Nations and refused to recognize aboriginal title (Cail 1974). Small reserves were set aside by executive act. The legal position of First Nations has changed dramatically in the last few decades with the constitutional recognition of aboriginal and treaty rights in 1982 (Constitution Act, 1982, s.35) and the judicial recognition of aboriginal title (*Delgamuukw v. British Columbia*, 1997, 3 S.C.R. 1010; *Tsilhqot'in Nation v. British Columbia*, 2014 S.C.C. 44).

10.2.2 Engineering the River

Engineered transformation of the river began with construction of locks for navigation in the late 1800s, followed by a combination of public works projects during the Great Depression and federal reclamation projects for irrigation. This period saw a dramatic decline in the salmon population, leading to the development of over 200 hatcheries (Cosens and Fremier 2014; White 1995; Mouat 2012; Hirt and Sowards 2012; USBR Grand Coulee Dam, no date; USBR Columbia Basin Project, no date; Bonneville Power Administration History, no date; Reisner 1987; Billington et al. 2005; Barton and Ketchum 2012; Fiege 1999; White 2012; Shurts 2012; Peery 2012; Foundation for Water and Energy Education, no date). The major federal investment in dams fueled growth and economic stability within the basin, but it was recognized that further control of the river required partnership with Canada.

The International Joint Commission formed by the 1909 Boundary Waters Treaty between the United States and Canada was directed to study the possibility of storage within Canada to provide benefits to both countries (Mouat 2012; Shurts 2012). This process was accelerated when, in 1948, a slightly higher-than-average snowpack met a rapid warming event. The resulting flood peaked at an estimated flow of >1 million ft.^{3} per second (28,317 m^3/s) at the confluence with the Willamette River (average peak flows are less than half that rate) (Barton and Ketchum 2012), causing extensive damage to property and destroying the town of Vanport, Oregon. At the time, storage capacity on the Columbia River was approximately 6% of the average annual flow (White 2012). With the best remaining water storage sites located in Canada or requiring territory in Canada (e.g., Libby Dam), serious negotiations began.

10.2.3 The Columbia River Treaty

The CRT between the United States and Canada entered into force on September 16, 1964, for the purpose of developing the river for hydropower and flood control. Dams built in Canada under the CRT release water on an agreed-upon schedule and the incremental increase in hydropower production in the United States are divided equally between the United States and Canada. The CRT provided for appointment of operating entities by the United States and British Columbia (BC), and both selected technical managers rather than political appointees (Exec. Order No. 11,177, 29 Fed. Reg. 13097, September 16, 1964) (Barton and Ketchum 2012), a common approach in the mid-twentieth century (Bankes and Cosens 2014). The entities adjust dam operation to changes in snowpack and runoff, and have reached supplemental agreements to accommodate environmental values when mutual benefits were possible (Bankes and Cosens 2012).

Under the terms of an agreement between BC and the Canadian government, both the benefits and control of dams under the CRT lie with the province. Referred to as the

Canadian Entitlement, the hydropower benefits are sold back to US utilities by BC and have resulted in a substantial revenue stream. To make this sale possible, Congress authorized construction of the Pacific Northwest–Pacific Southwest Intertie, allowing transmission of power to utilities in the southwestern United States, with a preference for sale to Northwest utilities (Pacific Northwest Consumer Power Preference Act, 16 U.S.C. § 837. 2006). This contributed to an interconnected North American power grid and reliance on the capacity of the basin's hydropower outside the basin.

With the prospect of assured flood control expiring on September 16, 2024, the United States Entity and the province of BC initiated review processes (US Army Corps of Engineers 2009, 2013; British Columbia 2013, 2014). The US regional review included formation of a Sovereign Review Team advisory body with 5 representatives of the 15 US basin tribes and a representative of each of the 4 main basin states and included multiple avenues for input from the public (US Army Corps of Engineers, no date). The Canadian process was led by the Provincial Ministry of Energy and Mines and included a process of public sessions throughout the basin and multiple avenues for public comment, and recognized the changed status of First Nations by providing separate consultation (British Columbia, no date). To understand why the scope of review expanded well beyond the expiring flood control provisions, it is necessary to examine changes affecting the basin since 1964 and to consider the added implications of climate change.

10.3 Changes in the Columbia River Social–Ecological System

Over the life of the CRT, change affecting the basin did not unfold as predicted. The following sections describe changes that affect treaty purposes, changes that have led to a broader scope of review, changes that impact the adaptability of basin water management, and the predicted results of climate change that are already being felt in the basin.

10.3.1 Changes Affecting the Purposes of the Columbia River Treaty

Energy markets have not evolved as anticipated in 1964, and continue to challenge forecasting. Planners expected the rapid growth in power demand following World War II to continue, and new thermal generation was expected to replace hydropower as the dominant source of energy in the Pacific Northwest (White 1995; Shurts 2012). Conservation nationwide in the wake of the 1973 oil embargo eliminated this need and continues to slow demand growth, with hydropower remaining the dominant energy source in the region (Hirt and Sowards 2012; Northwest Power and Conservation Council 2010). Correspondingly, the system's value has grown. The result is that the Canadian Entitlement is much greater than either party anticipated in the 1960s, ranging in value as high as $350 million/year in recent years although currently lower due to the impact on energy prices from increased production of natural gas. Currently, the need to balance rapid fluctuation in supply by new wind generation is taxing the flexibility of the hydropower system (Northwest Power and Conservation Council 2013).

10.3.2 Changes Leading to a Broadening in the Scope of Treaty Review

The rising expectation of public involvement in decision making in democratic societies globally and empowerment of formerly marginalized communities within the basin since

1964 led to a demand for a broader CRT review. New US domestic laws since 1964 parallel the global increase in the demand for public access to information and participation in governmental decision making (Hirt and Sowards 2012). In the United States, the indications of this trend began with the passage of the Freedom of Information Act in 1966 (5 U.S.C. § 552. 2006) and the National Environmental Policy Act in 1970 (Pub. L. No. 91-190, 83 Stat. 852. 1970; current version at 42 U.S.C. § 4321. 2006) (Hirt and Sowards 2012). Developments in Canada followed this same track with the adoption of federal environmental assessment procedures in 1977 (revised 1984 as the Environmental Assessment and Review Process Guidelines Order, SOR/84-467) and endangered species legislation in 2002 (Species at Risk Act, SC 2002, c. 29) (Boyd 2003). Among the key issues identified by stakeholders in the basin during the CRT review process is the desire for public input in both negotiation and implementation of transboundary water management (McKinney et al. 2010).

Complementing the demand for participation, the capacity of local communities and Native American and First Nation governments to participate in basin management has increased. This has been accomplished through litigation to establish rights, and legislation recognizing rights and delegating authority as follows:

1. Legal recognition of indigenous treaty fishing rights ultimately led to capacity building that elevated certain of the tribes, organized as the Columbia River Inter-Tribal Fish Commission, to status as comanagers of the basin's fisheries (*United States v. Washington*, 384 F.Supp. 312 [D.Wash., 1974], aff'd by 520 F.2d 676 [9th Cir. 1975], cert denied *Washington v. U.S.*, 423 U.S. 1086 [1976]; *Washington v. Washington State Commercial Passenger Fishing Vessel Association*, 443 U.S. 658 [1979]) (Columbia River Inter-Tribal Fish Commission, no date; Cosens 2012). Upper-basin tribes within the United States whose land was blocked from anadromous fish migration organized as the Upper Columbia United Tribes (UCUT, no date). In 2005, UCUT and its member tribes entered a memorandum of understanding with Bonneville Power Administration recognizing the sovereign role of the tribes in management of, among other things, fish and water resources (UCUT 2005). Although Native American tribes had no voice in negotiation of the CRT, with this increased capacity, all 15 tribes in the US portion of the basin came together to develop a common position in the CRT review calling for recognition of cultural and ecological values and the sovereign rights of tribes (Columbia Basin Tribes 2010). The 15 tribes joined together to select five representatives to the Sovereign Review Team invited to advise the US regional review process (US Army Corps of Engineers, no date).
2. The Northwest Power and Conservation Council (NWPCC), composed of state representatives in the United States portion of the basin, was authorized by Congress in 1980 and established by interstate compact (Pacific Northwest Electric Power Planning and Conservation Act, Pub. L. No. 96-501, 94 Stat. 2697). The council is charged with energy and fisheries restoration planning, increasing the capacity of the states in these areas. The council has played a major role in education of the public and has funded informal transboundary dialogues facilitated by representatives of public universities in the basin during the review (UCCRG, no date). In addition, one member of the council from each state was also chosen to serve on the Sovereign Review Team (US Army Corps of Engineers, no date).
3. The rights of First Nations in Canada were afforded constitutional protection in 1982. Since then, a series of court decisions have confirmed that provincial and federal governments must consult and accommodate First Nations when developing

policies or making decisions that may have an effect on aboriginal or treaty rights (*Haida Nation v. British Columbia* [*Minister of Forests*], 2004, 3 S.C.R. 511). This duty applies when considering possible changes to the CRT.

4. First Nations in Canada formed the Canadian Columbia River Intertribal Fisheries Commission (CCRIFIC) in 1993. Along with UCUT, First Nations seek reintroduction of salmon to the river main stem in Canada (Columbia Basin Tribes and First Nations 2015).
5. The Columbia Basin Trust (CBT), initiated through a grassroots effort to assert the rights of communities and First Nations whose lands were flooded by treaty dams, was formally recognized by BC in 1995 (Cosens 2010, 2012). The CBT receives hydropower revenue and is charged with providing education on water and economic development in the basin, and CBT activities include community planning for adaptation to climate change and education on the CRT (CBT, no date). The CBT has played a major role in education during the review process and joined the NWPCC in funding the university-led transboundary dialogues.

These changes provide the local capacity base to make participation by the basin citizenry a reality. Participation of these emerging voices in the CRT review has highlighted the impact of optimization of hydropower and flood control on basin resilience and broadened the scope of review.

10.3.3 Changes in Basin Resilience

Dramatic change in the status of the basin ecosystem is reflected in the decline of populations of anadromous fish. Non-CRT dams block anadromous fish from 37% of their former spawning grounds, including all of the river main stem in Canada, and even where fish passage is provided, slack water behind dams slows out-migration, reducing juvenile survival (Columbia River Inter-Tribal Fish Commission 1995). Salmon runs, estimated at between 6 and 16 million in the early 1880s (Hirt 2008; Landeen and Pinkham 2008; Peery 2012), are closer to 1 million today, with 80–90% estimated to be from hatchery production (National Research Council 1996; Peery 2012). Twentieth-century decision makers knew dams would impact salmon (Bottom et al. 2009), but the value placed on that resource has changed.

New laws addressing the substance of environmental management, including the US Endangered Species Act (ESA), adopted in 1973, which forbids federal actions that jeopardize listed species (16 U.S.C. §§ 1531–1544. 2006), show increased diversity in the ecosystem services society values. Eight salmon and four steelhead species that rely on habitat within the basin have been listed in the United States.* Although numerous factors affect these

* Current listings of salmon species found in the Columbia Basin: Snake River sockeye (endangered), Upper Willamette River chinook (threatened), Lower Columbia River chinook (threatened), Upper Columbia River spring-run chinook (endangered), Snake River fall-run chinook (threatened), Snake River spring/summer-run chinook (threatened), Lower Columbia River coho (threatened), Columbia River Chum (threatened). Final listing determinations for 16 ESUs of West Coast salmon, 70 Fed. Reg. 37160, 37193 (June 28, 2005). Note that four Evolutionarily Significant Units (ESU) of steelhead are also currently listed: 69 Fed. Reg. 33105 (June 14, 2004) and 71 Fed. Reg 5178 (February 1, 2006). However, these listings are currently in litigation. See, e.g., *Trout Unlimited v. Lohn*, No. CV06-0483-JCC 2007 WL 1795036 (W.D. Wash. June 13, 2007), *aff'd in part, rev'd in part* 559 F.3d 946 (9th Cir. 2009); see also West Coast NOAA Fisheries, *Status of ESA Listings & Critical Habitat Designations for West Coast Salmon & Steelhead*, http://www.westcoast.fisheries.noaa.gov/publications/protected_species/salmon_steelhead/status_of_esa_salmon_listings_and_ch_designations_map.pdf.

species, operation of the Federal Columbia River Power System (the part of the hydropower system at federal dams in the US portion of the basin) has been the subject of numerous lawsuits under the ESA (e.g., *National Wildlife Federation v. National Marine Fisheries Service*, 524 F.3d 917 [9th Cir. 2008]). The need to comply with the ESA has led the US Entity to negotiate supplemental operating agreements with BC to alter flows (Bankes and Cosens 2012). The result is that the actual production of hydropower from the US dams is less than the calculated production used to determine the Canadian Entitlement (Shurts 2012). This combined with the unexpected continued high value of hydropower is reflected in the call for reduction in the Canadian Entitlement in the US review (US Army Corps of Engineers 2013).

The decline of the iconic salmon populations is just one lens through which the impact of optimization for hydropower and flood control may be viewed. Optimization reduces the space available for adaptation (Walker and Salt 2006); thus, a river developed to its maximum capacity for any specific societal goal will be vulnerable to change in water provisioning and/or demand. This brings us to the overarching factor affecting the basin's water supply—climate change.

10.3.4 Potential Impacts of Climate Change on the Columbia River Basin

Engineered storage capacity in the basin is now 40% of the average annual flow (in comparison, the Colorado River has storage capacity for 400–500% of the average annual flow) (Barton and Ketchum 2012). Thus, the water management of the basin is heavily dependent on snowpack storage. Most of the region's precipitation falls from October to March at higher elevations, where temperatures have been conducive to snowpack development (e.g., Mote et al. 2005). Spring warming begets snowmelt that synchronizes well with dry season and increased water demands of the system. Flexibility in operational planning envisioned by the CRT depends on seasonal and yearly variation that can be forecast within the degrees of historical variability (Barton and Ketchum 2012). Unfortunately, climate change predictions may exceed the historical range of variability (Hamlet 2003).

Widespread observations already show decline in mountain snowpack (e.g., Mote et al. 2005), advancement in the timing of runoff (e.g., Stewart et al. 2005; Hamlet et al. 2007), and reduction in annual streamflow in the driest years (Luce and Holden 2009) since 1950. While these changes may be regionally linked to decline in mountain precipitation (e.g., Luce et al. 2013), notable warming in winter and spring can reduce snow volume and shift the timing of runoff to earlier in the year. While natural climate variability exacerbates and masks changes tied to increased concentrations of atmospheric greenhouse gases (Abatzoglou et al. 2014), many of these changes appear attributable to long-term climate change (e.g., Pierce et al. 2008).

Projected changes in climate for the basin include a nominal change in annual precipitation, though with more expected to fall in the cool season and less in the warm season (Dalton et al. 2013). However, the projection of temperature change is robust and suggests widespread warming of 2–4°C by the mid-twenty-first century (Dalton et al. 2013) that will dramatically accelerate the transition from snow- to rain-dominated watersheds (Elsner et al. 2010; Klos et al. 2014). The consequences of this shift in the dominant form of precipitation include earlier peak runoff; reduced incidence of flood (as the result of reduced snowpack, particularly in snow-dominated watersheds); lower summer and fall contributions to the natural hydrograph; and higher water temperature (Hamlet et al. 2005). In addition, scientists are beginning to consider secondary impacts of climate change in the basin, including increased demand for summer electric power for air conditioning within

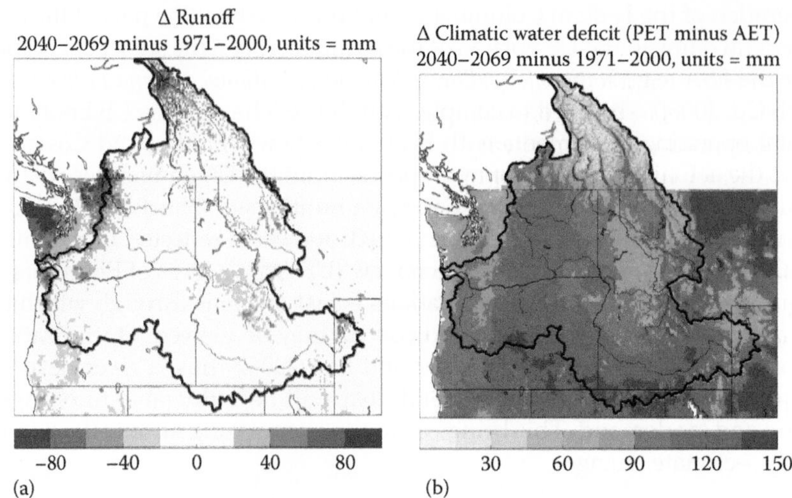

FIGURE 10.2
Projected change in (a) annual runoff and (b) climatic water deficit (right) between mid-twenty-first-century (2040–2069) climate under a high-emissions (RCP 8.5) scenario and late-twentieth-century (1971–2000) climate. The mean change as simulated by twenty climate models is shown only in areas where at least two-thirds of models agreed on the sign of change. Climate model output was downscaled to a 6 km horizontal resolution and then used to drive a simple one-dimensional modified Thornthwaite water balance model. The water balance model uses monthly temperature, precipitation, and potential evapotranspiration (PET), with water holding capacity of soils to estimate monthly soil moisture, actual evapotranspiration (AET), and snow water equivalent. The model estimates runoff as the excess precipitation or snowmelt not used by AET or to recharge soils, and the climatic water deficit as the unmet water demand by vegetation, or PET minus AET. The asymmetric timing of precipitation and PET across much of the western United States yields seasonally contrasting patterns of change in runoff (increasing in winter/spring) and climatic water deficit (dominated by summer drying). Projected warming across the region not only increases PET, primarily through increased vapor pressure deficits, but also advances the timing of snowmelt in mountain watersheds and increases the period over which PET exceeds AET. Collectively, this results in widespread increases in climatic water deficit across the region between 50 and 175 mm per year. Models project some increase in annual runoff in the Olympic Peninsula and northern Cascades as well as the headwaters of the Columbia River in British Columbia, in the Idaho panhandle, and near Glacier National Park primarily due to increases in precipitation during the cool season, when PET plays a nominal role. Farther south as well as in drier watersheds, increases in water lost to PET supersede nominal changes in cool-season precipitation and yield declines in annual runoff. These changes appear most prominent in the Colorado River Basin.

the basin (Hamlet et al. 2010) and increased demand for irrigation due to changes in the growing-season evapotranspiration (Dalton et al. 2013).

Projected changes in annual runoff and climatic water deficit are shown in Figure 10.2. Climatic water deficit refers to the unmet vegetative water demand. Despite projected increases in annual runoff over parts of the Columbia Basin, dominated by winter and early spring flows, models show strong agreement predicting increased water deficit from a combination of increased demand by vegetation in the summer months due to warming and a decline in available water due to advancement in the timing of snowmelt and runoff. These projections are likely to result in increased irrigation demand and contribute to changes in the makeup and vitality of ecosystems across the region. In addition, increased winter and early spring runoff from the Canadian portion of the basin may have implications for future flood control operations.

Changes in solar energy and moisture manifest through moisture availability and climate water deficit (e.g., Figure 10.2) will likely result in redistribution of vegetation. Wildfire and bark beetle outbreaks have both notably increased in their expanse over the past several decades (e.g., Westerling et al. 2006; Meddens et al. 2012). Climate-driven changes in these disturbances will likely continue, causing secondary changes in runoff and erosion rates, and the concomitant impacts to ecosystem services. This feedback creates the potential for cascading effects on water temperature and river flow regime, as well as legacy effects like nutrient retention, which impact aquatic ecosystems (Wegner et al. 2011).

While observed and projected changes in climate are gradual relative to seasonal variability in climate itself, we should not expect a slow and smooth transition. Rather, the interplay between anthropogenic climate change and natural climate variability may provide for a turbulent transition to a warmer climate. For example, superposed warming trends coincident with a protracted drought similar to the one experienced in the 1930s may occur in the coming decades, thereby invoking widespread climate impacts far earlier than anticipated.

In sum, the direct impacts of climate change, the indirect ecologically mediated impacts, and pronounced change in climate variability indicate a system already in transition. Going forward, the types and rates of ecosystem services and the adequacy of engineered facilities will be affected in unpredictable ways.

10.4 Water Policy Changes for Adaptation to Change

The changes since 1964 and predicted changes going forward call for greater flexibility in management, and it is within the review process itself that this need has become apparent.

10.4.1 The Columbia River Treaty Review

In December of 2013, the US Entity forwarded its regional recommendation to the US Department of State calling for continuation of the underlying concept of shared benefits with modernization of the treaty (US Army Corps of Engineers 2013). In March of 2014, the province of BC released its decision on the CT calling for continuation of the treaty with negotiated improvements in the treaty framework (British Columbia 2014).

Common ground between the two positions includes recognition of the need for flexibility and adaptation in the face of uncertainty and the need for more open processes in both negotiation and implementation. The US review recommendation specifically recognizes the need for flexibility to address flood risk management in light of both changes in objectives and change in climate; accommodate intermittent sources such as wind; and account for both climate and legal changes (such as ESA delisting) affecting a proposed third treaty purpose—ecosystem function (a term that is undefined in the document). The BC decision is much more spare but does recognize the need for "adaptive mechanisms to address significant changes to key components and interests" and to address climate change in both planning and implementation (British Columbia 2014). Reflecting the greater indigenous voice in review, both processes seek regional and tribal/First Nation involvement in any new negotiation and potentially an avenue for input to implementation, and both reflect recognition of the value of ecosystem function, with the US regional review recommending elevation of ecosystem function to a third prong of the treaty and the BC review considering ecosystem function (again without definition) a separate issue (US Army Corps of Engineers 2013; British Columbia 2014). The reviews have also staked out positions on the current treaty purposes, differing substantially in the interpretation of the level of flood protection to which the United States is entitled after 2024, and with each seeking changes in the Canadian Entitlement—BC higher, United States lower (US Army Corps of Engineers 2013; British Columbia 2014).

The BC decision represents the final step in the review in Canada. The US regional recommendation initially met with silence from the Department of State, with the regional congressional delegation twice (once in 2014 and again in 2015) submitting a letter to President Obama asking that negotiations be initiated. On May 20, 2015, the Department of State responded to the delegation, stating,

> Based on the Recommendation, we have decided to include flood risk mitigation, ecosystem-based function, and hydropower generation interests in the draft U.S. negotiating position. We hope to approach Canada soon to begin discussions on modernization of the Treaty.

Before turning to possible avenues to bridge positional gaps between the US and Canada positions and an analysis of models for flexibility in transboundary agreements, it may inform that effort to look behind three aspects of the review positions: (1) the BC position that ecosystem function, while important, is not an appropriate matter for the treaty; (2) the gap between the US regional recommendation calling for elevation of ecosystem function to a third purpose and the lack of any analysis of impacts; and (3) why the US regional recommendation has met with initial silence from the Department of State.

10.4.1.1 British Columbia's Position on Ecosystem Function

There may be several reasons why BC favors considering ecosystem function as a separate issue from the treaty, but it is clear that the goals of basin residents with respect to ecosystem function and their definition of the term will vary from the estuary to the headwaters. In some instances, the ecosystem has been altered by conversion of riverine to lacustrine systems through the development of dams. International river management is a cumbersome means to address issues that require local tailoring. In Section 10.4.2.1, we discuss subnational mechanisms for transboundary coordination that may better serve this need. Identification of only those issues requiring transboundary coordination, in this case, flow and fish passage, would limit the number of issues requiring international agreement.

10.4.1.2 The Gap between the Regional Recommendation and Supporting Studies

Surprisingly, despite the recommendation of inclusion of ecosystem function from the US regional review, none of the modeling done either jointly by the US Entity and BC Hydro or separately for the review processes, attempted to look at management scenarios that include aspects of ecosystem function (US Army Corps of Engineers, no date). The combined capacity building of Native American tribes and First Nations in the basin have led to collaboration across the international border to propose a phased process for the development of fish passage and restoration of salmonids to the Canada section of the river (US Columbia Basin Tribes and Canadian First Nations 2015), and use of a water bank to provide the flexibility needed for flow coordination. Neither of these proposals has been met with a parallel effort to understand the effects on transboundary reservoir coordination. Despite the expiration of assured flood control in 2024, the US review did not undertake a study of alternative means to distribute flood risk management in the basin, nor did climate change model predictions play a major role in review modeling.

Providing robust scientific bases for, and modeling the consequences of, proposed changes are key elements to bridging gaps and building adaptive capacity in any negotiation over natural resources. Without them, parties are left to speculate and may resort to positional bargaining, preventing recognition of means to enhance shared benefits going forward.

10.4.1.3 Initial Silence from the US Department of State

Understanding the difficult position the US Department of State was in with regard to navigating a way forward for the Columbia River requires understanding how US domestic law treats US involvement in international law. Despite the provision of the US Constitution requiring the advice and consent of two-thirds of the Senate for entry into a treaty (US Constitution Article II § 2[2]), not all international agreements require that process (see Bankes and Cosens 2012). In fact, there is considerable room for modification under an existing treaty provided that the Senate, in consultation with the Department of State, does not object (US Department of State 1985; Bankes and Cosens 2012). Thus, had the regional recommendation represented regional consensus, it is possible the Department of State could have proposed that negotiations within executive authority under the current CRT (1) change the Canadian Entitlement, (2) change flood control, (3) change the entity composition, and (4) change the flow regime for mutual ecosystem benefits. However, despite participating in the regional review, interest groups currently receiving the greatest benefit from the CRT and perceiving loss of some of those benefits if the recommendation went forward turned to their congressional leaders (US House Committee on Natural Resources 2013). This placed the Department of State in an untenable position by reducing the likelihood that Senate representatives from the region will take a broad view of executive authority under the existing treaty. The decision in May of 2015 by the Department of State to move forward in a manner consistent with the regional recommendation suggests confidence in the legitimacy of the regional process.

Without deluding ourselves or our readers that we have a silver bullet, we do believe that as the negotiations commence, there are measures that can be taken to bridge the gaps in stated positions and increase flexibility and inclusiveness going forward.

10.4.2 Introducing Flexibility and a Broader Voice: Adaptive Water Governance

Governance is the means through which political actors identify goals, make decisions, and take action. Governance includes not only the laws, policies, regulation, institutions, and organizational structures that enable and constrain the process of governing, but also the informal norms and interactions that influence decisions, including those of private and nongovernmental actors (Folke et al. 2005; Huitema et al. 2009; Cosens et al. 2014). Adaptive governance is governance that allows adaptive processes to emerge (Chaffin et al. 2014). Recognition of adaptive governance in systems that have been studied suggests that under the right circumstances, it is a natural (*emergent* or *self-organizing*) response to the challenges of managing complex landscapes in the face of change and uncertainty in that change (Chaffin et al. 2014). It arises out of the recognition and, often, frustration that rigid systems of environmental management cannot respond and evolve quickly enough as climate change unfolds. Adaptive governance cannot be mandated; nevertheless, the right circumstances are frequently products of organizational and institutional design. Thus, there are measures that can be taken to facilitate adaptive capacity and make it more likely that environmental governance can adapt to change and surprise while maintaining *constrained* or *bounded* flexibility (Bankes and Cosens 2014).

When considering how to create space for emergence of adaptive governance in international law, it is first important to note that international law is itself adaptive. Customary international law, which comprises the unwritten norms through which nation-states interact out of an understanding that they are legally obligated to do so, evolves with changing circumstances. International courts, as required by international treaty, interpret the generic terms and broad concepts in treaties in light of the changing understanding of the relevant norms of customary law (Vienna Convention on the Law of Treaties, Article 31; International Court of Justice, 1997; Arbitral Tribunal, the Hague, 2005). Thus, the United States and Canada are not bound to an interpretation of the CRT that comports with the intent of the drafters. They may, for example, interpret even the existing CRT in light of new understandings of the rights of indigenous people or take an expanded view of their obligation to mitigate harm to the environment through coordination on flow and fish passage.

Nevertheless, it may be that the existing CRT is simply too narrow to place the weight of modernization on its shoulders (Shurts 2012). Should the basin seek a new framework, other international water agreements increasingly illustrate recognition of the need for flexibility and an adaptive approach (McCaffrey 2003; Hearns and Paisley 2013) and provide models for governance of the Columbia River Basin. To identify the components of these models that will foster adaptive governance, we use the three areas of inquiry identified by the Adaptive Water Governance Project to mine existing international agreements: (1) structure, (2) capacity, and (3) process.

10.4.2.1 Structure

Allowing for flexibility in response to the uncertain impacts of climate change requires the ability to respond to the same problem at different levels and scales, and at the same level from different perspectives (Huitema et al. 2009; Rijke et al. 2012). This flexibility to respond requires both the authority to respond at multiple levels and coordination across those levels. For purposes of the CRT, it is important to identify those areas that require transboundary coordination and assure that the authority to coordinate at the international level is provided. Yet to ensure detection of change, rapid response, and increased innovation and the use of local knowledge, it is important that the authority exists to make decisions as close as possible to the individual citizen—the concept of subsidiarity (see, e.g., TFEU, Article 69; Vischer 2001), while recognizing that the level that best fulfills the above criteria may differ depending on the problem.

In the context of the CRT, consideration of level and subsidiarity cautions against the tendency to elevate every water-related issue to the international level simply because that is the level of the current review process. Attention to these factors might bridge the gap between the United States and BC on flood control while at the same time injecting breathing space into basin water management. Thus, diversification of flood risk management by incorporating local measures for flood control may not only introduce some redundancy into the system but also allow operation of reservoirs for environmental flows. Spreading the management of flood risk across multiple levels of government requires coordination across those levels. For a river basin like the Columbia, it may require coordination from the international to the federal/provincial to the local level.

The concept of subsidiarity is of particular importance for the issue of ecosystem function. Possible reasons for BC's hesitation in elevating ecosystem function to an international treaty include the recognition that *ecosystem function* will require substantial tailoring of solutions to local biophysical conditions and social values. By separating those aspects that require local input (e.g., restoration of spawning habitat, operation of hatcheries) from those that require transboundary coordination (e.g., flow and transboundary fish passage), the gap between the US and BC reviews may be lessened. The Great Lakes Compact and Agreement (Great Lakes–Saint Lawrence River Basin Agreement 2005; Compact 2008), a nonbinding, subnational agreement among states and provinces, provides a model for those issues that require local input but may benefit from transboundary communication. The subnational status provides greater flexibility and discretion to tailor local solutions on implementation (Karkkainen 2013).

International models that incorporate flexibility in structure also do so through a politically appointed implementing body authorized to make adjustments (Boundary Waters Treaty 1909; Great Lakes Water Quality Agreements 1972, 1978 as amended 1983, 1988; Great Lakes Water Quality Protocol 2012; Pacific Salmon Treaty 1985; Treaty between the United States of America and Mexico 1944). Political appointment provides an avenue to bring tribes, First Nations, and states into treaty implementation. International experience suggests that the political/decision-making function should be assigned to a separate entity from the one charged with scientific study, data coordination, and technical implementation (see, e.g., Boundary Waters Treaty 1909; Great Lakes Water Quality Agreements 1972, 1978 as amended 1983, 1988; Great Lakes Water Quality Protocol 2012; Pacific Salmon Treaty 1985). The overlap in these roles may create problems with political accountability (see, e.g., Treaty between the United States of America and Mexico 1944; McCarthy 2011). Indeed, the current CRT establishes agency-level, technical implementing entities that have been able to exercise a degree of discretion within the bounds of the detailed provisions of the CRT, but the absence of political appointment and avenues for public input would impair accountability should managers be charged with decision making on issues with high levels of uncertainty in the goal (such as ecosystem function).

Political accountability and inclusiveness may also be enhanced through structural measures, such as the establishment of advisory bodies, which include participation from tribes and First Nations, and additional venues for input from local government and other interests (Great Lakes levels agreement discussion in Bankes and Cosens 2014; Willoughby 1972; Great Lakes Water Quality Agreements 1972, 1978 as amended 1983, 1988; Great Lakes Water Quality Protocol 2012; Pacific Salmon Treaty 1985). In particular, the Great Lakes Water Quality Agreements use advisory bodies that are made up of representatives from national and subnational agencies and governments. This approach of overlap between actual players at different levels of government may improve coordination and flow of information across levels.

10.4.2.2 Capacity

Capacity for adaptive governance has two prongs: (1) adaptive and (2) participatory capacity (Cosens et al. 2014). Adaptive capacity requires both the authority to respond to change and the ability and resources to learn. This authority is a key component in seeking models from international agreements and may include authority to monitor for change, alter implementation in response to change, and revisit goals from time to time (i.e., adaptive management) (Bankes and Cosens 2014). Building adaptive capacity in the basin begins with the negotiation stage and may help bridge the gaps between parties and various interests identified here.

Lack of information may lead to speculation on the impact of change that is out of proportion to reality. In particular, the failure to model flood risk distribution, climate change, and approaches to flow adjustment for ecosystem function may fuel polarized positions in the basin. Consider, for example, that loss of snowpack may reduce the risk of major floods resulting from rapid warming or rain on snow events, as occurred in 1948. Alternatively, consider that while irrigation interests are justifiably concerned with distribution of flood risk management that impacts the operation of storage they rely on, no effort has been made to model exactly what this effect would be. Nor has any study considered the potential for irrigation benefits such as the use of storage drawdown to recharge aquifers in areas such as the Snake River Plain and the mid-Columbia region. Finally, rather than place ecosystem function on the table in its entirety, no effort has been made to identify only those issues that require transboundary coordination (i.e., flow and fish passage). Although unsuccessful in the past when attempted on a basin-wide scale (Volkman and McConnaha 1993; Cosens and Williams 2012), adaptive management applied to local implementation of restoration measures needed to return salmon to the Upper Columbia may be feasible and entail less risk. On the transboundary issues, while the tribes and First Nations have proposed use of a water bank to address flow, modeling has not been done to determine the actual impact of that change. Consider that the climate impact of declining summer–fall flows could provide incentive for environmental interests and irrigators to combine forces to share in the benefits of a water bank approach. Thus, information builds adaptive capacity. It nevertheless falls short without the ability to use it.

Participatory capacity is primarily related to the ability of affected parties to participate in decision making. Provisions for public participation and involvement of subnational levels of government are the primary focus related to local capacity building in an international agreement. In addition, the use and funding of advisory bodies, discussed previously, facilitates participation. Use of a public scientific forum (e.g., Great Lakes Water Quality Protocol 2012) facilitates participatory capacity through education and sharing of information and could begin during the basin negotiation process. The capacity building among tribes, First Nations, states, and local government since 1964 bodes well for the Columbia River Basin.

10.4.2.3 Process

Process elements of an agreement help in tailoring the balance between flexibility and certainty in a manner acceptable to affected parties. Process necessarily incorporates elements of *good governance* focused on equity and justice to assure sustainability of the communities affected, and can be viewed through the lens of legitimacy and inclusiveness.

Society seeks legitimacy in the actions of those who govern (Franck 1988; Bodansky 1999; Esty 2006). The formal processes used in international law for negotiation, ratification, and implementation of an international agreement (Bankes and Cosens 2012) are designed to assure legitimacy. Management allowing for flexibility and adjustment over time challenges the traditional mechanisms for securing legitimacy (Cosens 2013). Models must be sought that place bounds on the exercise of discretion in operational flexibility; consider both biophysical and social/economic time frames in setting periods for adjustment; establish processes to ensure accountability in adjustment of goals; and provide an avenue for broad, inclusive public input (Cosens 2013).

Accountability is enhanced through structural measures such as separation of the political and technical roles and the use of advisory bodies and public input. Standing bodies for political decision making are appropriate conduits for a process of adjustment to change, including addressing new issues (Boundary Waters Treaty 1909; Great Lakes Water Quality Agreements 1972, 1978 as amended 1983, 1988; Great Lakes Water Quality Protocol 2012; Pacific Salmon Treaty 1985). Processes for recording new decisions and transmitting them to the respective governments for veto have proven effective in allowing adjustment while maintaining accountability (Great Lakes levels agreement in Bankes and Cosens 2014; Treaty between the United States of America and Mexico 1944).

Constraints or bounds on discretion to ensure legitimacy through stability must be provided in the agreement itself. Thus, the Boundary Waters Treaty limits the discretion of its standing body, the International Joint Commission, to instances in which one or both governments refer an issue to it for study or consideration. Discretion may also be limited to exercise within the purposes of the treaty (Great Lakes Water Quality Agreements 1972, 1978 as amended 1983, 1988; Great Lakes Water Quality Protocol 2012; Pacific Salmon Treaty 1985). A practice of transmitting decisions to the respective governments for veto (the minute process under the Treaty between the United States of America and Mexico 1944) assures a check on discretion and accountability at the highest level, provided the line of authority (i.e., who has the final authority for review) is clearly provided in the international agreement (McCarthy 2011).

Attention to the issues of structure, capacity, and process in both the negotiation and implementation phase may improve the prospects of a modern and adaptable management regime for the Columbia River Basin.

10.5 Conclusion

Traditional transboundary water governance, which has relied on the historic record of variability of water supply and human and vegetative demand, is not equipped to govern a nonstationary future. Governance must become adaptive, yet this challenges one of the key purposes of governance—to provide stability for economic pursuits. The struggle to balance stability with the need for flexibility given uncertainty in the water resource is apparent in the current review of the CRT between the United States and Canada. Review of the CRT not only has drawn attention to the fact that social and biophysical changes since 1964 combined with the impact of climate change might warrant a new or revised treaty, but also provides a window of opportunity for the incorporation of the structures and processes, and the building of capacity to meet the challenge of climate change. The emerging concept of adaptive water governance provides a lens for viewing approaches to international water agreements that may be appropriate for the Columbia River Basin and would increase capacity to respond and adapt to future surprises.

Acknowledgments

This chapter relies on research from two projects: (1) the pursuit on Social–ecological System Resilience, Climate Change, & Adaptive Water Governance, chaired by Barbara Cosens and Lance Gunderson and supported by the National Socio-Environmental Synthesis Center (SESYNC) under funding from the National Science Foundation (DBI-1052875), and (2) the project on Protocols for Adaptive Water Governance: The Future of the Columbia River, coauthored by Nigel Bankes and Barbara Cosens and supported by the Program on Water Issues at the Munk School of Global Affairs, University of Toronto.

References

Abatzoglou, J., D.E. Rupp, and P.W. Mote. 2014. Understanding seasonal climate variability and change in the Pacific Northwest of the United States. *J. Climate* 27: 2125–2142.

Bankes, N. and B. Cosens. October 2012. The Future of the Columbia River Treaty. Munk School of Global Affairs, University of Toronto. http://munkschool.utoronto.ca/research/the-future-of-the-columbia-river-treaty/.

Bankes, N. and B. Cosens. 2014. Protocols for Adaptive Water Governance: The Future of the Columbia River for the Program on Water Issues. Munk School of Global Affairs, University of Toronto. http://powi.ca/wp-content/uploads/2014/10/Protocols-for-Adaptive-Water-Governance-Final-October-14-2014.pdf.

Barbar, K. 2006. Indigenous Regulations of the Harvest, in Canneries on the Columbia: A New Western History, Narrative. *The Oregon History Project.* http://www.ohs.org/the-oregon-history-project/narratives/canneries-on-the-columbia/native-fishery-katrine-barber/indigenous-regulations-of-harvest.cfm.

Barton, J. and K. Ketchum. 2012. Columbia River Treaty: Managing for uncertainty. *In* B. Cosens, ed. *The Columbia River Treaty Revisited: Transboundary River Governance in the Face of Uncertainty.* Corvallis: Oregon State University.

Billington, D.P., D.C. Jackson, and M.V. Melosi. 2005. The History of Large Federal Dams: Planning, Design, and Construction in the Era of Big Dams. US Department of the Interior. http://www.usbr.gov/history/HistoryofLargeDams/LargeFederalDams.pdf.

Bodansky, D. 1999. The legitimacy of international governance: A coming challenge for international environmental law? *American Journal of International Law* 93:596–624. http://dx.doi.org/10.2307/2555262.

Bonneville Power Administration. History https://www.bpa.gov/news/AboutUs/Pages/History.aspx

Bottom, D., K. Jones, C. Simenstad, and C. Smith, eds. 2009. Pathways to resilient salmon ecosystems. *Ecology and Society* 14(1):34 http://www.ecologyandsociety.org/issues/view.php?sf = 34.

Boundary Waters Treaty (BWT) between the United States and Great Britain (Canada). 1909. January 11, 1909, 6 Stat. 2448.

Boyd, D. 2003. *Unnatural Law: Rethinking Canadian Environmental Law and Policy.* Vancouver: University of British Columbia Press.

British Columbia. Columbia River Treaty Review. http://blog.gov.bc.ca/columbiarivertreaty/.

British Columbia, Columbia River Treaty Review, Draft B.C. Recommendation, December 2013, http://blog.gov.bc.ca/columbiarivertreaty/files/2012/07/Columbia-River-Treaty-Draft-BC-Recommendation.pdf.

British Columbia, Columbia River Treaty Review, B.C. Decision, March 2014, http://blog.gov.bc.ca/columbiarivertreaty/files/2012/03/BC_Decision_on_Columbia_River_Treaty.pdf.

CBT. Columbia Basin Trust. Climate Change in the Columbia Basin. http://www.cbt.org/Initiatives/Climate_Change/.

Cail, R. 1974. *Land, Man and the Law: The Disposal of Crown Lands in British Columbia, 1871–1913.* Vancouver: University of British Columbia Press.

Chaffin, B., H. Gosnell, and B. Cosens. 2014. A decade of adaptive governance scholarship: Synthesis and future directions. *Ecology and Society* 19:56, http://dx.doi.org/10.5751/ES-06824-190356.

Columbia Basin Tribes. 2010. Common Views on the Future of the Columbia River Treaty. http://www.usea.org/sites/default/files/event-/Common%20Views%20statement%20NQ.pdf.

Columbia Basin Tribes and First Nations. 2015 Fish Passage and Reintroduction into the U.S. & Canadian Upper Columbia Basin, http://www.ucut.org/Fish_Passage_and_Reintroduction_into_the_US_And_Canadian_Upper_Columbia_River3.pdf.

Columbia River Inter-Tribal Fish Commission http://www.critfc.org/.

Columbia River Inter-Tribal Fish Commission. 1995 (updated 2014). Spirit of the Salmon: Tribal Restoration Plan. http://plan.critfc.org/vol-1/.

Cosens, B. 2010. Transboundary river governance in the face of uncertainty: Resilience theory and the Columbia River Treaty. *University of Utah Journal of Land Resources, and Environmental Law* 30:229.

Cosens, B. 2012. Changes in empowerment: Rising voices in Columbia basin resource management. *In* B. Cosens, ed. *The Columbia River Treaty Revisited: Transboundary River Governance in the Face of Uncertainty*. Corvallis: Oregon State University Press.

Cosens, B. 2013. Legitimacy, adaptation, and resilience in ecosystem management. *Ecology and Society* 18(1):3. http://dx.doi.org/10.5751/ES-05093-180103.

Cosens, B. and A. Fremier. 2014. Assessing system resilience and ecosystem services in large river basins: A case study of the Columbia River Basin. *Natural Resources and Environmental Law Edition of the Idaho Law Review* 51:91–125.

Cosens, B. and L. Gunderson. 2013. Social–Ecological System Resilience, Climate Change & Adaptive Water Governance. National Socio-Environmental Synthesis Center. University of Maryland. http://www.sesync.org/project/water-people-ecosystems/adaptive-water-governance. The Adaptive Water Governance Project is supported by the National Socio-Environmental Synthesis Center (SESYNC) under funding from the National Science Foundation DBI-1052875.

Cosens, B., L. Gunderson, C. Allen, and M.H. Benson. 2014. Identifying legal, ecological and governance obstacles, and opportunities for adapting to climate change. *Sustainability* 6(4):2338–2356; doi:10.3390/su6042338 URL: http://www.mdpi.com/2071-1050/6/4/2338.

Cosens, B., L. Gunderson, and B. Chaffin. 2014. The adaptive water governance project: Assessing law, resilience and governance in regional socio-ecological water systems facing a changing climate. *Natural Resources and Environmental Law Edition of the Idaho Law Review* 51:1–27.

Cosens, B. and M.K Williams. 2012. Resilience and water governance: Adaptive governance in the Columbia River Basin. *Ecology and Society* 17 (4):3. http://www.ecologyandsociety.org/vol17/iss4/art3/.

Dalton, M., P.W. Mote, and A.K. Snover, eds. 2013. *Climate Change in the Northwest: Implications for Our Landscapes, Waters, and Communities*. 224 pp. D.C.: Island Press.

Deur, D. 1999. Toward an Environmental Prehistory of the Northwest Coast. *In* D.D. Goble and P.W. Hirt, eds. *Northwest Lands, Northwest Peoples: Readings in Environmental History*. Seattle: University of Washington Press.

Elsner, M.M., L. Cuo, N. Voisin, J.S. Deems, A.F. Hamlet, J.A. Vano, K.E.B. Mickelson, S. Lee and D.P. Lettenmaier. 2010. Implications of 21st century climate change for the hydrology of Washington, State. *Climatic Change* 102:225–260. doi: 10.1007/s10584-010-9855-0.

Esty, D.C. 2006. Good governance at the supranational scale: Globalizing administrative law. *Yale Law J*. 115:1490.

Fiege, M. 1999. *Irrigated Eden: The Making of an Agricultural Landscape in the American West*. Seattle: University of Washington Press.

Folke, C., T. Hahn, P. Olsson, and J. Norberg. 2005. Adaptive governance of social–ecological systems. *Annual Review of Environment and Resources* 30:441–473. http://dx.doi.org/10.1146/annurev.energy.30.050504.144511.

Foundation for Water & Energy Education, Irrigation. http://fwee.org/environment/what-makes-the-columbia-river-basin-unique-and-how-we-benefit/irrigation/.

Franck, T.M. 1988. Legitimacy in the International System. *American Journal of International Law* 82:705.

Goble, D.D. 1999. Salmon in the Columbia basin: From abundance to extinction. *In* D.D. D.D. Goble and P.W. Hirt, eds. *Northwest Lands, Northwest Peoples: Readings in Environmental History*. Seattle: University of Washington Press.

Great Lakes–St. Lawrence River Basin Water Resources Compact (GL Compact), Pub. L. No. 110-342, 122 Stat. 3739 (2008), http://www.cglg.org/projects/water/Agreement-Compact.asp.

Great Lakes–Saint Lawrence River Basin Sustainable Water Resources Agreement (GL Agreement), Dec. 13, 2005, http://www.cglg.org/projects/water/Agreement-Compact.asp.

Great Lakes Water Quality Agreements between the United States and Canada, 1972 amended 1978, 1983, 1988, available at 1978 GLWQA as amended, http://epa.gov/grtlakes/glwqa/1978/index.html.

Great Lakes Water Quality Protocol of 2012, http://www.ijc.org/en_/Great_Lakes_Water_Quality.

Hamlet, A. 2003. The role of transboundary agreements in the Columbia River Basin: An integrated assessment in the context of historic development, climate, and evolving water policy. *In* H. Diaz and B. Morehouse, eds. *Climate and Water: Transboundary Challenges in the Americas.*

Hamlet, A.F., P.W. Mote, M.P. Clark, and D.P. Lettenmaier. 2005. Effects of temperature and precipitation variability on snowpack trends in the Western United States. *Journal of Climate* 18:4545–4561.

Hamlet, A.F., P.W. Mote, M.P. Clark, and D.P. Lettenmaier. 2007: 20th century trends in runoff, evapotranspiration, and soil moisture in the Western U.S. *J. of Climate* 20:1468–1486.

Hamlet, A.F., S.Y. Lee, K.E.B. Mickelson, and M.M. Elsner. 2010. Effects of projected climate change on energy supply and demand in the Pacific Northwest and Washington State. *Climate Change* 102:103–128.

Harrison, J. 2012. Northwest Power and Conservation Council, Hatcheries. http://www.nwcouncil.org/history/hatcheries

Hearns, G. and P.R. Kyle. 2013. Lawyers write treaties, engineers build dikes, gods of weather ignore both: Making transboundary waters agreements relevant, flexible and resilient in a time of global climate change. *Golden Gate U. Environmental Law J.* 6:259.

Hirt, P. 2008. Developing a plentiful resource: Transboundary rivers in the Pacific Northwest. *In* J.M. Whiteley et al., eds. *Water, Place, & Equity*. MIT Press Scholarship.

Hirt, P.W. and A.M. Sowards. 2012. The past and future of the Columbia River. *In* B. Cosens, ed. *The Columbia River Treaty Revisited: Transboundary River Governance in the Face of Uncertainty.* Corvallis: Oregon State University Press.

Huitema, D., E. Mostert. W. Egas. S. Moellenkamp, C. Pahl-Wostl, and R. Yalcin. 2009. Adaptive water governance: Assessing the institutional prescriptions of adaptive (co-) management from a governance perspective and defining a research agenda. *Ecology and Society* 14:26. http://www.ecologyandsociety.org/vol14/iss1/art26/.

Josephy, A.M. Jr. 1997. *The Nez Perce Indians and the Opening of the Northwest.* Complete and Unabridged edition. New York and Boston: Houghton Mifflin Co.

Karkkainen, B.C. 2013. The Great Lakes Water Resources Compact and Agreement: Transboundary normativity without international law. *William Mitchell Law Review* 39:997.

Klos, P.Z., T.E. Link, and J.T. Abatzoglou. 2014. Extent of the rain–snow transition zone in the western U.S. under historic and projected climate. *Geophysical Research Letters.* 41, doi:10.1002/2014GL060500.

Landeen, D. and A. Pinkham. 2008. *Salmon and His People: Fish and Fishing in Nez Perce Culture.* Lewiston: Confluence Press.

Luce, C.H. and Z.A. Holden. 2009: Declining annual streamflow distributions in the Pacific Northwest United States, 1948–2006. *Geophysical Research Letters* 36, 1–6.

Luce, C.H., J.T. Abatzoglou, and Z.A. Holden. 2013. The missing mountain water: Slower westerlies decrease orographic enhancement in the Pacific Northwest USA. *Science* doi: 10.1126/science.1242335.

McCaffrey, S. 2003. The need for flexibility in freshwater treaty regimes. *Natural Resources Forum* 27:156–162.

McCarthy, R.J. 2011. Executive authority, adaptive treaty interpretation, and the International Boundary and Water Commission. U.S.–Mexico. *University of Denver Water Law Review* 14:197.

McKinney, M., L. Baker, A.M. Buvel, A. Fischer, D. Foster, and C. Paulu, 2010. Managing transboundary natural resources: An assessment of the need to revise and update the Columbia River Treaty. *West–Northwest Journal of Environmental Law and Policy* 16(2):307–350.

Meddens, A.J.H., J.A. Hicke, and C.A. Ferguson. 2012. Spatiotemporal patterns of observed bark beetle-caused tree mortality in British Columbia and the Western US. *Ecological Applications* 22 (7):1876–1891. doi: 10.1890/11-1785.1.

Mote, P., A. Hamlet, M.P. Clark, and D.P. Lettenmaier. 2005. Declining mountain snowpack in Western North America. *Bulletin American Meteorological Society* 86:39–49.

Mouat, J. 2012. The Columbia exchange: A Canadian perspective on the negotiation of the Columbia River Treaty, 1944–1964. *In* B. Cosens, ed. *The Columbia River Treaty Revisited: Transboundary River Governance in the Face of Uncertainty*. Corvallis: Oregon State University Press.

National Research Council. 1996. Upstream: Salmon and society in the Pacific Northwest *Report on the Committee on Protection and Management of Pacific Northwest Anadromous Salmonids* for the National Research Council of the National Academy of Sciences, National Research Council, DC: National Academy Press.

Northwest Power and Conservation Council, *Sixth Northwest Power Plan, Plan Overview* 1–2, February 2010, http://www.nwcouncil.org/energy/powerplan/6/plan/.

Northwest Power and Conservation Council, *Sixth Power Plan: Mid-Term Assessment Report*, March 13, 2013, https://www.nwcouncil.org/media/6391355/2013-01.pdf.

Pacific Salmon Treaty: Treaty between the Government of Canada and the Government of the United States of American Concerning Pacific Salmon, 1985 as revised, 1999 and 2009. http://www.psc.org/about_treaty.htm.

Pearson, M.L. 2012. The river people and the importance of salmon. *In* B. Cosens, ed. *The Columbia River Treaty Revisited Transboundary River Governance in the Face of Uncertainty*. Corvallis: Oregon State University Press.

Peery, C. 2012. The effects of dams and flow management on Columbia River ecosystem processes. *In* B. Cosens, ed. *The Columbia River Treaty Revisited Transboundary River Governance in the Face of Uncertainty*. Corvallis: Oregon State University Press.

Pierce, D.W., T.P. Barnett, H.G. Hidalgo, T. Das, C. Bonfils, B.D. Santer, G. Bala, M.D. Dettinger, D.R. Cayan, A. Mirin, A.W. Wood, and T. Nozawa. 2008. Attribution of declining western U.S. snowpack to human effects. *J. Climate* 21:6425–6444.

Reisner, M. 1987. *Cadillac Desert: The American West and its Disappearing Water*. D.C.: Penguin.

Rijke, J., R. Brown, C. Zevenbergen, R. Ashley, M. Farrelly, P. Morison, and S. van Herk. 2012. Fit-for-purpose governance: A framework to make adaptive governance operational. *Environmental Science & Policy* 22:73–84. http://dx.doi.org/10.1016/j.envsci.2012.06.010.

Shurts, J. 2012. Rethinking the Columbia River Treaty. *In* B. Cosens, ed. *The Columbia River Treaty Revisited: Transboundary River Governance in the Face of Uncertainty*. Corvallis: Oregon State University Press.

Stewart, I.T., D.R. Cayan, and M.D. Dettinger. 2005. Changes toward earlier streamflow timing across Western North America. *J. Climate* 18:1136–1155.

Treaty Between Canada and the United States of America Relating to Cooperative Development of the Water Resources of The Columbia River Basin (1964 CRT), U.S.–Can., Jan. 17, 1961, entered into force September 16, 1964. http://www.ccrh.org/comm/river/docs/cotreaty.htm.

Treaty between the United States of America and Mexico for the Utilization of Waters of the Colorado and Tijuana Rivers and of the Rio Grande, February 3, 1944, Proclamation, http://www.ibwc.state.gov/Files/1944Treaty.pdf.

TFEU, Treaty for the Functioning of the European Union, Article 69 states National Parliaments ensure that the proposals and legislative initiatives submitted under Chapters 4 and 5 comply with the principle of subsidiarity, in accordance with the arrangements laid down by the Protocol on the application of the principles of subsidiarity and proportionality. http://eur-lex.europa.eu /resource.html?uri = cellar:ccccda77-8ac2-4a25-8e66-a5827ecd3459.0010.02/DOC_1&format = PDF.

UCCRG. Universities Consortium on Columbia River Governance. n.d. http://www.columbiarivergovernance.org/.

Upper Columbia United Tribes (UCUT). nd. http://www.ucut.org/index.ydev.

UCUT. 2005. In the Field. http://www.ucut.org/in_the_field.ydev#news_paragraph6.

US Columbia Basin Tribes and Canadian First Nations. 2015. Fish Passage and Reintroduction into the U.S. & Canadian Upper Columbia Basin. January 9, 2015. http://www.ucut.org/Fish_Passage_and_Reintroduction_into_the_US_And_Canadian_Upper_Columbia_River3.pdf.

US Army Corps of Engineers and Bonneville Power Admin., Columbia River Treaty: 2012/2024 Review Website. http://www.crt2014-2024review.gov/.

US Army Corps of Engineers and Bonneville Power Admin., Columbia River Treaty: 2012/2024 Review: Phase 1 Technical Studies (Apr. 2009), http://www.bpa.gov/corporate/pubs/Columbia_River_Treaty_Review__2_-_April_2009.pdf.

US Army Corps of Engineers and Bonneville Power Administration, *Columbia River Treaty 2014/2024 Review, Regional Recommendation*, December 13, 2013, http://www.crt2014-2024review.gov/RegionalDraft.aspx.

US Bureau of Reclamation (USBR), Columbia Basin Project. http://www.usbr.gov/projects/Project.jsp?proj_Name = Columbia+Basin+Project.

USBR, Grand Coulee Dam, Overview. https://www.usbr.gov/projects/Facility.jsp?fac_Name= Grand+Coulee+Dam.

USBR, History: Reclamation: Managing Water in the West. http://www.usbr.gov/history/.

US Department of State, Circular 175: Procedures on Treaties, Revised February 25, 1985, at 11 FAM 721.2(b).

US House Committee on Natural Resources, Oversight Field Hearing on The Future of the US–Canada Columbia River Treaty—Building on 60 years of Coordinated Power Generation and Flood Control Monday, December 9, 2013, http://naturalresources.house.gov/calendar/eventsingle.aspx?EventID = 363025.

Vischer, R.K. 2001. Beyond devolution. *Indiana Law Review* 35:103.

Volkman, J.M. and W.E. McConnaha. 1993. Through a glass, darkly: Columbia River salmon, the Endangered Species Act, and adaptive management. *Environmental Law* 23:1249–1272.

Walker, B. and D. Salt. 2006. *Resilience Thinking: Sustaining Ecosystems and People in a Changing World*. D.C.: Island Press.

Wegner, S.J., D.J. Isaak, C.H. Luce, H.M. Neville, K.D. Fausch, J.B. Dunham, and J.E. Williams. 2011. Flow regime, temperature, and biotic interactions drive differential declines of trout species under climate change. *Proceedings of the National Academy of Sciences of the United States of America, 108*(34):14175–14180. doi:10.1073/pnas.1103097108.

Westerling, A.L., H.G. Hidalgo, D.R. Cayan, and T.W. Swetnam. 2006: Warming and earlier spring increases western U.S. forest wildfire activity. *Science* 313:940–943. doi:10.1126/science.1128834.

White, A.C. 2012. The Columbia River: operation under the 1964 Treaty. *In* B. Cosens, ed. *The Columbia River Treaty Revisited: Transboundary River Governance in the Face of Uncertainty*. Corvallis: Oregon State University Press.

White, R. 1995. *The Organic Machine: The Remaking of the Columbia River.* New York: Hill and Wang.

Willoughby, W.R. 1972. The International Joint Commission's role in maintaining stable water levels. *Inland Seas* 28:109–118.

11

California, a State of Extremes: Management Framework for Present-Day and Future Hydroclimate Extremes

Jeanine Jones

CONTENTS

ABSTRACT California's extensive system of major statewide and regional-scale water infrastructure, together with its robust institutional infrastructure for water management, were developed in response to long-recognized needs to connect locations of water supplies with areas of water supply deficiencies and to deal with temporal variability in precipitation. Managing water in California is about managing for the extremes—droughts and floods. The historical context for development of the state's management capacity has set the stage for dealing with another stressor being added to an already significant list of water supply and flood risk reduction challenges—that of adapting to extreme events related to climate change. California has put in place a framework for climate change adaptation, one that will facilitate responding to expected long-term impacts such as loss of mountain snowpack and the storage capacity it provides for water supply and flood management purposes. All views expressed in this chapter are the author's; nothing herein represents views of the California Department of Water Resources or the State of California.

11.1 Introduction and Setting

California is a leader in many respects—most populous state, top agricultural production state, home of the nation's most extensive system of large-scale water infrastructure, and a pioneer in climate change policy. Water has been central to California's development, beginning with the discovery of gold in the American River that triggered the Gold Rush and quickly led to statehood. Students of California water history will recall that Governor Leland Stanford traveled to his inauguration in Sacramento in January 1862 via rowboat, due to the so-called Great Floods of the winter of 1861–1862. These were immediately followed by the Great Drought of 1862; together this pair of extreme seasons was credited with helping put the last nails in the coffin of California's Mexican-era cattle rancho system.

The state's water management framework was shaped in the early twentieth century by recognition of the geographic mismatch between locations of water availability and of water needs, fostering development of an unparalleled system of federal, state, and local water projects. This system was designed not only because of the need for spatial realignment of water supplies and water needs but also because of temporal variability of precipitation—at timescales ranging from subseasonal to seasonal. Managing water during the wet season in many parts of the state entails complicated operations to balance between the competing needs of immediate flood risk management and storing water for later use during the dry summer season.

California is a large state with diverse physiography, encompassing seven National Oceanic and Atmospheric Administration (NOAA) climate divisions. Proximity to Pacific Ocean moisture sets the stage for some of the nation's most extreme precipitation events (leading to severe flooding potential); atmospheric patterns that block this moisture give rise to major droughts. Orography is destiny with respect to the state's precipitation, and much of California's developed water supplies are fed by mountain snowpack. A warming climate will alter the amount and spatial distribution of precipitation falling as snow, contravening historical hydrologic criteria used for design of much of the state's major water infrastructure.

California has been establishing a legal and institutional framework for mitigation of, and adaptation to, climate change. As a practical matter, however, water management in California has always been about planning for wet and dry extremes, and much of the nearer-term adaptation work represents continuing efforts by water agencies in this vein. Nevertheless, by the end of the century, the response to projected declines in mountain snowpack will require an expanded level of effort to replace the snowpack water storage capacity that supports both water supply and flood risk management functions.

11.1.1 Setting

Most of the water vapor that provides California's precipitation comes from the Pacific Ocean; as the moist air moves over major mountain ranges such as the Sierra Nevada or Transverse Ranges (Figure 11.1), the air is lifted and cooled, resulting in condensation and rain or snow. Snowpack in the Cascade Range and Sierra Nevada contributes to the runoff in the state's largest rivers and to the groundwater basin recharge that support much of California's developed urban and agricultural water use. The average water content of Sierra and Cascade snowpack has historically augmented the collective storage capacity of the major reservoirs in the mountains' watersheds by about one-third (California Natural Resources Agency [CNRA] 2014).

Much of California experiences a Mediterranean-like climate, with dry summers that are warm or hot and wet winters that are cool or cold. Summers are characterized by a blocking high-pressure zone that diverts atmospheric moisture away from the state. On

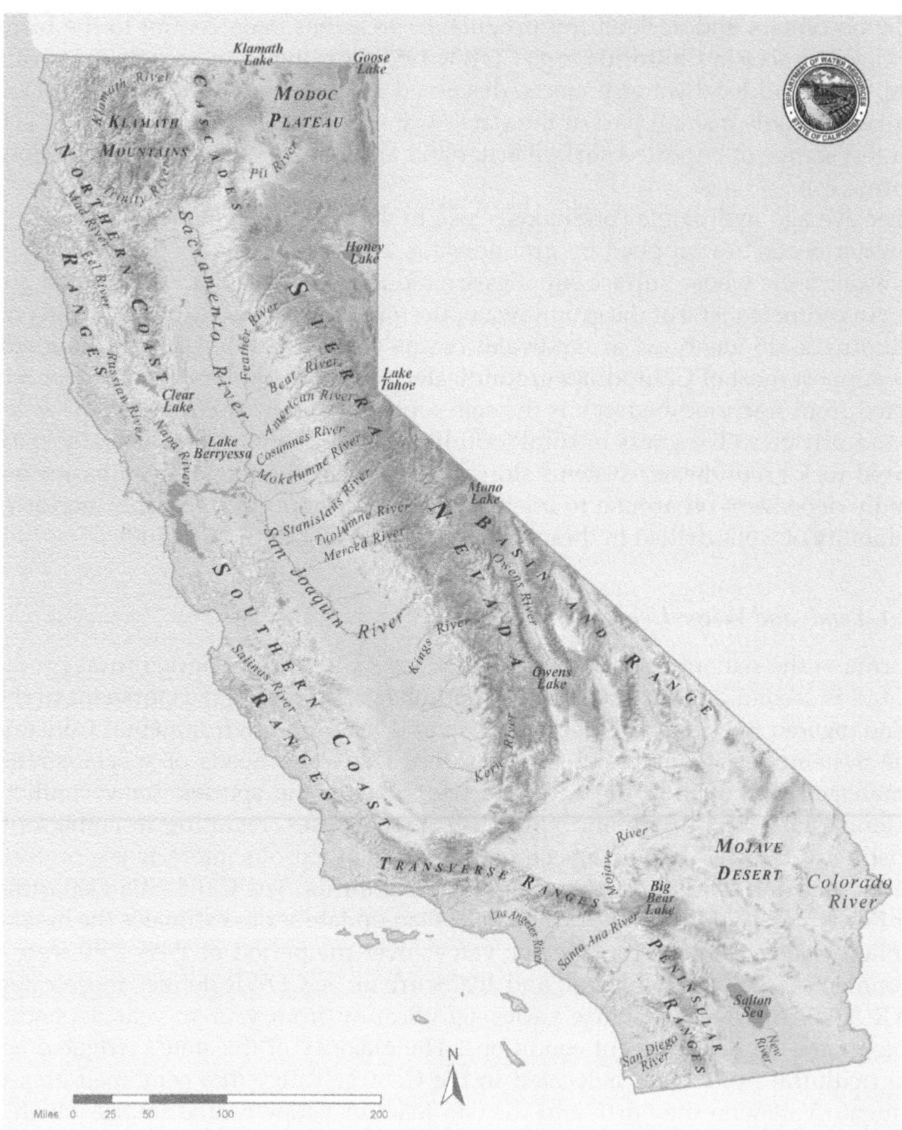

FIGURE 11.1
California location map.

average, about 75% of the state's average annual precipitation falls between November and March, with 50% occurring between December and February. The state's annual water budget is determined by a relatively small number of storms. A shortfall of a few major winter storms usually results in a dry year; conversely, a few very wet storms usually lead to a wet year (CDWR 2010). High annual variability in precipitation is characteristic of California (Dettinger et al. 2011).

Annual precipitation (and hence runoff) is greater in Northern California (north of the Sacramento–San Joaquin River Delta) than in Southern California. The fundamental imbalance between the availability of surface water supplies and the location of major population centers and agricultural production areas has been central to the history of water development in California (e.g., CDWR 1957), leading to the construction of major federal, state, and local water projects, discussed in this chapter. Imported surface supplies make up only a small part of the state's water budget. The Colorado River is by far the largest source of imported surface water and an important water supply for Southern California.

Under average hydrologic conditions, close to 40% of California's urban and agricultural water needs are supplied by groundwater, an amount that increases in dry years, when water users whose surface supplies are reduced increase their reliance on groundwater. An estimated 90% of the groundwater used in California is extracted from only 126 of California's 515 identified groundwater basins (CDWR 2015). Although large alluvial basins support most of California's groundwater use on a volumetric basis, groundwater extracted from fractured bedrock is the sole source of supply for many small water systems and private well owners in rural foothill and mountain areas. Generally speaking, fractured rock groundwater systems store far less water than do alluvial basins and are markedly dependent on annual to interannual precipitation for recharge, increasing the vulnerability of wells drilled in these formations to drought (CDWR 2010).

11.1.1.1 Land- and Water-Use Patterns

California is the nation's most populous state and the leading agricultural production state, and is second only to Hawaii in the number of species listed pursuant to the federal Endangered Species Act (ESA) (US Fish and Wildlife Environmental Conservation Online System website, accessed August 2014). The water needs of special-status species managed pursuant to the ESA—particularly aquatic species—may conflict with prior long-established urban and agricultural water uses, resulting in cutbacks to the preexisting water uses. Balancing competing needs of various interests is a central challenge for state and federal water managers in California. The California Department of Water Resources' (CDWR's) California Water Plan update series estimates the breakdown in applied water use by sector; average values over the period of 1998–2010 were 48.6% environmental, 40.8% agricultural, and 10.6% urban, as CDWR defines those categories (CDWR 2014). Applied water use varies significantly from year to year depending on hydrology and on management conditions. The majority of the state's irrigated acreage and agricultural production is located in the Central Valley (the combined area of the Sacramento Valley on the north and the San Joaquin Valley on the south), which is fed by the two rivers—the Sacramento and San Joaquin—that provide much of California's developed agricultural and urban water supplies (and represent the state's largest flood management challenges).

Despite its relatively short heyday, the Gold Rush era left a lasting legacy for California water management—creating the legal foundation for the prior appropriation doctrine of water rights (Attwater et al. 1988) and fostering development of the state's first large regional-scale flood control projects focused on the Sacramento Valley and the Sacramento–San Joaquin River delta. There, several factors spurred early development of what subsequently became a complicated system of Central Valley flood management infrastructure. These included flood damages caused by hydraulic mining debris from the Sierra Nevada mountains, large volumes of water from the major rivers, and local levee systems being constructed in response to federal and state legislation encouraging reclamation of swamplands for farming (Kelley 1989). In concept, the Sierran reservoirs—which are owned by federal, state, and local agencies and by private power companies—are operated as required by US Army Corps of Engineers (USACE) rule curves to reserve flood control storage capacity during the winter wet months; the reservoirs then fill with snowmelt runoff later in the spring season after the risk of large winter storms has passed. Reservoir operators manage winter flood events by holding back peak flood volumes long enough to be able to move releases through complex valley-floor infrastructure networks that include river channelization and levee projects, and flood control bypasses or floodways. If California's more than 13,000 mi. of flood control levees were laid end to end, they would stretch more than four times the distance between San Francisco and Miami, Florida.

The delta occupies a unique position in California's water supply and flood management systems; its physical setting and resource management challenges have been widely discussed (e.g., Lund et al. 2008). The delta's meandering waterways thread their way toward the San Francisco Bay among a maze of leveed islands used primarily for farming; most islands are below sea level. The delta's levees, largely constructed in the late 1800s, were not engineered structures, and some were constructed either with or upon fragile peat soils. Delta levee failures were historically common, on the order of more than 160 failures in the past 100 years (Suddeth et al. 2010). The modern delta serves as a de facto hub for much of the state's water supply system, and levee failures carry a risk beyond that of local flooding of agricultural land. Under worst-case scenarios, a single levee failure could cause a chain reaction of multiple levee breaches that would allow saline water to penetrate inland and contaminate freshwater urban and agricultural sources. Releases from upstream reservoirs are used to control salinity intrusion in the delta, in effect creating a hydraulic barrier to protect urban and agricultural diversions and to meet ecological water quality protection criteria. However, the volume of freshwater water stored in upstream reservoirs would be insufficient to overcome tidal inflow associated with large-scale levee failure, potentially rendering freshwater diversions inoperable for many months until levees can be repaired and hydraulic control reestablished.

11.2 Water Management Infrastructure

Just as orography is destiny with respect to California's precipitation, plumbing is everything when it comes to managing water. Most of the state's major water-use areas are linked through a complex network of water projects (Figure 11.2). The Central Valley's large water supply reservoirs also serve a dual role as key parts of the flood management infrastructure. The state's largest water projects are as follows:

- The US Bureau of Reclamation's (USBR's) Central Valley Project (CVP), which develops water in the Central Valley and delivers it primarily for agricultural use within the valley, with a smaller component of in-valley and San Francisco Bay Area urban use
- CDWR's State Water Project (SWP), which develops water in the Central Valley and serves it primarily for urban use in Southern California and parts of the Bay Area, with a smaller component of agricultural use, including in the San Joaquin Valley
- USBR's facilities that store and convey interstate Colorado River water to Southern California for urban and agricultural use

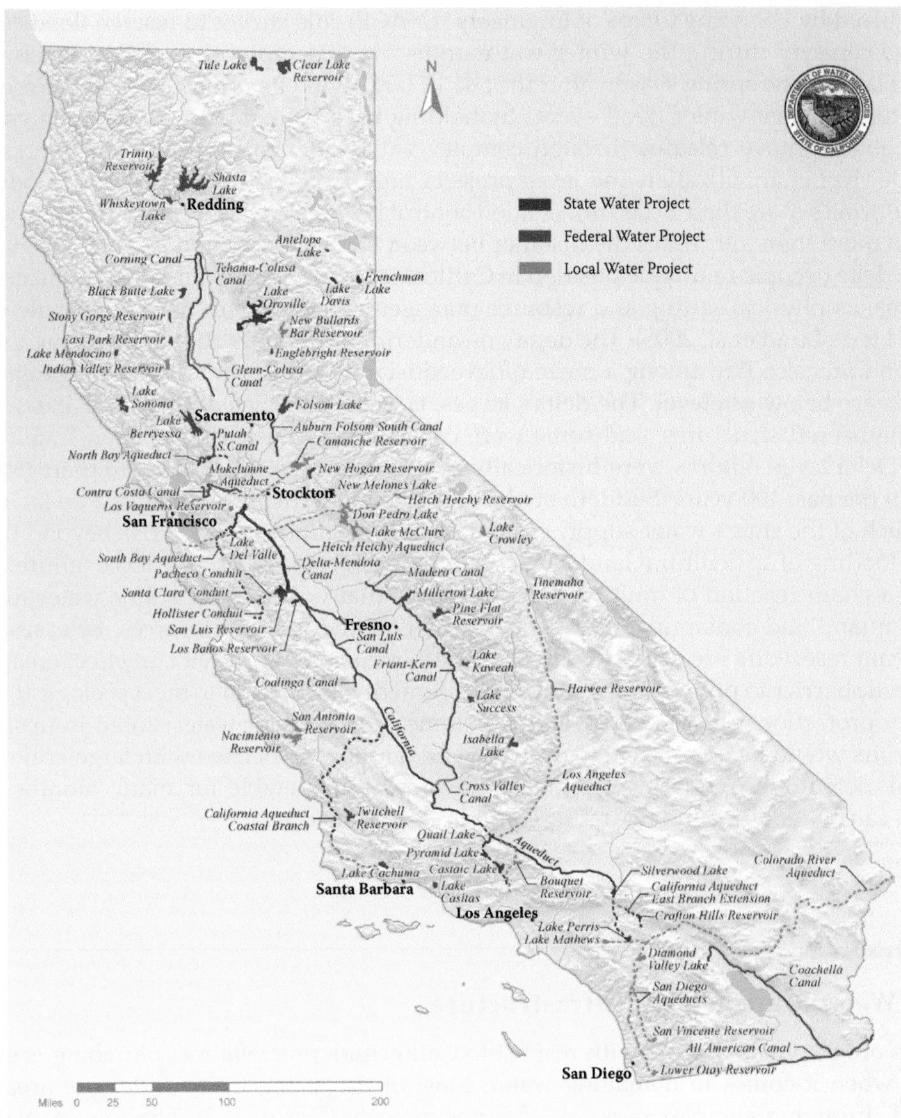

FIGURE 11.2
California's major water projects.

California has a long tradition of comprehensive, coordinated state water planning. Legislation enacted in 1878 provided for establishment of the Office of the State Engineer and tasking it to collect data and develop policy for irrigation development. William Hammond Hall, California's first state engineer, is credited with advancing the vision for systematic and coordinated development of Central Valley water resources, for both irrigation and flood control (e.g., Hall 1880). By the early 1930s, ongoing state planning had resulted in release of the State Water Plan, which called for development of facilities subsequently constructed as the federal CVP (California Department of Public Works 1930). Large local agency urban water supply projects were also being built during this time period (see the following list). As California's population boomed during World War II and the postwar period, continuing state planning culminated in CDWR's 1957 California Water Plan, a vision for optimized development of federal, state, and local water infrastructure (CDWR 1957). Subsequent construction of the SWP stemmed from this planning effort.

The following list shows the chronology of development of major water supply projects and of selected major institutional actions:

1850 California is admitted to the Union

1871 First reported construction of a dam on Lake Tahoe

1887 Legislature enacts the Wright Irrigation District Act, allowing creation of special districts

1902 Congress enacts the Reclamation Act, authorizing federal construction of water projects

1913 First barrel of Los Angeles Aqueduct completed

1922 Colorado River Compact signed

1929 Mokelumne River Aqueduct of East Bay Municipal Utility District completed

1931 Legislature enacts the Water Conservation Act of 1931, spurring formation of many new special districts (most for agricultural purposes)

1934 San Francisco's Hetch Hetchy Aqueduct completed

1940 All American Canal completed

1941 Colorado River Aqueduct completed

1945 Shasta Dam completed

1968 Oroville Dam completed

1968 Congress enacts National Wild and Scenic Rivers Act

1972 Legislature enacts California Wild and Scenic Rivers Act

1973 Congress enacts ESA

1978 State Water Resources Control Board (SWRCB) adopts Water Right Decision 1485 regarding CVP/SWP water operations criteria for the delta

1984 Legislature enacts California Endangered Species Act

1992 Congress enacts Central Valley Project Improvement Act

1999 SWRCB adopts Water Right Decision 1641 regarding CVP/SWP water operations criteria for the delta

Several important points about the development of this major water infrastructure are particularly important when considering long-term climate adaptation.

- Almost all of the state's significant reservoir storage is located in Central Valley watersheds. Local storage in urban Southern California, where about half of the state's population lives, is predominantly in the form of groundwater storage or terminal reservoirs for imported water supplies.
- Reservoir storage volume on most rivers tributary to the Central Valley is only a fraction of the rivers' average annual flow—in contrast to the Colorado River, where total system storage equals about four times the river's annual inflow to Lake Powell.
- Regionally, flood risks are greatest in the Central Valley due to the high potential flood volumes from the Sacramento and San Joaquin Rivers and the large spatial extent of the intensive agricultural and urban development located in the floodplains.

- Central Valley reservoirs are operated by a mix of federal, state, and local agencies (and private utilities) in a complicated seasonal balancing act between storing water for water supply and managing flood control releases to avoid downstream damages.
- Most of the state's major infrastructure was designed based on the observed hydrology of the first half of the twentieth century.
- Most of the state's major infrastructure was constructed prior to enactment or imposition of environmental regulatory statutes or administrative provisions that subsequently reduced water supplies historically available for urban or agricultural purposes in order to provide water for environmental purposes.

Hydroclimate by itself is only one of many factors shaping management and operations of water infrastructure. Federal and state agencies must often balance competing objectives such as water supply versus flood control or water contracts with urban and agricultural agencies versus regulatory requirements for special-status fish species. Two extreme events with similar hydrology that occur some years apart may not result in the same response actions or impacts, due to changes in other conditions—service-area water-demand amounts or patterns, watershed urbanization, or regulatory requirements for fish species protected pursuant to the ESA.

11.3 Institutional Framework for Extreme Events

Managing water in California is about managing for extremes, as the extreme conditions—droughts and floods—define the necessary performance criteria for water supply or flood management projects. Experience gained in managing historical extreme events described in this chapter has led to development of institutional tools to facilitate preparedness and response for future events. Additionally, matching California's long historical tradition of comprehensive, coordinated state planning for major water infrastructure is a tradition of using state financial assistance for local agencies to accomplish water development or management goals. Increasingly, such financial assistance is being conditioned by statutory or administrative requirements for eligibility designed to promote objectives such as water conservation, water-shortage contingency planning, or climate change adaptation, objectives that can also assist with managing for extremes.

11.3.1 Statutory Framework

California law recognizes both riparian and appropriative rights to surface water (and pueblo rights in some limited cases) (Attwater and Markle 1988). Rights to groundwater have not been administered at the state level, but authority was historically provided to local agencies for various forms of groundwater management through general legislative acts and acts specific to individual local agencies, as well as through case law. The Sustainable Groundwater Management Act, signed into law in 2014, now prescribes actions intended over the long term to achieve sustainable management in groundwater basins designated by CDWR as high- and medium-priority basins, through requiring formation of local groundwater sustainability agencies and authorizing the State Water Resources Control Board (the state agency that administers surface water rights) to step in if local agencies fail to act.

Numerous statutory provisions were put in place in response to historical droughts to facilitate water transfers, exchanges, and banking, such as provisions that a transfer of water did not constitute forfeiture or abandonment of the underlying water right or contractual right (CDWR 2000). Examples of other statutory provisions that facilitate drought response include the following:

- Water Code Sections 1810 et seq. are uniquely California provisions that require public agencies owning water conveyance facilities to wheel water for others, subject to availability of capacity and payment of compensation. This authority is fundamental to the execution of local water agency transfers that require transportation in the SWP or CVP.
- Water Code Sections 10610 et seq. require water purveyors serving "more than 3000 customers or supplying more than 3000 acre-feet of water annually" to prepare Urban Water Management Plans and to submit them to CDWR. Plans must be updated every 5 years and must include a water-shortage contingency element that sets forth how a supplier will respond to drought, including to a single-year supply cutback of up to 50%. This requirement applies to more than 400 water purveyors that provide almost all of the state's urban supplies, and compliance with it is a condition of eligibility for some state financial assistance programs.

11.3.2 Administrative Framework

Administratively, state financial assistance programs have been used to advance goals for water supply reliability and flood risk reduction. In recent years, voters have approved major water bonds providing grants for these purposes, including the following:

- Proposition 204 in 1996 for $995 million
- Proposition 13 in 2000 for $2.1 billion
- Proposition 50 in 2002 for $3.44 billion
- Proposition 84 in 2006 for $5.388 billion
- Proposition 1E in 2006 for $4.09 billion
- Proposition 1 in 2014 for $7.12 billion

One outcome of recent bond measures has been dedication of funding for local agency integrated regional water management (IRWM) planning and plan implementation. IRWM planning encourages local agencies to develop multiobjective, multibeneficiary projects that could, for example, combine water supply reliability benefits with flood risk reduction benefits. CDWR's guidelines for IRWM grants for Proposition 84 and Proposition 1E require applicants to meet a climate change standard that includes preparation of a regional vulnerability assessment.

11.3.3 Specifics regarding Climate Change

Formal incorporation of climate change science into California state agency programs has occurred in a remarkably short time. State policy activity began in the mid-2000s, with a Governor's Executive Order S-3-05 that established greenhouse gas emission (GHG) reduction targets and formed a state agency Climate Action Team (CAT), and also with the California Global Warming Solutions Act of 2006 that mandated GHG reduction.

Our Changing Climate, California's first climate change assessment, was published in 2006 (California Climate Change Center 2006), with subsequent assessments in 2009 and 2012. The then–Public Interest Energy Research program administered by the California Energy Commission funded substantial research on basic impacts of climate change specific to California, and facilitated the CAT's adoption of scenarios that state agencies would use for climate change planning. In 2008, Executive Order S-13-08 directed state agencies to plan for sea-level rise (SLR) and to commission a National Research Council (NRC) study to recommend ranges of SLR to use for state planning. California's Ocean Protection Council (COPC) took the lead in 2011 in adopting initial guidance to be used by state agencies whose programs involve planning for SLR for their own facilities or for permitting local development in coastal and estuarine zones (COPC 2011). The state's first California Adaptation Strategy was published in 2009 (CNRA 2009); a subsequent update was released in 2014.

11.4 Extreme Events—Historical Perspective and Lessons Learned

California is no stranger to droughts and floods, and even concerns about rising sea levels due to climate change represent an intensification of impacts historically observed during extreme storm events. Broadly speaking, extreme storms—their occurrence or lack thereof—fundamentally drive California water management activities.

California's most significant droughts of statewide geographic extent are those with the longest duration or driest hydrology. Three events in particular stand out in the historical record: the 6-year drought of 1929–1934, the 2-year drought of 1976–1977, and the 6-year event of 1987–1992 (CDWR 2014). It remains to be seen how the ongoing drought that began in 2012 and has, as of this writing, continued through 2015 will end up ranking in comparison to these significant historical events. This most recent event has stood out for its occurrence during record statewide warmth, a condition that might be expected to become more common in future droughts.

Occurring some 80 years ago, the 1929–1934 drought is difficult to place in context with modern conditions in terms of impacts. However, the drought was severe from a hydrologic perspective: within the 11-year period of water years 1924–1934, there were four extremely dry years, including 1924—holder of many site-specific hydroclimate records in California. In terms of extended duration of largely dry years, this historical event was on a par with the major events reconstructed in a millennium of Sacramento and San Joaquin River flows (described in this section). The hydrology of the 1929–1934 drought was widely used in design and analysis of much of the state's major water infrastructure.

Although the 1976–1977 drought was only a 2-year event, it included the driest single year of statewide runoff. The widespread impacts experienced in this drought were a wake-up call for water agencies statewide and highlighted vulnerabilities in some larger urban water suppliers' systems. California's most recent long-duration drought was 1987–1992; it provides the closest analog to present conditions of water management infrastructure. This event spurred enactment of legislation to facilitate water transfers and water banking, as well as to strengthen requirements for urban water-shortage planning. One outcome following both of these droughts was improvement in interconnections among the larger urban water agencies in most of the state's major population areas, facilitating water transfers and emergency response actions. The 1987–1992 drought also helped build momentum for development of new local agency large-scale managed groundwater banking projects (Jones 2003). These large historical droughts and the more recent events since then have, however, highlighted areas where improvements are needed: long-term stability for moving water across the delta and for water project diversions from the delta, reduction in water-shortage risks for small water systems in rural areas that depend on fractured rock groundwater sources, and improvement in subseasonal to seasonal climate forecasting.

Paleoclimate reconstructions have identified California droughts as far more severe than today's water institutions and infrastructure were designed to manage. Perhaps the earliest recognition of the relative severity of paleodroughts dates back to the modern drought of 1929–1934, when Lake Tahoe dropped below its natural rim and exposed tree stumps rooted in place on the lake bottom (Harding 1965). Later, studies of additional relict tree stumps rooted in place in Lake Tahoe and other central Sierra Nevada waterways identified chronic dry periods dating back to mid-Holocene times (e.g., Lindstrom 1990; Stine 1994; Kleppe et al. 2011). CDWR has funded streamflow reconstructions for the Sacramento, San Joaquin, and Klamath Rivers (Meko et al. 2014) to improve understanding of drought vulnerability. This work discovered that the Sacramento and San Joaquin Rivers share 1580 as their single driest year in the combined reconstructed and instrumental streamflow record; the reconstructed flow in 1580 was only about half of that of the driest year (1924) in the observed record. CDWR has likewise contributed to funding Colorado River streamflow reconstructions. Reconstructions of Lake Powell inflow show multidecadal periods when flows were below the long-term average, and the driest period in the observed record is greatly surpassed in severity by conditions in parts of the reconstructed record (Meko et al. 2007).

Paleoflood information is site-specific and difficult to generalize at a regional or statewide scale, so this approach does not lend itself to a broad examination of floods prior to the historical record. Within the historical record, the extent of flooding and the occurrence of flood damages are greatly influenced by the level of development in a given area and by human modification of the natural environment. In using the time period from the second half of the twentieth century onward as representative of modern conditions, the years distinguished as experiencing large-scale flooding across much of the state were 1950, 1955, 1964, 1969, 1986, and 1997. These major winter floods highlighted two important points—the limitations of flood frequency analyses given a relatively short historical record and the need to improve Central Valley levees and related infrastructure. With respect to the valley's levee systems, the widely used aphorism that there are only two kinds of levees—those that have failed and those that will fail—is attributed to State Engineer William Hammond Hall and likely reflected his familiarity with the construction techniques used for the early agricultural levees. Long-standing recognition of the need to repair and upgrade valley levees was given greater emphasis in the public consciousness in the aftermath of the damage in Gulf Coast states caused by Hurricane Katrina, spurring state emergency funding in the mid-2000s to begin critical repairs. Subsequently, voter approval of Proposition 1E in 2006 provided initial funding for a programmatic effort on flood risk reduction. The backlog of flood risk reduction work remains substantial, however, and substantial additional investments are needed, as is work on nonstructural risk management techniques such as local land-use planning.

The example of Folsom Dam on the American River upstream of Sacramento—the example sometimes facetiously used to point out that building dams causes floods of record—speaks to limitations associated with flood frequency analyses. Folsom Dam was authorized under the Flood Control Act of 1944 and was originally expected to provide a roughly 250-year level of protection for the downstream urban area. The USACE began dam construction in 1948, and in 1950, the largest flood of measured record occurred (its size relative to the Great Floods of 1861–1862 is uncertain, because streamflow data are not available for that early event). The 1950 flood was followed by an even larger flood in 1955, when the partially complete Folsom Dam was credited with saving much of Sacramento from flooding. Peak inflows to Folsom Lake in the subsequent 1964 flood surpassed those of the 1955 flood, and then in 1986, a new flood of record occurred in the watershed. A possible flood disaster was averted through a joint flood operations decision by USACE, USBR, and CDWR to release more water from the dam than the rated levee downstream capacity; fortunately, the levees held, and massive urban flooding was avoided. Subsequent hydrologic reanalysis placed the 1986 flood at a 70-year recurrence interval event, with Folsom Dam providing about a 63-year level of downstream protection at the rated levee capacity (USACE 1988). Another peak flood occurred in 1997; its 3-day flood volume was roughly equal to that of the 1986 event. There were thus five major floods of record in the second half of the twentieth century, as compared to the relatively less active first half of the century.

The American River example demonstrates how an essential flood response tool—weather forecasting—has dramatically improved over time. The modern era of weather satellites began in 1960 with launch of the Tiros I satellite. Weather forecasts available to reservoir operators for the floods of the 1950s and even the 1964 event provided only a fraction of the information now available from satellite data and modern weather models. The decisions made in preparation for American River operations in the 1986 flood would not have been possible but for long-lead weather observations and forecasts that were not available during the earlier events. Looking forward, further improvements in forecasting at the weather and subseasonal to seasonal climate timescales are an important tool for adaptation actions associated with reservoir operations for both drought and flood conditions.

One step on the path of improving these forecasts has been a substantial investment made in understanding and monitoring extreme precipitation in California through a research partnership between the NOAA's Hydrometeorology Testbed program and CDWR. Tracking water-vapor transport across the Pacific Ocean into California and subsequent research to understand storm dynamics revealed the importance of atmospheric river (AR) storms from both a flood management and a water supply perspective. On average, these storms are estimated to contribute about 40% of California's annual precipitation (Dettinger et al. 2011); when the storms stall at coastal or Sierra mountain ranges, conditions are favorable for intense orographic precipitation that often sets the stage for floods. Improving AR prediction would support forecast-informed reservoir operations, an active area of adaptation planning for CDWR. Because managing for either wet or dry extremes in California is fundamentally about managing for the presence or absence of large winter storms, improving understanding of these events is an important aspect of adaptation.

11.5 Extremes in a Changing Climate

Extensive material has been published about climate change in California and its expected impacts. The reader is directed to the California climate assessments (CNRA 2009) and to the Southwest climate assessment (Garfin et al. 2013) for detailed discussion of this subject. Briefly, global climate model studies generally show good agreement with respect to projections of warmer temperatures. Although agreement among climate model studies is not as good for precipitation as it is for temperature, one of the best areas of model agreement is for increased aridity in the US Southwest, including Southern California. Northern California, in contrast, is projected to become slightly wetter than the present. Expected climate change impacts include increased water demands due to warmer temperatures and loss of Sierra Nevada and Cascades snowpack. Warming in the Sierra Nevada, expressed in Figure 11.3 in the form of a time series of the freezing level for an example location, is already being observed. The elevation of the freezing level is important in a snowpack-dominated hydrologic system, at the timescale both of an individual storm (how much precipitation will run off immediately and must be managed for flood risk reduction) and of a winter season (how much snowpack will be available to fill reservoirs in the spring). Climate models generally show very pronounced impacts due to warming—such as loss of half or more of Sierra Nevada snowpack—by the end of the century, with notable impacts being observed by midcentury (California Climate Change Center 2006).

The Southwest climate assessment summarizes some expected outcomes of climate change related to extreme events, listed here.

- Winter precipitation extremes are expected to become more frequent and more intense.
- Winter floods in Sierra Nevada watersheds are expected to increase in intensity, while spring snowmelt floods are expected to diminish in frequency and intensity. Flood risks from large AR storms may increase, but more investigation of the relationship between climate change and occurrence of these storms is needed.
- Drought in the Colorado River Basin is expected to become more frequent and more intense. Northern Sierra watersheds may become somewhat less drought-prone due to increased precipitation.

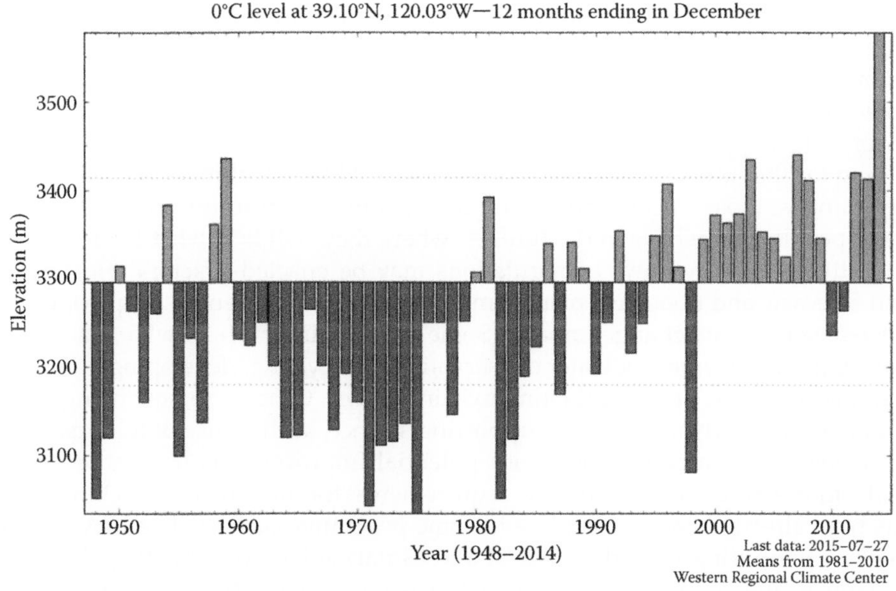

FIGURE 11.3
Annual elevation of freezing at Lake Tahoe.

Over time, rising sea levels will increase the risk of inundation for coastal and estuarine low-lying areas, a risk particularly associated with winter storms during El Niño conditions. Sea cliffs and dune-backed beaches constitute about 80% of California's open coast, reducing the area topographically vulnerable to inundation (California Climate Change Center 2009). Much of the shoreline of the San Francisco Bay estuary, however, is low-lying and topographically vulnerable to inundation. Historically, flood damages due to coastal inundation have been associated with winter storms when high tides, storm surges, and large waves combine; especially damaging events occurred during the strong El Niño events of 1982–1983 and 1997–1998 (when observed sea levels exceeded regional mean sea levels projected for 2100) (NRC 2012). The incidence of observed water levels many feet above historical mean sea level in the San Francisco Bay Area is expected to increase over time, going from a historical rate of 9 h per decade to 12,000 h per decade by the century's end (NRC 2012).

11.6 Discussion

The aphorism that the future will be like the past but more so is applicable when considering climate change and extreme events in California, at least within reasonable time frames for conventional public agency water planning by (e.g., not beyond the end of the century). California's water infrastructure and institutional framework were designed for managing water under conditions of shortage (droughts) and excess (floods), although much of the state's water infrastructure is aging and investments are needed to maintain and expand its capacity and to incorporate new technologies for increasing water-use efficiency. Setting aside climate change impacts, existing systems are already being managed to respond to the challenges of population growth, increased urban development in floodplains, new water and habitat requirements for special-status species, and unanticipated natural variability in the climate system. Climate change adds an additional challenge to this list and further emphasizes needs to upgrade the state's aging infrastructure, which the American Society of Civil Engineers (ASCE) grades as "D" for levees/flood control and as "C" for water supply (ASCE 2012).

Planning for the future always holds uncertainty. Although climate change science has its uncertainties, those uncertainties can pale in comparison to uncertainties associated with how people will behave in the future—where they will live, what land-use patterns will look like, what new laws or regulations may be enacted—factors affecting water-demand forecasts and flood risk management needs. Many potential adaptive tools exist for water-sector climate change impacts. Some entail doing more of what water agencies are already doing—increasing water conservation or recycling, developing more groundwater storage capacity, or strengthening existing levees. Others are now being evaluated in conjunction with advances in the supporting science, such as use of forecast-informed reservoir operations in conjunction with potential improvements of weather-to-climate timescale forecasting. Administrative requirements for preparing for climate change impacts have already been put in place in some programs, such as state IRWM grants or state agency permitting related to coastal and estuarine land use. Public policy is inherently adaptive, and changes may be made to the institutional framework in response to observed needs, as was the case with statutory changes made to facilitate water transfers during the 1987–1992 drought. Adapting to major reductions in Sierra and Cascades snowpack will arguably be, for example, one of the more difficult institutional challenges that California must face, but the long period over which this impact will occur offers lead time for putting the elements of a response in place and financing needed actions.

References

ASCE. 2012. California Infrastructure Report Card, 2012. http://www.ascecareportcard.org/.

Attwater, W.R. and J. Markle. 1988. Overview of California Water Law, *Pacific Law Journal* 19(4), 1988.

California Climate Change Center. 2006. *Our Changing Climate: Assessing the Risks to California.* Sacramento: California Energy Commission, July.

California Climate Change Center. 2009. *The Impacts of Sea Level Rise on the California Coast*. Sacramento: California Energy Commission, May.

California Department of Public Works, Division of Water Resources. 1930. Bulletin No. 25, Report to the Legislature of 1931 on State Water Plan. Sacramento.

CDWR. 1957. *Bulletin No. 3, the California Water Plan*. Sacramento.

CDWR. 2000. *Preparing for California's Next Drought: Changes Since 1987–92*. Sacramento, July.

CDWR. 2010. *California's Drought of 2007–2009, An Overview*. Sacramento, November.

CDWR. 2014. *California Water Plan Update 2013*. Sacramento, October.

CDWR. 2015. *California's Most Significant Droughts: Comparing Historical and Recent Conditions.* Sacramento, February.

CNRA. 2014. *Safeguarding California: Reducing Climate Risk, an Update to the 2009 California Climate Adaptation Strategy*. Sacramento, July.

CNRA. 2009. *2009 California Climate Adaptation Strategy: A Report to the Governor of the State of California in Response to Executive Order S-13-08*. Sacramento.

COPC. 2011. *Proposed Resolution of the California Ocean Protection Council on Sea-level Rise*, http://opc.ca.gov/webmaster/ftp/pdf/agenda_items/20110311/12.SLR_Resolution/20110311OPC-SLR-Resolution.pdf., March.

Dettinger, M.D., F.M. Ralph, T. Das, P.J. Neiman, and D.R. Cayan. 2011. Atmospheric rivers, floods, and the water resources of California. *Water*, 3, 455–478.

Garfin, G., A. Jardine, R. Merideth, M. Black, and S. LeRoy, eds. 2013. *Assessment of Climate Change in the Southwest United States: A Report Prepared for the National Climate Assessment.* Washington, D.C.: Island Press.

Hall, W.H. 1880. *Report of the State Engineer to the Legislature of California, Session of 1880, Part 4, Irrigation of the Plains*. Sacramento.

Harding, S.T. 1965. *Recent Variations in the Water Supply of the Western Great Basin*. Berkeley, CA: University of California Berkeley, Water Resources Center Archives.

Jones, J. 2003. Groundwater Storage—The Western Experience. *Journal, American Water Works Association.*

Kelley, R. 1989. *Battling the Inland Sea: Floods, Public Policy, and the Sacramento Valley*. Berkeley: University of California Press.

Kleppe, J.A., D.S. Brothers, G.M. Kent, S. Jensen, and N.W. Driscoll. 2011. Duration and severity of medieval drought in the Lake Tahoe Basin. *Quaternary Science Reviews* 30(23–24), 3269–3279.

Lindström, S. 1990. Submerged tree stumps as indicators of Mid-Holocene aridity in the Lake Tahoe Region. *Journal of California and Great Basin Anthropology*, 12(2). Retrieved from: http://www.escholarship.org/uc/item/4s95f878.

Lund, J., E. Hanak, W. Fleenor, R. Bennett, J. Howitt, P. Mount, and P. Moyle. 2008. *Comparing Futures for the Sacramento–San Joaquin Delta*. Sacramento: Public Policy Institute of California.

Meko, D.M., C.A. Woodhouse, and R. Touchan. February 2014. *Klamath/San Joaquin/Sacramento Hydroclimatic Reconstructions from Tree Rings, Report to the California Department of Water Resources, Agreement 4600008850.*

Meko, D.M., C.A. Woodhouse, C.A. Baisan, T. Knight, J.J. Lukas, M.K. Hughes, and M.W. Salzer. 2007. Medieval drought in the upper Colorado River Basin. *Geophysical Research Letters* 34, L10705.

National Research Council. 2012. *Sea-Level Rise for the Coasts of California, Oregon, and Washington: Past, Present, and Future.* Washington: National Academies Press.

Stine, S. 1994. Extreme and persistent drought in California and Patagonia during mediaeval time. *Nature* 369, 546–549.

Suddeth, R., J. Mount, and J. Lund. August 2010. Levee decisions and sustainability for the Sacramento–San Joaquin Delta. *San Francisco Estuary and Watershed Science* 8(2), 1–23.

USACE. 1988. *American River Watershed Investigation, California: Reconnaissance Report*. Sacramento, January.

12

California's Sacramento–San Joaquin Delta: Reflections on Science, Policy, Institutions, and Management in the Anthropocene

Richard B. Norgaard*

CONTENTS

ABSTRACT The Sacramento–San Joaquin Delta is a critical node in California's water system, a node through which many complex social and environmental feedbacks loop. Understanding likely futures of the Delta requires a historical, complex socioecological system perspective of its dynamics. Appropriate institutional organization to promote effective science and management of the feedbacks has eluded California political and policy processes. Climate change and sea level rise increase the complexity, adding to the political and policy difficulties of finding effective institutional structures for science and management. The challenges in this Delta provide broad insights into the difficulties of water and ecological management in the Anthropocene.

12.1 Introduction

The socioecological story of the Sacramento–San Joaquin Delta (hereafter, Delta) is an excellent regional example of humankind's global predicament. We are now in the Anthropocene, an era in which people are the primary drivers of environmental change

* While acknowledging that much of my understanding stems from my engagement with Delta scientists, managers, policy makers, and stakeholders, the details selected, interpretations provided, and judgments expressed in this document are strictly my own unless otherwise attributed.

with complex feedbacks on a global scale (Crutzen 2002). Water management in California provides a similar tale of strong feedbacks at a regional level with the Delta at the center. The water technologies used over the past century and a half and other interventions with what was once an unspoiled environment have had unexpected environmental and social consequences. These instigated corrective interventions that, in turn, have had their own unexpected consequences. Delta water management already almost entirely entails dealing with problems we ourselves have created. Now, with human-driven global environmental dynamics also driving regional environmental change, there are yet more complex socioenvironmental feedbacks with which to contend.

Given the relatively gradual pace at which we come to understand our environment and the fact that the environment is now changing more rapidly, the limitations of our environmental understanding are considerably greater than we have historically thought they were (Healey Dettinger, and Norgaard 2008; NRC 2012). This in itself suggests that more scientific humility and less technological and developmental hubris is in order. The problem, however, is not simply a matter of rates of change and learning. Scientific research has yet to be framed so as to incorporate the critical characteristic of the Anthropocene: that people are the primary drivers of environmental change and the change is ongoing. This is true with respect to understanding the Delta (Norgaard, Kallis, and Kiparsky 2009), and also true globally (Millennium Ecosystem Assessment 2005; IPCC 2014). The California Delta experience provides a vivid example of regional-scale challenges to environmental science and governance. These challenges are also emerging with increasing frequency around the globe.

The natural and social aspects of history in the Anthropocene tightly unfold together. To understand the Anthropocene, we need to see how people and nature constantly affect each other. Unfortunately, text consists of a sequence of words, while words are laden with historic meanings that are difficult to reuse for new purposes. Given these limitations, Section 12.2 emphasizes the water and ecological systems and their modification by people. Section 12.3 emphasizes the social portion, centering now on the political and institutional responses to what are perceived as water, environmental, or ecological problems but in fact are human problems—the consequences of earlier human interventions guided by earlier scientific framings of our relationships with nature. This bare-bones background provides a basis for reflecting on current scientific and institutional quandaries in Section 12.4. Key points are brought out in Section 12.5, the conclusion, through commentary on points made at a Delta science conference.

12.2 Water and Environmental History

The Delta forms as the Sacramento and San Joaquin Rivers and several smaller tributaries converge, slow, and deposit their silt, creating an inland delta before continuing on to San Francisco Bay, itself an inland bay. Historically, 60% of California's runoff to the sea drained through the Delta. Most of the water originates in the Sierra Nevada mountain range. Major reservoirs store precipitation that falls mainly as snow at the higher elevations in winter, with delays in melting moderating some of the seasonal concentration in precipitation. The reservoirs are managed so as to moderate annual variation in precipitation and, thereby, historic flows through the Delta. The Delta is the hub of California's complex water distribution system. The federal government's Central Valley Project (CVP) and California's State Water Project (SWP) distribute northern California water southward from the southern end of the Delta. Regional irrigation and municipal water districts also take water from the Delta. About 25 million Californians, nearly two out of three, get some of their water from the Delta or use water intercepted before it otherwise would have arrived there.

Humans have heavily influenced the Delta environment since the second half of the nineteenth century. Delta marshlands were reclaimed for agriculture through the construction of levees. Hydraulic mining in the Sierra Nevada foothills left debris in the spawning grounds of salmon, and their populations declined dramatically. Salmon populations declined further with the construction of upstream dams. Striped bass were introduced for sport fishing in the early twentieth century, crowding out other species, and then went into decline themselves. Agricultural, municipal, and industrial pollutants soon became significant. The environmental consequences of these earlier transformations became more critical when compounded by the export of northern California water from the Delta to the south in the 1950s for the CVP, and then more so with the export in the 1970s for the SWP.

The map of the Sacramento–San Joaquin Delta in Figure 12.1 shows the labyrinth of islands formed by levees built to transform marshland into farmland. The Sacramento

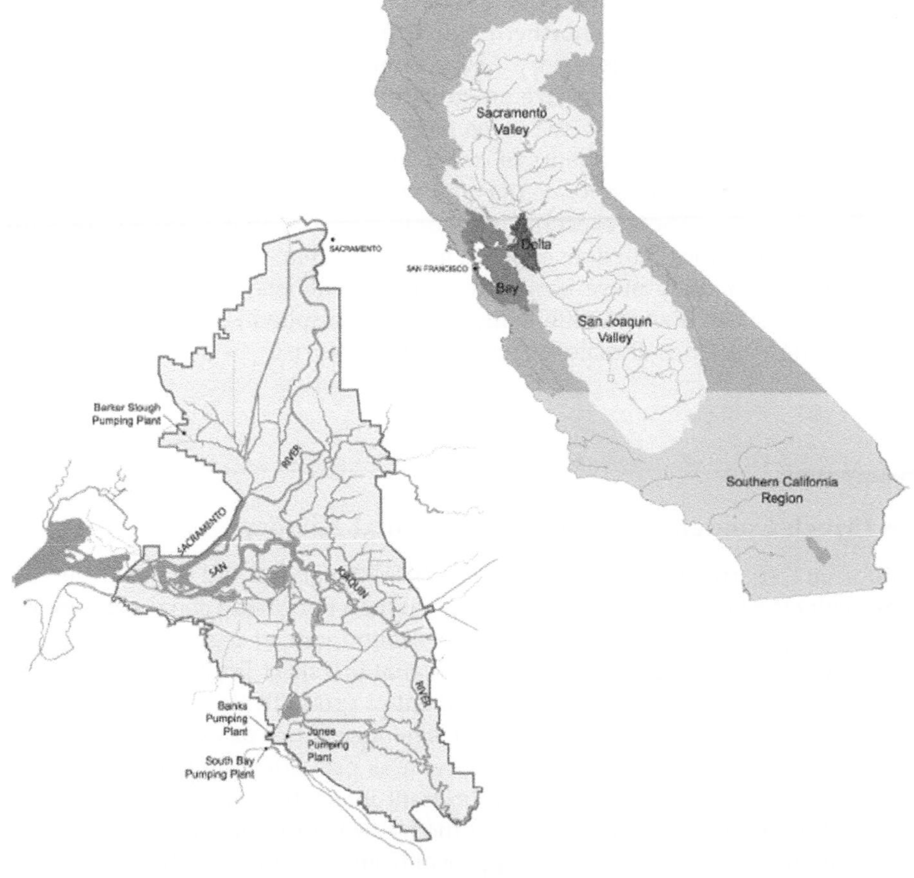

FIGURE 12.1
Regions of California, with detail of the Sacramento–San Joaquin Delta. (Figure courtesy of California Department of Water Resources.)

River enters the Delta from the north. Water is exported from the southern end of the Delta into the canals of both the federal and state water projects for use in San Joaquin Valley agriculture and for municipal use in southern California.

Water engineers have always seen the Delta as an obstacle, starting with the first advocates of grand plans to bring water to southern California in the nineteenth century. Encouraging water to flow from north to diversion pumps in the south of the Delta, through a maze of islands through which water naturally flows west, was long recognized as a bad idea. Building a peripheral canal around the Delta, however, also raised political and financial problems that repeatedly got in the way of having one built. Diverting water around the Delta would leave Delta island farmers with lower-quality water; fish and other wildlife would also suffer (Hundley 2001; Kallis, Kiparsky, and Norgaard 2009; Norgaard, Kallis, and Kiparsky 2009). Perhaps more importantly, Delta farmers and environmentalists feared that a canal would facilitate greater water exports from the Delta, compounding existing problems. Rather than establishing trust between in-Delta and other stakeholders, federal and state agencies exported water from the southern end of the Delta, subsequently complicating Delta problems and politics with pollutants returning from newly irrigated land in the San Joaquin Valley. Delta political conflicts have evolved over time and have mostly become more difficult as the environmental consequences in the Delta of marshland conversion, water exports, and pollution have unfolded and compounded.

With sea level rise, diverting northern California water around the Delta by some means will someday become necessary. Levee failure is more likely with sea level rise, and levee failure would bring massive amounts of salty water into the Delta to fill the area reclaimed for agriculture by the levee. Such failures will endanger the quality of water for Delta farmers and the quality of water diverted southward. Eventually, sea level rise alone will significantly deteriorate the quality of water that can be exported from the existing diversion sites. These concerns, of course, presume that there will be sufficient water in northern California to deliver south as climate changes. Furthermore, some irrigated crop production may move northward with climate change to avoid temperature extremes. Nevertheless, it seems safe to argue that a peripheral canal or tunnel of some size is increasingly likely as climate changes and the seas rise.

12.3 A (Much Condensed) Delta Institutional History

The California Department of Water Resources (DWR) planned and was actively engaged in the difficult effort to obtain legislative approval for a peripheral canal beginning in the 1950s. DWR and the first Jerry Brown administration gained legislative ground in the late 1970s and early 1980s by attaching to the canal bill strict limits on further environmental degradation in the Delta, the management of groundwater, and the protection of north coast streams from water development. The legislature passed a complex package of promises in mid-1980. Rather than being seen as providing something for everyone, the bill set off a multifronted storm with different interests protesting different components of the bill. A ballot measure to repeal the bill was formulated, vigorously debated, and passed in June 1982, garnering 63% approval from a complex combination of interests north and south. Diverse interests had clearly not come to a shared understanding as to the nature of Delta water-environmental dynamics and how to intervene in them so as to set them on an acceptable course. Indeed, to give a peek into the direction of this chapter,

Delta scientists have not had such a dynamic, integral socioecological framework for conducting and communicating their research.

It is important to keep in mind that all of this is taking place in the context of increasing distrust in government combined with a growing sense that somehow, private firms and markets can solve all problems. At the same time, the awareness of environmental problems is rising and broadening from regional to global framings. While the expertise of agency scientists advanced to meet new challenges, state and federal agency budgets were steadily being cut, reducing the number of agency scientists and their ability to do research. Consultants, from independent entrepreneurs to employees of multinational consulting corporations, do more and more science on contract, leaving agency scientists to increasingly serve as science administrators.

In 1994, California and the US Department of the Interior signed an agreement, the Bay–Delta Accord, to facilitate the collaboration of water and environmental agencies to resolve water conflicts related to the Delta and to coordinate future management. The structure and process, known as CALFED, initiated a science program with a small coordinating staff, a lead scientist, and an Independent Science Board. The science program proved fairly effective at reviewing, apparent strengthening, and, though to a lesser extent, coordinating Delta science. As an administrative reform that would lead to better management of California water and the Delta, however, CALFED proved ineffective at setting a course of action and was officially declared dead in 2005 (Little Hoover Commission 2005). CALFED's administrative failure, however, had been becoming increasingly clear as Delta stakeholders thought it advantageous to do battle in the courts rather than await administrative clarity. This led to the courts declaring what constituted best science with respect to water management and the protection of endangered species. Once again, the institutional structure proved inadequate to formulate sufficient shared understanding for effective management (Kallis, Kiparsky, and Norgaard 2009).

With the demise of CALFED, two parallel efforts began at the same time. First, the DWR, with funding provided by water agencies south of the Delta, initiated the Bay Delta Conservation Plan (BDCP). The BDCP soon settled on a proposal to tunnel under the Delta to assure water reliability, with the stress on reliability and water quality rather than a higher level of imports. The BDCP also includes some 150,000 acres (60,000 ha)* of ecological restoration in the Delta to assure habitat and food for critical and endangered species. The entire BDCP effort is billed as a habitat conservation plan in exchange for which *take permits*, permissions to kill a limited number of *now* better-protected endangered species, would be issued by the state's Department of Fish and Wildlife. BDCP would take 7 years to elaborate.

Simultaneously with the emergence of BDCP, Governor Schwarzenegger established a Delta Vision Blue Ribbon Task Force in 2006 to work with Delta stakeholders to come up with a vision for what the Delta is and could be and how the Delta should best be managed. Their efforts laid the groundwork for the Delta Protection Act passed by the state legislature in 2009 (State of California 2009). This act formed a Delta Stewardship Council (DSC), an independent agency whose members are appointed. The DSC is specifically charged with devising and maintaining a Delta Plan. A Delta Science Program (DSP) and a Delta Independent Science Board (DISB), both remnants from CALFED, support the DSC. The DSP, continuing its operation under CALFED, strives to coordinate and provides review services for Delta research programs undertaken across all of the relevant agencies. The

* 1 acre = 0.4 ha.

DISB reviews agency science programs to see that they are using the *best available science.* In recognition of the increasingly complex environmental dynamics, the Delta Reform Act also mandates the use of adaptive management.

To coordinate the array of scientists working in water and environmental agencies, the DSP initiated a Delta Science Plan as part of the Delta Plan. Scientists from multiple agencies as well as the DSC and DISB vetted the plan. Through the Delta Science Plan, operating under the logo "one Delta, one science," scientists from multiple agencies grapple with research priorities, coordinate and improve monitoring and analysis, and strengthen the connections between researchers and policy makers. This newly initiated science plan will evolve as it is implemented and hopefully prove increasingly effective over time.

By the end of 2013, BDCP consisted of a 17,000-page draft plan and 22,000-page draft environmental impact report (EIR) prepared under the auspices of the DWR with the aid of other key agencies, but with much of the work done by ICF International, a management and engineering consultancy. The Delta Protection Act specifies that the DISB shall review the EIR prepared for the BDCP. It further states that BDCP will be incorporated into the DSC's Delta Plan if and when it becomes approved by the state's Department of Fish and Wildlife as an adequate habitat conservation plan. Federal approval also complicates adoption of the BDCP.

The DISB reviewed the plan and its EIR during early 2014 and reported to the DSC on May 15, "The potential effects of climate change and sea-level rise on the implementation and outcomes of BDCP actions are not adequately evaluated" (Delta Independent Science Board 2014, p. 3).

With respect to adaptive management, the DISB found the following:

> Details of how adaptive management will be implemented are left to a future management team without explicit prior consideration of (a) situations where adaptive management may be inappropriate or impossible to use, (b) contingency plans in case things do not work as planned, or (c) specific thresholds for action. (Delta Independent Science Board 2014, p. 3)

Then on August 26, 2014, the US Environmental Protection Agency stated the following in the transmittal letter for its review of BDCP's Draft EIS:

> EPA supports the Draft EIS's recognition that climate change and sea level rise would likely result in decreased freshwater flows into and through the Delta and increased salinity intrusion, however, the assumption that in the face of diminished overall water supply due to climate change, diversions north of the Delta would be allowed to increase seems unrealistic. (US Environmental Protection Agency 2014, p. 19)
>
> EPA fully supports the stated purpose of the BDCP effort: to produce a broad, long-term planning strategy that would meet the dual goals of water reliability and species recovery in this valuable ecosystem, and we recognize the potential benefits of a new conveyance facility. However, we are concerned that the actions proposed in the Draft EIS may result in violations of Clean Water Act water quality standards and further degrade the ecosystem. (US Environmental Protection Agency 2014, p. 1)

The BDCP is a major collaborative effort, albeit driven by water users south of the Delta, to resolve some of California's most difficult water issues. Climate change and sea level rise are making the existing system of getting water through the Delta ever more inappropriate. The Delta will continue to change, but Californians can help determine how it will change as well as be in a better position to adapt to change. Yet a plan of action designed for diverse Delta stakeholders and the electorate as a whole, one that would lead to a sufficient consensus, seems as elusive as ever.

12.4 Understanding, Institutions, and the Anthropocene

Let me proffer a broad explanation as to why not only Californians but modern societies in general have difficulties addressing how to understand and manage complex, dynamic environmental systems such as the Delta, made more complex by climate change. Historically, the role of science in policy has been to more clearly define alternative choices and their likely impacts. Choices need to be made; science makes those choices more explicit. With the aid of science, the facts can be agreed upon. The facts themselves are not subject, or at least much less subject, to the influence of political manipulation. This expectation for the role of science is imbedded in existing environmental legislation, agency structure, and our understanding of the science–policy interface.

Science has never quite lived up to this expectation of how it is supposed to work, however. The fractured structure of science; the difficulties of predicting the future; and the ongoing tendency of stakeholders to influence what politicians, policy makers, and the public see as relevant information have all contributed to this failure. Nevertheless, this vision of the role of science persists (Sarewitz, Pielke, and Byerly 2000). The California Delta provides an informative example of how this view of science in the policy process is further challenged by climate change. Indeed, new complexities can provide new screens behind which political maneuvering with respect to the facts takes place.

Let me briefly provide some background on the organization of science with respect to complexity.

12.4.1 From Complicated to Complex

Both natural and social systems theorists have suggested that there are critical distinctions between *complicated* and *complex* (Waldrop 1992; Poli 2013). A system consisting of knowable parts connected in relatively simple, knowable ways becomes more complicated as the number of parts and interactions increases. For example, automobiles have become more complicated over time, partly by adding new parts such as air-conditioning. Even though modern braking systems and air-conditioning systems depend on power from the propulsion system, the three systems can be thought of pretty much independently. The interdependencies can be stated quite simply and kept in mind when necessary because each is relatively simple. The hydrological system leading to and through the Delta, even with water diversions, is complicated: there are many parts and relations, but the parts are knowable, and the hydrological relationships are relatively straightforward.

A system with many parts becomes complex when the parts are less easily understood and are interdependent in multiple ways that are also not easily understood. Feedbacks in relationships, especially positive feedbacks, make understanding more difficult. The system becomes especially complex when parts come and go (think of invasive species and local extinctions), resulting in wholly new relationships among all of the parts. In nature, and in society too, existing parts and relations evolve and develop wholly new characteristics over time.

This distinction between complicated and complex is a state of mind that guides how we understand, while reality is what it is, unfolding as it will. The structural inertia of the institutions of science—the disciplinary departments of universities and the administrative divisions between agencies—can keep the scientists studying the parts of a complex system separately, as if the parts fit together simply, ignoring the more challenging and beguiling complexity. In fact, the success of western science has come predominantly through reductionism, by assuming that systems are merely complicated. Implicit in western science is the assumption that the parts will fit together when each part is known well enough. In the process, science becomes unified. New environmental and social problems arise, however, precisely because our understanding of whole systems is so poor. Technological and organizational changes are introduced without adequate systemic environmental and social understanding that then result in unexpected systemic environmental and social problems. Since the 1960s, there have been increased efforts to understand environmental systems. There have been parallel, though less ambitious, efforts to understand social systems as whole systems. Now, to the extent that we understand that we are in the Anthropocene, we increasingly need to think in terms of combined socioecological systems operating across different spatial and temporal scales (Crutzen 2002; Folke et al. 2005).

The framework suggested in the Millennium Ecosystem Assessment (MA) shown in Figure 12.2 integrates social and natural systems. It provides a visual example of interaction, with reminders that the processes pictured are working across different spatial and

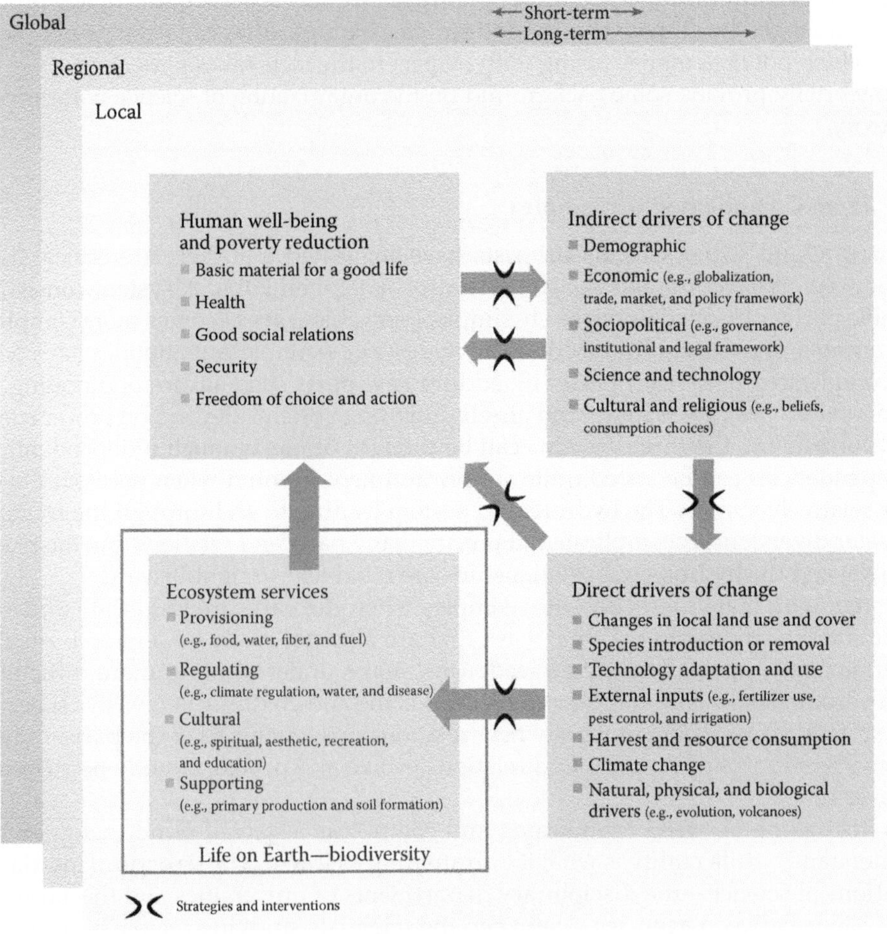

FIGURE 12.2
Framework, Millennium Ecosystem Assessment, 2005.

temporal scales (Millennium Ecosystem Assessment 2005). While the diagram appears to be merely complicated, both the components and relations between them are difficult to define, different in different places and at different spatial and temporal scales, and evolving over time, making the system truly complex. This framework is an excellent guide to how we might think more clearly in the future about social and natural systems' interaction. The framework, however, was of relatively little help in evaluating the literature existing at the time of the MA. Scientists prior to the MA did not have this framework in mind when they identified the questions they were asking, designed their research, and interpreted their findings (Norgaard 2008); hence, their work did not readily fit the new framework devised for the assessment. The MA framework has since become influential only to the extent that it has framed some new research.

12.4.1.1 Delta Complexities

Looking simply at water flows associated with the Delta, we see a system that is definitely complicated. Most of the water, from the Southern Sierra that would have drained through the Delta is delivered directly to San Joaquin Valley farmers. Some of that water recharges groundwater basins; a small amount makes its way back into the San Joaquin River. San Francisco and East Bay cities dammed Central Sierra streams, diverting this water into major pipelines that cross the Delta to municipal and industrial users. Much of this water returns to the San Francisco Bay after treatment. Northern California water, mostly from the Sacramento and Feather Rivers, irrigates the Sacramento Valley, provides water for Sacramento and other cities, and flows into the Delta. A significant percentage of the water that flows into the Delta is pumped at its southern end and sent south for San Joaquin Valley agriculture as well as municipal use in southern California. Central Californians also tap water for municipal use directly from the Delta. Lastly, Delta residents farming reclaimed marshland are also significant users of water in the Delta. There are also environmental uses of water in order to maintain habitats and protect species at risk before and within the Delta. The water system is complicated, the stakeholders diverse, but computer models of the hydrological system, basically spatial accounting models, reproduce historic patterns and guide management reasonably well.

Other factors complicate the Delta socioenvironmental system, bringing it close to complexity. Agricultural, industrial, and municipal users of freshwater are simultaneously putting pollutants into the water system. The Delta levees are vulnerable to breaking now and will become more so with sea level rise. The levees are maintained by a diverse mix of agencies backed by an equally heterogeneous mix of stakeholders. Urban encroachment transforms the margins of the Delta landscape, limiting the options for ecological restoration and bringing new stakeholders into Delta politics. Politics inevitably entails questions of fairness, but it is difficult to discuss the allocation of surface water among users when groundwater is not allocated and has been exploited freely by many of the same users. Groundwater is only now coming under regulation so that its use can be coordinated with surface water (State of California 2014).

The Delta becomes truly complex when we put species and their interactions into the system. The biological system responds to highly variable hydrological conditions, some natural, some man-made, as well as the human transformation of the landscape. The California Bay-Delta system is one of the most heavily invaded estuaries in the world (Cohen and Carlton 1998; Cohen 2011). Introduced species have transformed food chains. People are active agents of change and are constantly responding to the consequences of the changes they have initiated. Delta research needs to be done with an overarching framework in mind, comparable to that in Figure 12.2, but framings to date have yet to include people as active agents.

12.4.1.2 Complexity, Science, and Policy Processes

Acknowledging that the Sacramento–San Joaquin Delta system is complex rather than merely complicated raises critical issues about the organization of science and decision making (Dryzek 1987; Mitchell 2009) that have not yet been openly broached by Delta scientists and policy makers. Table 12.1 suggests the organizational needs for understanding and managing complicated and complex systems. The breakdown suggests how conventional bureaucratic systems can handle complication reasonably well because parts of environmental problems can be seen as separable and can be studied separately by separate agencies and scientific teams. Complexity suggests a quite different approach, one based on much more sharing of information and ideas among scientists, managers, and policy makers (Norgaard and Baer 2005a,b). Working with complexity also suggests less distinction between learning through science and learning through management. For complex systems, there needs to be flexibility with respect to goals and research agendas to accommodate learning and a changing environment. Such flexibility, however, raises serious issues about the role of scientists as merely objectively carrying out democratically determined goals versus being setters of goals (Ezrahi 1990).

Standard descriptions of adaptive management have elements in the "Complicated System" column—being responsive to new information, learning more systematically from information generated through management, and being able to switch to new goals as either the system changes or knowledge about the system changes. Adaptive management has been central to Delta science since the Delta Reform Act (2009), but discussions about its implications for organizational structure have mostly entailed layering additional, but informal, linkages into the existing structure rather than designing from scratch and seeking legislative approval for structural change.

Adaptive management presents a significant quandary. One cannot redesign the proposed tunnels once they have been built. Managers can adaptively manage how major water

TABLE 12.1

Organizational Forms Best Suited for Complicated and Complex Systems

Complicated System	Complex System
Broad research agenda can be "parted out" and tackled by relatively independent research programs for use by separate managers who communicate little with scientists or each other.	Frequent communication among researchers and managers and between researchers and managers is necessary to maintain understanding.
Hierarchical structure and chain of command works effectively.	Loosely coupled communities of thinkers, scientists, and managers with the voices of those with more experience making wise judgments are given more attention.
Answers among the separate issues exist by themselves; best answers exist; best decisions can be made.	Collective interpretation leads to new insights into the system and suggests new ways it might be managed.
Knowing—decide and tell others what to do.	Learning—act/monitor/learn/plan together, all at the same time.
Staying the course—align and maintain focus.	Notice emergent directions—build on what works but be ready to change.

Source: Modified from http://learningforsustainability.net/sparksforchange/complicated-or-complex-knowing-the-difference-is-important-for-the-management-of-adaptive-systems/#sthash.FioUKvZH.dpuf.

infrastructure is used to change the environmental consequences but not change the infrastructure itself. Multiple small and diverse fixed investments will always be more appropriate in a highly complex and hence uncertain world than a single fixed investment. In this respect, investing in the tunnels is somewhat like investing in a large nuclear power plant instead of investing in multiple smaller and dispersed wind and solar facilities that can be removed, and possibly even moved to new locations, if conditions change. When making large, fixed investments, our understanding of the future needs to be especially clear. On the other hand, there are economies of scale that favor large projects, leaving us in yet another quandary.

The DSP has been a positive force in bringing scientists together to help coordinate efforts. Now, the DSP, spurred in part by a critical analysis of Delta science and management undertaken by the National Research Council (NRC 2012), is working with the Delta scientific community to develop a Delta Science Plan to further coordinate Delta science. The idea for a science plan originated with the DISB and was included as an element of the overall Delta Plan. The science plan calls for the unification of Delta science and scientists getting right answers, while it also calls for encouraging discourse among scientists and managers to arrive at more robust answers. This is somewhat between the left and right columns of Table 12.1 but is a clear effort to overcome the divisions in science so characteristic of academic and governmental bureaucracies. The DSP and DSC encourage agency heads, who meet as the Delta Interagency Implementation Committee, to align agency research and management efforts with the Delta Science Plan.

12.4.2 Scientific Lags and Institutional Constraints

While some scientists are actively struggling to think in a socioecological system framework, the vast majority of scientists and scientist-managers are only slowly changing. The careers of many of today's biologists started with intellectual revolutions. Conservation biology arose in the 1970s to bring modern ecological understanding into wildlife and landscape management. Today's conservation biologists fought the status quo during the 1980s, transforming wildlife and landscape academic departments and management agencies around the globe. Conservation biology replaced management approaches emphasizing predator control dating from the nineteenth century that had become institutionalized with the creation of water and wildlife agencies in the beginning of the twentieth century. Restoration ecologists emerged a decade after conservation biology, arguing that damaged environments should not be abandoned and showing how human-influenced environments could be improved. Restoration ecologists had to drag environmental scientists and environmental movements' efforts away from relatively unspoiled natural environments to clearly spoiled, unnatural environments. It is important to keep in mind that many of the scientists we now see as doing rather conventional science and management were once radicals or carry the radical banners of their professors.

The realization that people have become significant drivers of global environmental change does not suddenly transform science and management. Few of the once-radical conservation biologists or restoration ecologists are mentally prepared for the Anthropocene. They were trained in ecology, not socioecology, a field only beginning to arise in the academe. Equally importantly, they occupy now well-defined positions within environmental agencies that were established by legislative and regulatory decisions, forward looking at the time, that now constrain what they can do. It is uncomfortable to think thoughts that are not consistent with intellectual commitments already made or to contemplate engaging in decisions on which one is not actually empowered to act.

Among conservation biologists, what might someday be considered transformative theoretical battles to understand what it means to think and work within the Anthropocene are taking place over practical management questions. One example is whether biologists can and should engage in *assisted migration* or *assisted colonization* (for examples of this debate, see McLachlan, Helman, and Schwartz 2007; Minteer and Collins 2010; Dawson et al. 2011; Lunt et al. 2013). Conservation biologists and restoration ecologists have heretofore held up recent ecosystems, or those that existed within some relevant historical period, as ideal. Species that have been in the area for quite some time are deemed natural and thereby good, whereas species that have arrived and become established recently, frequently with the help of human activity, are *invasives* and unnatural and thereby bad. The Endangered Species Act assumes that species persistence is to be facilitated in current locations, and the act has not been modified to include relocation of species as a remedy.

Of course, species have been moving around for a long time. With climate change, however, established species may need to move more quickly to survive, and some species, not necessarily seen as desirable, may be especially adept at taking advantage of rapidly changing conditions. This changes the whole idea of what constitutes an invasive species.

Conservation biologists have long advocated for corridors so that species can move from place to place; now the corridor idea has another rationale and is being advocated on a larger scale. Some species, however, are moving up mountainsides, and it is difficult to build corridors from southern peaks to more northern foothills. Similarly, it is difficult for freshwater fish to change watersheds. Should biologists assist species in migrating to more appropriate habitats to enhance their likelihood of surviving climate change? Someday, the answer in many cases will be yes, but at this time, even asking the question clashes with the philosophical foundations of many biologists. The science behind taking such action is also not yet developed. And it is to the credit of biologists that they are humbly saying so. We need to be looking forward, rather than backward to some previously natural world, more so than ever with climate change, but the science and legislation lag the pace of change of reality. Attempts to restore the past may eventually be off the table because the past will simply have been eradicated. If the salmon are gone, the goal of managing salmon populations has little practical meaning.

Conservation biologists are also struggling with the question of whether species should be taken off the endangered species list if they are *conservation reliant*, i.e., if they are surviving successfully, not in the wild but with ongoing human effort. In 2014, salmon, already being reared in hatcheries, were trucked from northern California hatcheries nearly to the sea because the Sacramento River had insufficient water. The *conservation reliant* discussions also have recognized that the Endangered Species Act is not suitable for a world of climate change or the Anthropocene generally (Scott et al. 2005; Goble et al. 2012, 2014; Rohlf, Carroll, and Hartl 2014a,b).

Similarly, in the face of climate change, the dominant management discourse is about *restoring* habitats largely using criteria drawn from their past rather than "futurating" habitats. I have only vague ideas as to what futurating ecosystems would actually mean, but I think it would include preparing them for a range of possible futures that might unfold, as well as facilitating desirable new species, assisted or not. The BDCP is a conservation plan, not a futuration plan. The science is not there, let alone the enabling legislation to support futuration science and management. So we have a plan that looks forward 50 years and tries to address accelerating climate change and sea level rise while remaining constrained by existing science and environmental legislation that, to a great extent, still look backward.

The words we use, the ways conservation biologists and the public think, as well as the requirements of existing legislation to manage for a predictable future are inconsistent with the uncertainties of the future that lies ahead (for a lay synthesis, see Marris 2011). The contradictions, in my judgment, left the preparers of the BDCP and its EIR with both dilemmas and exploitable wiggle room. They seemed inclined to assume greater certainty in predicting a favorable future when doing so favored the plan. This approach led the DISB to critique, "Many of the impact assessments hinge on overly optimistic expectations about the feasibility, effectiveness, or timing of the proposed conservation actions, especially habitat restoration" (Delta Independent Science Board 2014, p. 3).

But when the uncertainties overwhelmed tight analysis, and it was more convenient to ignore uncertainties, they were indeed ignored. This led the DISB to conclude, "The project is encumbered by uncertainties that are considered inconsistently and incompletely; modeling has not been used effectively to bracket a range of uncertainties or to explore how uncertainties may propagate" (Delta Independent Science Board 2014, p. 3).

12.5 Conclusions

In late October 2014, some 1200 scientists from government, academia, and consulting firms along with managers and key policy makers gathered in Sacramento for the 8th Biennial Bay-Delta Science Conference. "Making Connections" was the theme of the 3-day conference, and in this fairly scholarly setting, scientists indeed connected with each other and with managers and policy makers to openly discuss the quandaries of managing the Delta. Difficult topics touched upon in scientific presentations were elaborated in discussion periods and deliberated more intensely in hallway conversations. Several of the presenters made serious efforts to speak to the dynamics of the Delta as a whole environmental system. Many noted that this conference works as an important integrating effort for collective scientific understanding. Several speakers noted that the Delta Science Plan seemed to already have had some impact in focusing the presentations in this year's conference.

Still, few speakers came close to seriously discussing the long-run implications of the Anthropocene. The concept of ecosystem resilience seemed to have gained ground in this conference over the conference 2 years ago, but resilience was proffered as a way of protecting ecosystems from the stresses of climate change, as if climate change were a temporary perturbation. None of the presenters whom I heard raised questions about how resilience might be a feature of ecosystems continually adapting to ongoing climate change. One speaker commented that in rare instances in the future, it might be necessary to move a species to save it. Apparently, it is better to give than to receive, for no one in the sessions I attended mentioned how the Delta might be better prepared to receive assisted migrants from other areas. Discussion of invasive species pretty much continued within its historic framework of seeing invasives as problems rather than acknowledging that some invasives will be necessary for species survival.

The science–policy–management interface was often mentioned. One speaker noted that scientists are expected to prepare habitat conservation plans that will work. Permission to incidentally kill Delta smelt, for example, is not likely to be given in exchange for a habitat conservation plan that officially acknowledges that if the plan does not work, scientist-managers will adaptively manage until they find something else that does work...at least for a while. Current politics and institutions frown on balancing certain, ongoing losses against a plan that basically says "leave it to us, we will figure it out." Promising to adaptively manage through a future of surprises makes habitat conservation plans look weak;

it is an insufficient way to balance the already existing consequences of historic Delta transformations including those resulting from water development projects.

At least one speaker at the conference argued that the individual scientists from the diverse agencies are much more in tune with each other's knowledge and hence more in tune with the Delta as a whole system than are their respective agency leaders. Many scientists thought much more effort will be needed to bring the agency leaders to work together. One suggested that the existing process of checks and balances in governance needs to be dramatically accelerated or the system will have to be scrapped to deal with the rapid dynamics of the Anthropocene. One presenter suggested that the governor needed to appoint a Delta tsar with the power to force agencies to work together. Problems with the current science–policy–management interface were on the minds of many.

Most of the members of the DISB also attended the conference, and at the board's meeting the following day, there was general agreement that the conference provided a critically important way to share knowledge among scientists. Board members also noted that the thinking of the scientists seemed to be well ahead of the agendas of their agencies.

The Sacramento–San Joaquin Delta has had a key and controversial role in California water politics, policy, and management for decades. The Delta has also been difficult for scientists to understand while also being little known by Californians. Climate change and sea level rise make the Delta more important, more complex and difficult to know, and more controversial too. The natural sciences are not easily integrated, and the difficulties of framing and studying ongoing human-driven change are immense. The scientific difficulties are compounded by the fact that environmental science does not work apart from the existing institutions that structure how science is undertaken; how findings are shared among scientists; and how information is communicated to politicians, policy makers, and the public. The Delta Science Plan and the Delta science conferences are helping to meet the challenge, but institutional changes are also needed to help science address the nature of the Anthropocene.

Acknowledgments

Lindsay Correa, Jessica Goddard, and John Wiens provided very helpful, substantive and editorial suggestions but are neither corroborators in the judgments expressed nor responsible for any remaining errors.

References

Cohen, A.N. 2011. *The Exotics Guide to Non-Native Marine Species of the North American Pacific Coast.* Center for Research on Aquatic Bioinvasions: Richmond, CA and San Francisco Estuary Institute. http://www.exoticsguide.org.

Cohen, A.N. and J.T. Carlton. 1998. Accelerating invasion rate in a highly invaded estuary. *Science* 279(5350, 23 January):555–558

Crutzen, P.J. 2002. Geology of mankind. *Nature* 415:23.

Dawson, T.P. S.T. Jackson, J.I. House, I.C. Prentice, and G.M. Mace. 2011. Beyond predictions: Biodiversity conservation in a changing climate. *Science* 332(1 April):53–58.

Delta Independent Science Board. 2014. Review of the Draft BDCP EIR/EIS and Draft BDCP, May 15, 2014. http://deltacouncil.ca.gov/sites/default/files/documents/files/Attachment-1-Final-BDCP-comments.pdf.

Dryzek, J.S. 1987. *Rational Ecology: Environment and Political Economy*. New York: Basil Blackwell.

Ezrahi, Yaron. 1990. *The Descent of Icarus: Science and the Transformation of Contemporary Democracy.* Cambridge: Harvard University Press.

Folke, C., T. Hahn, P. Olsson, and J. Norberg. 2005. Adaptive governance of socio-ecological systems. *Annual Review of Environment and Resources* 30: 441–473.

Glacken, C.J. 1967. *Traces on the Rhodian Shore.* Berkeley: University of California Press.

Goble, D.D., J.A. Wiens, J.M. Scott, T.D. Male, and J.A. Hall. 2012. Conservation reliant species. *BioScience* 62:869–873.

Goble, D.D. J.A. Wiens, T.D. Male, and J. M. Scott. 2014. Response to conservation reliant species: Toward a biology-based definition. *BioScience* 64(10):857–858.

Healey, M.C., M. D. Dettinger, and R.B. Norgaard (eds.). 2008. *The State of Bay-Delta Science, 2008.* Sacramento, California: CALFED Science Program.

Hundley Jr., N. 2001. *The Great Thirst: Californians and Water: A History*. Revised Edition. Berkeley: University of California Press.

IPCC, 2014. *Climate Change 2014: Synthesis Report*. Contribution of Working Groups I, II and III to the Fifth Assessment Report of the Intergovernmental Panel on Climate Change Core Writing Team, R.K. Pachauri and L.A. Meyer (eds.). IPCC, Geneva, Switzerland, 151 pp.

Kallis, G., M. Kiparsky, and R.B. Norgaard. 2009. Adaptive governance and collaborative water policy: California's CALFED Bay-Delta Program. *Environmental Science and Policy* 12(6):631–643.

Little Hoover Commission. 2005. *Still Imperiled, Still Important: The Little Hoover Commission's Review of the CALFED Bay-Delta Program.* Sacramento, CA: Commission on California State Government and Organization Economy.

Lunt, I.D. et al. 2013. Using assisted colonisation to conserve biodiversity and restore ecosystem function under climate change. *Biological Conservation* 157:172–177.

Marris, E. 2011. *Rambunctious Garden: Saving Nature in a Post-Wild World.* New York: Bloomsbury.

Marsh, G.P. 1864 (1965). *Man and Nature, or Physical Geography as Modified by Human Action*. Cambridge: Harvard University Press.

McLachlan, J.S. J.J. Hellman, and M.W. Schwartz. 2007. A framework for debate of assisted migration in an era of climate change. *Conservation Biology* 21(2):297–302.

Millennium Ecosystem Assessment. 2005. http://www.millenniumassessment.org/en/index.html.

Minteer, B.A. and J.P. Collins. 2010. Move it or lose it: The ecological ethics of moving species under climate change. *Ecological Applications* 20(7):1801–1804.

Mitchell, S.D. 2009. *Unsimple Truths: Science, Complexity, and Policy*. Chicago: University of Chicago Press.

National Research Council. 2012. *Sustainable Water and Environmental Management in the California Bay-Delta*. Committee on Sustainable Water and Environmental Management in the California Bay-Delta, Water Science and Technology Board. D.C.: National Academies Press.

Norgaard, R.B. 2008. Finding Hope in the Millennium Ecosystem Assessment. *Conservation Biology* 22(4):862–869.

Norgaard, R.B., G. Kallis, and M. Kiparsky. 2009. Collectively engaging complex socio-ecological systems: Re-envisioning science, governance, and the California Delta. *Environmental Science and Policy* 12(6):644–652.

Norgaard, R.B. and P. Baer. 2005a. Collectively seeing complex systems: The nature of the problem. *BioScience* 55(11):953–960.

Norgaard, R.B. and P. Baer. 2005b. Collectively seeing climate change: The limits of formal models. *BioScience* 55(11):961–966.

Poli, R. 2013. A note on the difference between complicated and complex social systems. *Cadmus* 2(1):142–147.

Rohlf, D.J., C. Carroll, and B. Hartl. 2014a. Conservation-reliant species: Towards a biology based definition. *BioScience* 64:601–611.

Rohlf, D.J.C. Carroll, and B. Hartl. 2014b. Reply to Goble and colleagues. *BioScience* 64(10):859–860.

Sarewitz, D., R.A. Pielke, and R. Byerly. (eds.). 2000. *Prediction: Science, Decision Making, and the Future of Nature*. Washington D.C.: Island Press.

Scott, J.M., D.D. Gobble, J.A. Wiens, D.S. Wilcove, M. Bean, and T. Male. 2005. Recovery of imperiled species under the Endangered Species Act: The need for a new approach. *Frontiers in Ecology and the Environment* 3:383–389.

State of California. 2009. Sacramento–San Joaquin Delta Reform Act of 2009, SBX7 1.

State of California. 2014. Sustainable Groundwater Management Act (AB1739, SB1168, and SB1319), Sacramento.

US Environmental Protection Agency. 2014. Letter to Will Stelle, Regional Administrator West Coast Region National Marine Fisheries Service, August 26, 2014. http://www.ewccalifornia.org/reports/epa-bdcp-deis-comments-8-26-2014.pdf.

Waldrop, M.M. 1992. *Complexity: The Emerging Science at the Edge of Order and Chaos*. New York: Simon and Schuster.

13

California's Climate Change Response Strategy: Integrated Policy and Planning for Water, Energy, and Land

Robert C. Wilkinson

CONTENTS

ABSTRACT It is important to understand the water/energy/land nexus in order to design effective climate change response strategies. California is using such understanding to craft integrated strategies to reduce greenhouse gas emissions (mitigation of the root problem) and build resilience (adaptation) in the face of climate impacts. Climate change poses serious challenges for California's $2.32 trillion economy and 38.8 million residents. Water systems are a critical impact area, as precipitation patterns shift, intensity of weather events increases, and snow processes change. Water is a large user of energy, so it is also a significant cause of emissions driving climate change. This chapter addresses the challenges and opportunities in water/energy/land system management, and it discusses prospects for developing climate change response strategies that would have multiple benefits.

13.1 Introduction: The Climate Challenge and Integrated Response Strategies

California is taking climate change seriously. Addressing the issue of climate change at the United Nations World Environment Day in 2005, then-governor Arnold Schwarzenegger famously declared, "The debate is over. We know the science. We see the threat. And we know the time for action is now" (Schwarzenegger 2005). While there may still be debate in some quarters, California has moved forward with an increasingly integrated set of strategies to better understand the implications of climate change, reduce greenhouse gas (GHG) emissions, and adapt to changes. Governor Jerry Brown declared in 2015, "California has the most far-reaching environmental laws of any state and the most integrated policy to deal with climate change of any political jurisdiction in the Western Hemisphere" (Brown 2015).

Governor Brown opened an unprecedented fourth term in January 2015 with some remarkable statements regarding climate change. In his inaugural address, he said, "Neither California nor indeed the world itself can ignore the growing assault on the very systems of nature on which human beings and other forms of life depend. We must demonstrate that reducing carbon is compatible with an abundant economy and human wellbeing. So far, we have been able to do that. In fact, we are well on our way to meeting our AB 32 goal of reducing carbon pollution and limiting the emissions of heat-trapping gases to 431 million tons by 2020" (Brown 2015).

13.2 Climate Change Impacts

California's $2.31 trillion economy is the largest in the United States and ranks as one of the top 10 economies in the world (USBEA 2015; LAO 2015; World Bank 2015). It is also the most populous state in the nation, with 38.8 million residents confronted with serious challenges from climate change and variability (US Census 2015; CDF 2015; USGCRP 2014; IPCC 2014; NAS 2014; Little Hoover Commission 2014). The opening sentence of California's Global Warming Solutions Act (AB32) states, "Global warming poses a serious threat to the economic well-being, public health, natural resources, and the environment of California" (AB32, Section 38501(a) 2006). Surface water and groundwater sources in California are stressed, with many systems overallocated and overtapped. About 80% of California's developed water goes to agriculture, which produces 1–2% of the state's gross product (CDWR 2014; Hanak et al. 2012).

Climate change and related impacts have been the subject of research and assessments since at least the 1970s (NAS 1975; US Senate 1979; Revelle and Waggoner 1983), and the findings consistently confirm that water, energy, and land impacts are critical concerns (Stewart et al. 2005; Dettinger and Cayan 1994; CCCC 2006, 2008, 2009; NRC 2001, 2002; Gleick 2000; Wilkinson 2002; USGCRP 2014; IPCC 2013, 2014; Plattner 2014; NAS 2014). Building on the science, California has developed a number of policies and response strategies, ranging from the Global Warming Solutions Act and other laws and policies to reduce GHG emissions, to state and local agency mitigation plans and adaptation strategies (AB32 2006; California Water Action Plan 2014; CDWR 2008, 2009, 2012, 2014a; CNRA 2013, 2015; CEPA 2013; CARB 2014; CRPS 2015). In the language of climate science and policy, the term *mitigation* refers to actions that address the root problem—reductions in GHG concentrations in the atmosphere, typically by reducing emissions. *Adaptation* refers to efforts to cope with the impacts of climate change such as extreme weather events and sea level rise. Integrated response strategies combine the two and focus on multiple benefits from policies and investments.

Climate variability—including floods, droughts, and heat waves—is not new. While greater variability is expected, it is not clear whether California will trend wetter or drier in the future. Response strategies therefore need to be robust and resilient under both possibilities in order to be effective. Higher temperatures, however, are expected to increase the intensity of wet events (storms and floods), dry conditions (drought), and extreme heat events (heat waves) (USGCRP 2014; IPCC 2014). Increased temperatures also cause a reduction in snowpack (a critically important and valuable source of free water storage in California), leading to earlier spring runoff, reduced hydropower production, and shifts in the timing of hydropower availability. Other projected impacts include increased intensity of precipitation events, increased evaporation, and increased demand for water (Dettinger and Cayan 1994; Cayan et al. 2010; Seager and Hoerling 2014; IPCC 2013, 2014; USGCRP 2014). Recent temperature and precipitation extremes may be indicative of what the future will be like under these projections. The year 2014, for example, was the hottest (the average temperature in California in 2014 was 61.5°F, which was 4.1°F warmer than the twentieth-century average) and among the driest in the instrumental record for California (NCDC 2015; WRCC 2015; Griffin and Anchukaitis 2014).

13.3 Integrated Response Strategies

Integrated strategies need to address both sides of the climate issue (mitigation and adaptation), because water, energy, and land will be impacted by climate change, and our management strategies for each will affect the climate (Wilkinson 2011; Pacific Council on International Policy 2010; IPCC 2014). There are important links between water, energy, and land that involve both vulnerabilities and opportunities. For example, water is one of the largest users of energy in California (CEC 2005; GEI 2010).

California has developed a number of integrated resource management strategies over the past several decades to address energy, water, land management, and climate (CEC 2005; CDWR 2014a,b; CARB 2014; CDWR 2009, 2012). The logic of integrated planning is increasingly being adopted in state and local policy. Building on the pioneering work in the energy sector with *integrated resources planning* (IRP), the California Energy Commission (CEC) established a biennial Integrated Energy Policy Report. In 2005, it provided a seminal contribution to the water/energy discussion with estimates of the total electricity, gas, and diesel fuel used by the state's water systems (CEC 2005).

For water policy and investment programs, *integrated water management* and variants like *integrated regional water management* provide a basis for policy strategies and funding programs run by the State Water Resources Control Board and the Department of Water Resources (Integrated Regional Water Management Planning Act of 2008; CDWR 2014a,b, 2015). The Department of Water Resources has made integrated management, defined as "a comprehensive and collaborative approach for managing water to concurrently achieve social, environmental, and economic objectives," the overarching theme of the state water plan (CDWR 2014a, p. H-2). The governor's Water Action Plan follows the theme of promoting integration and multiple benefits, and the California Air Resources Board's latest update of the Scoping Plan for implementation of AB32, the California Global Warming Solutions Act, is structured on integrated efforts across state agencies (California Water Action Plan 2014; CARB 2014). As the Air Resources Board notes in the 2014 update to the Scoping Plan, "California is taking a proactive approach to climate change policy, through integrated policy and planning that will build a higher-quality, resilient economy while continually reducing GHG emissions" (CARB 2014, p. 25).

In January 2015, the governor proposed ambitious new goals for the state as part of a more integrated approach in California: "It is time to establish our next set of objectives for 2030 and beyond. Toward that end, I propose three ambitious goals to be accomplished within the next 15 years: Increase from one-third to 50 percent our electricity derived from renewable sources; Reduce today's petroleum use in cars and trucks by up to 50 percent; Double the efficiency of existing buildings and make heating fuels cleaner. We must also reduce the relentless release of methane, black carbon and other potent pollutants across industries. And we must manage farms and rangelands, forests and wetlands so they can store carbon. All of this is a very tall order. It means that we continue to transform our electrical grid, our transportation system and even our communities. I envision a wide range of initiatives: more distributed power, expanded rooftop solar, micro-grids, an energy imbalance market, battery storage, the full integration of information technology and electrical distribution and millions of electric and low-carbon vehicles. How we achieve these goals and at what pace will take great thought and imagination mixed with pragmatic caution. It will require enormous innovation, research and investment. And we will need active collaboration at every stage with our scientists, engineers, entrepreneurs, businesses and officials at all levels" (Brown 2015).

This chapter examines challenges and opportunities at the nexus of water, energy, and land in the context of climate change and variability. It then identifies cross-sector links between them and options for further development of integrated response strategies.

13.4 The Water–Energy Nexus

One of the most important cross-sector links in the context of climate change is between water and energy (Wilkinson 2011; Kenney and Wilkinson 2011). Water is one of the largest users of energy in California, accounting for approximately 19% of total electricity use, about 33% of the non–power plant natural gas, and significant amounts of diesel fuel for pumping (CEC 2005). The CEC concluded that energy used for water presents large untapped opportunities for cost-effective energy-efficiency improvements and GHG emission reductions, commenting, "The Energy Commission, the Department of Water Resources, the CPUC, local water agencies, and other stakeholders should explore and pursue cost-effective water efficiency opportunities that would save energy and decrease the energy intensity in the water sector" (CEC 2005, p. 7). This aligns well with the findings of the California Department of Water Resources, which has indicated in each of the past three state water plans, including the most recent in 2014, that water-use efficiency is the largest new water supply over the next several decades (CDWR 2014). Energy systems also use large amounts of water. Across the United States, thermal power plants are the largest user of water in total withdrawals (including fresh and saline sources), and they are roughly tied with irrigation at 38% each as the largest users of fresh water (Maupin et al. 2014). Actual water consumed by energy systems is considerably less than the amout withdrawn.

The water–energy nexus is the relationship between the use of water to extract, convert, and use energy, and the use of energy to extract, treat, deliver, and use water. They are inherently interrelated; energy is usually required throughout the water-use cycle, and water is required for many (though not all) energy system processes. It is possible to significantly reduce energy use and related emissions in water systems and water use in energy systems. The potential multiple benefits, such as cost savings, improved reliability, resource efficiency, reduced environmental impacts, and GHG emission reductions, of the integrated management of water and energy are important aspects of the water–energy nexus (AWE/ACEEE 2013; Wilkinson 2011; Cooley et al. 2013).

Innovations in water, energy, and land management systems present interesting and important potential for synergies (Cooley and Donnelly 2013; Wilkinson 2011; Cooley et al. 2013). The water–energy nexus is an opportunity for innovation that yields multiple benefits. Figure 13.1, developed for an analysis of innovation opportunties in California for water management by the California Council on Science and Technology, identifies the steps in the process and important linkages between them (CCST 2014). The water cycle methodology was developed originally to determine the energy inputs to water (Wilkinson 2000). It was then adapted for the CEC's seminal 2005 Integrated Energy Policy Report (CEC 2005). The methodology is applicable to water sources ranging from surface water and groundwater supplies to ocean desalination and water recycling. It has now been used as the basic approach to calculating the energy intensity of water supplies by

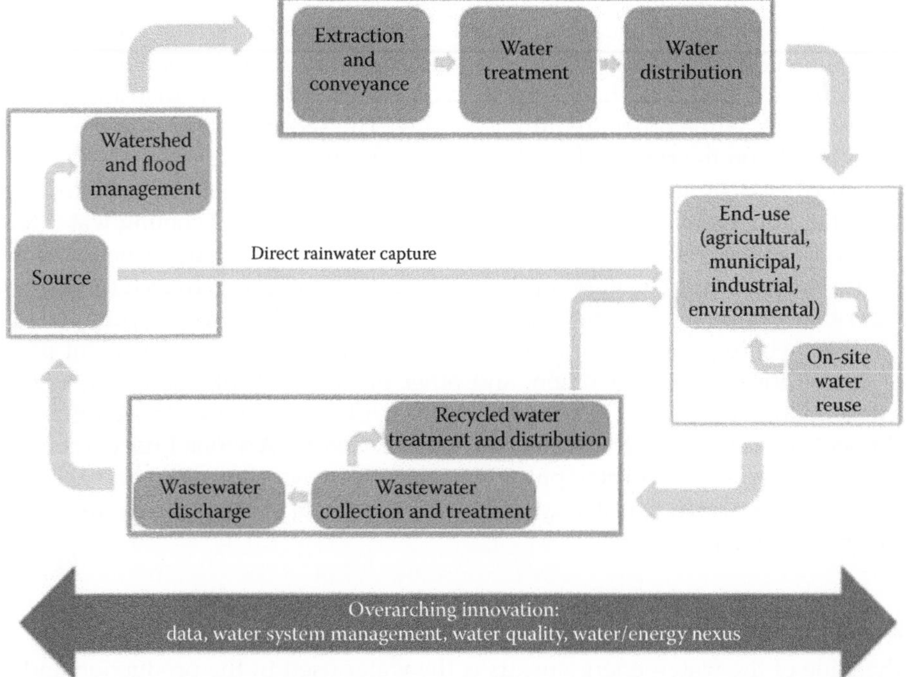

FIGURE 13.1
Water-use cycle. (From California Council on Science and Technology. 2014. *Achieving a Sustainable California Water Future through Innovations in Science and Technology*, California Council on Science and Technology, http://www.ccst.us/news/2014/0409water.php, p. ES 1.)

a number of entities in the United States and in other countries. The Pacific Institute and the Bren School at the University of California–Santa Barbara used funding from the CEC, the Canadian government, and the WateReuse Research Foundation to develop an open-access computer model based on the methodology (Cooley and Wilkinson 2012).

13.4.1 Energy Inputs to Water Systems

Water systems are often energy intensive. Moving large quantities of water over long distances with significant elevation changes, treating and distributing it within communities, using the water, and finally collecting and treating wastewater together account for a major use of energy (Wilkinson 2000; CEC 2005; GEI 2010). The *energy intensity* of water (or *embedded energy*) is the total amount of energy required to make a unit of water available at a particular place, and it varies with location, source, and use. Pumping water at each stage is often energy intensive. Other important energy inputs include thermal energy (heating and cooling) at the point of use and aeration in wastewater treatment processes.

As depicted in Figure 13.1, there are three broad categories of energy use in water systems that correspond directly to the water-use cycle. The first is water extraction, conveyance, treatment, and distribution (shown in the box at the top). Extracting and conveying water can be highly energy intensive (Wilkinson 2000, 2011; CEC 2005). The largest user of electricity in California, for example, is the State Water Project, and the largest single facility consuming electricity in the state is the Edmunston Pumping Plant—part of that system (Wilkinson 2011). Surface water and groundwater pumping require significant amounts of energy depending on the depth of the source. Where water is stored in intermediate facilities, energy is usually required to store and then recover the water. Within local service areas, water is treated, pumped, and pressurized for distribution, with energy use determined by local conditions and sources. For example, some distribution systems are gravity driven, while others require pumping. The second category of energy inputs (shown in the box on the right) relates to on-site water use and includes activities such as pumping, further treatment (e.g., softeners, filters, etc.), circulation and pressurization of water supplies (e.g., building circulation pumps), and heating and cooling water for various purposes. Finally, wastewater collection, treatment, and discharge (shown in the box on the bottom) each require energy inputs. Wastewater is collected, treated (unless a septic system or other alternative is being used), and discharged. Wastewater is pumped to treatment facilities where gravity flow is not possible, and the standard treatment processes require energy for pumping, aeration, and other processes (Wilkinson 2011; CEC 2005). Ocean desalination and some interbasin water supply systems like the California State Water Project, the Colorado River Aqueduct, and the Central Arizona Project require large amounts of energy. Groundwater pumping and water recycling are less energy intensive than these interbasin transfer systems, and water-use efficiency requires no energy (Wilkinson 2008, 2011).

13.4.2 Water Inputs to Energy Systems

The other side of the water–energy nexus is the water used in the production and use of energy. The US Geological Survey (USGS) estimates in its most recent analysis that 38% of all US freshwater withdrawals were used for thermoelectric power (Maupin 2014, p. 1). The water intensity of energy is the total amount of water, calculated on a whole-system basis, required to produce and use a given amount of energy.

Water inputs to energy systems are highly variable. They depend on the primary energy source and on conversion technologies employed at each step in the process. For example, primary fuels such as fossil fuels—oil, gas, coal—often require water for production, and they sometimes produce water of varying quality as a by-product of extraction. Biofuels often require water for irrigation of crops as well as for production processes. There is even a significant consumptive water use by hydroelectric systems when evaporation from surface water impoundment is taken into consideration. It is important to note that both renewable and nonrenewable energy sources can be either water thrifty or water intensive depending on a number of factors, including technologies deployed. Water inputs at each step need to be accounted for to develop a comprehensive water-intensity metric.

Water is a limiting factor for thermal power plant siting and operation primarily due to cooling requirements (EPRI 2008; Averyt et al. 2011; Rogers et al. 2013; Davis and Clemmer 2014). Cooling systems account for the majority of water used in power generation, but water is also used to mine, extract, process, and/or transport fuels (e.g., coal slurry lines), and these processes have impacts on water resources. Some energy systems are highly dependent on large volumes of water and are vulnerable to disruptions, while other energy sources are relatively independent of water (USDOE 2006; EPRI 2008; Sandia 2015; Davis and Clemmer 2014). Water use for renewable forms of energy varies substantially. Solar photovoltaic panels and wind turbines use essentially no water to produce electricity. Some geothermal and concentrating solar power systems that employ dry cooling technology, and landfill gas-to-energy projects, have minimal water use (NREL 2014; Rogers et al. 2013). In contrast, irrigated bioenergy crops like corn consume substantial amounts of water per unit of energy. Finally, although reservoirs often have multiple purposes (e.g., flood control, water storage, and recreation), evaporative losses from hydroelectric facilities per unit of electricity are higher than many other forms of generation.

13.5 The California Context and Key Trends for Water, Energy, and Land Management

We have designed cities and built water and energy systems in California responding to opportunities and constraints—physical, economic, social, and environmental—as perceived at different stages of history. Some of yesterday's solutions are today's challenges (Delta Stewardship Council 2013). This historical context and the identification of new trends are important to crafting effective climate response strategies.

Existing stresses, such as water scarcity, air pollution, and ecosystem impacts, will be exacerbated by climate change (USGCRP 2014; IPCC 2014; Wilkinson 2002). Overallocated and overtapped surface water and groundwater supplies, vulnerable energy systems, and some land-use approaches present challenges to long-term sustainability. The decline in aquatic ecosystems and species impacts, including listings by the state and federal governments, are indicators of stress.

The focus of technology development and policy for much of the past century was on the supply side of both the energy and water equations, with large-scale power plants and dams being the options of choice. Large centralized systems are less reliable than anticipated. The State Water Project announced 0% and then 5% deliveries in 2014. The San Onofre Nuclear Generating Station shut down permanently in 2013, ahead of schedule, due to multiple problems. Since the 1970s, technological innovation has increasingly been applied to the demand side. Dramatic improvements have been made in the ability to secure the *services and benefits* desired from each unit of water and energy. Various technologies, from lighting and electric motors to pumps and plumbing fixtures, have vastly improved efficiencies. It is clear that substantial economic and environmental benefits can be cost-effectively achieved through further efficiency improvements in water and energy systems.

13.5.1 Dramatic Increases in Energy and Resource Efficiency

While California's economy and population have been growing rapidly and are the largest in the United States, both water and energy use have *decoupled* from total product (USBEA 2015; CDF 2015; Hanak et al. 2012). Electricity use per capita has remained level at around 7000 kWh per person in California since the mid-1970s, while the US figure is around 13,000 kWh per person (USEIA 2012) (Figure 13.2).

Water use has been declining in California and across the United States in all use sectors over the past several decades (DWR 2014; Maupin 2014). Total freshwater withdrawals peaked in the 1980s and then declined by 15% between 2005 and 2010 across the United States (Maupin 2014). Technical innovations, price signals, and policy measures have enabled California to *quadruple* its gross domestic product (GDP) per unit of water between the 1960s and 2010 (Hanak et al. 2012) (Figure 13.3).

Water and energy efficiency are explicit goals of both local and state governments (CDWR 2014; CEC 2005; California Water Action Plan 2014; Brown 2015; LADWP 2011). Significant potential exists to increase efficiency in all water-use areas (CDWR 2014; Pacific Institute and NRDC 2014). Mayor Garcetti of Los Angeles explicitly linked the two in a policy directive in 2014: "Just as water conservation is how we will get through our drought and control our water costs, energy conservation is how we will address climate change and keep our power bills low.... investing in efficiency is three to four times cheaper than building new power plants and cleans the air. The cheapest and cleanest way to ensure we have enough electricity to keep the lights on and power our economy is through energy efficiency" (Garcetti 2014b).

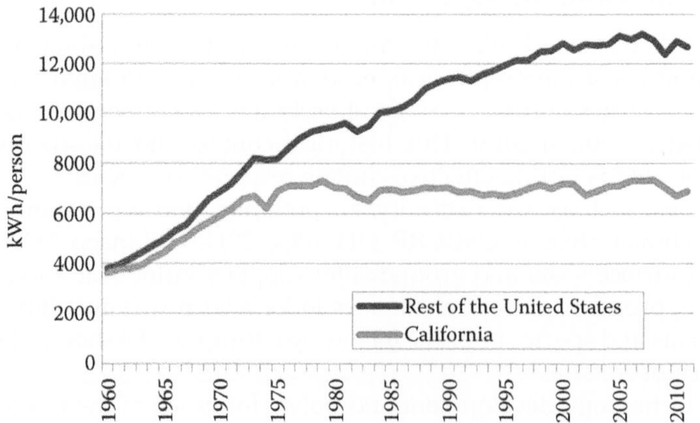

FIGURE 13.2
Per capita electricity consumption: California versus rest of nation. (From United States Energy Information Administration, State Energy Data System, All Consumption Estimates, in Physical Units, ESTCP and TPOPP, 1990–2010 (June 2012), http://www.eia.gov/state/seds/seds-data-complete.cfm#Consumption, Data plot from NRDC, 2013, California's Energy Efficiency Success Story: Saving Billions of Dollars and Curbing Tons of Pollution, 2012, web.)

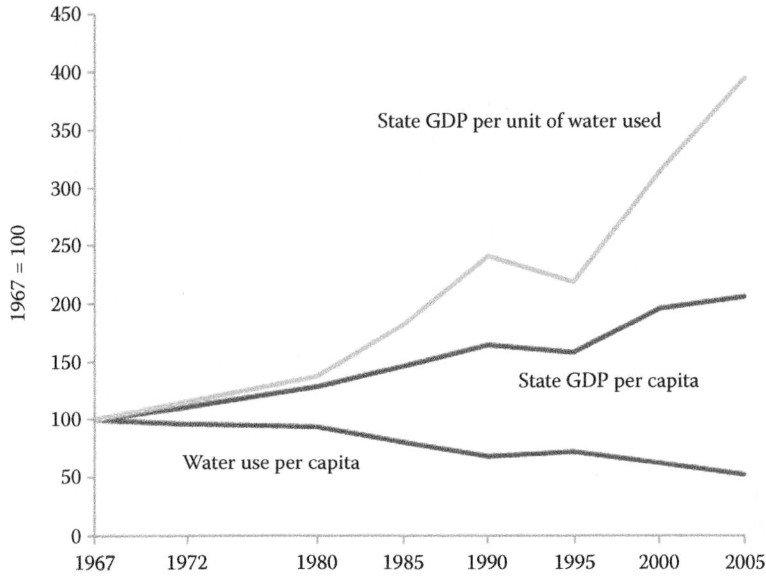

FIGURE 13.3
California GDP and population per unit of water used. (From Hanak, E. et al., Water and the California Economy, Public Policy Institute of California, http://www.ppic.org/main/publication.asp?i=1015, 2012.)

Major cities like Los Angeles are, in fact, using less water now than they were a quarter century ago, even with an increase in population of over 1 million people (LADWP 2011). In urban Southern California as a whole, the population grew by about 5 million over 25 years leading up to 2015 (to about 19 million), yet overall water demand declined by 20% according to Jeff Kightlinger, general manager of the Metropolitan Water District of Southern California (MWD) (Rogers 2015).

Under California law enacted in 2009, statewide urban per capita water use is to be cut by 20% by the year 2020 (SBX7-7 2009). Water managers are responding in part with incentives like rebates for turf removal, high-efficiency clothes washers and toilets, multistream rotating nozzles for sprinklers, and weather-based irrigation controllers (CUWCC 2014). In Southern California, the Metropolitan Water District estimates that over 1.1 billion m^3 (900,000 acre-feet) of water was saved in 2013 through these programs along with water efficiency codes and rate systems designed to encourage efficient use (MWD 2014). In terms of reduced GHG emissions, the improved efficiency involves a dramatic reduction in the energy and emissions that would have been required to deliver and use the water (Wilkinson 2011).

13.5.2 Increasing Reliance on Local and Decentralized Water and Energy Sources

There is a strong trend in California and elsewhere toward increased reliance on decentralized, local supplies and solutions, and away from large-scale centralized water and energy systems (Ruys and Hogan 2014; Lovins and Palazzi 2014; CDWR 2014; MWD 2014). Virtually every water agency in the state is shifting to greater reliance on local solutions, including efficiency improvements—like fixing leaks and replacing old plumbing fixtures and appliances—and recycling water and capturing and using storm water. These actions, when aggregated at the state level, will provide the largest new supplies of water for California (CDWR 2014). In some areas, ocean desalination is also being developed as a local option. A 189,250 m^3/day (50 million gal./day) facility is currently under construction in Carlsbad, just north of San Diego.

The Metropolitan Water District of Southern California developed one of the early applications of IRP for water in 1996. Over the past few decades, "Metropolitan has moved from securing reliability by emphasizing imported supplies to embracing local solutions and a diverse portfolio approach" (MWD 2014, p. 6). Though the 1996 IRP called for imported supplies to comprise 60% of the overall water supply in its service area, MWD found that "that has dramatically changed over time" as the 2010 IRP has reduced Southern California's reliance on imported Colorado River and State Water Project deliveries to 36% of the overall supply mix (MWD 2014, p. 3).

In most cases (but not all), shifting to local water sources substantially reduces energy use and GHG emissions (Park et al. 2008; CARB 2014). Ocean desalination is an exception, as it requires about 3.569 kWh/m^3 (4400 kWh/acre-foot) (Wilkinson 2011). The trend to local options is driven by a number of factors, including economics, technology innovation, reliability, and environmental considerations (CCST 2014). Local resources dominate the investments being made in water systems in California. Hanak et al. (2012) found that local entities in California are annually investing almost 10 times more ($30.3 billion) in water systems than the state and federal governments ($3.6 billion) combined (p. 4).

Los Angeles is investing billions of dollars to develop local supply options and reduce its purchases of imported water. The city plans to reduce purchased imports 50% by 2024 under an executive directive from Mayor Garcetti and as part of the city's integrated climate response plan (Garcetti 2014a). Local supplies, including efficiency, rainwater harvesting, and reusing wastewater, are to be substantially increased (Garcetti 2014a).

Indirect and direct potable reuse is also moving forward in leading urban centers in addition to Los Angeles, including San Diego, Silicon Valley, and a number of other communities throughout the state. A precedent-setting program in San Diego for direct potable reuse is moving forward with a $3.5 billion investment. With unanimous support from the city council and mayor, the project, called Pure Water, will start producing water in less than a decade and generate 314,155 m^3 (83 million gal.) per day at three water treatment plants by 2035 (San Diego 2014). San Diego's Climate Action Plan explicitly calls for the city to reduce dependence on imported water and energy in order to build resilience and increase the reliability of supplies (San Diego 2014). The region plans to shift from 95% reliance on imported water in 1990 to 40% local water supplies by 2020 (San Diego 2014).

The Orange County Water District and the Orange County Sanitation District developed one of the most advanced recycled wastewater systems in the world, and the agencies have been augmenting groundwater supplies since 2008 to serve 2.4 million people. In 2015, Orange County boosted production from 264,979 m^3 (70 million gal.) per day to 378,541 m^3 (100 million gal.) per day—twice the capacity of the Carlsbad ocean desalination facility by comparison (Groundwater Replenishment System 2014). Important drivers for expansion are reliability and cost. Recycling the water, even with advanced treatment, uses "less than half the energy required to pump imported water from Northern California to Orange County and other parts of Southern California, and less than one-third the energy it takes to desalinate ocean water" (Groundwater Replenishment System 2014). In Silicon Valley, the Santa Clara Valley Water District decided in 2014 to build water treatment facilities for potable reuse for the city of Sunnyvale and its service areas in western Santa Clara County (Santa Clara Valley Water District 2014).

Electricity generation is also transitioning toward smaller-scale distributed technologies and away from large centralized facilities like nuclear power plants. Once thought to be "too cheap to meter" (Strauss 1954), only one operating nuclear plant is left in California. Two of California's three large nuclear facilities have been permanently shut down, with the San Onofre Nuclear Generating Station in Southern California formally retired in 2013 (USNRC 2015). No large conventional power plants of any kind have been built in decades, or are even planned in the state, while smaller-scale systems are taking the place of the power plants that are being retired.

Electricity is increasingly coming from renewable sources. California law now requires that 33% of the electricity sold in the state must come from renewable sources by 2020, and the governor has called for it to be increased to 50% (CRPS 2015; Brown 2015). In addition to technology advances, an important driver of this trend is the dramatic cost reductions, particularly in wind energy, solar photovoltaic panels, and batteries (Lazard 2014; UBS 2014). In place of large, centralized fossil and nuclear power plants, California is increasingly turning to *micropower* systems. These systems, as Amory Lovins notes, are "sources of electricity that are relatively small, modular, mass-producible, quick-to-deploy, and hence rapidly scalable—the opposite of cathedral-like power plants that cost billions of dollars and take about a decade to license and build" (Lovins and Palazzi 2014).

Investment analysts indicate that batteries and solar photovoltaic technology may be "at the tipping point" as electricity users become generators, and they suggest, "Solar systems and batteries will be disruptive technologies for the electricity system" (UBS 2014, p. 1). The implications are significant: "In this decentralized electricity world, the key utilities' assets will be smart distribution networks, end customer relationships and small-scale backup units. Utilities should be able to extract more value in (highly competitive) supply activities, as customer needs will be more complex. Large-scale power generation, however, will be the dinosaur of the future energy system: Too big, too inflexible, not even relevant for backup power in the long run" (UBS 2014, p. 1).

Both of these trends—increased efficiency and increased reliance on local and renewable water and energy sources—have two important outcomes in the context of climate change: reduced emissions (mitigation in climate terms) and increased resilience in the face of impacts (adaptation). The full costs of energy and water include climate change and other environmental impacts. Those costs are not too cheap to meter.

13.6 Redesigning Our Activities on the Landscape

The management of land, and our activities on it, from restoration and stewardship of high mountain meadows and forest systems to new urban design in cities, has important implications and impacts in the context of climate change. Recharging groundwater aquifers, preventing erosion and flooding, and beneficially harvesting rainwater on the land are important elements of this land management opportunity relating to water and energy. Practices such as *low-impact development* (LID), a land planning and engineering design approach to storm water management, yield both water quantity and quality benefits while also reducing energy requirements and associated GHG emissions (Garrison et al. 2009, 2014; EPA 2007; Stoner et al. 2006; CARB 2014). Successful rainwater harvesting and management practices include maximizing infiltration, which recharges local and regional groundwater systems; providing retention areas and slowing runoff, which reduce flooding and erosion; minimizing the impervious footprint of a project through reducing paved surfaces; directing runoff from impervious areas onto landscaping; and capturing runoff in rain barrels or cisterns for beneficial use (Garrison et al. 2009).

These practices are highly cost-effective (USEPA 2007) and can yield significant water supplies while at the same time reducing pollution, flood damage, energy use, and GHG emissions (Garrison et al. 2009, 2014). The National Association of Home Builders states, "Ever wish you could simultaneously lower your site infrastructure costs, protect the environment, and increase your project's marketability? With LID techniques, you can" (NAHB 2002).

Rainwater harvesting, along with water-use efficiency and reuse, are included as GHG reduction measures in California's Scoping Plan for implementation of AB32 (CARB 2014). Land-use planning authorities and organizations have embraced these water management and design concepts and they have created guidelines and policies for implementation. The Local Government Commission, for example, developed the Ahwahnee water principles in 2005 to guide local land-use authorities in designs for LID and water management (LGC 2005).

California has also developed land-use policy to address GHG emissions and sustainable-communities issues. The California Air Resources Board notes that the Sustainable Communities and Climate Protection Act of 2008 (SB 375, 2008) "developed a critical, unique policy mechanism for reducing transportation-sector GHG emissions" (CARB 2014). Regional and local planning agencies are responsible for developing Sustainable Communities Strategies as part of the federally required Regional Transportation Plan, and they are also responsible for developing state-required general plan housing elements to help meet these targets. "The goal of SB 375 is to reduce GHG emissions from passenger vehicles through better-integrated regional transportation, land use, and housing planning that provides easier access to jobs, services, public transit, and active transportation options" (CARB 2014, p. 49; SB 375, 2008).

13.6.1 Restoration of Natural System Functions as Part of Integrated Strategies

Restoration of natural system functions, like recharging of groundwater aquifers, is an important part of integrated response strategies. Restoration efforts include reforestation and revegetation, meadow and floodplain restoration, erosion control, the appropriate reintroduction of fire, and removal of barriers like obsolete dams. One key goal of this work is to allow natural systems to function like a sponge, absorbing water during high-precipitation events to recharge soil moisture and aquifers, and retaining water for use during dry times. The state has already recognized measures like mountain meadow restoration as significant water supplies in its assessment of future water supply options (CDWR 2014). Water quality and ecosystem services can be enhanced in this way, and pressures on species and natural systems can be reduced. Investments are now being made in these activities, and more will be needed. Recent analysis indicates that the economic benefits of measures like forest treatment (thinning and controlled burning) are two to three times the costs (Sierra Nevada Conservancy 2014).

Barriers to the functioning of natural systems are being removed to restore physical and ecosystem processes. The largest dam removal project in California history is taking place on the Carmel River. The 32.3 m (106 ft.) concrete San Clemente Dam owned by California American Water is being removed at a cost of over $84 million to reduce risk and restore fish passage (CalAm 2015). The dam sat on an earthquake fault and was filled with debris and mud, no longer providing water benefits. Costs are being shared by the owner and various public and private sources. With increased variability in precipitation and greater intensity of storm events expected with climate change, the dam removal is a risk reduction measure and an environmental restoration opportunity. A number of other dams, such as the Matilija Dam near Ojai on the central coast, are also slated for removal as funding becomes available.

Groundwater aquifers are being restored to provide valuable water supply options. In Los Angeles, a major project is under way to clean up and restore the functions of the San Fernando aquifer. The city plans to use the aquifer, in conjunction with rainwater harvesting and water recycling strategies, as a key part of its plan to reduce purchased imported water by 50% in the next decade. The Los Angeles effort is an excellent example of an integrated effort with multiple benefits. The reduced reliance on imported water takes some pressure off a seriously overallocated State Water Project and damaged delta ecosystem. The local supplies of rainwater and recycled water require far less energy, even with pumping and treatment, so GHG emissions are reduced. And the capture and use of rain and recycled water reduces impacts from flooding during high-precipitation events.

13.6.2 Cross-Sector Links, Coordination, and Integrated Approaches

Integration of science, policy, planning, and management at all levels of governance—local, state, and federal—and with nongovernmental organizations (NGOs), business, and other stakeholders will be required to tap opportunities and deal with challenges posed by climate change. The California Council on Science and Technology noted in a recent analysis, "California has a long history of success in leveraging innovations in science, technology, management and implementation strategies to improve its resource management, including its continued leadership in energy efficiency. The state's best strategy for dealing with its water challenges, both current and future, lies in taking a system management approach to water similar to the approach used for energy. Also, as with energy, innovative water technologies represent a sound business opportunity for California" (CCST 2014, p. 1).

Leadership is emerging at all levels, and there are exciting and inspiring examples of successful efforts involving integrated response strategies. For example, in 2006, the Sonoma County Water Agency committed to the goal of operating a net carbon-free water system by 2015 by reducing its energy use through efficiency and by diversifying its energy portfolio with renewable energy production. The agency achieved 97% of the goal within 5 years and reached the goal ahead of schedule. The water agency stated that it "has a duty to secure water for future generations of Sonoma County residents. The potential impacts of climate change will make it more difficult to meet that mission. The more the Water Agency can do to mitigate the impact of climate change, the more secure the source of water will be for the future" (Sonoma County Water Agency 2015).

In the Los Angeles basin (Los Angeles, San Gabriel, and Santa Ana River watersheds), in San Diego and Silicon Valley, and in counties including Ventura, Sonoma, Alameda, and many more jurisdictions, water, flood control, and wastewater agencies are working with land-use authorities, NGOs, and universities to improve water-use efficiency, capture and recharge storm water, and reuse wastewater to meet water supply needs and prevent pollution. Many of the same partners are seeking to intelligently relandscape communities to improve water-use efficiency and groundwater recharge. Indeed, we need a statewide effort to relandscape California with attractive and appropriate designs and materials that are water efficient and that function like a sponge—absorbing rainwater and retaining moisture (Council for Watershed Health 2015; Bay-Friendly Landscaping Coalition 2015; Tree People 2015).

California has developed new approaches to integrated planning, including interagency efforts to respond to climate change. The Water–Energy Team of the state's Climate Action

Team (WETCAT), for example, has effectively facilitated policy coordination and integration across state agencies and with the academic and stakeholder communities (WETCAT 2015; CARB 2014; Spanos 2012). The University of California—systemwide with all campuses involved—is planning to reach net neutrality for on-campus emissions (scope 1) and for emissions related to purchased electricity (scope 2) by 2025. Other entities, including cities, are embracing similar goals.

State and federal agencies are developing new approaches to better understand and quantify critical dimensions of the water–energy–land nexus in the context of climate change. For example, the Department of Water Resources and the National Aeronautics and Space Administration's (NASA's) Jet Propulsion Lab (JPL) are developing new ways to measure snowpack from airborne observations. Mountain snow is a critically important water supply source in many parts of the world, including California. The two most critical properties for understanding snowmelt runoff and timing, the amount of water stored in snowpacks and when it will melt and become runoff, are the spatial and temporal distributions of snow water equivalent (SWE) and snow albedo. As Tom Painter of JPL notes, "Despite their importance in controlling volume and timing of runoff, snowpack albedo and SWE are still largely unquantified in the U.S. and not at all in most of the globe, leaving runoff models poorly constrained. NASA/JPL, in partnership with the California Department of Water Resources, has developed the Airborne Snow Observatory, an imaging spectrometer and scanning lidar system, to quantify SWE and snow albedo, generate unprecedented knowledge of snow properties for cutting edge cryospheric science, and provide complete, robust inputs to water management models and systems of the future" (NASA 2015). This kind of collaborative approach will be critical to integrated response strategies.

Federal and state policy regarding both energy and water has addressed embedded energy use for decades. For example, in recognition of the energy impacts of water use, the US Energy Policy Act of 1992 set standards for the maximum water use of toilets, urinals, showerheads, and faucets (Energy Policy Act 1992). As the policy makers anticipated, the energy savings resulting from water efficiency are significant.

New initiatives and efforts to coordinate water and energy policy are being proposed. At the federal level, the Nexus of Energy and Water for Sustainability Act was proposed in 2014 and reintroduced in 2015 (S 1971). It would require the director of the Office of Science and Technology Policy to establish either a committee or subcommittee on energy–water nexus for sustainability under the National Science and Technology Council (NSTC), cochaired by the secretary of the Department of Energy (DOE) and secretary of the interior. The act would require the committee or subcommittee to serve as a forum for developing common federal goals and plans on energy–water nexus issues; promote coordination of the related activities of federal departments and agencies; coordinate and develop capabilities for data collection, categorization, and dissemination of data from and to other federal departments and agencies; and engage in information exchange between federal departments and agencies.

13.7 Conclusion: Tapping Multiple Benefits with Integrated Response Strategies

A quarter century ago, Roger Revelle and Paul Waggoner made the following prescient observation as part of the Climate and Water Panel of the American Association for the Advancement of Science: "Among the climatic changes that governments and other public bodies are likely to encounter are rising temperatures, increasing evapotranspiration, earlier melting of snowpacks, new seasonal cycles of runoff, altered frequency of extreme events, and rising sea level. *Governments at all levels should reevaluate legal, technical, and economic procedures for managing water resources in the light of climate changes that are highly likely"* (Revelle and Waggoner 1990, p. 4, italics in original). This process of reevaluation is taking place as Revelle and Waggoner advised, and integrated response strategies are emerging.

A diverse portfolio of options, including technologies and techniques, is available to address the challenge of climate change and variability. Multiple-benefit opportunities exist that can improve resilience and adaptive responses, reduce GHG emissions, and improve quality of life. In many cases, these opportunities are available at less cost than doing nothing. California is increasingly embracing these options. Integrated response strategies involving investments and actions that both reduce emissions and improve resilience and adaptive capacity are being made. Better information and coordination is needed, and California should continue to integrate water-use, energy-use, and land-use planning and management at all levels.

The state's key water and energy management agencies have made important strides in identifying areas where water-use, energy-use, and land-use planning and management can be integrated, and plans are increasingly incorporating the nexus. This work should be enhanced and expanded. The water–energy–land nexus should be further incorporated into local and regional water and energy plans and assessments. Urban and agricultural water management plans, for example, are beginning to incorporate energy and emissions data and analysis. This should become standard practice in planning processes.

Opportunities are available to substantially reduce energy inputs (and costs) of water systems, and water use in energy systems. We have cost-effective options to improve resilience, environmental quality, and security. A standardized multiple-benefit analysis methodology should be developed to determine the cost-effectiveness of investments in both water and energy systems. This would support stronger cofunding strategies for water-use, energy-use, land-use, and other entities, and generate supportive policy structures to enable water and energy entities to tap linked water–energy improvement opportunities.

Governor Brown declared, "Taking significant amounts of carbon out of our economy without harming its vibrancy is exactly the sort of challenge at which California excels. This is exciting, it is bold and it is absolutely necessary if we are to have any chance of stopping potentially catastrophic changes to our climate system" (Brown 2015). The Los Angeles Times opined, "California may lose its wager on climate change, but the cost of not trying is too great" (Los Angeles Times 2015).

References

Alliance for Water Efficiency (AWE) and American Council for an Energy-Efficient Economy (ACEEE). 2013. Water–Energy Nexus Research: Recommendations for Future Opportunities. http://www.allianceforwaterefficiency.org/1Column.aspx?id=8514&terms=energy+water+nexus.

Assembly Bill 32. 2006. California Global Warming Solutions Act of 2006. http://www.arb.ca.gov/cc/docs/ab32text.pdf.

Averyt, K., J. Fisher, A. Huber-Lee, A. Lewis, J. Macknick, N. Madden, J. Rogers, S. Tellinghuisen. 2011. Freshwater Use by U.S. Power Plants: Electricity's Thirst for a Precious Resource. Union of Concerned Scientists. http://www.ucsusa.org/assets/documents/clean_energy/ew3/ew3-freshwater-use-by-us-power-plants.pdf.

Bay-Friendly Landscaping Coalition. 2015. http://www.bayfriendlycoalition.org/.

Brown, E.G. Jr., 2015. Governor's Inaugural Address, Remarks as Prepared, January 5, 2015. http://gov.ca.gov/home.php.

California Air Resources Board (CARB). 2014. Climate Change Scoping Plan, First Update. http://www.arb.ca.gov/cc/scopingplan/2013_update/first_update_climate_change_scoping_plan.pdf.

California American Water, California Coastal Conservancy, National Oceanographic and Atmospheric Administration (CalAm). 2015. http://www.sanclementedamremoval.org/.

California Climate Change Center (CCCC). 2006. Our Changing Climate, Assessing the Risks to California http://www.energy.ca.gov/2006publications/CEC-500-2006-077/CEC-500-2006-077.PDF.

California Climate Change Center (CCCC). 2008. The Future Is Now: An Update on Climate Change Science Impacts and Response Options for California. http://www.energy.ca.gov/2008publications/CEC-500-2008-071/CEC-500-2008-071.PDF.

California Climate Change Center (CCCC). 2009. Using Future Climate Projections to Support Water Resources Decision-Making. http://wwwdwr.water.ca.gov/pubs/climate/using_future_climate_projections_to_support_water_resources_decision_making_in_california/usingfutureclimateprojtosuppwater_jun09_web.pdf.

California Climate Change Portal http://www.climatechange.ca.gov/.

California Council on Science and Technology (CCST). 2014. Laspa, J., K. Longley, S. Sorooshian, R. Wilkinson, D. Zoldoske, M.D. DeCillis, A. Michelson, B. Hannegan. Achieving a Sustainable California Water Future through Innovations in Science and Technology. California Council on Science and Technology. http://www.ccst.us/news/2014/0409water.php.

California Department of Finance (CDF). 2015. http://www.dof.ca.gov/research/demographic/.

California Department of Water Resources (CDWR). 2008. Managing an Uncertain Future, Climate Change Adaptation Strategies for California's Water. http://www.water.ca.gov/climatechange/docs/ClimateChangeWhitePaper.pdf.

California Department of Water Resources (CDWR). 2009. Possible Impacts of Climate Change to California's Water Supply. http://baydeltaoffice.water.ca.gov/climatechange/ClimateChangeSummaryApr09.pdf.

California Department of Water Resources (CDWR). 2012. Final Climate Action Plan Phase I: Greenhouse Gas Emissions Reduction Plan. http://wwwdwr.water.ca.gov/climatechange/CAP.cfm.

California Department of Water Resources (CDWR). 2013. Strategic Plan for the Future of Integrated Regional Water Management in California. http://wwwdwr.water.ca.gov/irwm/stratplan/.

California Department of Water Resources (CDWR). 2014a. The California Water Plan. Bulletin 160-13 California Department of Water Resources. http://www.waterplan.water.ca.gov.

California Department of Water Resources (CDWR). 2014b. Integrated Regional Water Management. http://www.water.ca.gov/irwm/grants/index.cfm.

California Department of Water Resources (CDWR). 2015. Integrated Regional Water Management (IRWM). http://www.water.ca.gov/irwm/grants/index.cfm and http://www.water.ca.gov/irwm/grants/guidelines.cfm.

California Energy Commission (CEC). 2005. Integrated Energy Policy Report, November 2005, CEC-100-2005-007-CMF.

California Environmental Protection Agency (CEPA), Office of Environmental Health Hazard Assessment. 2013. Indicators of Climate Change in California. http://oehha.ca.gov/multimedia/epic/pdf/ClimateChangeIndicatorsReport2013.pdf.

California Legislative Analyst's Office (LAO) 2015. http://www.lao.ca.gov/LAOEconTax/Article/Detail/90.

California Natural Resources Agency (CNRA). 2013. California Climate Adaptation Strategy. http://climatechange.ca.gov/adaptation/index.html.

California Natural Resources Agency (CNRA). 2015. Cal-Adapt Website. http://resources.ca.gov/climate_adaptation/science/cal-adapt.html.

California Renewables Portfolio Standard (CRPS), 2015. (Established in 2002 under Senate Bill 1078, accelerated in 2006 under Senate Bill 107 and expanded in 2011 under Senate Bill 2.) http://www.cpuc.ca.gov/PUC/energy/Renewables/.

California Urban Water Conservation Council (CUWCC), 2014. http://www.cuwcc.org/Resources/Best-Management-Practices-BMPs.

California Water Action Plan: Actions for Reliability, Restoration and Resilience. 2014. Governor's plan prepared by the California Natural Resources Agency, the California Environmental Protection Agency, and the California Department of Food and Agriculture. http://resources.ca.gov/california_water_action_plan/docs/Final_California_Water_Action_Plan.pdf.

Cayan, D., T. Das, D.W. Pierce, T.P. Barnett, M. Tyree, A. Gershunov. 2010. Future dryness in the southwest US and the hydrology of the early 21st century drought. *PNAS* 107: 21271–21276. http://www.pnas.org/content/107/50/21271.abstract.

Cooley, H., R. Wilkinson. 2012. Implications of Future Water Supply Sources for Energy Demands, and Computer Model with WESim User Manual, Pacific Institute and Bren School, University of California, Santa Barbara, for WateReuse Research Foundation, the California Energy Commission, and the Canadian Mortgage and Housing Corporation. http://www.pacinst.org/publication/wesim/.

Cooley, H., K. Donnelly, N. Ajami. 2013. Energizing Water Efficiency: California Energy Sector Experiences Can Advance State's Water Conservation and Efficiency. http://pacinst.org/publication/energizing-water-efficiency/.

Cooley, H., K. Donnelly. 2013. Water–Energy Synergies: Coordinating Efficiency Programs in California. http://pacinst.org/publication/water-energy-synergies/.

Council for Watershed Health. 2015. http://www.watershedhealth.org/jobboard.aspx.

Davis, M., S. Clemmer. 2014. *Power Failure. Union of Concerned Scientists.* www.ucsusa.org/powerfailure.

Delta Stewardship Council. 2013. The Delta Plan: Ensuring a Reliable Water Supply for California, a Healthy Delta Ecosystem, and a Place of Enduring Value. http://deltacouncil.ca.gov/sites/default/files/documents/files/DeltaPlan_2013_CHAPTERS_COMBINED.pdf.

Dettinger, M.D., D.R. Cayan. 1994. Large-scale atmospheric forcing of recent trends toward early snowmelt runoff in California. *Journal of Climate* 8, 606–623.

Electric Power Research Institute (EPRI). 2008. Water Use in Power Generation.

Energy Policy Act of 1992. (102nd Congress H.R.776.ENR.) http://thomas.loc.gov/cgi-bin/query/z?c102:H.R.776.ENR:.

Garcetti, E. 2014a. Executive Directive #5: Emergency Drought Response—Creating a Water Wise City. http://www.lamayor.org/executive_directive_5_emergency_drought_response_creating_a_water_wise_city.

Garcetti, E. 2014b. Policy Statement, Office of Los Angeles Mayor Eric Garcetti. November 10, 2014. http://www.lamayor.org/mayor_garcetti_highlights_ladwp_s_new_15_energy_efficiency_goal_which_leads_nation.

Garrison, N., R.C. Wilkinson, R. Horner. 2009. A Clear Blue Future: How Greening California Cities Can Address Water Resources and Climate Challenges in the 21st Century. Natural Resources Defense Council and Water Policy Program Bren School of Environmental Science and Management, University of California, Santa Barbara. http://www.nrdc.org/water/lid/default.asp.

Garrison, N., J. Sahl, A. Dugger, R. Wilkinson. 2014. Stormwater Capture Potential in Urban and Suburban California. http://pacinst.org/publication/ca-water-supply-solutions/.

GEI Consultants/Navigant Consulting. 2010. Embedded Energy in Water Studies. Study 1: Statewide and Regional Water–Energy Relationship. http://www.cpuc.ca.gov/PUC/energy/Energy+Efficiency/EM+and+V/Embedded+Energy+in+Water+Studies1_and_2.htm; and Study 2: Water Agency and Function Component Study and Embedded Energy–Water Load Profiles. California Public Utilities Commission, Energy Division, Managed by California Institute for Energy and Environment, August 31, 2010. http://www.cpuc.ca.gov/PUC/energy/Energy+Efficiency/EM+and+V/Embedded+Energy+in+Water+Studies1_and_2.htm.

Gleick, P. 2000. Water: The Potential Consequences of Climate Variability and Change for the Water Resources of the United States. Report of the Water Sector Assessment Team of the National Assessment of the Potential Consequences of Climate Variability and Change for the US Global Change Research Program. http://pacinst.org/wp-content/uploads/sites/21/2013/02/natl_assessment_water3.pdf.

Griffin, D., K.J. Anchukaitis. 2014. How unusual is the 2012–2014 California drought? *Geophysical Research Letters* 41(24), 9017–9023. doi: 10.1002/2014GL062433.

Groundwater Replenishment System (Orange County). 2014. http://www.gwrsystem.com/index.php?option=com_content&view=article&id=2&Itemid=26.

Hanak, E., J. Lund, A. Dinar, B. Gray, R. Howitt, J. Mount, P. Moyle, B. Thompson. 2011. Managing California's Water: From Conflict to Reconciliation. Public Policy Institute of California. http://www.ppic.org/main/publication.asp?i=944.

Hanak, E., J. Lund, B. Thompson, W.B. Cutter, B. Gray, D. Houston, R. Howitt et al. 2012. Water and the California Economy. Public Policy Institute of California. http://www.ppic.org/main/publication.asp?i=1015.

Intergovernmental Panel on Climate Change (IPCC). 2013. Climate Change 2013: The Physical Science Basis: Summary for Policymakers. Contribution of Working Group I to the Fifth Assessment Report of the Intergovernmental Panel on Climate Change. http://www.ipcc.ch/report/ar5/wg1/.

Intergovernmental Panel on Climate Change (IPCC). 2014. Fifth Assessment Report of the Intergovernmental Panel on Climate Change. http://www.ipcc.ch/.

Integrated Regional Water Management Planning Act of 2008. http://www.water.ca.gov/irwm/grants/fundsource_legis.cfm.

Kenney, D.S., R. Wilkinson eds. 2011. *The Water–Energy Nexus in the Western United States*. Cheltenham: Edward Elgar Publishing.

Lazard. 2014. Levelized Cost of Energy Analysis—Version 8.0, August 2014. http://www.lazard.com/PDF/Levelized percent20Cost percent20of percent20Energy percent20- percent20Version percent208.0.pdf.

Little Hoover Commission. 2014. Governing California through Climate Change. Report Number 221. http://www.lhc.ca.gov/studies/221/report221.html.

Local Government Commission (LGC). 2005. The Ahwahnee Water Principles. http://www.lgc.org/resources/water.

Los Angeles Department of Water and Power (LADWP). 2011. Los Angeles Department of Water and Power. 2010 Urban Water Management Plan. http://www.water.ca.gov/urbanwatermanagement/2010uwmps/Los%20Angeles%20Department%20of%20Water%20and%20Power/LADWP%20UWMP_2010_LowRes.pdf.

Los Angeles Times. 2015. California's bold attack on climate change. L.A. Times Editorial 1/8/15. http://www.latimes.com/opinion/editorials/la-ed-climate-change-jerry-brown-20150109-story.html.

Lovins, A., T. Palazzi. 2014. Micropower's Quiet Takeover. Forbes. http://www.forbes.com/sites/amorylovins/2014/09/19/micropowers-quiet-takeover/.

Maupin, M.A., J.F. Kenny, S.S. Hutson, J.K. Lovelace, N.L. Barber. 2014. Estimated Use of Water in the United States in 2010. USGS Circular: 1405. http://pubs.er.usgs.gov/publication/cir1405.

Metropolitan Water District of Southern California (MWD). 2014. Regional Progress Report to the California State Legislature. http://www.mwdh2o.com/mwdh2o/pages/yourwater/SB60/archive/SB60_2014.pdf.

National Aeronautics and Space Administration (NASA), Airborne Snow Observatory. 2015. http://aso.jpl.nasa.gov/.

National Academy of Sciences (NAS). 1975. *Understanding Climatic Change: A Program for Action*. Washington, D.C.: National Academy of Sciences.

National Academy of Sciences (NAS). 2014. What We Know. http://whatweknow.aaas.org/get-the-facts/.

National Academies; The Royal Society. 2014. Climate Change: Evidence and Choices. http://www.nap.edu/catalog.php?record_id=18730.

National Association of Home Builders (NAHB). 2002. Builder's Guide to Low Impact Development. http://www.lowimpactdevelopment.org/lid%20articles/Builder_LID.pdf.

National Oceanic and Atmospheric Administration's National Climatic Data Center (NCDC). 2015. NOAA. http://www.ncdc.noaa.gov/cag/and for California 2014 temperatures http://www.ncdc.noaa.gov/temp-and-precip/climatological-rankings/index.php?periods[]=12¶meter=tavg&state=4&div=0&month=12&year=2014#ranks-form.

National Renewable Energy Lab (NREL). 2014. Making Sustainable Energy Choices: Insights on the Energy/Water/Land Nexus (Technical Report), Analysis Insights. http://www.nrel.gov/docs/fy15osti/62566.pdf.

National Research Council (NRC). 2001. Climate Change Science: An Analysis of Some Key Questions. D.C.: National Academy Press. http://www.nap.edu.

National Research Council (NRC). 2002. *Abrupt Climate Change: Inevitable Surprises.* D.C.: National Academy Press. http://www.nap.edu/catalog/10136.html.

Natural Resources Defense Council (NRDC). 2014. Drought Recommendations to the State Water Resources Control Board. February 26, 2014. http://www.swrcb.ca.gov/waterrights/water_issues/programs/drought/docs/workshops/nrdc_drought_recommend.pdf.

Pacific Council on International Policy. 2010. Preparing for the Effects of Climate Change—A Strategy for California. A Report by the California Adaptation Advisory Panel to the State of California. Los Angeles, CA. http://www.pacificcouncil.org/document.doc?id=183.

Pacific Institute, Natural Resources Defense Council, R. Wilkinson. 2014. The Untapped Potential of California's Water Supply. www.pacinst.org/publication/ca-water-supply-solutions.

Park, L., B. Bennett, S. Tellinghuisen, C. Smith, R. Wilkinson. 2008. The Role of Recycled Water in Energy Efficiency and Greenhouse Gas Reduction. California Sustainability Alliance. http://sustainca.org/programs/water_energy/recycled_water_study.

Plattner, G.K. 2014. Highlights of the New IPCC Report. Director of Science, AR5 IPCC WGI TSU. http://www.ipcc.ch/pdf/unfccc/cop19/cop19_pres_plattner.pdf.

Revelle, R.R., P.E. Waggoner. 1983. *Changing Climate.* D.C.: National Academy Press.

Revelle, R.R., P.E. Waggoner. 1990. Effects of a carbon dioxide-induced climatic change on water supplies in the western United States. In P.E. Waggoner, ed. 1990. *Climate Change and U.S. Water Resources.* New York: John Wiley and Sons.

Rogers, J., K. Averyt, S. Clemmer, M. Davis, F. Flores-Lopez, D. Kenney, J. Macknick et al. 2013. *Water-Smart Power: Strengthening the U.S. Electricity System in a Warming World.* Cambridge: Union of Concerned Scientists. http://www.ucsusa.org/clean_energy/our-energy-choices/energy-and-water-use/water-smart-power.html.

Rogers, P. 2015. California drought: State residents increase conservation but still fall far short of governor's goal. *San Jose Mercury News.* http://www.mercurynews.com/science/ci_27269399/california-drought-state-residents-increase-conservation-but-still.

Ruys, J., M. Hogan. 2014. Consumers at the Gate. Has energy reached 'peak centralization'? Rocky Mountain Institute Blog. http://blog.rmi.org/blog_2014_09_10_consumers_at_the_gate.

San Diego County Water Authority. 2014. Climate Action Plan. http://www.sdcwa.org/climate-action-plan.

Sandia National Laboratory. 2015. Energy–Water Nexus. http://www.sandia.gov/energy-water/.

Santa Clara Valley Water District. 2014. http://www.valleywater.org/Services/RecycledWater.aspx.

Schwarzenegger, A. 2005. United Nations World Environment Day Conference, June 1, 2005, San Francisco.

Seager, R., M. Hoerling. 2014. Atmosphere and Ocean Origins of North American Droughts. *Journal of Climate* 27, 4581–4606. doi: http://dx.doi.org/10.1175/JCLI-D-13-00329.1.

Senate Bill (SB) 375. The Sustainable Communities and Climate Protection Act of 2008. Steinberg, Chapter 728, Statutes of 2008. http://www.arb.ca.gov/cc/sb375/sb375.htm.

Senate Bill (SB) X7-7. Water Conservation Act of 2009 (SBX7-7). http://www.water.ca.gov/wateruseefficiency/sb7/.
S.1971—113th Congress (2013–2014). Nexus of Energy and Water for Sustainability Act of 2014. https://www.congress.gov/bill/113th-congress/senate-bill/1971.
Sierra Nevada Conservancy, California Natural Resources Agency. 2014. Mokelumne Watershed Avoided Cost Analysis. http://www.sierranevada.ca.gov/our-work/mokelumne-watershed-analysis.
Sonoma County Water Agency. 2015. Carbon Free Water. http://www.scwa.ca.gov/carbon-free-water/.
Spanos, K.A. 2012. The climate has changed: Now what? Integrated Regional Water Management and climate change planning a coincidental or inevitable union? 30th Annual Water Law Conference. *American Bar Association*. http://www.water.ca.gov/climatechange/articles.cfm.
Stewart, I., D. Cayan, M. Dettinger. 2005. Changes toward earlier streamflow timing across western North America. *Journal of Climate* 18, 1136–1155.
Stoner, N., C. Kloss, C. Calarusse. 2006. Rooftops to Rivers: Green Strategies for Controlling Stormwater and Combined Sewer Overflows. http://www.nrdc.org/water/pollution/rooftops/contents.
Strauss, L. 1954. The phrase "too cheap to meter" comes from the Chairman of the United States Atomic Energy Commission, Lewis Strauss, in a 1954 speech to the National Association of Science Writers.
Tree People. 2015. http://www.treepeople.org/.
UBS. 2014. Will solar, batteries and electric cars re-shape the electricity system? UBS Global Research, August 20, 2014. http://knowledge.neri.org.nz/assets/uploads/files/270ac-d1V0tO4LmKMZuB3.pdf.
United States Bureau of Economic Analysis (USBEA). 2015. http://www.bea.gov/regional/index.htm.
United States Census. 2015. http://quickfacts.census.gov/qfd/states/06000.html.
United States Department of Energy (USDOE). 2006. Energy Demands on Water Resources Report to Congress on the Interdependency of Energy and Water. December 2006. http://www.sandia.gov/energy-water/congress_report.htm.
United States Energy Information Administration (USEIA). 2012. State Energy Data System, All Consumption Estimates, in Physical Units, ESTCP and TPOPP, 1990–2010 (June 2012). www.eia.gov/state/seds/seds-data-complete.cfm#Consumption. Data plot from NRDC, 2013. California's Energy Efficiency Success Story: Saving Billions of Dollars and Curbing Tons of Pollution.
United States Environmental Protection Agency (USEPA). 2007. Reducing Stormwater Costs through Low Impact Development (LID) Strategies and Practices. Fact sheet number 841-F-07-006.
United States Environmental Protection Agency, California Department of Water Resources, Resources Legacy Fund, and the US Army Corps of Engineers, (USEPA). 2011. Climate Change Handbook for Regional Water Planning. http://wwwdwr.water.ca.gov/climatechange/CCHandbook.cfm.
United States Global Change Research Program (USGCRP). 2014. National Climate Assessment. http://www.globalchange.gov/what-we-do/assessment/draft-report-information.html; and United States Global Change Research Program (USGCRP). http://library.globalchange.gov/products/annualreports.
United States Nuclear Regulatory Commission (USNRC). 2015. http://www.nrc.gov/info-finder/reactor/songs/decommissioning-plans.html.
United States Senate. 1979. Carbon Dioxide Accumulation in the Atmosphere, Synthetic Fuels and Energy Policy: A Symposium. Committee on Governmental Affairs. 96th Congress, 1st Session. US Government Printing Office, Washington, D.C.
Water–Energy Team of the California Climate Action Team (WETCAT). 2015. http://www.climatechange.ca.gov/climate_action_team/water.html.

Wilkinson, R. 2000. Methodology for Analysis of the Energy Intensity of California's Water Systems, and an Assessment of Multiple Potential Benefits through Integrated Water–Energy Efficiency Measures. Exploratory Research Project, Ernest Orlando Lawrence Berkeley Laboratory, California Institute for Energy Efficiency. http://www.es.ucsb.edu/faculty/wilkinson.php.

Wilkinson, R. 2002. Preparing for a Changing Climate: The Potential Consequences of Climate Variability and Change for California. The California Regional Assessment Report of the California Regional Assessment Group for the U.S. Global Change Research Program, National Center for Geographic Information Analysis, and the National Center for Ecological Analysis and Synthesis. University of California, Santa Barbara (Sponsored by the National Science Foundation). http://www.ncgia.ucsb.edu/products.html.

Wilkinson, R.C. 2008. Invited Testimony to Congress: Water Supply Challenges for the 21st Century. Committee on Science and Technology, United States House of Representatives. http://www.bren.ucsb.edu/people/Faculty/wilkinson_more.htm.

Wilkinson, R. 2011. The Water–Energy Nexus: Methodologies, Challenges, and Opportunities. In Kenney, D.S. and R. Wilkinson eds. *The Water–Energy Nexus in the Western United States*. Cheltenham: Edward Elgar Publishing.

World Bank. 2015. GDP. http://data.worldbank.org/indicator/NY.GDP.MKTP.CD.

Western Regional Climate Center (WRCC). 2015. http://www.wrcc.dri.edu/.

14

California's Irrigated Agriculture and Innovations in Adapting to Water Scarcity

Heather Cooley

CONTENTS

ABSTRACT California is one of the most productive agricultural regions in the world. The state, however, is prone to multiyear droughts, with wide-ranging impacts, including on the state's agricultural sector. Climate change, combined with continued population and economic growth, will make meeting water demands even more difficult. The good news is that water efficiency improvements, water recycling and reuse, and groundwater recharge can promote the long-term sustainability of the state's agricultural sector.

14.1 Introduction

California is one of the most productive agricultural regions in the world, producing more than 400 different farm products. The state is the nation's largest agricultural producer, supplying both US and international markets. In 2013, California farm output was valued at a record $46.4 billion. Additionally, California is the nation's largest agricultural exporter, with exports reaching a record $21.2 billion in 2013 (CDFA 2015).

California's agricultural production has been made possible, in part, by irrigation supplied by a vast and integrated infrastructure network. While actual water use remains largely unknown due to a lack of consistent measurement and reporting, estimates show that California's agricultural sector uses the majority of California's developed water supply. An average of 44 million acre-feet (MAF) (54.25 BCM)* of water was withdrawn annually from surface water and groundwater aquifers between 2001 and 2010. Approximately 80% of that water, or 35 MAF, was used for agriculture, and the remaining 20% was used in homes, businesses, and institutions in urban areas (DWR 2014).

California has the most variable climate in the United States as measured by the coefficient of variation (standard deviation/mean) of water-year (October–September) precipitation (Dettinger et al. 2011). This variability is driven by the fact that a small number of winter storms typically account for the bulk of annual precipitation, so that small changes in the number or intensity of these storms can be the difference between a wet and dry year. Moreover, the state is prone to multiyear droughts. In the twentieth century, for example, California experienced droughts in 1929–1934, 1976–1977, and 1987–1992. More recently, California experienced a relatively modest drought in 2007–2009 and, as of this writing, is in the midst of a much more severe drought beginning in 2012, extending into 2015 and possibly longer. Indeed, in a recent analysis, Griffin and Anchukaitis (2014) found that 2012–2014 was the worst 3-year drought in the last 1200 years in California and that 2014 was the worst single drought year during this period.

Droughts have wide-ranging social, economic, and environmental impacts. Several studies have documented the impacts of these droughts on the state's agricultural sector (e.g., Gleick and Nash 1991; Michael et al. 2010; Christian-Smith et al. 2015). It is important to note, however, that determining the impact of a drought is challenging, as there is no standard methodology, data are often lacking, and it is difficult to isolate drought from other factors.

Across California, drought and its impacts have become exacerbated by growing water scarcity. Water scarcity is a man-made phenomenon resulting from water demand exceeding the natural renewable availability. It is characterized by "a permanent and continued degradation of water ecosystems and reduced water availability for other (economic) functions" (Schmidt and Benitez-Sanz 2013). California's thirst for water exceeds supply in even an average year, as evidenced by collapsing freshwater ecosystems in the Sacramento–San Joaquin Delta and widespread groundwater overdraft. Climate change, combined with continued economic and population growth, will exacerbate these concerns. This chapter examines growing evidence of water scarcity in California and identifies a set of adaptation strategies to support the long-term sustainability of California agriculture.

14.2 Signs of Water Scarcity

California's water system is out of balance, and human demands for water exceed volumes that can be sustainably extracted. As one example, an analysis of the state's water rights database found that post-1914, appropriative water rights allocations are approximately five times the state's mean annual runoff (Grantham and Viers 2014). Other forms of water rights, e.g., pre-1914 and riparian water rights, were not included in this analysis but would further increase surface water rights allocations. While water can be used multiple times as it flows downstream and there is an incentive for permit holders to overstate their water use, the findings "provide strong evidence that the state has overallocated water in many, if not most river basins." Evidence of the overallocation of surface

* 1 million acre-feet (MAF) = 1.233 billion cubic meters (BCM).

water resources can be found across California. For example, the San Joaquin River—the state's second longest river—was, until recently, completely dewatered along a 60 mi. stretch in all but the wettest years.* Likewise, the Colorado River—an important water source for Southern California—has not regularly flowed into the Sea of Cortez for more than 50 years. Moreover, in a recent analysis of the Sacramento–San Joaquin Delta—the hub of the state's water system—the State Water Resources Control Board notes that "the best available science suggests that current flows are insufficient to protect public trust resources" (SWRCB 2010).

In addition to the overallocation of California's surface water resources, groundwater use has been unregulated in most parts of California. This has fostered overpumping, leading to declining groundwater levels across major parts of the state. According to the California Department of Water Resources (CDWR 2014), groundwater levels have dropped to all-time lows in most areas of the state. In many areas of the San Joaquin Valley, groundwater levels are more than 100 ft. below previous historic lows (Figure 14.1). While some groundwater recharge occurs in wet years, it is more than offset by pumping in dry and even average years, with over 50 MAF of groundwater having been lost in the Central Valley over the last half century (UCCHM 2014). While the California Department of Water Resources (DWR) has been estimating, with considerable uncertainty, overdraft of 1 and 2 MAF/year (CDWR 2003), there are strong indications that groundwater overdraft is worsening.† Recent data show that the Sacramento and San Joaquin river basins alone collectively lost over 16 MAF of groundwater between October 2003 and March 2010, or about 2.5 MAF/year (Famiglietti 2014). Overdraft in some areas has already led to subsidence, threatening water conveyance infrastructure and reducing long-term groundwater storage capacity.

14.3 Strategies for Adapting to Water Scarcity

A variety of strategies are available to adapt to growing water scarcity. Water efficiency improvements—defined as measures that reduce water use without affecting the benefits water provides—have been shown to be cost-effective and flexible tools for adaptation. Several research studies have shown that agricultural water withdrawals could be reduced by 5.6–6.6 MAF/year, or by about 17–22%, while maintaining agricultural production (CALFED 2000, 2006; Cooley et al. 2009). Part of these savings are reductions in consumptive use, ranging from 0.6–2 MAF/year, representing additional supply that can be allocated to other beneficial uses. The remainder of the savings comes from a reduction in water required to be taken from rivers, streams, and groundwater, which would provide improvements in water quality, in-stream flow, and energy savings (Gleick et al. 2011). Moreover, these efficiency improvements could improve the reliability of existing supplies and reduce vulnerability to water scarcity and drought. Additionally, expanding water reuse and promoting groundwater recharge can help to boost supplies. The following section describes some of these strategies in detail.

* Following an 18-year legal battle, water was released to restore river flows and fish populations beginning in 2009.

† A comprehensive statewide assessment of groundwater overdraft has not been conducted since 1980, and there are major gaps in groundwater monitoring.

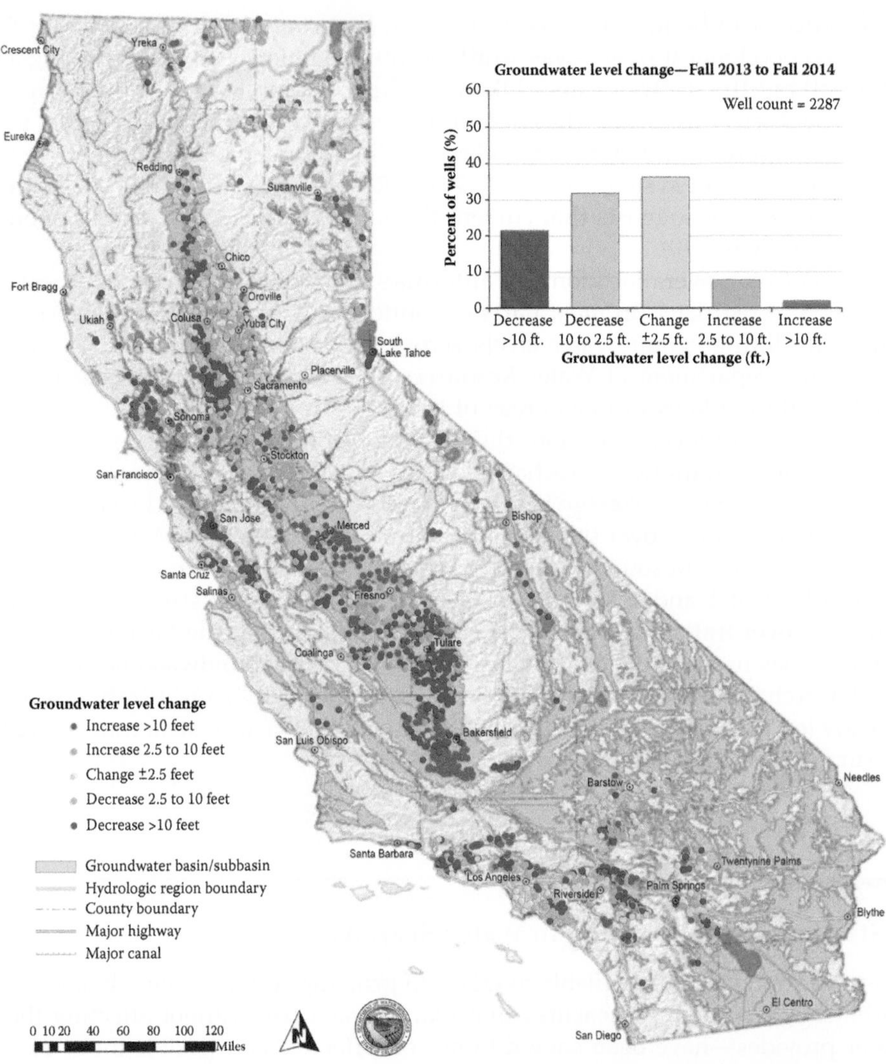

FIGURE 14.1
Groundwater levels in drought low spring (2008–2014) compared to historical low spring (1990–1998). Note: Groundwater level change determined from water level measurements in wells. Map and chart based on available data from the DWR Water Data Library as of September 15, 2014. Data subject to change without notice. (From DWR, Statewide Water Balances, 1998–2010, Sacramento, California, 2014.)

14.3.1 Efficient Irrigation Methods

Numerous irrigation methods are currently available to deliver water where and when it is needed. These methods are typically divided into three categories: flood, sprinkler, and drip irrigation. The oldest form of irrigation is flood irrigation, which refers to the application of water by gravity flow to the surface of the field. The entire field may be flooded (by uncontrolled flood or basin irrigation), or the water is fed into small channels (furrows) or strips of land (borders). Sprinkler irrigation, introduced in the 1930s, delivers water to the field through a pressurized pipe system and distributes water via rotating sprinkler heads, spray nozzles, or a single gun-type sprinkler. Drip irrigation refers to the slow application of low-pressure water from plastic tubing placed near the plant's root zone. Water is applied through drip emitters placed aboveground or belowground, referred to as surface and subsurface drip, respectively.

With proper design, installation, operation, and maintenance, drip irrigation is the most efficient irrigation method, where irrigation efficiency is defined as the volume of irrigation water beneficially used by the plant divided by the volume of irrigation water applied minus change in storage of irrigation water. By this definition, flood irrigation is the least efficient because of the larger volumes of unproductive evaporative losses, water application to nontargeted surface areas, and the propensity for deep percolation. The efficiency of sprinkler irrigation varies with the type of system but generally falls between flood and drip irrigation, although newer drop-head center-pivot installations come close to the water-use efficiency of drip systems. Potential irrigation efficiencies for flood irrigation systems range from 60% to 85%, compared to 70–90% for sprinklers and 88–90% for drip systems (Salas et al. 2006).

In addition to reducing water use, drip irrigation provides a number of other benefits. Drip irrigation allows for the precise application of water and fertilizer to meet crop needs, which can increase crop yield and/or quality. Additionally, improved water and chemical management "can benefit water quality, reduce potential runoff, and reduce potential leaching of nutrients and chemicals" (Evans et al. 1998). Moreover, with drip irrigation, diseases are also less likely to develop because water does not come into contact with crop leaves, stems, or fruit (Shock 2006).

One of the major disadvantages of converting to drip is the relatively high initial investment required. However, these costs can be offset with a reduction in operation costs and/or increase in crop revenue. In addition to the high upfront cost, there are a number of other disadvantages associated with drip irrigation. For example, drip requires management to ensure that emitters do not leak or become clogged by silt, chemical deposits, or even algal growth in the drip lines. Farmers may switch to using groundwater because of its consistency in quality and availability, which may further exacerbate groundwater overdraft. Rodents can also be a problem, especially where the drip line is buried. Moreover, irrigation technologies are only methods to distribute water, not measures of efficiency. Thus, effective management is essential for achieving the water savings of an efficient irrigation system (Lewis et al. 2008).

California farmers have made considerable investments in more efficient irrigation technologies over the past several decades (Figure 14.2). In 1991, for example, flood irrigation was employed on more than 6.1 million acres (2.44 million ha)*, nearly 70% of the state's irrigated land. Drip irrigation, by contrast, was used on less than 1.3 million acres, 14% of the state's irrigated land. By 2010, less than two decades later, flood-irrigated land declined by 1.8 million acres, while drip-irrigated land increased by 2.1 million acres. During this period, the land area irrigated with sprinklers remained fairly constant. Despite these improvements, flood remains the most common irrigation method. Continued adoption of more efficient irrigation methods would allow farmers to respond to water scarcity concerns.

* 1 acre = 0.4 ha.

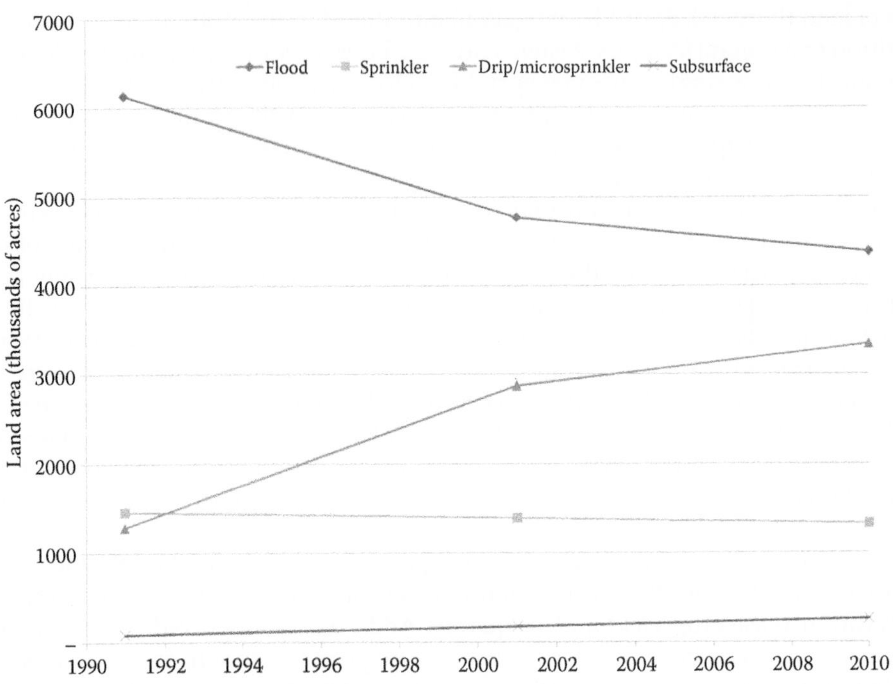

FIGURE 14.2
Total irrigated land area under each irrigation method, 1990–2010. Note: These data do not include rice acreage, which is grown using flood irrigation. If rice acreage were included, the percentage of cropland using flood irrigation would be higher. (Data from Tindula, G.N. et al., *ASCE J. Irrigation Drainage Eng.*, 139, 233–238, 2013.)

14.3.2 Weather-Based Irrigation Scheduling

Crop water requirements vary throughout the crop life cycle and depend on weather and soil conditions. Irrigation scheduling provides a means to evaluate and apply an amount of water sufficient to meet crop requirements at the right time. The California Irrigation Management Information System (CIMIS), for example, is an integrated network of automated weather stations throughout the state that provides information needed to estimate crop water requirements. Since its inception in 1982, the CIMIS network has expanded to include more than 145 automated weather stations across the state. A 1997 study found that using CIMIS increased crop yields by 8% and reduced applied water by 13%, on average (CDWR 1997). These results are consistent with other studies, which generally find that weather-based irrigation scheduling reduces applied water by 11–20% (Kranz et al. 1992; Buchleiter et al. 1996; Dokter 1996).

Despite its benefits, weather-based irrigation scheduling is practiced by a small but growing number of California farmers. A survey of growers in the late 1990s estimated a statewide total of approximately 364,000 agricultural acres under CIMIS (Parker et al. 2000), representing about 4% of the state's irrigated acreage. More recent data from the United States Department of Agriculture show that the majority of farmers still rely primarily on visual inspection or personal experience to determine when to irrigate (USDA 2008) (Table 14.1). Soil or plant moisture sensors, computer models, daily evapotranspiration (ET) reports, and scheduling services, which have long been proven effective, are still fairly uncommon, suggesting that there is significant room for improvement.

TABLE 14.1

Method Used by California Farmers to Decide When to Irrigate, 2003 and 2008

Method	2003	2008
Condition of crop	71%	66%
Feel of soil	36%	45%
Personal calendar schedule	27%	32%
Scheduled by water delivery organization	11%	10%
Soil moisture sensing device	10%	14%
Daily ET reports	8%	12%
Other	6%	5%
Commercial or government scheduling service	5%	10%
When neighbors irrigate	4%	6%
Plant moisture sensing device	3%	3%
Computer simulation model	1%	3%

Source: Data from Table 36 in USDA, Farm and Ranch Irrigation Survey, 2003 and 2008.
Note: Many farmers use more than one method when deciding when to irrigate; thus, the total of all methods exceeds 100%.

14.3.3 Regulated Deficit Irrigation

The traditional irrigation strategy is to provide crops with sufficient water to transpire at their maximum potential, thereby meeting the full crop ET requirements throughout the growing season. Alternative methods whereby water is applied below full crop ET requirements—referred to as deficit irrigation—can be an effective tool to reduce water use. Regulated deficit irrigation (RDI) is an irrigation strategy whereby watering is restricted to levels below crop ET requirements during certain drought-tolerant growth stages, such as the vegetative stages and the late ripening period. RDI provides a means to optimize crop productivity per unit of water rather than simply maximize total crop productivity, and thus has been examined in water-scarce areas (Fereres and Soriano 2007; Geerts and Raes 2009).

Response to water stress can vary considerably by crop and growth stage, and thus, RDI may be limited to certain crops. To date, RDI has been more successfully applied on tree crops and vines than on field crops because (1) crop quality, rather than total yield, is an important determinant of economic returns for these crops and (2) the yield-determining processes in many trees and vines are less sensitive to water stress during certain growth stages (Fereres and Soriano 2007). Goldhamer and Fereres (2005) estimate that applying RDI techniques to major tree crops and wine grapes in California would provide annual reductions in water consumption of 1.0–1.5 MAF without reducing grower profits.*

Recent studies suggest that RDI may also have applications on other crops. In some cases, deficit irrigation improves crop quality. For example, deficit irrigation has been found to reduce tomato yield but improve fruit quality compared to fully irrigated treatments (Favati et al. 2009; Ozbahce et al. 2010; Shahein et al. 2012). In other cases, deficit irrigation may simply help growers minimize impacts associated with water shortages. In field trials in 2003 in the Klamath Basin and in the Sacramento Valley of California, Orloff et al. (2004) examined alfalfa yield under three irrigation treatments, which included normal full-season irrigation and two deficit irrigation regimes. The authors found that deficit irrigation reduced applied water as well as alfalfa yield and quality. However, the authors

* Major tree crops evaluated included almonds, citrus, pistachios, prunes, peaches, olives, apples, and pears.

conclude that it "could provide a partial solution to water shortages in drought years." In particular, deficit irrigation allowed some forage production and economic viability of alfalfa in the face of water shortage and could allow farmers to voluntarily transfer the water saved, providing some compensation to offset production losses.

14.3.4 Water Recycling and Reuse

Water reuse is practiced on California farms, although it could be expanded. California farms can capture runoff from their fields and reuse that water on site. In addition, wastewater from a nearby community can undergo treatment and then be used to irrigate crops or recharge groundwater that is subsequently used for irrigation. This is commonly referred to as recycled water (or municipal recycled water).

Title 22 of the California Water Code allows for the use of recycled water on all types of food and nonfood crops, although with varying water quality criteria for different crop types. Recycled water can be applied directly to the crop or blended with some other water source. Recycled water is a safe, reliable water supply that reduces dependence on limited surface water and groundwater supplies; reduces pollution in rivers and oceans from wastewater discharge; and even provides essential crop nutrients, thereby reducing fertilizer costs.

California's agricultural sector has been using recycled water for more than 100 years, and it currently represents the single largest user of recycled water in the state. In 1910, recycled water was used for agriculture at nearly three dozen sites, and by 1970, agriculture accounted for two-thirds of the estimated 175,000 acre-feet of municipal wastewater beneficially reused annually (Newton et al. 2012). The most recent statewide recycled water survey, conducted in 2009, identified the annual reuse of 670,000 acre-feet of municipal wastewater, of which 245,000 acre-feet (37%) was for irrigation (Newton et al. 2012). Additionally, some amount of recycled water is used for groundwater recharge in agricultural areas, and some of this may be used for irrigation.

The direct use of recycled water for agriculture is widely practiced in 36 of the state's 58 counties. Despite wide application, its use for agriculture is concentrated in the San Joaquin Valley (122,000 acre-feet), parts of Southern California (57,000 acre-feet), and the Central Coast (19,000 acre-feet). The city of Bakersfield is the single largest producer of recycled water for agriculture in the state (Newton et al. 2012) and has been providing treated wastewater to a nearby farm for more than 80 years (Wu et al. 2009). In 2009, the facility provided 35,000 acre-feet of recycled water to irrigate alfalfa, hay, cotton, and other crops (Newton et al. 2012; City of Bakersfield 2014). The state has adopted aggressive recycled water goals (SWRCB 2013), and undoubtedly, California agriculture will play a role in achieving those goals.

14.3.5 Groundwater Recharge

Groundwater is an important water source for California farmers, accounting for nearly 40% of irrigation withdrawals (USGS 2014). State totals, however, hide regional dependence on groundwater. Groundwater accounts for more than 90% of irrigation withdrawals in 10 counties, most of which are located along the coast. Large volumes of groundwater are also used for irrigation in Tulare, Kern, and Fresno Counties, which account for 27% of the state's irrigated area. During drought years, when surface supplies are limited, groundwater becomes an increasingly important stopgap measure for farmers. For example, Howitt et al. (2014) projected that groundwater extraction in 2014 would increase by 5 MAF statewide, largely offsetting the estimated 6.6 MAF decline in surface water availability and reducing the economic impacts of the drought on the state's agricultural sector. Yet, as described previously, the current use of groundwater is unsustainable, as evidenced by declining groundwater levels across large parts of the state, saltwater intrusion and other water quality impacts, land subsidence, lost storage, and increased energy costs, among other adverse impacts.

California's overdrafted aquifers provide significant water storage opportunities and could help the state respond to climate change, particularly to reductions in snowpack due to warmer temperatures. Snowpack represents California's largest reservoir, capturing precipitation during the winter months when water demand is low and releasing it through the spring and summer when demand is high. Scientists forecast that warming will reduce total snowpack by as much as 70% by the end of this century (Hayhoe et al. 2004), increasing winter streamflows while reducing summer flows. With proper management, California's groundwater aquifers could help capture some of this water, reducing the risk of floods in the winter and drought in the summer. Additionally, excess surface water in wet years, treated wastewater, and storm water runoff could be used to recharge groundwater.

California's groundwater recharge potential is not well known. While preliminary estimates suggest that the total groundwater storage capacity ranges from 890 million to 1.3 billion acre-feet (CDWR 1994), only a fraction of the storage capacity is available for recharge. A recent analysis of grant applications submitted to the DWR suggests that the recharge potential is at least 785,000 acre-feet/year—although this is considered an underestimate because less than one-third of the grant applications submitted were available for review (Choy et al. 2014). Moreover, the authors find that the median cost of groundwater recharge, including capital and operation and maintenance costs, is $390/acre-foot, less than many other water supply options.

14.4 Policy Instruments

While California farmers are implementing strategies to adapt to water scarcity in varying degrees, they face a number of barriers that hinder more widespread adoption. These barriers are wide ranging and may include financial, legal, institutional, informational, and educational barriers. For example, implementing these strategies can be costly. While reductions in operating costs and/or increases in crop revenue may make these investments cost-effective in the long term, the initial investment may be cost prohibitive.

Moreover, the price of water for irrigation in large parts of the state is too low, failing to provide an incentive to conserve water. This is particularly true for the Central Valley Project (CVP), which provides water to more than 3 million acres of farmland and 1 million households in California. The federal government invested heavily in the construction of the CVP, and these investments were to be paid by those who benefited from the projects. Under the original contracts, which were negotiated and signed in the late 1940s, the capital costs for the project were to be paid off in 40 years. By the 1970s, however, the contractors had made little progress in repaying the capital costs and had also failed to pay enough to cover annual operating and maintenance costs. In 1986, Congress enacted legislation (the Coordinated Operations Act) that set 2030 as a firm repayment deadline. However, prices for water from the CVP remain too low, and as a result, CVP contractors are behind on repaying project costs. A 2013 analysis by the Office of the Inspector General estimated a shortfall of $330–390 million in repaying the capital investment in the CVP irrigation facilities if current water delivery trends continue (US DOI 2013). For comparative purposes, the shortfall in 2012 was about $600 million.

Additionally, in some areas, water is not available on demand. In California, water is predominantly delivered through gravity-fed canals designed and constructed in the early and middle twentieth century, and a recent survey found that nearly 80% of these water systems fail to provide water to farmers on demand (AWMC 2008). Rather, water is primarily available on an arranged ordering system, limiting the irrigator's ability to respond to changing weather conditions: one-third of those surveyed must place orders 24–48 h in advance, and about 5% of those surveyed were delivered water based on a fixed rotation.* The relatively inflexible delivery systems that still characterize many California irrigation districts limit effective and efficient water resource management.

Several policy instruments have been developed and implemented in California to help farmers overcome some of these barriers. These instruments include grants and loans, as well as water markets. In 2000, for example, California voters approved Proposition 13, which authorized the California DWR to make $28 million in loans to local public agencies and incorporated mutual water companies to finance agricultural water conservation and efficiency projects and programs. Additional funding was made available for other water projects, such as groundwater recharge and storage, which could also benefit the agricultural community. Shortly thereafter, in 2002, California voters approved Proposition 50, which included grants for agricultural water-use efficiency improvements. Some of the projects funded include lining canals, automating delivery systems, and installing tailwater recovery systems.

For several decades, California has maintained a water market that allows for the temporary, long-term, or permanent transfer of the right to use water in exchange for compensation. Water transfers in California are not new, dating back to the Gold Rush era (LAO 1999). Early state efforts to facilitate water markets began in the late 1970s in response to a severe drought, and both state and federal efforts were greatly expanded in the 1990s. Initially, market activity was slow, averaging 100,000 acre-feet annually in the early 1980s. Between 2003 and 2011, however, water market activity was more active, with an average of 2.1 MAF committed and 1.4 MAF actually transferred annually (Hanak and Stryjewski 2012a). The volume of water traded was 3.2% of statewide water use during that period, an order of magnitude less than in Chile's Limarí Valley and Australia's Murray–Darling Basin, where an estimated 30% of the entitlements available are traded (Grafton et al. 2010). The end users of the water transferred in California were largely municipal and industrial (39%), followed by farmers (36%), the environment (20%), and mixed purposes (5%) (Hanak and Stryjewski 2012b). While early market activity was dominated by short-term transfers, recent activity has been dominated by long-term and permanent transfers, and most of these have been purchases by cities.

The impacts of water markets on California's agricultural sector are not well established and require further analysis. Temporary water transfers among farmers can allow for a response to drought or other short-term water supply constraints and help moderate the economic impacts of those constraints on the agricultural sector, e.g., by fallowing land and selling that water for use on higher-value crops. Long-term and permanent transfers, by contrast, allow for changes in water usage and demand patterns, and in almost all cases, these transfers have shifted water away from agriculture to other uses. While this may promote economic efficiency for society, the impacts on California's agricultural

* Rotational deliveries are rigid delivery systems where water is delivered in fixed amounts at fixed intervals.

sector are more uncertain and depend on how water was made available for these transfers, i.e., whether it was from fallowing land or from changing the type of crops grown (from more water intensive to less water intensive) or improving the efficiency with which these crops were grown. This information is not readily available, and as a result, it is difficult to assess the effect these transfers have had on the agricultural sector. Moreover, water transfers can result in socioeconomic and environmental impacts that are not always well understood.

Finally, another notable and important policy instrument was recently adopted by the California legislature: the Sustainable Groundwater Management Act of 2014. Previously, groundwater usage was largely unregulated, as evident in severe groundwater overdraft in some parts of the state. The act provides a framework for local authorities to manage groundwater supplies but allows for state intervention if necessary to protect groundwater resources. It requires the formation of local agencies by mid-2017 and requires those agencies to adopt and implement local basin management plans to achieve long-term groundwater sustainability by 2022. Moreover, it requires basins to achieve groundwater sustainability goals by 2040 in medium- and high-priority basins in critical overdraft and by 2042 in medium- and high-priority basins. While achieving more rational and sustainable use of California's groundwater resources is still decades away, passage of the Sustainable Groundwater Management Act is an important and essential step in the right direction.

14.5 Conclusions

California is one of the most productive agricultural regions in the world, and that productivity has been made possible by a vast and integrated water infrastructure network that provides large volumes of water to the agricultural sector. The state, however, is prone to multiyear droughts, with wide-ranging impacts, including on its agricultural sector. Across California, drought and its impacts have become exacerbated by increasing water scarcity, whereby human demands for water greatly exceed volumes that can be sustainably extracted from surface waters and groundwater aquifers, especially in dry years. Climate change, combined with continued population and economic growth, will make meeting water demands even more difficult.

A variety of strategies are available to help California farmers adapt to water scarcity. Water efficiency improvements—defined as measures that reduce water use without affecting the benefits water provides—have been shown to be cost-effective and flexible tools to adapt to water scarcity. Studies show that while California farmers are much more efficient than they were a decade ago, significant efficiency potential remains. Moreover, boosting local supplies through water reuse and groundwater recharge can further promote the long-term sustainability of the state's agricultural sector.

While California farmers are implementing strategies to adapt to water scarcity in varying degrees, they face a range of financial, legal, institutional, informational, and educational barriers that hinder more widespread adoption. Several policy instruments have been introduced in California to overcome these barriers, including grants and low-interest loans, new groundwater management requirements, as well as water markets and transfers. Additional effort is needed to assess the effectiveness of these efforts, determine what other instruments may be needed, and evaluate the degree to which these instruments will perform as climate impacts intensify.

References

Agricultural Water Management Council (AWMC). 2008. Efficient Water Management: Irrigation District Achievements. http://www.agwatercouncil.org/images/stories/pdfs/AWMC_final.pdf.

Buchleiter, G.W., D.F. Heermann, and R.J. Wenstrom. 1996. Economic analysis of on-farm irrigation scheduling. *Evapotranspiration and Irrigation Scheduling: Proceedings of the International Conference.* November 3–6, 1996. San Antonio, Texas.

California Department of Food and Agriculture (CDFA). 2015. California Agricultural Statistics 2013 Crop Year. US Department of Agriculture National Agricultural Statistics Service. http://www.nass.usda.gov/Statistics_by_State/California/Publications/California_Ag_Statistics/2013cas-all.pdf.

California Department of Water Resources (CDWR). 2014. Groundwater Basins with Potential Water Shortages and Gaps in Groundwater Monitoring. Public Update for Drought Response. http://www.water.ca.gov/waterconditions/docs/Drought_Response-Groundwater_Basins_April30_Final_BC.pdf.

California Department of Water Resources (CDWR). 2003. California's Groundwater Bulletin 118. http://www.water.ca.gov/groundwater/bulletin118/update_2003.cfm.

California Department of Water Resources (CDWR). 1997. *Fifteen Years of Growth and a Promising Future: The California Irrigation Management Information System.* Sacramento: State of California Department of Water Resources.

California Department of Water Resources (CDWR). 1994. California water plan update. Sacramento. *Bulletin 160-94*. 2 v.

California Department of Water Resources (CDWR). 1975. California's groundwater. *Bulletin 118.* Sacramento. P. 135.

CALFED Bay-Delta Program. 2000. Water Use Efficiency Program Plan. Final Programmatic EIS/EIR Technical Appendix. http://calwater.ca.gov/content/Documents/library/307.pdf.

CALFED Bay-Delta Program. 2006. Water Use Efficiency Comprehensive Evaluation Final Report. CALFED Bay-Delta Program Water Use Efficiency Element. http://www.calwater.ca.gov/content/Documents/library/WUE/2006_WUE_Public_Final.pdf.

Christian-Smith, J., M. Levy, and P.H. Gleick. 2015. Maladaptation to drought: A case report from California, USA. *Sustainability Science* 10(3):491–501.

Choy, J., G. McGhee, and M. Rohde. 2014. Recharge: Groundwater's Second Act. Water in the West. Stanford Woods Institute for the Environment and the Bill Lane Center for the American West.

City of Bakersfield. 2014. Treatment Plant 2. Public Works: Wastewater Division. Website http://www.bakersfieldcity.us/cityservices/pubwrks/wastewater/plant2/index.htm.

Cooley, H., J. Christian-Smith, and P.H. Gleick. 2009. *Sustaining California Agriculture in an Uncertain Future.* Oakland: Pacific Institute. http://pacinst.org/wp-content/uploads/sites/21/2014/04/sustaining-california-agriculture-pacinst-full-report.pdf.

Department of Water Resources (DWR). 2014. *Statewide Water Balances, 1998–2010.* Sacramento, California.

Dettinger, M.D., F.M. Ralph, T. Das, P.J. Neiman, and D.R. Cayan. 2011. Atmospheric rivers, floods and the water resources of California. *Water* 3:445–478.

Dokter, D.T. 1996. AgriMet—The Pacific Northwest Cooperative Agricultural Weather Station Network. Evapotranspiration and Irrigation Scheduling: *Proceedings of the International Conference.* November 3–6, 1996. San Antonio, Texas.

Evans, R.O., K.A. Harrison, J.E. Hook, C.V. Privette, W.I. Segars, W.B. Smith, D.L. Thomas, and A.W. Tyson. 1998. Irrigation Conservation Practices Appropriate for the Southeastern United States. Georgia Department of Natural Resources Environmental Protection Division and Georgia Geological Survey. Project Report 32.

Famiglietti, J. 2014. Epic California drought and groundwater: Where do we go from here? *Water Currents. National Geographic Blog*. http://voices.nationalgeographic.com/?s= epic+california+drought.

Favati, F., S. Lovelli, F. Galgano, V. Miccolis, T. Di Tommaso, and V. Candido. 2009. Processing tomato quality as affected by irrigation scheduling. *Sci. Hort.* 122:562–571.

Fereres, E. and M.A. Soriano. 2007. Deficit irrigation for reducing agricultural water use. *Journal of Experimental Botany* 58(2):147–159.

Geerts, S. and D. Raes. 2009. Deficit irrigation as an on-farm strategy to maximize crop productivity in dry areas. *Agricultural Water Management* 96(9):1275–1284.

Gleick, P.H., J. Christian-Smith, and H. Cooley. 2011. Water-use efficiency and productivity: Rethinking the basin approach. *Water International* 36(7):784–798.

Gleick, P.H. and L. Nash. 1991. The Societal and Environmental Costs of the CContinuing California Drought. Pacific Institute. http://pacinst.org/wp-content/uploads/sites/21/2014/05/societal-enviromental-cost-drought.pdf.

Goldhamer, D. and E. Fereres. 2005. The promise of regulated deficit irrigation in California's orchards and vineyards. *The California Water Plan Update. Bulletin 160-05*, vol. 4.

Grafton, R.Q., C. Landry, G.D. Libecap, S. McGlennon, and R. O'Brien. 2010. An integrated assessment of water markets: Australia, Chile, China, South Africa and the USA. *National Bureau of Economic Research.*

Grantham, T.E. and J.H. Viers. 2014. 100 years of California's water rights system: Patterns, trends, and uncertainty. *Environmental Research Letters* 9:10.

Griffin, D. and K.J. Anchukaitis. 2014. How unusual is the 2012–2014 California drought? *Geophysical Research Letters* 41(24):9017–9023.

Hanak, E. and E. Stryjewski. 2012a. *California's Water Market, by the Numbers: Update 2012*. Public Policy Institute of California. San Francisco.

Hanak, E. and E. Stryjewski. 2012b. *Technical Appendix of California's Water Market, by the Numbers: Update 2012*. Public Policy Institute of California. San Francisco.

Hayhoe, K., D. Cayan, C.B. Field, P.C. Frumhoff, E.P. Maurer, N.L. Miller, S.C. Moser, S.H. Schneider, K. Nicholas Cahill, E.E. Cleland, L. Dale, R. Drapek, R.M. Hanemann, L.S. Kalkstein, J. Lenihan, C.K. Lunch, R.P. Nielson, S.C. Sheridan, and J.H. Verville. 2004. Emission pathways, climate change, and impacts on California. *Proceedings of the National Academy of Sciences.* 101(34):12422–12427.

Howitt, R., J. Medellín-Azuara, D. MacEwan, J. Lund, and D. Sumner. 2014. Economic Analysis of the 2014 Drought for California Agriculture. UC–Davis Center for Watershed Sciences. https://watershed.ucdavis.edu/files/content/news/Economic_Impact_of_the_2014_California_Water_Drought.pdf.

Kranz, W.L., D.E. Eisenhauer, and M.T. Retka. 1992. Water and energy conservation using irrigation scheduling with center-pivot irrigation systems. *Agricultural Water Management* 22:325–334.

Legislative Analyst's Office (LAO). 1999. The Role of Water Transfers in Meeting California's Water Needs. Sacramento, CA. http://www.lao.ca.gov/1999/090899_water_transfers/090899_water_transfers.html.

Lewis, D.J., G. McGourty, J. Harper, R. Elkins, J. Christian-Smith, J. Nosera, P. Papper, R. Sanford, L. Schwankl, and T. Prichard. (2008). Meeting irrigated agriculture water needs in the Mendocino County portion of the Russian River. University of California Cooperative Extension Mendocino County, University of California Davis, Department of Land Air and Water Resources, and University of California Kearny Agricultural Center.

Michael, J., R. Howitt, J. Medellin-Azuara, and D. MacEwan. 2010. A Retrospective Estimate of the Economic Impacts of Reduced Water Supplies to the San Joaquin Valley in 2009. http://forecast.pacific.edu/water-jobs/sjv_rev_jobs_2009_092810.pdf.

Newton, D., D. Balgobin, D. Badyal, R. Mills, T. Pezzetti, and H.M. Ross. 2012. Results, Challenges, and Future Approaches to California's Municipal Wastewater Recycling Survey. State Water Resources Control Board (SWRCB) and California Department of Water Resources (CDWR): Sacramento, CA.

Orloff, S., D. Putnam, B. Hanson, and H. Carlson. 2004. Controlled deficit irrigation of alfalfa (*Medicago sativa*): A strategy for addressing water scarcity in California. New directions for a diverse planet. R.A. Fischer ed. *Proceedings of the 4th International Crop Science Congress*. Brisbane, Australia, September 26, October 1, 2004.

Ozbahce, A. and A.F. Tari. 2010. Effects of different emitter space and water stress on yield and quality of processing tomato under semi-arid climate conditions. *Agric. Water Manage.* 97:1405–1410.

Parker, D., D.R. Cohen-Vogel, D.E. Osgood, and D. Zilberman. 2000. Publicly funded weather database benefits users statewide. *California Agriculture* 54(3):21–25.

Salas, W., P. Green, S. Frolking, C. Li, and S. Boles. 2006. *Estimating Irrigation Water Use for California Agriculture: 1950s to Present*. California Energy Commission, PIER Energy-Related Environmental Research. CEC-500-2006-057.

Schmidt, G. and C. Benitez-Sanz. 2013. How to distinguish water scarcity and drought in EU water policy? *GWF Discussion Paper 1333. Global Water Forum, Canberra, Australia*. http://www.globalwaterforum.org/2013/08/26/how-to-distinguish-water-scarcity-and-drought-in-eu-water-policy/.

Shahein, M.M., M.E. Abuarab, and A.M. Hassan. 2012. Effects of regulated deficit irrigation and phosphorous fertilizers on water use efficiency, yield and total soluble solids of tomato. *American–Eurasian J. Agric. & Environ. Sci.* 12(10):1295–1304.

Shock, C. 2006. Drip Irrigation: An Introduction. Oregon State University Extension Service. http://extension.oregonstate.edu/umatilla/mf/sites/default/files/Drip_Irrigation_EM8782.pdf.

State Water Resources Control Board (SWRCB). 2013. Recycled Water Policy. Revised January 22, 2013. State Water Resources Control Board. Sacramento (CA), p. 39.

State Water Resources Control Board (SWRCB). 2010. Development of Flow Criteria for the Sacramento–San Joaquin Delta Ecosystem. Sacramento, CA. http://www.swrcb.ca.gov/waterrights/water_issues/programs/bay_delta/deltaflow/docs/final_rpt080310.pdf.

Tindula, G.N., M.N. Orang, and R.L. Snyder. 2013. Survey of irrigation methods in California in 2010. *ASCE Journal of Irrigation and Drainage Engineering* 139:233–238.

United States Department of Agriculture (USDA). 2008. Farm and Ranch Irrigation Survey. Washington, D.C.

United States Department of Agriculture (USDA). 2003. Farm and Ranch Irrigation Survey. Washington, D.C.

United States Department of the Interior (US DOI). 2013. Central Valley Project, California: Repayment Status and Payoff. Report No. WR-EV-BOR-0003-2012. http://www.doi.gov/oig/reports/upload/WR-EV-BOR-0003-2012Public.pdf.

United States Geological Survey (USGS). 2014. California water use estimates for 2010. Accessed October 5, 2014 at http://ca.water.usgs.gov/water_use/.

University of California Center for Hydrologic Modeling (UCCHM). 2014. Water Storage Changes in California's Sacramento and San Joaquin River Basins from GRACE: Preliminary Updated Results for 2003–2013. UCCHM Water Advisory #1.

Wu, L., W. Chen, C. French, and A. Chang. 2009. Safe Application of Reclaimed Water Reuse in the Southwestern United States. Publication 8537. University of California Division of Agriculture and Natural Resources.

15

Responses of Southern California's Urban Water Sector to Changing Stresses and Increased Uncertainty: Innovative Approaches

Celeste Cantú

CONTENTS

ABSTRACT The Santa Ana River Watershed Project Authority (SAWPA) was formed to resolve conflicts over increasing water use and water quality deterioration in a large, rapidly developing and water-stressed Southern California watershed. This chapter chronicles SAWPA's experience in becoming a pioneer in the development of a collaborative planning process that addresses all aspects of water resources in the watershed. SAWPA spearheaded articulation of a vision for watershed-wide collaboration through the One Water One Watershed (OWOW) process. This deliberative process facilitates the translation of desired outcomes to cooperative actions. The foundation for the process starts with the definition of two types of objectives: *problem solving* and *creating anew*, in other words, bringing a desired goal into reality. The OWOW collaborative process is evolving to place greater emphasis on this latter positive perspective. OWOW's guiding principles, organizational structure, and achievements can provide valuable lessons for communities in other watersheds that are facing challenges from uncoordinated land-use development, growing water demands, water quality degradation, and increasing climatic variability.

15.1 Introduction

The first decades of the twenty-first century have seen mounting pressures on California's watersheds and water resources. A conjunction of record warmth and recurrent droughts has led to a series of destructive forest fires, with 12 of the 20 largest fires recorded in California since 1932 occurring after the year 2000 (California Department of Forestry and Fire Protection 2014). Damaged watersheds and droughts have impaired both water quantity and quality in many parts of the state. In addition, California residents have become increasingly concerned about climate change, sea-level rise, crashing ecosystems, projected population growth, development, and financial lean times. They are demanding a more efficient and synergistic approach to the management of their water resources (California Department of Finance 2014).

In the Santa Ana Watershed in Southern California, growing pressures on the local environment and changing citizen demands for management of the area's water resources have been met with an innovative collaborative planning process that provides a successful example of how to transition from twentieth-century water management strategies to a new approach that meets twenty-first-century needs. This new approach is responsive to the demands of ratepayers who do not want to pay for water three times over. They are no longer willing to pay their water supplier top dollar for water imported from far away with a huge carbon footprint, while they pay flood managers to protect people and property by channeling what naturally falls on their community out of it as fast as possible, and then also pay dearly for the water they have used to be highly treated and dumped in the river. The argument for adopting a strategic path to water management has never been clearer. Integrated water resource management (IWRM) can deliver more bang for the buck, facilitate regulatory compliance, protect the environment, and manage a reliable water supply during this time of change (Global Water Partnership Technical Advisory Committee 2000; Kemper et al. 2007). As scarcity and costs increase, the argument for the integrated approach is even stronger (National Research Council 1999; UN-Water 2012). We need to integrate the management of each water drop through multiple stages and multiple uses. And we need to view the drop as finite but never ending. The Santa Ana Watershed Project Authority's (SAWPA's) One Water One Watershed's (OWOW's) IWRM plan does just that (Santa Ana Watershed Project Authority 2013).

15.2 The Watershed

The San Bernardino and San Gabriel Mountains create the headwaters for the Santa Ana River, which drains the 2650 mi.^{2} (6864 km^2) terminating at the Pacific Ocean at Huntington Beach (Figure 15.1). The Santa Ana River flows for 100 mi. (160 km),* draining the largest coastal stream system in Southern California, which includes parts of Orange, Riverside and San Bernardino Counties, as well as a sliver of Los Angeles County. The watershed drops in elevation as much as the Mississippi but is only 100 mi. from the headwaters to the coast. The total length of the river with its major tributaries is about 700 mi.

* 1 mile (mi) = 1.6 kilometers (km).

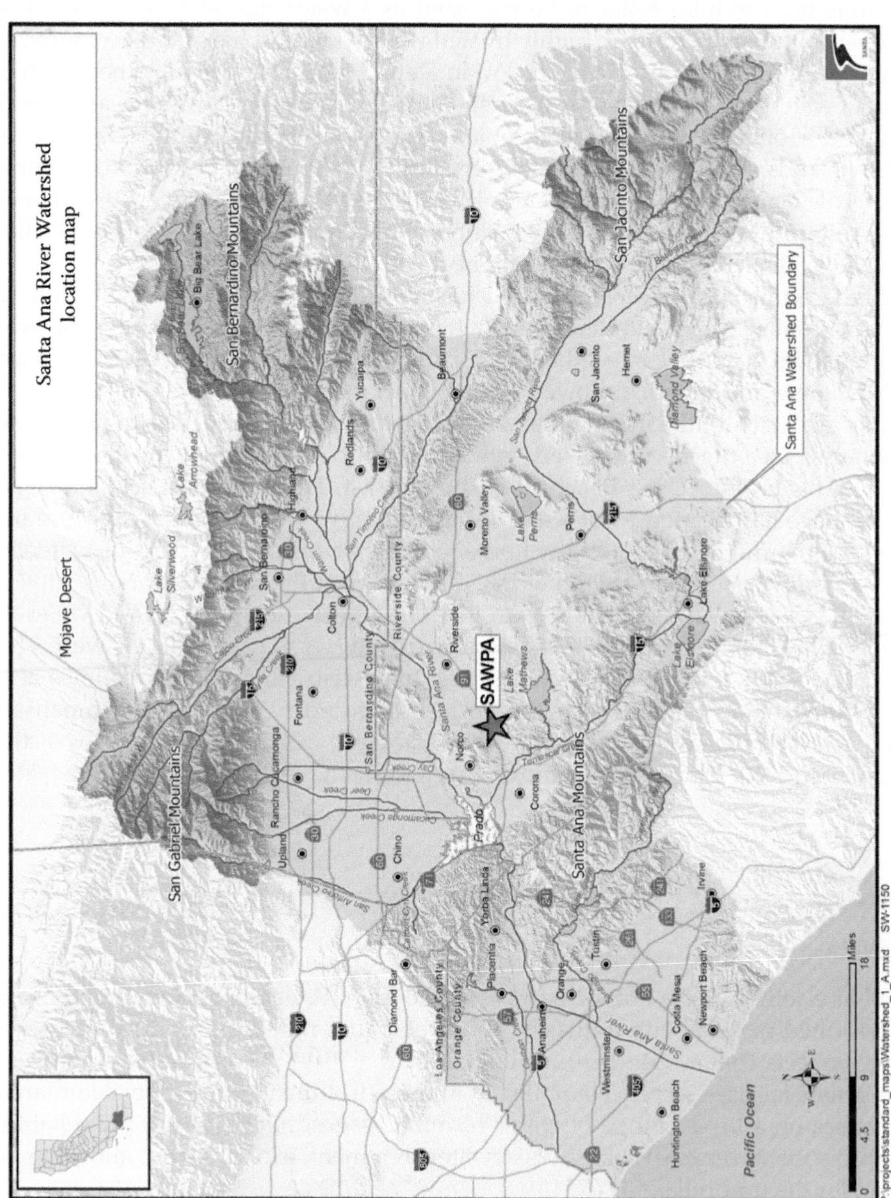

FIGURE 15.1
Santa Ana River Watershed, California.

Surface water and groundwater replenishment come from snowmelt and storm water runoff from the mountainous upper zones of the watershed. The Santa Ana River Watershed has always been characterized by flashy runoff, causing dangerous flood risks. The river is tamed by two dams. The Seven Oaks Dam constructed by the US Army Corps of Engineers in 1999 is the 10th largest dam in the United States. Originally built as a flood control dam, today it also is used to hold water to be managed as a water supply. Below Seven Oaks Dam, the river continues picking up highly treated wastewater discharge, urban runoff, and inflow from groundwater until it reaches Prado Dam. Prado Dam divides the watershed into the upper watershed and the lower watershed. Prado Dam is twice as large as the Seven Oaks Dam and is operated for flood control and water conservation. Prado Dam has created fertile wetlands, which today are among the largest in Southern California, sustaining a complex fertile ecosystem. The San Jacinto River starts in the San Jacinto Mountain and runs to the west until it terminates in Lake Elsinore. Only once in about 15 years does a flood event cause Lake Elsinore to overflow and connect with the Santa Ana River.

The semiarid Santa Ana River Watershed has a Mediterranean climate with hot, dry summers and cooler, wetter winters, which has been the pattern to which the water resource managers have become adapted. Historically, precipitation occurs primarily between November and March, followed by a long drier season.

This classic watershed, like many others, developed from early ranches; to farms, dairies, and wineries; to a highly urbanized, industrialized transportation corridor from the ports of Los Angeles and Long Beach to all points east. Huge warehouses and related development have hardscaped what was once a productive working landscape and have interrupted the natural hydrology. Now home to almost 6 million people, this watershed is projected to grow faster than any other in California over the next 50 years.

The watershed has always had more than its share of water challenges, and water districts responded by hiring some of the most innovative water managers in the west. These water managers worked to protect and optimize their districts' water, sometimes at the expense of downstream users. The resulting lack of coordination and uncompensated adverse impacts led to the watershed's largest civil lawsuit, which was resolved by a stipulated judgment in 1969. Among other things, this judgment created the SAWPA.

15.3 SAWPA

Established as a joint-powers authority in 1968, the SAWPA began as a planning agency and was reformed in 1972 with a mission to develop and maintain regional plans, programs, and projects that would protect the Santa Ana River Basin's water resources to maximize beneficial uses in an economically and environmentally responsible manner. SAWPA focuses on a broad range of water resource issues, including supply reliability, quality improvement, recycled water, wastewater treatment, groundwater, and salt management, under the umbrella of IWRM.

The SAWPA Commission is composed of five member agencies—Eastern Municipal Water District, Inland Empire Utilities Agency, Orange County Water District, San Bernardino Valley Municipal Water District, and Western Municipal Water District—which together represent the majority of the water management authorities and stakeholders within the watershed. They are wholesale and retail water agencies that manage groundwater production, desalination, resource management, wastewater collection and treatment, and regional water recycling. Water demands in the watershed outgrew the local water supply, and imports from the Colorado River and the San Joaquin–Sacramento River Delta made up the difference.

15.4 Twenty-First-Century Challenges

Taming the Santa Ana has never been easy, but there is now a changing hydroclimatic regime for which interpretation and adaptation will require work. The river was once declared "the greatest flood hazard in the US, west of the Mississippi," and the dams were built, but today, we are experiencing climate variability that compounds other stresses on water resources, amplifying the need for a watershed-based integrated approach to water management. In 2015, California entered a fourth year of a drought, prompting the governor to declare a drought emergency. In the midst of this enduring drought, the Santa Ana Basin experienced dramatic changes in weather patterns.

On New Year's Eve 2014, an unprecedented low-elevation snow fell in Western Riverside County, breaking avocado trees and ancient live coast oaks, and freezing citrus. A foot of snow was recorded in 5 h, a first in recorded history. Six months later, when traditionally, it never rains, bone-dry Southern California had flash flood warnings. Dozens of Californian cities set all-time rainfall records for the month of July with this unprecedented midsummer rainfall washing out major bridges. One National Weather Service meteorologist called it "super historic." Uncontained wildfires jumped a freeway, igniting cars and trucks. These unexpected climatic extreme events lend further weight to the argument that we need to change our water management routines. We cannot use the past to model the future. A study conducted by the US Bureau of Reclamation (2013) of the Santa Ana Watershed shows that weather extremes will become the norm as we move to drier decades. We need to work together to capture the elusive once-in-every-5-years deluge and have the storage to contain it during the long dry years. The rhythm and patterns are changing; the challenges continue. We see catastrophic floods, and debris flows on the heels of catastrophic fires. These challenges are bigger than can be addressed by any one community or water management entity. They can only be addressed by coming together to understand the whole system.

15.5 One Water One Watershed

In 2007, SAWPA convened the OWOW process. OWOW creates a venue where water managers can see a picture bigger than their specific districts. They can see how working together can address challenges on the system level. To guide the development of the OWOW plan, SAWPA staff working with the OWOW Steering Committee established a vision along with goals and objectives for the watershed that would incentivize a holistic approach to resource management. OWOW adopted guiding principles from Peter Senge's *The Necessary Revolution* (2008), where he frames two concepts: problem solving and creating anew. Problem solving is about making what you do not want go away. Creating anew involves bringing something you care about into reality, such as resiliency, sustainability, water reliability, and quality. We are shifting the conversation from the familiar—avoiding something bad—to doing something positive and new. We are shifting the conversation from problems to possibilities.

15.5.1 Guiding Principles

The following OWOW principles guide our plan.

15.5.1.1 Take a Systems Approach

- See the Santa Ana River Watershed as a hydrologic whole. The planning process must be watershed-wide and bottom-up to allow for a holistic, inclusive approach to watershed management.
- Working in concert with nature is cost-effective. Gravity flow, percolation, wetlands, meadows, and other green infrastructure perform important functions and often have lower life-cycle costs.
- See all problems as interrelated; seek efficiencies and synergies. The OWOW plan and projects must pursue multiple objectives beyond the traditional objective of reliable water, including to ensure high-quality water for all users; preserve and enhance the environment; promote sustainable water solutions; manage rainfall and highly treated water as a drinking water resource; preserve open space and recreational opportunities; maintain quality of life, including the needs of disadvantaged communities; provide economically effective solutions; and improve regional integration and coordination.

15.5.1.2 Create Anew

- OWOW is a shared vision for the watershed, one that requires water managers to understand and respect each other's needs.
- Breakthrough innovations are needed to answer emerging challenges. We understand that working harder within the old paradigms will not address the challenges emerging today.
- Establish a water ethic: everyone knows where their water comes from, how much of it they use, what they put into it, and where it goes after it leaves them.
- We asked stakeholders to move from problem solving to creating a new shared vision and to realize breakthrough innovations. The change we are seeking is not to try harder to maintain the status quo of the twenty-first century. We are shifting the conversation from the familiar—avoiding something bad—to doing something positive and new.

15.5.1.3 Collaborate across Boundaries

- As citizens of the watershed, create solutions. We asked everyone to check their identity at the door. All other loyalties, agendas, and priorities were to be secondary to those of the holistic view of the Santa Ana River Watershed. We asked stakeholders who feel strongly to let go of cherished beliefs and views so that they could allow something bigger than themselves to develop. We asked a lot.
- No one person can do it alone. No one has enough understanding, credibility, or authority to connect the larger networks of people and organizations to do this work. We have to do it together. There is no reason to assume that when each

agency seeks to optimize results within its jurisdiction, the results will be a solution that is optimum overall. In fact, we know that just the opposite is often the case. We first optimize in the aggregate and later implement at a smaller district scale.

- Think big. The plan must improve conditions throughout the watershed, ensuring that an improvement in the welfare of one area is not at the expense of others and that when such expenses are unavoidable, compensation is found. Quality of life must be protected, and economic impacts must be understood.

With these established principles, the OWOW Steering Committee conveyed a sense of urgency and direction to produce a plan that was more aggressive in how the watershed was managed.

15.5.2 Organization

To manage the technical and planning work, the stakeholders organized into separate work groups, each designated as a *pillar* within the plan's framework. The pillars identify and vet creative ideas, conduct brainstorming, and assist with regional coordination, outreach efforts, gathering and reviewing data, and developing and reviewing analysis, resulting in the OWOW chapters. The pillars included the following:

- *Water-Use Efficiency*, focusing on waste reduction and increased conservation
- *Water Resource Optimization*, addressing water reliability, supply, and security
- *Beneficial-Use Assurance*, focusing on water quality
- *Energy and Environmental Impact Response*, addressing the water–energy nexus and climate change
- *Natural Resources Stewardship*, addressing the environment, habitat, parks, recreation, and open space
- *Land-Use and Water Planning*, addressing the water–land nexus
- *Storm Water*, including both resource opportunities and risk management
- *Operational Efficiency and Water Transfers*, looking for opportunities to optimize water management within and among the agencies
- *Disadvantaged and Tribal Communities*, addressing issues particular to these communities
- *Government Alliance*, cochaired by Reclamation and the Los Angeles Regional Office of the US Army Corps of Engineers and including representatives from nine federal agencies, five state and local agencies, and two tribes, as well as the emergency support services of all three counties within the watershed

Each pillar consisted of 10 to 60 volunteers, including participants from local agencies, special districts, nonprofit organizations, universities, Native American tribes, and private citizens. Expert volunteer cochairs, responsible for facilitating the work-group process, led each pillar group. The pillars were asked to view watershed resources and problems from a multidisciplinary perspective that extended beyond their topic area, while considering other pillars' perspectives. For example, the Water Reliability pillar considered environmental and habitat-restoration issues when developing its strategies. Through this process, synergies were developed, and multibenefit programs were identified. For example, through this approach, it was possible to incorporate the understanding that many downstream water resource and water-quality problems could be more effectively and efficiently addressed upstream, at the source, thus requiring collaboration with other entities. Over time, this collaboration among the pillar groups provided a more unified vision, resulting in new integrated and multibeneficial solutions to water resource challenges, which increased collaboration among jurisdictions and geographies. To encourage collaboration between pillars, the responsibilities of each were designed to overlap.

SAWPA staff provides administrative and facilitative assistance to the pillars and the OWOW Steering Committee for overall OWOW plan development. In addition, SAWPA provides decision support tools to assist the Steering Committee and pillars in decision-making processes, provides planning documents to allow pillars to build upon existing plans, and performs significant public outreach and education about the integrated planning approach for the Santa Ana River Watershed. In addition to the OWOW planning process itself, SAWPA administers 10 to 15 multiagency task forces that support OWOW. These task forces range from surface water and groundwater quality to threatened-species preservation and restoration, and are integrated with water resources. These task forces bring together stakeholders from over 100 different agencies and organizations in the watershed. The work of these task forces—often involving retail and wholesale water agencies, groundwater management agencies, wastewater agencies, nongovernmental organizations (NGOs), businesses, universities, and other organizations—has been integrated into the OWOW planning process.

The resulting 2013 OWOW 2.0 Plan advances a paradigm change from a strictly water-supply to an IWRM approach, moving from a mission of providing abundant high-quality water at the lowest possible cost to one in which water resources are managed in a sustainable manner and with regard for the needs of the environment and those downstream. OWOW 2.0 seeks a proactive approach that is lighter on the land, and protects habitat and a resilient future.

Engaging stakeholder involvement in any large, diverse watershed is challenging, particularly when one considers every resident of the watershed a stakeholder. OWOW was designed to be a "bottom-up meets top-down" process. By encouraging the participation of different groups of people and those holding varying viewpoints from throughout the watershed, OWOW seeks to reach a larger number of stakeholders. The OWOW pillar groups represent an effective means to ensure public involvement. The list of stakeholders involved is one of the most extensive ever taken by any regional water management group and includes over 4000 representatives from 120 water agencies, including flood control, water conservation districts, and wastewater and water supply agencies. It also includes representatives from the 63 incorporated cities, including mayors, key department heads, city council members, and planning commissioners. Also included are representatives from county, state, and federal government; Native American tribes; the real estate community; members of the environment and environmental justice, agricultural, and development communities; consultants; trade associations; academia; nonprofit organizations; and others simply interested in water.

SAWPA has hosted multiple workshops, forums, and presentations, including a TEDx Talk, to discuss the benefits of collaboration and multibenefit watershed projects. The annual OWOW watershed conferences attract over 400 attendees. The conferences serve as an opportunity to invite the public to become involved in OWOW, and to discuss OWOW plan development to date. They also serve to reinforce the OWOW plan goals to encourage a watershed focus, and encourage collaboration in developing multibenefit projects.

Social media is a component of SAWPA's overall public outreach to provide leadership and information to stakeholders in reaching the goal of a sustainable Santa Ana River Watershed. In its social media presence, SAWPA provides a virtual venue to invite collaboration and encourage interaction from others, to inspire and educate watershed residents, and to provide web-based information. SAWPA's website is considered *home base.*

15.5.3 OWOW Projects

The pillars and the OWOW plan's cross-disciplinary and organizational approach developed many projects. Some were new; others were changed to be synergistic and comprehensive. The projects are horizontally and vertically integrated. *Horizontally* means that water districts across the landscape worked together with a common goal, and *vertically* means that water districts worked with flood control, water recovery agencies (new name for wastewater managers), and habitat and environmental groups. Here are a few of those projects.

15.5.3.1 Water-Use Efficiency and Water Quality

The largest crop cultivated the United States is ornamental lawn, which feeds no one but uses more water than any other crop—water that is scarce. The Santa Ana River Watershed developed a watershed-wide water-use efficiency comprehensive program. Funding coming from the California Proposition 84 bond, the Metropolitan Water District, and local water agencies exceeds $22 million and is available throughout the watershed.

Key to the program is the idea that if water users understood how much water they use and what they really needed for personal in-home and outside use, they would be much better at conserving water and could achieve efficiency. Conservation-based rates set a price signal by charging a reasonable amount for water needed for personal use and a little more for water needed outside the home in the yard, but a lot more if that budget is exceeded. An even higher rate is charged if so much water was wasted that it ran off the yard and created pollution. We wanted water districts with traditional rate structures to be able to adopt these conservation-based rate structures if they wanted to. The California State Water Resource Control Board encouraged adoption with a new mandate that California reduce retail water consumption by 25% overall, with many districts given a reduction mandate as high as 36%. A University of California Riverside study (Baerenklau et al. 2013) documents that implementing this conservation-based rate structure achieved a 15% water-use reduction. This program is providing tools and funding to retail water agencies in the watershed to assist the implementation of this rate structure. It creates incentives for water-use efficiency, and charges the lowest prices for the most essential uses, by establishing tiers that reflect the agency's true water service costs. In order to set an accurate water budget for each ratepayer, we had to determine how much landscape each ratepayer needed to irrigate. We mapped the urban areas of the watershed, using the latest available technology, through infrared imaging that can distinguish between vegetated and non-vegetated areas. The digital mapping data will be provided to water agencies throughout the watershed. Also provided is a web based tool that will allow retail water agencies in the watershed to connect with their customers and encourage conservation. Finally, a financial incentive is provided to create a market-based transformation from turfgrass to drought-tolerant landscaping. The incentive, which will be in the form of a rebate, will be available to homeowner associations and public agencies who will replace ornamental turf with drought-tolerant landscaping in highly visible areas across the watershed.

Changing rate structures is often politically dangerous, but we have collectively moved in directions that afford some protection from backlash. Customers with conservation-based rates report that they are happy and that this is a fair and equitable rate system because no longer does one ratepayer subsidize a large user when the water supplier has to find expensive water on the margin. We have seen a great groundswell of public awareness among concerned individuals and organizations, and water consumption and water runoff have been reduced. But this strategy has recently come under scrutiny in the recent court of appeals decision in *Capistrano Taxpayers Association v. City of San Juan Capistrano* (http://www.courts.ca.gov/opinions/documents/G048969.PDF). In 1996, Proposition 218 (http://www.cvwd.org/DocumentCenter/View/2220) changed the California Constitution to require that property-related fees cannot exceed the reasonable or the proportional cost of providing a service to the parcel on which the fee is imposed. Water rates are subject to these requirements. In the Capistrano case, the court struck down the city's water budget–based rate structure because it found no evidence in the record connecting differences in cost of service to the tiers in the rate structure or the price for each tier. The court made clear, however, that tiered rates could be constitutional if the nexus between the tiers and water usage and cost of service were established in the record. The court noted, for example, that a water supplier might have multiple sources of supply, each with unique costs, which could be allocated to different levels of water use. While the Capistrano court did not discuss other kinds of costs of service, such as the cost of implementing required conservation programs, system costs associated with peaking, and the costs of developing new sources of supply, those costs would be part of the reasonable cost of providing water service. Proposition 218 does not restrict a water agency's ability to use other unrestricted funds, such as ad valorem tax revenues, to support separate water rate tiers. Other California courts have interpreted Proposition 218's proportionality requirements differently, and further clarification may come forth in the future. In the meantime, tiered rates are supportable with an appropriate cost-of-service study that establishes the nexus between costs and tiers.

Investment in water-use efficiency inside the home resulted in an average per capita use below the statewide average. Inland, the hotter days and larger lots mean that up to 80% of total water consumption occurs in the yard. In coastal areas, this could be closer to 40%. About half the water used to irrigate residential landscape is not needed. It is wasted and runs off the property, carrying pet waste, fertilizers, and salt into the storm water collection system, eventually hitting the receiving water body with a slug of toxicity.

Cities are required to implement expensive *total maximum daily loads* (TMDLs) meant to protect water bodies for fish and people. Taking the systems approach to solving problems, we can see that if we immediately improve our irrigation practices and ultimately improve our landscaping practices by removing turf and planting water-efficient plants, we can simultaneously improve water quality and water reliability. Waste at one end of the watershed causes water-quality problems in the river or the ocean at the other end. The disconnect is that water suppliers are not accountable for pollution resulting from water use, but cities are. Not until we get all the players from different corners of the watershed in a room can we find solutions. We could save as much as 40% of highly treated drinking water, and if we learned to be better landscapers, using water-efficient plants and less turf, we could save closer to 60% of all treated potable water! Working together, we can see that adopting a budget-based water rate will improve water quality. Water-use efficiency ultimately is not only about incentives or mandates but must also be driven by a water ethic held by the consumer. We will never have the regulatory structure or the funding to fully incentivize water-use efficiency; it must also arise from the ethics of the user. This can only happen if the user has a relationship with water, understanding both its limitations and its value.

15.5.3.2 Forest First

By managing the forest with a commitment to water quality and supply, this new partnership will deliver an integrated response that shaves the peaks off catastrophes. The San Bernardino and Cleveland National Forests encompass approximately 33% of the Santa Ana Watershed's land mass and receive 90% of its annual precipitation. Runoff from the forests directly affects the amount and quality of water received downstream. Yet, historically, there has been little, if any, relationship between forest and water management. The fire-suppression practices of the twentieth century have resulted in an unnaturally dense forest, reducing water supply and increasing fuel, resulting in more intense and catastrophic fires. Meadows, nature's sponge, retain water for groundwater recharge but dry up due to natural and manmade channelization, blocking the potential benefit of this resource. Hard lessons were learned after the forests experienced devastating fires in the early 2000s, and the aftermath of those fires directly impaired the quality of water downstream. It took years to recover. The 2013 Rim Fire, the largest ever in the Sierra Nevada and the third largest in California history, was a direct result of these past practices. Today, SAWPA and its members have partnered with the US Forest Service to develop the Forest First partnership to correct this disconnect. The joint goal is to manage the forest strategically, focusing on the areas that are most vulnerable to catastrophic fires, which could affect the downstream watershed. When trees are thinned, more water is released to the lower watershed; fire threats are lessened; and a natural fire regime, faster-moving and not as hot, prevails.

Thinning the forest to mimic a time before it was so heavily occupied by people serves to reduce fires and to protect people and the environment, and it also serves water supply, quality, and flood missions. Restored meadows attenuate floods, allowing them to absorb water and slowly percolate it into aquifers that store this high-quality sweet water. This partnership is being pursued among downstream groundwater management agencies, flood control and water conservation districts, water supply agencies, resource agencies, and the Forest Service. The goal is to find agreeable projects that can be executed in specific areas within the forest and that will have a direct effect in preserving and enhancing the quality and quantity of water resources from the source or headwaters, contributing to the overall health of the watershed. Evidence from a quantitative cost–benefit analysis is being sought to validate investment in projects that will help the Forest Service keep the forest healthy and, in turn, promote water quantity and quality. This investment in green infrastructure reduces carbon footprint, is less expensive than building a treatment plant, and has Mother Nature doing work for us. Integrating water and disaster management, *from prevention to cure*, learning from the aftermath of fires to establish cooperation for prevention, brings two traditionally separate communities together.

15.5.3.3 Salt Management

Almost a century of agriculture and industry has resulted in salts and other constituents infiltrating aquifers and streams. Crops irrigated by imported Colorado River water are harvested, but they leave behind a salt legacy accumulating in the groundwater. Water quality in the Santa Ana River has improved in recent years thanks to brackish-groundwater desalination and water-quality planning, but challenges remain. Drought exacerbates these challenges. Without sweet rain or Sacramento–San Joaquin Delta water to blend with native water, which is higher in salt, the watershed is closer to maximum salt limits, causing a double whammy of water loss due to both drought and salt.

Technologically advanced wastewater-control infrastructure has been rigorously employed, and work continues on minimizing the negative impacts from agricultural runoff. Nevertheless, the existing salts and contaminants present in the watershed from past practices still need to be removed, as improving water quality is inextricably linked to improving water supplies and implementing a comprehensive groundwater storage program. As part of the solution to the salt issues within the watershed, SAWPA constructed almost 100 mi. of pipeline to convey high-saline brine out of the watershed. Desalters throughout the upper watershed remove and concentrate salt from brackish groundwater, which is collected and conveyed to the coast, where it is treated before being discharged to the sea. The result is sweet water that can be used as supply, while aquifers, the river, and the people who use that water are protected from salt contamination. The lower watershed invested in the brine line in the upper watershed (outside their jurisdiction) to protect the quality of water that would ultimately reach them. This commitment continues as the lower watershed invests in infrastructure in their area to accommodate brines from upper watershed desalters and investments from the upper watershed to maintain a water quality for the protection of the lower. This practice has been key in helping the watershed survive the drought.

15.6 Conclusion

SAWPA has practiced successful IWRM for decades, and still, it is not easy. Success does not lie only in the establishment of a regional agency like SAWPA but, rather, in a regional perspective that builds upon existing organizations and relationships working from the group up and from the inside out. SAWPA has benefited from a stable revenue stream garnered from the five member agencies and from fees for services and grants. Funding helps support efforts, but it is not the most important factor. It can also run counter to the effort. If each agency had all the funds required to address its own problems and needs, they might not be motivated to cooperate with others. The severity of water resource problems also creates opportunities for IWRM to flourish. This is a strong factor in SAWPA's success, because the watershed does not produce sufficient local water to serve current and expected future uses, and agencies had to band together and cooperate.

The west is ripe for integrated water management planning and successful collaboration across a broad range of stakeholders with differing perspectives. This is borne out in SAWPA's history, beginning with the five members, who are distinct geographically, and in their original missions (groundwater management, water importer, and wastewater treatment). As time went on, each of them further diversified and worked closely with the others, and with other missions such as flood control. The latest iteration of OWOW brought in far more diverse stakeholders, including those outside of traditional water managers.

SAWPA enjoys strong support from federal and state agencies as well. While we do not have perfect regulatory alignment, the agencies are aware of the need to examine this and are willing to see if alignment can be achieved. There is increasing skepticism about the ability of higher agencies at the state or federal level to render viable long-term solutions to our complex water-quality and water resource problems on their own. Success is closer to our grasp when there is alignment of objectives and decision criteria throughout the hierarchy of regulatory and other resource agencies. The US Council on Environmental Quality's updated *Principles and Requirements for Federal Investments in Water Resources* (2013) provides a common framework for federal agencies' funding decisions and promotes investment in integrated, multibenefit solutions. Even before this direction, the US Bureau of Reclamation's Southern California Area Office and the US Army Corps of Engineers, Los Angeles District, have demonstrated strong leadership, support, and commitment to integration. These two organizations cochair SAWPA's OWOW Government Alliance pillar and are working hand in glove to support the process. Reclamation invested $1 million, matched by the California Department of Water Resources and SAWPA, to conduct several studies to support OWOW, including climate change adaptation and other regional efforts. The Corps has created a pilot to support watershed-based budgeting. To avoid the piecemeal approaches of the past, the Corps has developed a strategic plan that incorporates environmental operating principles mirroring IWRM principles. The US Environmental Protection Agency has also reiterated support for IWRM with their storm water combined permit. The California legislature passed the Integrated Regional Water Management grant program, funding $1 billion to incentivize IWRM plans, processes, and multibenefit projects for California. The State Water Resources Control Board and the Regional Water Quality Control Boards (RWQCBs) are organized along watershed hydrologic boundaries and have long been committed to basin planning. But this support itself is not nearly enough to fuel a successful effort on the regional watershed level that asks all water resource managers to check their agencies' priorities at the door and to come to the table with a commitment to the greater good of the watershed. We are asking a lot, but it is increasingly clear that the path to resiliency in today's challenges must be predicated in IWRM.

References

Baerenklau, Kenneth A., Kurt A. Schwabe and Ariel Dinar. 2013. *Do Increasing Block Rate Water Budgets Reduce Residential Water Demand? A Case Study in Southern California.* UC Riverside Water Science and Policy Center Working Paper.

California Department of Finance. 2014. *Total Population Projections for California and Counties: July 1, 2015 to 2060 in 5-Year Increments.* http://www.dof.ca.gov/research/demographic/reports/projections/P-1/.

California Department of Forestry and Fire Protection. 2014. http://www.fire.ca.gov/communications/downloads/fact_sheets/20LACRES.pdf.

Global Water Partnership Technical Advisory Committee. 2000. *Integrated Regional Watershed Management.* TAC Background Papers, no 4. Stockholm, Sweden. www.gwp.org/Global/GWP-CACENA_Files/en/pdf/tec04.pdf.

Kemper, Karen E., Ariel Dinar, and William Blomquist, eds. 2007. *Integrated River Basin Management Through Decentralization.* New York: Springer.

National Research Council. 1999. *New Strategies for America's Watersheds.* Washington, DC: National Academy of Sciences.

Santa Ana Watershed Project Authority. 2013. *One Water, One Watershed Plan 2.0.* http://www.sawpa.org/owow-2-0-plan-2/.

Senge, Peter. 2008. *The Necessary Revolution.* New York: Crown Business.

UN-Water. 2012. *Status Report on the Application of Integrated Approaches to Water Resources Management.* United Nations Environment Programme. www.un.org/waterforlifedecade/pdf/un_water_status_report_2012.pdf.

US Bureau of Reclamation. 2013. *Technical Memorandum No. 86-68210-2013-02 Climate Change Analysis for the Santa Ana River Watershed Santa Ana Watershed Basin Study, California Lower Colorado Region* http://www.sawpa.org/wp-content/uploads/2014/01/Appendix-F2-Technical-Memorandum-No.-86-68210-2013-02-Climate-Change-Analysis-for-the-Santa-Ana-River-Watershed-Santa-Ana-Watershed-Basin-Study-California-Lower-Colorado-Region.pdf.

US Council on Environmental Quality. 2013. *Principles and Requirements for Federal Investments in Water Resources.* www.whitehouse.gov/sites/default/files/final_principles_and_requirements_march_2013.pdf.

16

Climate Change and Allocation Institutions in the Colorado River Basin

Jason Anthony Robison

CONTENTS

ABSTRACT Climate change has profound implications for allocation institutions in the Colorado River Basin. After providing an overview of current and projected hydrological conditions within the basin, this chapter analyzes salient issues posed by climate change for the Colorado River Compact, Upper Colorado River Basin Compact, Supreme Court decree in *Arizona v. California*, and water rights held by American Indian tribes. Apparent from the analysis are two overarching policy and management priorities: (1) a compelling need for clarity regarding the specific boundaries of water-use entitlements established by the allocation institutions, and (2) an associated need for parties holding these entitlements to foster measures and commitments aimed at facilitating levels of water use that realistically align with their respective water budgets.

16.1 Introduction

At the core of the elaborate body of laws governing water allocation and management in and around the Colorado River Basin—colloquially called the *Law of the River*—are a host of interconnected institutions that apportion the use of this water among federal, state, and tribal sovereigns and water users. These institutions have shaped, and will continue to shape, the essential character of the modern US Southwest. Climate change holds a variety of implications for these institutions—and, in turn, for economic, environmental, political, and social conditions within this region—and it is an analysis of these implications that constitutes the focus of this chapter. The broad pattern apparent from this analysis is one in which the projected impacts of climate change on basin-wide water supplies—arguably more so than at any earlier point in the Law of the River's history—necessitate sustained efforts by sovereigns and water users (1) to comprehend the precise boundaries imposed by their water-use entitlements and (2) to foster associated commitments and measures aimed at facilitating levels of water use that comport with these boundaries. In short, these parties' water budgets must be clarified, planned around, and ultimately lived within. Two main sections comprise this chapter and shed light on this pattern. The discussion begins with a thumbnail sketch of the basin focusing primarily on historical and projected water supplies and demands and related impacts of climate change. Against this backdrop, the remainder of the chapter provides an analysis of pressing issues associated with the foregoing pattern. Two limitations in the analysis should be flagged at the outset—one hydrological, one institutional. First, the analysis focuses almost exclusively on the implications of climate change for surface water allocation under existing institutions, despite recent findings of extensive groundwater depletion in the basin and advocacy for conjunctive surface water and groundwater management under the Law of the River (Castle et al. 2014). Second, given the vast expanse of the Law of the River, the allocation instruments encompassed within the analysis account for a relatively modest scope of institutions potentially affected by climate change within the Colorado River Basin. More far-reaching and complementary analyses are warranted.

16.2 The Basin of Contention in the Twenty-First Century

Dubbed the "basin of contention" by its premier western historian (Hundley 2009, p. 352), the Colorado River Basin encompasses roughly 244,000 mi. (631,960 km^2)* of territory that falls within portions of seven US states (Arizona, California, Colorado, Nevada, New Mexico, Utah, and Wyoming) and two Mexican states (Baja California and Sonora). As depicted in Figure 16.1, the mainstream of the Colorado River meanders nearly 1400 mi. (2240 km)† from its headwaters in the Rocky Mountains of northern Colorado to its terminus in the Gulf of California in Mexico (MacDonnell 2009). Its major tributaries include the Green River, San Juan River, and Gila River. Legally and politically, the basin is bifurcated into two "basins," an Upper Basin and Lower Basin, at a point along the Colorado River mainstream in northern Arizona called Lee Ferry. Colorado, New Mexico, Utah, and Wyoming are commonly referred to as the *Upper Basin states*, and Arizona, California, and Nevada are commonly referred to as the *Lower Basin states*. Whereas the Upper Basin consists of high desert and alpine terrain, a significant portion of the Lower Basin is of low desert character, particularly in the southernmost areas of Arizona and California. Diverse federal lands pervade the Upper and Lower Basins (e.g., Grand Canyon National Park), and the two largest Indian reservations in the United States likewise are located within the basin: the Navajo Nation in the Four Corners area and the Uintah and Ouray Reservation in eastern Utah.

Hydrological conditions in and around the Colorado River Basin have changed markedly over the past century. As illustrated by Figure 16.2, a gradual pattern is apparent across this period involving an increasingly disproportionate relationship between water

* 1 square mile (mi^2) = 2.59 square kilometers (km^2).

† 1 mile (mi.) = 1.6 kilometers (km).

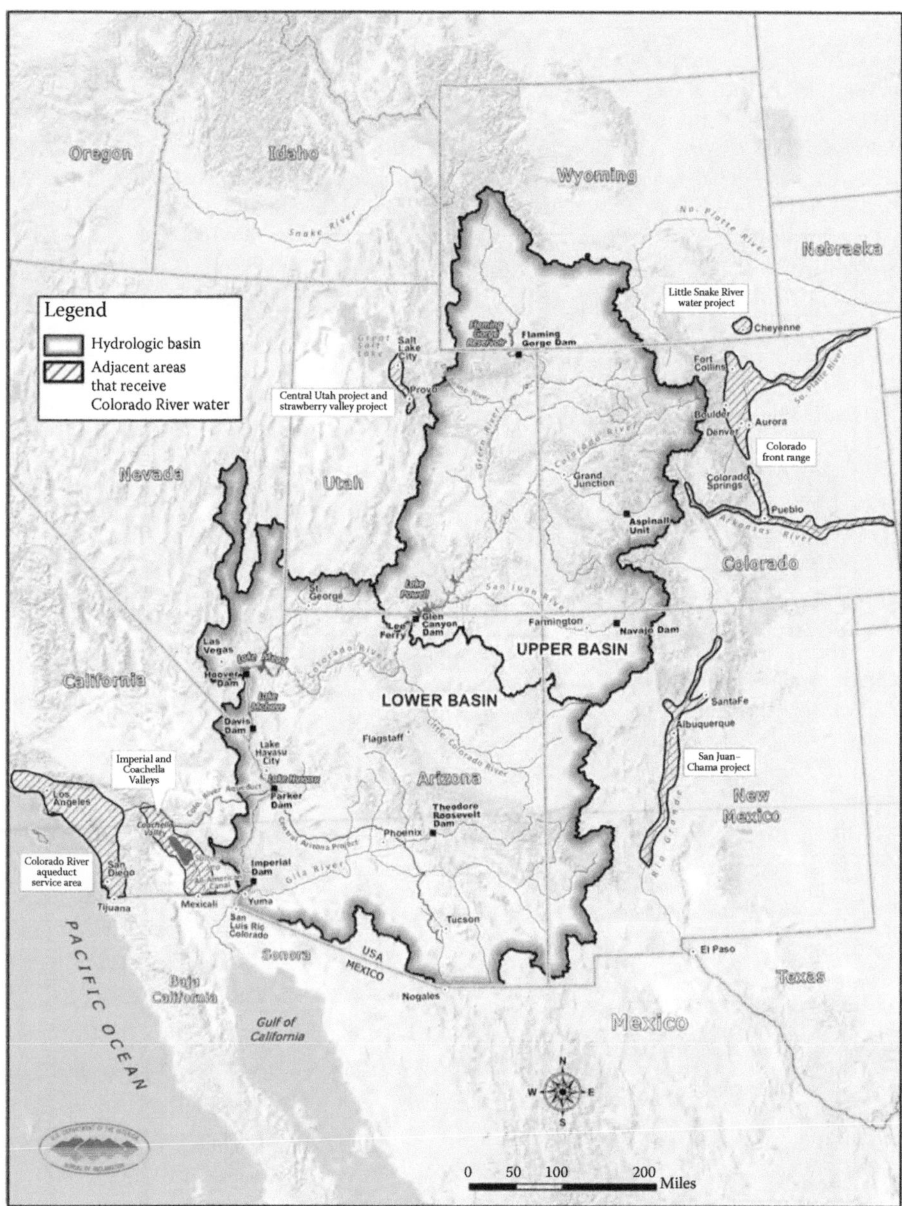

FIGURE 16.1
Colorado River Basin and adjacent export areas within United States.

supplies and demands. The Bureau of Reclamation examined this supply–demand imbalance in a landmark study released in December 2012 hailed as "the most comprehensive basin-wide analysis ever undertaken within the Department of the Interior": the Colorado River Basin Water Supply and Demand Study (Basin Study) (Connor 2013). According to the Basin Study, basin-wide water demands exceeded water supplies on average for the first time in the roughly 100-year historical record during the preceding decade, and this

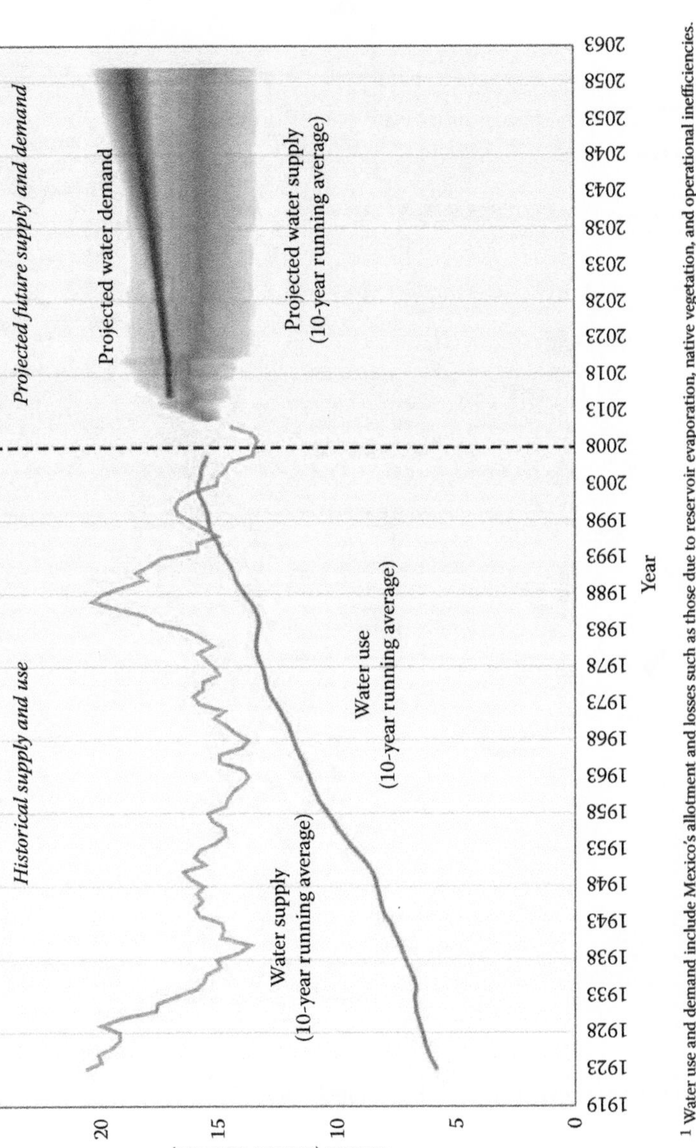

[1] Water use and demand include Mexico's allotment and losses such as those due to reservoir evaporation, native vegetation, and operational inefficiencies.

FIGURE 16.2
Historical supply and use[1] and projected future supply and demand[1] in million acre-feet (1 million acre-feet [MAF] = 1.233 billion cubic meters [BCM]).

imbalance is projected to widen to an annual shortfall of about 3.2 million acre-feet (MAF)* (equivalent to more than 1 trillion gallons of water) by 2060 (US Bureau of Reclamation 2012a). Notably, this projected 3.2 MAF annual shortfall is the *mean* supply–demand projection. It should also be highlighted that this imbalance affects the lives and livelihoods of nearly 40 million people in the United States who currently rely on the basin's water (US Bureau of Reclamation 2012a).

To elaborate briefly on the demand side of this imbalance, historical consumptive uses and losses in the Colorado River Basin grew from approximately 13.0 to 15.0 MAF between 1971 and 2010, with this figure falling slightly above 15.0 MAF in the latter year. Figure 16.3 reveals this trend. Drawn from the Basin Study, these figures unfortunately do not account for consumptive uses and losses along tributaries in the Lower Basin, including those associated with the primary Lower Basin tributary, the Gila River. Figure 16.4 evidences how Gila River uses and losses ranged from slightly above 2.5 MAF to slightly below 4.5 MAF between 1971 and 2005 and were approximately 3.5 MAF during the latter year (the most recent for which data are provided).

This pattern of increasing water demands in the Colorado River Basin is not projected to abate in coming decades according to the Basin Study. Rather, as illustrated by Figure 16.5, basin-wide consumptive uses and losses are anticipated to range from 17.7 to 20.1 MAF by 2060, although again these figures do not account for uses and losses along the Lower Basin tributaries. A wide scope of related figures addressing projected water demands of the Upper Basin, Lower Basin, individual states, and American Indian tribes appear throughout the discussion in this chapter.

Running in the opposite direction is the projected water supply trend for the Colorado River Basin. As described in the Basin Study, approximately 92% of the basin's natural flow—as measured at Imperial Dam upstream of the Gila River's confluence with the Colorado River—originates within the Upper Basin (US Bureau of Reclamation 2012b). Historical and projected flow levels for the Colorado River mainstream at Lee Ferry—again, the dividing point between the Upper and Lower Basins—are therefore critically important. The hydrological figures denoted as *Lee Ferry flows* here and elsewhere in the chapter consist of reconstructed natural flows at this dividing point. Looking backward, annual Lee Ferry flows had a mean of 15.0 MAF and ranged from 5.6 to 25.2 MAF from 1906 to 2007 (US Bureau of Reclamation 2012b). Looking forward, variable projections exist for future Lee Ferry flows as influenced by climate change (US Bureau of Reclamation 2011, 2012b, 2013; Vano et al. 2014). One of four water supply scenarios examined in the Basin Study, the downscaled global circulation model (GCM) scenario reflected in Figure 16.6, includes such projections (US Bureau of Reclamation 2012b). Future Lee Ferry flows are projected to range from 4.2 to 44.3 MAF per year in this scenario. Its mean projection anticipates an 8.7% decrease between now and 2060 in the 15.0 MAF mean annual Lee Ferry flows from the historical record. Such a decrease translates to average annual flows of 13.7 MAF. Its median projection is described textually in the Basin Study as "nearly 1.0 MAF lower (annual flow of around 12.7 MAF) than the mean"—a roughly 15.3% decrease from the historical average—although this value is subsequently noted in tabular form as 13.6 MAF (US Bureau of Reclamation 2012b, B-66). For purposes of this analysis, the mean and lower of the two median figures in the GCM scenario—involving future average annual Lee Ferry flows of 13.7 MAF and 12.7 MAF, respectively—are the focal point of the following assessment of water policy and management issues posed by climate change.

* 1 million acre-feet (MAF) = 1.233 billion cubic meters (BCM).

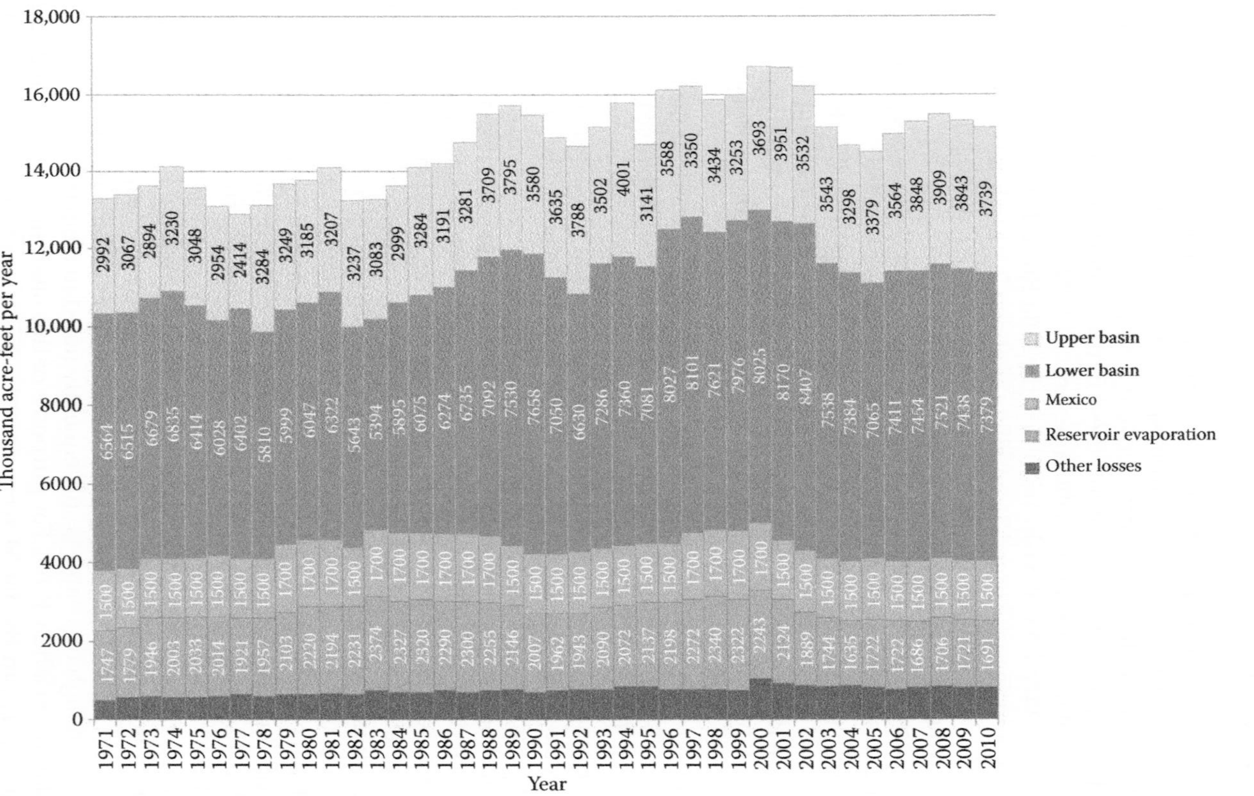

[1] Excluding consumptive use in Lower Basin tributaries.
[2] Reservoir evaporation losses are accounted differently in the Upper and Lower Basin. In the Upper Basin, reservoir evaporation losses are accounted as part of each state's total uses. In the Lower Basin, reservoir evaporation losses are accounted separately from each state's uses. Reservoir evaporation losses from Upper and Lower Basin reservoirs have been aggregated for this presentation.
[3] Phreatophyte and operational inefficiency losses.

FIGURE 16.3
Historical Colorado River water consumptive use[1] and loss by basin, Mexico, reservoir evaporation,[2] and other losses,[3] 1971–2010.

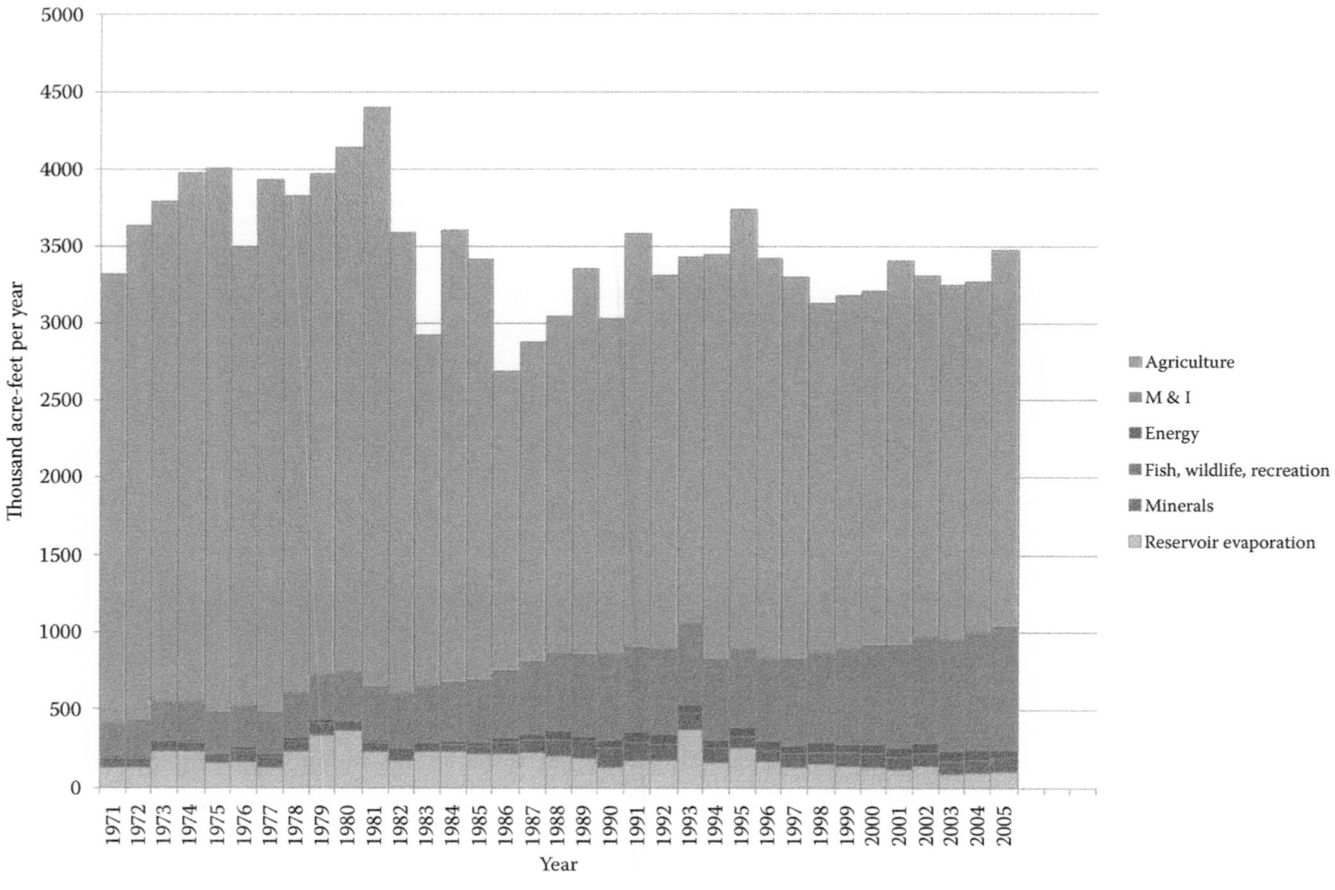

FIGURE 16.4
Historical consumptive uses and losses for the Gila River, 1971–2005. M & I, municipal and industrial.

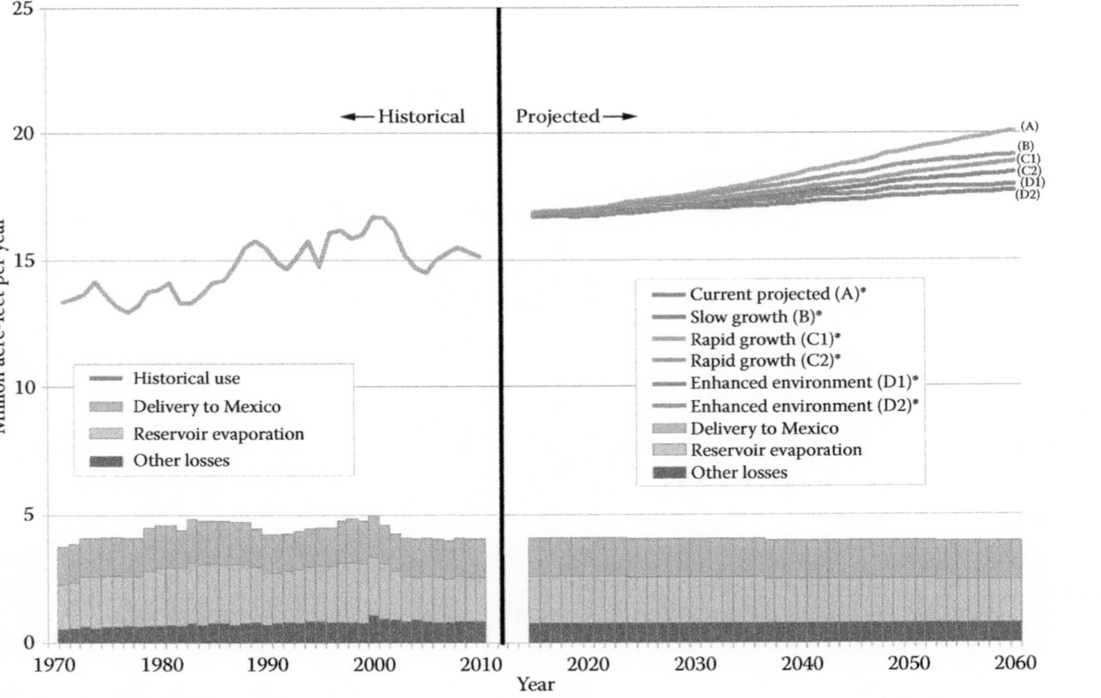

* *Quantified demand scenarios have been adjusted to include Mexico's allotment and estimates for future reservoir evaporation and other losses.*

1 Excluding consumptive use in Lower Basin tributaries.

2 Assumed 1.5 MAF delivery to Mexico 2012–2060. Modeling to support the system reliability analysis will project future deliveries to Mexico in accordance with the 1944 treaty.

3 Median value of Colorado River Simulation System (CRSS)-simulated reservoir evaporation across supply and demand scenarios.

4 Other losses include phreatophyte and operational inefficiency losses. Future phreatophyte losses are computed by assuming 1995–2008 average of 632,000 acre-feet (kaf). Future operational inefficiency losses are computed as the sum of 109 kaf (the 1990–2010 average bypass of return flows from the Welton–Mohawk Irrigation and Drainage District to the Cienega de Santa Clara in Mexico) and 7 kaf (computed by assuming the 1964–2010 historical average annual volume of nonstorable flows delivered to Mexico [excluding flood years] is reduced by 90% due to the operation of Warren H. Brock Reservoir).

FIGURE 16.5
Colorado River Basin historical use[1] and projected future demand,[1] delivery to Mexico,[2] reservoir evaporation,[3] and other losses.[4]

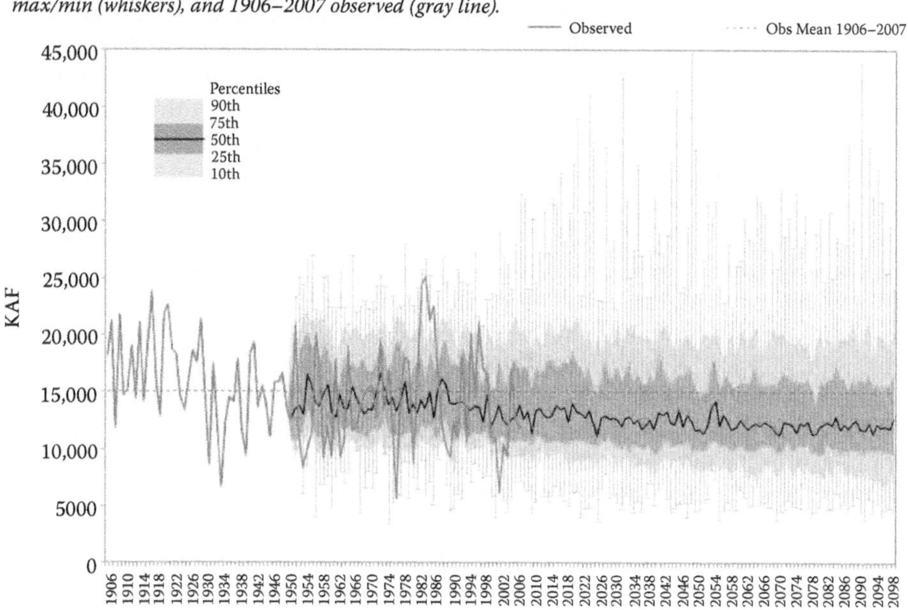

FIGURE 16.6
Lee Ferry natural flow statistics for downscaled GCM scenario as compared to observed flow.

16.3 Implications of Climate Change for Allocation Institutions

Encompassed within the vast legal framework that has facilitated the foregoing water demands, and that governs allocation and management of the foregoing water supplies, are literally dozens of laws that have come into existence in an iterative manner over the past century to collectively form the Law of the River. The analysis in this chapter focuses selectively on the implications of climate change for allocation institutions that govern water use within the US portion of the Colorado River Basin. These institutions include (1) the Colorado River Compact ("Compact") at the basin-wide level, (2) the Upper Colorado River Basin Compact ("Upper Basin Compact") in the Upper Basin, and (3) the supreme court's decree in *Arizona v. California* governing water use from the Colorado River mainstream in the Lower Basin. Also briefly touched on are climate change's implications for quantified and unquantified water-use entitlements held by tribal sovereigns throughout the basin. As identified in the introduction, the overarching pattern apparent from the various issues posed by climate change for these institutions essentially entails two types of challenges: (1) those associated with understanding the precise boundaries of water-use entitlements established by the institutions and (2) those associated with entitlement holders ensuring their respective levels of water-use comport with the water budgets prescribed by their entitlements. The analysis begins by considering this pattern in relation to the Colorado River Compact.

16.3.1 US Basin-Wide Apportionment

Regulating water use throughout the US portion of the Colorado River Basin is the Compact—a roughly 90-year-old document forged by federal and state members of the Colorado River Commission in 1922 that has been aptly characterized as the Law of the River's "constitution" (Adler 2008, p. 21). It is the Compact that frames the basin-wide apportionment scheme in the United States. Climate-based reductions in average Lee Ferry flows like those outlined previously pose numerous issues for this scheme. Prominent among these issues are those concerning (1) the Compact's equity with respect to the relative levels of water availability in the Upper and Lower Basins as influenced by climate change and (2) the Compact's interpretation in various particulars that bear directly on the perceived equity of the basin-wide apportionment.

A brief survey of the apportionment scheme established in Article III of the Compact sets the stage for these two issues. Article III divides the Colorado River Basin into an Upper Basin and Lower Basin at Lee Ferry, as identified previously, and authorizes groups of states with portions of territory in these two subbasins to collectively use set amounts of water from the *Colorado River system*—defined as "that portion of the Colorado River and its tributaries within the United States of America." Specifically, the Compact, in Article III(a) and (b), authorizes "beneficial consumptive use" of 7.5 and 8.5 MAF of water in the Upper and Lower Basins, respectively, per annum. It then goes on to impose pivotal flow obligations in Article III(c) and (d). The former provision, Article III(c), obligates the Upper and Lower Basins to each contribute half of the flows needed to satisfy Mexico's 1.5 MAF entitlement under a 1944 treaty if *surplus* water is not available to supply these flows in a given year. This flow obligation is regarded as the first priority of the Law of the River (MacDonnell 2009). Dovetailing with this obligation, the Upper Basin states also owe a flow obligation to the Lower Basin under Article III(d), which appears in the form of a prohibition against the Upper Basin states causing flows at Lee Ferry from being depleted below 75.0 MAF during any consecutive 10-year period.

Stemming from the particular design of the Compact's scheme, an initial issue posed by climate change for the US basin-wide apportionment concerns its equity—or, put differently, the extent to which the Compact is capable of realizing its fundamental purpose of effecting an *equitable apportionment* in the Colorado River Basin (Robison and Kenney 2012). It has been common knowledge for decades that the quantified entitlements and flow obligations identified previously were founded on erroneous overestimates of average annual Lee Ferry flows (Hundley 2009). What remains to be further examined, however, is precisely how the Upper and Lower Basins will respectively fare under the Compact's framework given the projected decreases in Lee Ferry flows. Stated broadly, the crux of this equitable apportionment issue concerns a distributional disparity resulting from the application of the Compact flow obligations in Article III(c) and (d) to the water available for consumptive use from the Upper and Lower Basins' respective 7.5 and 8.5 MAF entitlements in Article III(a) and (b) (Robison and Kenney 2012).

A few figures serve to illustrate this equitable apportionment issue. If Lee Ferry flows were to average 13.7 MAF per year in the future—again, the mean projection in the Basin Study's GCM scenario—the Upper Basin states potentially would only be able to rely on 5.45 MAF of their ostensible 7.5 MAF entitlement on average per year after satisfying the Article III(c) and (d) flow obligations. To be clear, this 5.45 MAF postflow obligation remainder would need to suffice to cover both consumptive uses *and* reservoir evaporation and other losses in the Upper Basin. The remainder would drop to roughly 4.45 MAF if Lee Ferry flows were to decline further to 12.7 MAF. Consider, by way of contrast, the Lower Basin states' situation in the same circumstances. These states would be able to rely on two sources of flows to cover their consumptive uses and reservoir evaporation and other losses. The first would be mainstream flows supplied by the Upper Basin at Lee Ferry—specifically, an assumed annualized average of 7.5 MAF pursuant to Article

III(d)—and the second would be flows from tributaries of the Colorado River within the Lower Basin. It is impossible to provide a precise annual average for prospective tributary flows. That said, the Bureau of Reclamation's reports for the 1971–2005 period—which contain the most recent data as of this writing—indicate that annual consumptive uses and losses along the Lower Basin tributaries averaged 3.52 MAF across this period and ranged from 1.99 to 5.23 MAF (US Bureau of Reclamation 2015a). Apparently due to methodological differences, these figures unfortunately do not fully align with counterpart figures contained in a Lower Basin tributaries appendix to the Basin Study, particularly those noted earlier identifying annual consumptive uses and losses along the Gila River alone of slightly above 2.5 MAF to slightly below 4.5 MAF from 1971 to 2005 (US Bureau of Reclamation 2012i). Whatever the annual average may be for combined mainstream and tributary flows in the Lower Basin going forward, however, the takeaway here is hopefully plain. The more pronounced the climate-based reductions in Lee Ferry flows, the smaller the amount of water legally available for use in the Upper Basin under the Compact, and the greater the deviation from its original 7.5/8.5 MAF equitable apportionment set forth in Article III(a) and (b) (CRGI 2010).

Accompanying this equitable apportionment issue are longstanding yet unresolved matters of compact interpretation. These interpretive issues revolve around the contested meaning of Article III(c) and (d) and the details of their respective flow obligations. Regarding the former, the precise method for determining whether surplus water exists within the Colorado River system to supply Mexico's treaty entitlement has been disputed (CRGI 2012b). Does this surplus determination require accounting for, or refraining from considering, water in the Gila River and other Lower Basin tributaries? In a similar vein, does surplus consist of water being consumed in the Lower Basin in excess of that subbasin's 8.5 MAF entitlement? Or, conversely, should the existence of surplus be determined by assessing whether water exists in the Colorado River system above and beyond the collective 16.0 MAF associated with the Upper and Lower Basins' entitlements? Dovetailing with these questions is a related interpretive issue involving Article III(d) (CRGI 2012a). To what extent, if any, might future climate-based reductions in Lee Ferry flows alter the Upper Basin's obligation not to deplete these flows below 75.0 MAF every 10 years? Simply put, is this flow obligation fixed or potentially contingent? Varied answers have been offered to these questions, and these answers significantly influence the extent to which the basin-wide apportionment scheme is viewed as equitable.

One final question is appropriate in drawing this initial section to a close: What is the optimal way for dealing with interpretive conflicts over the Compact in the years ahead? These conflicts bear directly on the Upper and Lower Basins' relative abilities to use water afforded by their entitlements. Further, as highlighted in greater detail in subsequent sections, water demands in both the Upper and Lower Basins are projected to increase considerably over the next 50 years, while Lee Ferry flows are again projected to decrease (US Bureau of Reclamation 2012a). It appears inevitable in these circumstances that conflicts over the interpretive issues will need to be managed carefully going forward. At present, consultation is the mandatory process for addressing interpretive disputes concerning the Compact, pursuant to an agreement forged by the basin states in 2007 contemporaneous with the adoption of the Interim Guidelines (i.e., Colorado River Interim Guidelines for Lower Basin Shortages and the Coordinated Operations for Lake Powell and Lake Mead). These guidelines will remain in effect until 2026, and it is possible their coordinated operating regime for Lake Powell and Lake Mead will ameliorate tensions stemming from the interpretive issues. If that proves not to be the case, or if an alternate type of resolution seems preferable upon expiration of the guidelines, options for addressing the interpretive issues include arbitration and supreme court litigation as provided for in Articles VI and IX of the Compact. Although the future course remains to be seen, the closing message is that ongoing management of these conflicts will remain an important priority.

16.3.2 Upper Basin Apportionment

Apparent from the previous section is the fact that the Upper Basin is in a position of creating a future from "leftovers" under the Compact—at least with regard to the role played by consumptive use, as opposed to nonconsumptive use, of water from the Colorado River system within that future (Getches 1985). Numerous policy and management priorities emerge given this state of affairs, including the need for ongoing and realistic assessments of water availability within the Upper Basin as a whole, and within Upper Basin states individually, as these matters are governed by the flow obligations imposed by the Compact and derivative entitlements prescribed by the Upper Basin Compact. Interconnected with this priority is the need for policy measures aimed at promoting efficient and flexible water use by Upper Basin water users, including potential interstate water banking. The material that follows considers these subjects after briefly sketching out the Upper Basin Compact's allocation framework.

Future availability of Colorado River system water for consumptive use in the Upper Basin hinges as much on the composition of the apportionment schemes established by the Compact and Upper Basin Compact as it does on projected climate-based reductions in Lee Ferry flows. The latter apportionment scheme is set forth in Article III of the Upper Basin Compact. It generally calls for allocating the water remaining available for consumptive use within the Upper Colorado River system after the Upper Basin states have satisfied their Compact flow obligations (i.e., the leftovers) on an annual percentage basis among these states. The specific percentages of this water that the Upper Basin states are authorized to consume are as follows: Colorado, 51.75%; New Mexico, 11.25%; Utah, 23%; and Wyoming, 14%. Arizona is also entitled to consume 50,000 acre-feet [61.65 million cubic meters (Mm^3)]* of this water per year. Notably, the Upper Basin Compact treats reservoir evaporation and other losses as consumptive use under Articles V and VI—an important distinction from the *Arizona v. California* decree as identified in the next section. Overall, given the nested relationship between the Compact and Upper Basin Compact, the core insight is that just as the Compact flow obligations dictate the amount of water available for consumptive use within the Upper Basin collectively, so too do the Upper Basin Compact's entitlements control the amount of water available for consumptive use within the Upper Basin states individually. Future climate-based reductions in Lee Ferry flows portend to diminish both water budgets so to speak as these budgets are circumscribed by the nested compacts.

Figures from the Basin Study tangibly illustrate this state of affairs and the derivative need for realistic contemporary analyses of water availability and consumptive use in the Upper Basin. Taken together, Upper Basin consumptive uses and reservoir evaporation losses fell somewhere between 4.24 and 4.49 MAF in 2010, consisting of 3.74 MAF of uses and approximately 500,000–750,000 acre-feet of losses (US Bureau of Reclamation 2012c). Projecting across the next 50 years, annual consumptive uses in the Upper Basin are anticipated to hover around 5.0 MAF by 2035 and to increase to somewhere between slightly below 5.0 and 6.0 MAF by 2060 (US Bureau of Reclamation 2012c). Combining these projections with projected reservoir evaporation losses of 805,000 acre-feet per year throughout

* 1 acre-foot (af) = 1233.48 cubic meters (m^3).

the Upper Basin, the resulting figures entail annual consumptive uses and losses of approximately 5.8 MAF by 2035 and 5.8–6.8 MAF by 2060 (US Bureau of Reclamation 2012d,e,f,g).

A tenuous relationship exists between these projected figures and those in the previous section addressing future water availability in the Upper Basin after the Compact flow obligations have been satisfied. To reiterate, if Lee Ferry flows were to average 13.7 MAF in the future, then after backing out Arizona's 50,000 acre-foot entitlement under the Upper Basin Compact, perhaps an annual average of roughly 5.4 MAF would be available to cover consumptive uses and reservoir evaporation and other losses in the Upper Basin. This figure would drop to 4.4 MAF if Lee Ferry flows were to average 12.7 MAF. These projections raise serious questions about how realistic it is to expect Upper Basin consumptive uses to increase to the levels forecast in the Basin Study—or, stated differently, how much additional consumptive use should be considered viable in water planning processes given the constraining relationship between the Upper Basin's Compact flow obligations and its ostensible 7.5 MAF entitlement. Neither the 5.8 MAF projection for 2035 nor the 5.8–6.8 MAF projected range for 2060 appear feasible looking ahead given that an annual average of only 5.4 MAF will be available for consumptive uses and losses in the Upper Basin if future Lee Ferry flows average 13.7 MAF. Further, if an annual average of only 4.4 MAF ends up being available for consumptive uses and losses in the Upper Basin—because Lee Ferry flows average 12.7 MAF per year—it appears questionable whether any additional uses and losses would be feasible in excess of the 4.24–4.49 MAF range during 2010. The bottom line is that diligent efforts need to be undertaken to realistically assess the amount of Colorado River system water potentially available for consumption in the Upper Basin in coming years.

This perspective applies with equal force to water planning and management activities within the individual Upper Basin states. Tables 16.1 and 16.2 integrate a variety of figures to convey this point. Table 16.1 identifies the states' projected water budgets under the Upper Basin Compact in the event that future Lee Ferry flows average 13.7 or 12.7 MAF. These calculations assume that (1) 8.25 MAF will flow from the Upper Basin to the Lower Basin and Mexico on average per year stemming from the Compact flow obligations and (2) 50,000 acre-feet will be consumed in Arizona annually extending from that state's Upper Basin Compact entitlement. In turn, Table 16.2 addresses the Upper Basin states' historical and projected consumptive uses and losses. The historical figures are based on provisional data in the Bureau of Reclamation's Consumptive Uses and Losses Report for 2011–2015 and account for consumptive uses and reservoir evaporation losses in 2012 (the most recent year covered in the report). As for the projected figures, they are taken from appendices to the Basin Study and identify anticipated consumptive uses and reservoir evaporation losses for each Upper Basin state in 2035 and 2060.

TABLE 16.1

Projected Water Budgets under Upper Colorado River Basin Compact

	Colorado	New Mexico	Utah	Wyoming
Compact entitlement	51.75%	11.25%	23.0%	14.0%
Water budget: 13.7 MAF Lee Ferry flows (MAF)*	2.795	0.608	1.242	0.756
Water budget: 12.7 MAF Lee Ferry flows (MAF)	2.277	0.495	1.012	0.616

* 1 MAF = 1.233 BCM.

TABLE 16.2

Historical and Projected Consumptive Uses and Losses (MAF) of Upper Basin States

		Colorado	New Mexico	Utah	Wyoming
2012	Total	2.658	0.434	1.033	0.480
	Uses	2.268	0.359	0.837	0.389
	Losses	0.390	0.075	0.196	0.091
2035	Total	2.876–3.160	0.743–0.828	1.223–1.331	0.648–0.752
	Uses	2.446–2.730	0.673–0.758	1.033–1.141	0.533–0.637
	Losses	0.430	0.070	0.190	0.115
2060	Total	2.964–3.460	0.753–1.049	1.274–1.467	0.691–0.884
	Uses	2.534–3.030	0.683–0.979	1.084–1.277	0.576–0.769
	Losses	0.430	0.070	0.190	0.115

Source: Data from US Bureau of Reclamation, Colorado River Basin Water Supply and Demand Study, Technical Report C—Water Demand Assessment, 2012d,e,f,g, http://www.usbr.gov/lc/region/programs/crbstudy/finalreport/Technical%20Report%20C%20-%20Water%20Demand%20Assessment/TR-C-Water_Demand_Assessmemt_FINAL.pdf, accessed July 27, 2015; US Bureau of Reclamation, Colorado River System Consumptive Uses and Losses Reports, 2015a, http://www.usbr.gov/uc/library/envdocs/reports/crs/crsul.html, accessed July 27, 2015.

A number of conclusions can be drawn by comparing the figures in these tables, but the ultimate message from the comparison again concerns hydrological realism. For sake of brevity, three points should be highlighted. First, if future Lee Ferry flows average 13.7 MAF annually, all of the Upper Basin states appear to have some cushion for additional consumptive use in light of the 2012 figures. Only in Utah and Wyoming, however, does this cushion appear adequate, partly or wholly, for consumptive use at levels comparable to the 2035 projections. Second, Wyoming is the only Upper Basin state whose cushion appears sufficient to enable consumptive use to increase to levels on par with the 2060 projections—albeit only through the lower end of the projected range. Third, if future Lee Ferry flows average 12.7 MAF annually, two of the Upper Basin states, Colorado and Utah, apparently would need to decrease their consumptive use to levels below the 2012 figures. Further, New Mexico and Wyoming would be unable, in this circumstance, to increase their consumptive use to levels contemplated by either the 2035 or 2060 projections. In sum, although the gravity of this issue varies from state to state, the projected Lee Ferry flow reductions throw into question how much additional consumptive use, if any, actually will be possible given the Upper Basin states' water budgets (CRRG 2014).

A natural question arises from the quantitative perspective on future Upper Basin water availability offered here. Which demand management or supply augmentation measures should the Upper Basin states employ as they continue to consumptively use water afforded by their compact entitlements? Formulation and implementation of such measures is a necessity given the Upper Basin's position under the Colorado River Compact of creating a future from leftovers. One notable measure submitted for consideration in conjunction with the Basin Study is an interstate water bank within the Upper Basin (US Bureau of Reclamation 2012j; CRGI 2013). It would empower water users in Upper Basin states to use their entitlements in an efficient and flexible manner. They could choose to lease portions of these entitlements via voluntary exchanges in whatever way might make sense given their goals as lessors and lessees. A number of design-related issues would need to be ironed out for this water bank to take shape, however, including avoiding potential legal obstacles posed by the Upper Basin Compact, which was not drafted with interstate water banking in mind (Getches 1997). One potential legal issue concerns a prohibition in Article III of the Upper Basin Compact against "countenancing average uses by any signatory State in excess of its apportionment." This prohibition might be triggered if banking activities in an Upper Basin state were to cause consumptive use levels to exceed that state's entitlement on a recurrent basis. This detail and many others would need to be navigated, but the design of an Upper Basin water bank is a noteworthy aspect of the policy frontier.

16.3.3 Lower Basin Apportionment

As beneficiary of one of the two flow obligations imposed by the Colorado River Compact, the Lower Basin's situation with regard to consumptive use of Colorado River system water may initially appear distinct from that of the Upper Basin. While anchored in the legal framework, however, this distinction obfuscates a host of similarities that exist concerning allocation issues posed by climate change for the two subbasins. Closely resembling the state of affairs in the Upper Basin, the Lower Basin states currently face major challenges involving reconciling their existing and projected water demands with limitations imposed by the Compact's apportionment scheme—particularly, constraints on mainstream supplies in the Lower Basin extending from the Compact flow obligations. Other notable priorities in this realm of the Law of the River include the need for technical studies of water uses and losses along the Lower Basin tributaries and the potential expansion of existing interstate transfer schemes for mainstream water.

Regulating consumptive uses and losses of water from the Colorado River mainstream in the Lower Basin is the decree issued by the supreme court in *Arizona v. California*. This decree establishes a *sliding-scale* apportionment scheme whose features reflect the court's interpretation of the Boulder Canyon Project Act in this landmark case. Of critical importance among these features is the central role played by the secretary of the interior in determining the amount of water available for consumptive use along the Lower Colorado River annually. If the secretary determines that 7.5 MAF of consumptive use is possible, this amount is to be apportioned 4.4 MAF to California, 2.8 MAF to Arizona, and 0.3 MAF to Nevada, and "normal conditions" are said to exist. Conversely, if the secretary determines that more than 7.5 MAF of consumptive use can be made, such surplus is to be apportioned 50% to California, 46% to Arizona, and 4% to Nevada. Finally, in the event of shortage conditions, wherein less than 7.5 MAF of consumptive use is possible in a given year, the secretary has discretion to apportion this consumptive use among the Lower Basin states subject to an order of priority for present perfected rights that exist within these states and a 4.4 MAF limitation on California's consumptive use. The scope of the Lower Basin states' entitlements under the decree thus hinges on an annual secretarial determination of available mainstream supplies. The secretary takes into account reservoir evaporation and other losses when making this determination; however, these losses are *not* counted against the states' entitlements as consumptive use. Nor does the decree regulate consumptive uses and losses along tributaries to the Lower Colorado River, with the lone exception of New Mexico's consumptive use from the Gila River. The decree notably treats "water drawn from the mainstream by underground pumping" as consumptive use.

Extensive reliance interests and future expectations attach to the mainstream water governed by the *Arizona v. California* decree. A few figures from the Basin Study illustrate this point. Consumptive uses and losses along the Lower Colorado River were at least 8.9 MAF in 2010, consisting of approximately 7.4 MAF of consumptive uses and 1.5 MAF of reservoir evaporation and phreatophyte losses (US Bureau of Reclamation 2012c). Even more broad-based expectations are projected for the Lower Colorado River going forward. The

Basin Study does not include figures for projected losses, but annual consumptive uses alone are anticipated to approach 8.5 MAF in several scenarios by 2035 and to range from roughly 8.5–10.0 MAF by 2060 (US Bureau of Reclamation 2012c). In short, the Lower Basin states are heavily dependent on mainstream water and projected to become even more so.

Projected reductions in Lee Ferry flows due to climate change cast doubt on the extent to which the Lower Basin states will be able to maintain existing reliance interests or to realize additional expectations associated with the Lower Colorado River. The amount of mainstream water available for consumptive use in the Lower Basin under the *Arizona v. California* decree—again, as determined annually by the secretary of the interior—is primarily a function of the volume of Lee Ferry flows supplied by the Upper Basin as releases from Lake Powell in fulfillment of the Compact flow obligations. Although tributary inflows into the Lower Colorado River and holdover storage in Lake Mead and other mainstream reservoirs also influence these supplies, it is primarily the Lee Ferry flows sent downstream from the Upper Basin as Lake Powell releases that control annual mainstream water availability within the Lower Basin. Ultimately, as Lee Ferry flows decrease and Upper Basin consumptive uses increase in line with the projections discussed earlier, it appears inevitable that the amount of water supplied by the Upper Basin extending from its flow obligations will approach the minimum required by the Compact. This trajectory gives rise to two questions of much importance vis-à-vis the existing reliance interests and future expectations highlighted previously: (1) what exactly is the minimum amount of Lee Ferry flows the Upper Basin is obligated to supply under the Compact, and (2) precisely how much mainstream consumptive use can this minimum sustain within the Lower Basin?

Tables 16.3 and 16.4 contain some thought-provoking figures in relation to these questions. It should be clarified at the outset that, although the supreme court has never interpreted the Colorado River Compact, the secretary of the interior historically has maintained a minimum objective annual release of 8.23 MAF from Lake Powell based on the Compact flow obligations. Most commentators view this release as consisting of 7.50 MAF per Article III(d), 750,000 acre-feet per Article III(c), and a 20,000 acre-foot deduction for Paria River inflows below Glen Canyon Dam (Schiffer et al. 2007). Assuming for sake of analysis that Lake Powell releases will average at least 8.23 MAF per year in the future to comply with the Compact flow obligations—i.e., despite the fact that the Interim Guidelines call for annual releases of *less* than 8.23 MAF in certain circumstances (as a successor instrument also might do)—the two tables aim to assess how much mainstream consumptive use average annual Lake Powell releases of 8.23 MAF might enable within the Lower Basin. Table 16.3 contains figures for the 15-year period from 2000 to 2014, including rounded averages, for annual Lake Powell releases, Lake Mead inflows, Lake Mead releases (including treaty flow releases), and Lower Basin mainstream consumptive use. In turn, Table 16.4 tracks the decline in Lake Mead's storage over this period. Notable figures apparent from this latter table include (1) an overall decline of 12.344 MAF in Lake Mead's live storage, (2) a corresponding decline in Lake Mead's storage from 82% to 39% of capacity, and (3) an average annual rate of decline in Lake Mead's storage of 963,900 acre-feet.

As with the Upper Basin figures in Tables 16.1 and 16.2, a number of conclusions can be drawn from Tables 16.3 and 16.4 regarding how much mainstream consumptive use might be possible in the Lower Basin if annual releases from Lake Powell average 8.23 MAF in the future. Three points are especially worth considering.

First, average annual releases from Lake Powell will need to be greater than 8.23 MAF for the Lower Basin states to consumptively use 7.5 MAF of mainstream water—again, their normal entitlement under the *Arizona v. California* decree—on average per year. Table

TABLE 16.3

Lake Powell Releases, Lake Mead Inflows and Releases, and Lower Basin Mainstream Consumptive Use, 2000–2014

Year	Lake Powell Release	Lake Mead Inflow	Lake Mead Release	Lower Basin Mainstream Consumptive Use
2000	9.40 MAF	n/a	10.993 MAF	8.025 MAF
2001	8.23 MAF	n/a	10.492 MAF	8.17 MAF
2002	8.23 MAF	n/a	10.50 MAF	8.407 MAF
2003	8.23 MAF	n/a	9.46 MAF	7.538 MAF
2004	8.23 MAF	n/a	9.635 MAF	7.384 MAF
2005	8.23 MAF	10.07 MAF	7.941 MAF	7.065 MAF
2006	8.23 MAF	8.931 MAF	9.395 MAF	7.411 MAF
2007	8.231 MAF	8.906 MAF	9.452 MAF	7.454 MAF
2008	8.978 MAF	9.894 MAF	9.531 MAF	7.521 MAF
2009	8.23 MAF	8.888 MAF	9.211 MAF	7.438 MAF
2010	8.23 MAF	9.164 MAF	9.26 MAF	7.379 MAF
2011	12.52 MAF	13.68 MAF	9.80 MAF	7.317 MAF
2012	9.47 MAF	10.10 MAF	9.42 MAF	7.444 MAF
2013	8.23 MAF	9.06 MAF	9.04 MAF	7.478 MAF
2014	7.48 MAF	8.16 MAF	9.76 MAF	7.571 MAF
Average	8.677 MAF	9.685 MAF	9.593 MAF	7.573 MAF

Source: Data from US Bureau of Reclamation, Colorado River Basin Water Supply and Demand Study, Technical Report C—Water Demand Assessment, 2012, http://www.usbr.gov/lc/region/programs/crbstudy/finalreport/Technical%20Report%20C%20-%20Water%20Demand%20Assessment/TR-C-Water_Demand_Assessmemt_FINAL.pdf, accessed July 27, 2015; US Bureau of Reclamation, Upper Colorado Region, Water Resources Group, Colorado River Basin Annual Operating Plans, 2015, http://www.usbr.gov/uc/water/rsvrs/ops/aop/, accessed July 27, 2015; US Bureau of Reclamation, Lower Colorado Region, Lower Colorado River Water Accounting, 2015, http://www.usbr.gov/lc/region/g4000/wtracct.html, accessed July 27, 2015.

16.3 reveals that average annual Lake Powell releases of approximately 8.68 MAF were sufficient to enable average annual mainstream consumptive use of approximately 7.57 MAF in the Lower Basin from 2000 to 2014. Table 16.4, however, demonstrates that Lake Mead's storage contemporaneously declined at an average annual rate of 963,900 acre-feet. This analysis will not delve further into the precise volume of average annual Lake Powell releases needed to facilitate 7.5 MAF of average annual consumptive use from the Lower Colorado River. Whatever this mark may be, however, the 8.23 MAF historical minimum release plainly falls short of it, creating what has been appropriately described as a "structural deficit" (CRRG 2014, p. 3). That is, even if Lake Powell releases were to average 8.23 MAF in coming years, these figures suggest the Lower Basin states would nonetheless face chronic shortages under the *Arizona v. California* decree.

Second, as a corollary to the preceding point, the Basin Study's full range of annual projections for Lower Basin mainstream consumptive use identified previously—i.e., 8.0–8.5 MAF by 2035, 8.5–10.0 MAF by 2060—cannot be reconciled with the prospect of average annual releases of 8.23 MAF from Lake Powell in the future. It again goes beyond the scope of this discussion to evaluate precisely what volumes of average annual Lake Powell releases might be necessary to facilitate these projected levels of consumptive use. If average annual Lake Powell releases of 8.23 MAF will be clearly inadequate to satisfy the Lower Basin states' 7.5 MAF normal decree entitlement, however, as suggested by Tables

TABLE 16.4
Lake Mead Storage, 2000–2014

Year	Live Storage	Percent of Capacity	Change in Storage
2000	22.444 MAF	82%	−2.148 MAF
2001	19.873 MAF	73%	−2.571 MAF
2002	17.093 MAF	66%	−2.769 MAF
2003	15.618 MAF	60%	−1.475 MAF
2004	13.937 MAF	54%	−1.681 MAF
2005	15.22 MAF	59%	+1.282 MAF
2006	13.89 MAF	54%	−1.332 MAF
2007	12.51 MAF	48%	−1.382 MAF
2008	12.01 MAF	46%	−0.492 MAF
2009	10.933 MAF	42%	−1.080 MAF
2010	10.1 MAF	39%	−0.841 MAF
2011	13.0 MAF	50%	+2.885 MAF
2012	13.1 MAF	50%	+0.158 MAF
2013	12.4 MAF	47%	−0.773 MAF
2014	10.1 MAF	39%	−2.24 MAF

Source: Data from US Bureau of Reclamation, Upper Colorado Region, Water Resources Group, Colorado River Basin Annual Operating Plans, 2015, http://www.usbr.gov/uc/water/rsvrs/ops/aop/, accessed July 27, 2015.

16.3 and 16.4, such releases obviously will not enable mainstream consumptive use at annual levels of 500,000 acre-feet to 2.5 MAF higher.

Third, the viability of existing reliance interests and future expectations attached to the Lower Colorado River would be cast further into doubt, of course, if average annual Lake Powell releases were to drop below 8.23 MAF in the future. As reflected in Table 16.3, 2014 was the first year since Lake Powell's filling in the 1960s that the annual release fell to such a level—7.48 MAF pursuant to the Interim Guidelines (Bureau of Reclamation 2014, p. 3). As a matter of law, the prospect of a recurring pattern of this type hinges at bottom on how the Compact flow obligations are navigated. Interpretive conflicts surround these obligations—particularly, Article III(c)—as highlighted earlier. Ultimately, to the extent that the Compact *might* be construed (or at least administered) as permitting average annual Lake Powell releases of less than 8.23 MAF, the resulting diminution in Lee Ferry flows would proportionately reduce the amount of mainstream consumptive use possible within the Lower Basin under the *Arizona v. California* decree.

Overlapping with the preceding material regarding future mainstream water availability in the Lower Basin are two related policy and management priorities that should be noted before turning to the final section on tribal water rights. First, as has been suggested throughout this chapter, future technical studies of consumptive uses and reservoir evaporation and other losses along the Lower Basin tributaries need to be undertaken. Underlying this priority is this author's admittedly personal view that the Colorado River Compact indeed governs these uses and losses, and that the existence of accurate figures for them is essential to diligent administration of the Compact's apportionment scheme, including Article III(c)'s flow obligation. It is respectfully frustrating to resort in 2015 to use and loss figures from 2005 to convey a "current" sense of hydrological conditions along the Lower Basin tributaries. Second, as in the Upper Basin, a suite of novel demand management and supply augmentation measures assuredly will emerge in the Lower Basin in coming decades. Notable among these measures is the potential expansion of existing programs aimed at enabling efficient and flexible water use by entitlement holders to mainstream water—specifically, a water banking program established in 1999 via federal regulations and an intentionally created surplus (ICS) program adopted under the Interim Guidelines (US Bureau of Reclamation 2012j; CRGI 2013). Both programs are framed around the unique provisions of the *Arizona v. California* decree—in particular, Article II(B)(6)—and it is worth considering to what extent they might be suited for conceptually and geographically expanded offshoots.

16.3.4 Tribal Water Rights

Climate change also holds important implications for water rights held by American Indian tribes residing on 28 reservations located throughout the Colorado River Basin. Figure 16.7 depicts these reservations. As noted, they include the two largest in the United States: the Navajo Nation in the Four Corners region and the Uintah and Ouray Reservation in eastern Utah. Stated broadly, the overarching challenge facing these tribes with regard to their water rights involves translating what are legal rights on paper ("paper rights") into physical water supplies that can be utilized to create viable homelands ("wet water"). Because of the prevalent senior status of tribal water rights—which carry a higher allocation priority vis-à-vis junior water rights under the basin states' prior appropriation systems—this translation has the potential to profoundly affect non-Indian junior water users. Significant issues in the realm thus include the need for quantification of as-yet unquantified tribal water rights, additional funding for tribal water infrastructure, and full integration of tribal water rights into planning and management efforts aimed at assessing water availability and promoting equitable reallocation measures.

American Indian tribes hold quantified water rights to Colorado River system water under two types of legal instruments: adjudication decrees and negotiated settlements. Underpinning the adjudicated rights are the seminal supreme court decisions of *Winters* and *Arizona v. California,* the latter of which established and quantified water rights held by five tribes within the Lower Basin to water from the Colorado River mainstream (Cordalis and Cordalis 2015). In lieu of adjudications, 14 tribes in the basin up to this point have elected to secure and to quantify their water rights by entering into negotiated settlements, the earliest in 1978 and the most recent in 2010 (NARF 2013). Table 16.5 highlights 24 tribes identified in the Basin Study as holding quantified rights in one of these two forms (US Bureau of Reclamation 2012c). For accounting purposes under the Upper Basin Compact and the *Arizona v. California* decree, consumptive uses undertaken in conjunction with these tribes' water rights must be charged against the respective entitlement(s) of the basin state(s) in which the uses occur.

The total amount of consumptive use authorized by these tribes' quantified rights is substantial. Overall, quantified tribal diversion rights throughout the basin are about 2.9 MAF annually, with approximately 1.36 MAF in the Upper Basin and 1.58 MAF in the Lower Basin, including 1.4 MAF in Arizona (US Bureau of Reclamation 2012c). The Navajo Nation alone holds diversion and depletion rights to 606,660 and 325,670 acre-feet, respectively, of water from the San Juan River in New Mexico (US Bureau of Reclamation 2012h). Similarly, the Ute Indian tribe of the Uintah and Ouray Reservation holds diversion and depletion rights to 480,594 and 258,943 acre-feet, respectively, of Upper Basin water pursuant to a revised compact with the state of Utah currently pending reratification (US Bureau of Reclamation 2012h). A host of additional examples exist to illustrate the large volume of tribal water rights in the basin.

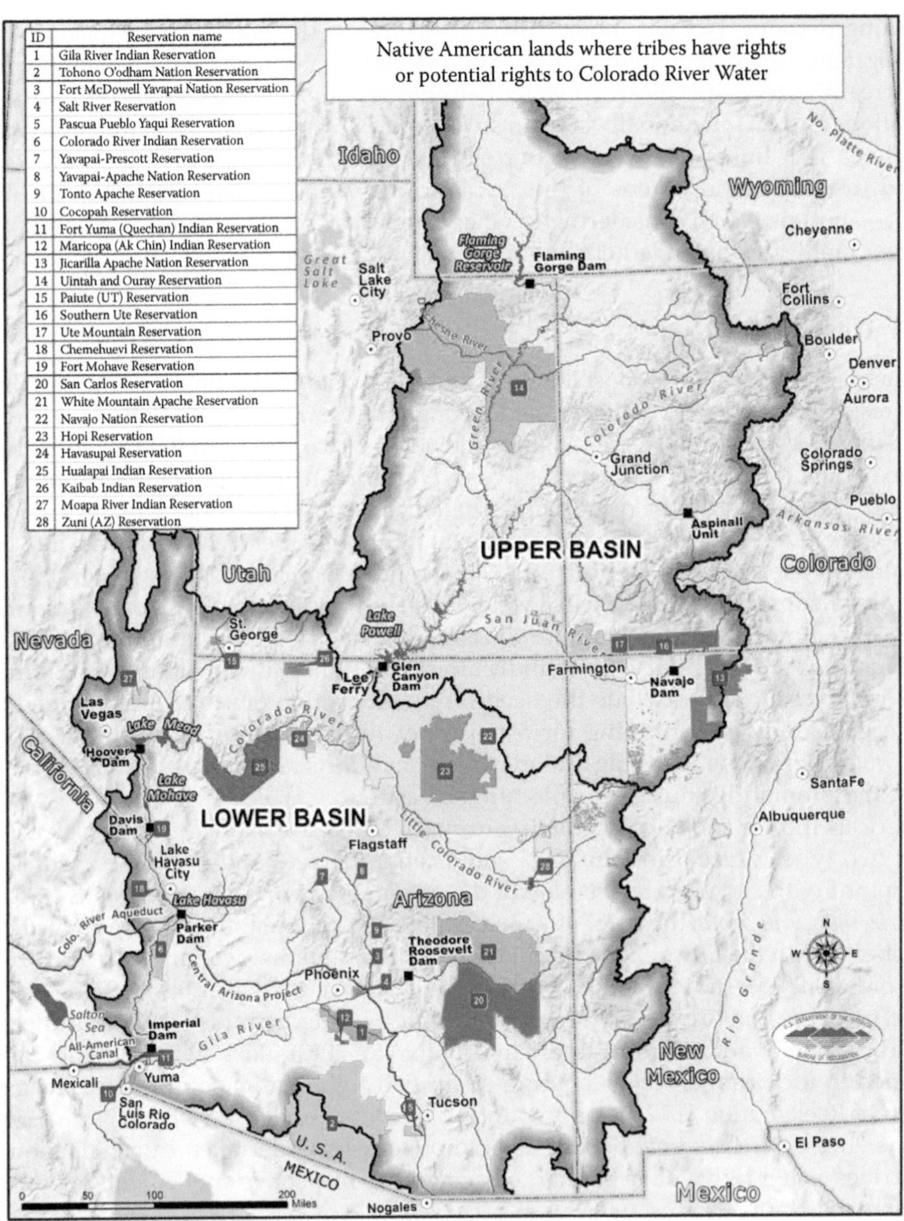

ID	Reservation name
1	Gila River Indian Reservation
2	Tohono O'odham Nation Reservation
3	Fort McDowell Yavapai Nation Reservation
4	Salt River Reservation
5	Pascua Pueblo Yaqui Reservation
6	Colorado River Indian Reservation
7	Yavapai-Prescott Reservation
8	Yavapai-Apache Nation Reservation
9	Tonto Apache Reservation
10	Cocopah Reservation
11	Fort Yuma (Quechan) Indian Reservation
12	Maricopa (Ak Chin) Indian Reservation
13	Jicarilla Apache Nation Reservation
14	Uintah and Ouray Reservation
15	Paiute (UT) Reservation
16	Southern Ute Reservation
17	Ute Mountain Reservation
18	Chemehuevi Reservation
19	Fort Mohave Reservation
20	San Carlos Reservation
21	White Mountain Apache Reservation
22	Navajo Nation Reservation
23	Hopi Reservation
24	Havasupai Reservation
25	Hualapai Indian Reservation
26	Kaibab Indian Reservation
27	Moapa River Indian Reservation
28	Zuni (AZ) Reservation

FIGURE 16.7
Tribal lands in the Colorado River Basin.

Basin tribes do not anticipate idling their water rights over the next 50 years—quite the opposite. The Basin Study projects basin-wide tribal diversions to range from approximately 2.7 to 3.4 MAF by 2060 (US Bureau of Reclamation 2012c). These projected diversions fall between 1.2 and 1.7 MAF in the Upper Basin and 1.5 and 1.7 MAF in the Lower Basin. Again focusing solely on the Navajo Nation, it anticipates making full use of its 606,660 acre-foot diversion right and 325,670 acre-foot depletion right to water from the San Juan River in New Mexico at this time. The same can be said for the plans of the Ute

TABLE 16.5

Tribes with Quantified Rights to Colorado River System Water

Tribe	Reservation Location
Ak-Chin Indian Community	Arizona
Chemhuevi Indian tribe	California
Cocopah Indian tribe	Arizona
Colorado River Indian tribes	Arizona, California
Fort McDowell Yavapai Nation	Arizona
Fort Mojave Indian tribe	Arizona, Nevada, California
Gila River Indian Community	Arizona
Hopi tribe	Arizona
Jicarilla Apache Nation	New Mexico
Moapa Band of Paiutes	Nevada
Navajo Nation	Arizona, New Mexico, Utah
Pascua Yaqui tribe	Arizona
Quechan Indian tribe	Arizona, California
Salt River Pima–Maricopa Indian Community	Arizona
San Carlos Apache tribe	Arizona
Southern Ute Indian tribe	Colorado
Tohono O'odham Nation	Arizona
Tonto Apache tribe	Arizona
Ute Indian tribe of the Uintah and Ouray Reservation	Utah
Ute Mountain Ute tribe	Colorado, New Mexico, Utah
White Mountain Apache tribe	Arizona
Yavapai–Apache Nation	Arizona
Yavapai-Prescott tribe	Arizona
Zuni Indian tribe	Arizona

Source: US Bureau of Reclamation, Colorado River Basin Water Supply and Demand Study, Technical Report C—Water Demand Assessment, 2012, http://www.usbr.gov/lc/region/programs/crbstudy/finalreport/Technical%20Report%20C%20-%20Water%20Demand%20Assessment/TR-C-Water_Demand_Assessmemt_FINAL.pdf, accessed July 27, 2015.

Indian tribe of the Uintah and Ouray Reservation regarding its 480,594 acre-foot diversion right and 258,943 acre-foot depletion right to Upper Basin water in Utah (US Bureau of Reclamation 2012h). This projected trend in increased tribal demands is a significant basin-wide policy dynamic (CRGI 2013; CRRG 2014).

Further complicating the allocation issues in this realm is the fact that a variety of tribes currently hold *unquantified* water rights to Colorado River system water. As listed in Table 16.6, the Basin Study identifies 12 tribes within this category. The Bureau of Reclamation has acknowledged that water demands associated with these unquantified rights "will be a factor impacting Basin-wide water availability" (US Bureau of Reclamation 2012c). The absence of concrete figures, however, makes it difficult to assess the precise role these rights will play in future water planning and management. That said, the basic notion that the scope of these unquantified rights needs to be ascertained—for the sake of both Indian and non-Indian communities—appears commonsensical (CRGI 2013; Nania et al. 2014; Cordalis and Cordalis 2015).

To synthesize the preceding tribal material, the projected impacts of climate change on water availability in the Colorado River Basin bring to the fore at least three key issues that, although relevant irrespective of climate change, are made more pressing by it. First,

TABLE 16.6

Tribes with Unquantified Rights to Colorado River System Water

Tribe	Reservation Location
Havasupai tribe	Arizona
Hopi tribe	Arizona
Hualapai tribe	Arizona
Kaibab Band of Paiute Indians	Arizona
Navajo Nation	Arizona, New Mexico, Utah
Pascua Yaqui tribe	Arizona
San Carlos Apache tribe	Arizona
San Juan Southern Paiute tribe	Arizona
Tohono O'odham Nation	Arizona
Tonto Apache tribe	Arizona
Ute Mountain Ute tribe	Colorado, New Mexico, Utah
Yavapai–Apache Nation	Arizona

Source: US Bureau of Reclamation, Colorado River Basin Water Supply and Demand Study, Technical Report C—Water Demand Assessment, 2012, http://www.usbr.gov/lc/region/programs/crbstudy/finalreport/Technical%20Report%20C%20-%20Water%20Demand%20Assessment/TR-C-Water_Demand_Assessmemt_FINAL.pdf, accessed July 27, 2015.

ongoing efforts to quantify as-yet unquantified tribal water rights appear critical given water supply and demand conditions in the basin and the derivative need for proactive water planning and management. Second, related efforts to secure funding for tribal water infrastructure—which historically has been included in negotiated settlements in various forms—likewise need to be prioritized to enable tribes to realize developmental goals. Third, prospective water planning and management efforts need to account for tribal water rights and associated water demands in a manner that is on par with the treatment afforded non-Indian water users and their entitlements. In particular, because of the often senior status of tribal water rights, close attention needs to be paid in these efforts to potential impacts of increased tribal water use on non-Indian junior water rights holders that historically have relied on water resources to be newly used by tribes. Tribal water marketing arrangements may be useful mechanisms in this regard—several variations of which were submitted for consideration in the Basin Study (US Bureau of Reclamation 2012j; CRGI 2013). At the same time, however, tribes should carefully assess whether leasing provisions in negotiated settlements may unduly restrict more broad-based (and remunerative) marketing activities potentially made possible by future interstate schemes (McCool 2002).

16.4 Conclusion

Allocation institutions in the Colorado River Basin face a historically unprecedented imbalance between water supplies and demands that they contributed mightily to facilitate. They again constitute only one strand of the intricate framework that is the Law of the River, but the issues posed by climate change for them are monumental, both for the institutions themselves as well as for the basin that the Law of the River shapes. Will the Colorado River Compact be capable of fulfilling its core purpose of equitable apportionment given projected climate-based reductions in Lee Ferry flows? Much of the answer seems to hinge on how the Compact is interpreted and which approaches are utilized toward this end. Construed in one form or another, the Compact's framework inevitably will confine the Upper and Lower Basin states to their respective entitlements under the Upper Basin Compact and the *Arizona v. California* decree. What kind of future can the Upper Basin states make with their leftovers, and exactly how much of a cushion actually exists in their water budgets? Climate change poses similarly pressing questions for the Lower Basin, given the role played by the Compact flow obligations in channeling water to the Lower Colorado River, yet on a far grander scale due to the scope of reliance interests and future expectations. Potential evolution of interstate transfer schemes in both contexts will be an important pattern to watch. So, too, will the future treatment of quantified and unquantified water rights held by tribal sovereigns throughout the basin. Climate change accentuates the need to address these legal rights in earnest in future planning and management efforts, and to craft equitable allocation arrangements that enable tribes to create viable homelands. Whether the existing allocation institutions maintain their current seams, bend marginally at these seams as hydrological conditions tighten, or potentially reweave the seams fundamentally is perhaps the essential question posed by climate change within the Colorado River Basin.

Acknowledgments

Funding for this chapter was generously provided by the George Hopper and Carl M. Williams faculty research funds.

References

Adler, R.W. 2008. Revisiting the Colorado River Compact: Time for a Change? *Journal of Land, Resources, & Environmental Law* 28:19–47. Available at http://epubs.utah.edu/index.php/jlrel/article/viewFile/97/87 (accessed July 27, 2015).

Castle, S.L. Thomas, B.F., Reager, J.T., Rodell, M., Swenson, S.C., and Famiglietti, J.S. 2014. Groundwater Depletion During Drought Threatens Future Water Security of the Colorado River Basin. *Geophysical Research Letters* 41:5904–5911. Available at http://onlinelibrary.wiley.com/doi/10.1002/2014GL061055/pdf (accessed July 27, 2015).

Colorado River Governance Initiative (CRGI). 2010. Rethinking the Future of the Colorado River: Draft Interim Report of the Colorado River Governance Initiative. Available at http://www.waterpolicy.info/archives/docs/CRGI-Interim-Report.pdf (accessed July 27, 2015).

Colorado River Governance Initiative (CRGI). 2012a. Does the Upper Basin Have a Delivery Obligation or an Obligation Not to Deplete the Flow of the Colorado River at Lee Ferry? Available at http://www.waterpolicy.info/archives/docs/Delivery%20Obligation%20memo.pdf?p=1693 (accessed July 27, 2015).

Colorado River Governance Initiative (CRGI). 2012b. Respective Obligations of the Upper and Lower Basins Regarding the Delivery of Water to Mexico: A Review of Key Legal Issues. Available at http://www.waterpolicy.info/archives/docs/Obligations%20Regarding%20the%20Delivery%20of%20Water%20to%20Mexico.pdf?p=1689 (accessed July 27, 2015).

Colorado River Governance Initiative (CRGI). 2013. Cross-Boundary Water Transfers in the Colorado River Basin: A Review of Efforts and Issues Associated with Marketing Water Across State Lines or Reservation Boundaries. Available at http://www.waterpolicy.info/docs/CrossJurisdictionalWaterMarketingCRBJune2013.pdf (accessed July 27, 2015).

Colorado River Research Group (CRRG). 2014. The First Step in Repairing the Colorado River's Water Budget: Technical Report. Available at http://www.coloradoriverresearchgroup.org/uploads/4/2/3/6/42362959/crrg_technical_report_1_updated2.pdf (accessed July 27, 2015).

Connor, M.L. 2013. Statement of Michael L. Connor, Commissioner, U.S. Department of the Interior, Before the Energy and Natural Resources Committee Subcommittee on Water and Power, U.S. Senate, on Colorado River Basin Water Supply and Demand Study. Available at http://www.usbr.gov/newsroom/testimony/detail.cfm?RecordID=2421 (accessed December 11, 2015).

Cordalis, A. and Cordalis, D. 2015. Indian Water Rights: How *Arizona v. California* Left An Unwanted Cloud Over the Colorado River Basin, *Arizona Journal of Environmental Law & Policy* 5:333–362. Available at http://www.ajelp.com/wp-content/uploads/Cordalis_final.pdf (accessed July 27, 2015).

Getches, D.H. 1985. Competing Demands for the Colorado River, *University of Colorado Law Review* 56:413–479. http://digitool.library.colostate.edu///exlibris/dtl/d3_1/apache_media/L2V4bGlicmlzL2R0bC9kM18xL2FwYWNoZV9tZWRpYS8xMjUwNjc=.pdf (accessed July 27, 2015).

Getches, D.H. 1997. Colorado River Governance: Sharing Federal Authority as an Incentive to Create a New Institution. *University of Colorado Law Review* 68:573–658. Available at http://digitool.library.colostate.edu///exlibris/dtl/d3_1/apache_media/L2V4bGlicmlzL2R0bC9kM18xL2FwYWNoZV9tZWRpYS8xMjUxMjM=.pdf (accessed July 27, 2015).

Hundley, N. 2009. *Water and the West: The Colorado River Compact and the Politics of Water in the American West*. Berkeley and Los Angeles: University of California Press.

MacDonnell, L.J. 2009. Colorado River Basin. In *Water and Water Rights*, ed. A.K. Kelley, 5–54. New York: LexisNexis.

McCool, D.C. 2002. *Native Waters: Contemporary Indian Water Settlements and the Second Treaty Era*. Tucson: University of Arizona Press.

Nania, J., Cozzeto, K., and Tapp, A.M. 2014. Chapter 4—Water Resources. In *Considerations for Climate Change and Variability Adaptation on the Navajo Nation*. Available at http://www.colorado.edu/law/sites/default/files/Considerations%20For%20Climate%20Change%20and%20Variability%20Adaptation%20on%20the%20Navajo%20Nation.vf_.pdf (accessed July 27, 2015).

Native American Rights Fund (NARF). 2013. Settlements Approved by Congress. Available at http://www.indian.senate.gov/sites/default/files/upload/files/5.20.15%20Witness%20Testimony%20-%20Steven%20Moore.pdf (accessed July 27, 2015).

Robison, J.A. and Kenney D.S. 2012. Equity and the Colorado River Compact. *Environmental Law* 42:1157–1209. Available at http://papers.ssrn.com/sol3/papers.cfm?abstract_id=2137761 (accessed July 27, 2015).

Schiffer, P.S., Guenther, H., and Carr, T.G. 2007. From a Colorado River Compact Challenge to the Next Era of Cooperation among the Seven Basin States. *Arizona Law Review* 49:217–233. Available at http://arizonalawreview.org/pdf/49-2/49arizlrev217.pdf (accessed July 27, 2015).

US Bureau of Reclamation. 2011. SECURE Water Act Section 9503(c)—Reclamation Climate Change and Water 2011. Available at http://www.usbr.gov/climate/SECURE/docs/SECUREWaterReport.pdf (accessed July 27, 2015).

US Bureau of Reclamation. 2012a. Colorado River Basin Water Supply and Demand Study, Study Report. Available at http://www.usbr.gov/lc/region/programs/crbstudy/finalreport/Study%20Report/CRBS_Study_Report_FINAL.pdf (accessed July 27, 2015).

US Bureau of Reclamation. 2012b. Colorado River Basin Water Supply and Demand Study, Technical Report B—Water Supply Assessment. Available at http://www.usbr.gov/lc/region/programs/crbstudy/finalreport/Technical%20Report%20B%20-%20Water%20Supply%20Assessment/TR-B_Water_Supply_Assessment_FINAL.pdf (accessed July 27, 2015).

US Bureau of Reclamation. 2012c. Colorado River Basin Water Supply and Demand Study, Technical Report C—Water Demand Assessment. Available at http://www.usbr.gov/lc/region/programs/crbstudy/finalreport/Technical%20Report%20C%20-%20Water%20Demand%20Assessment/TR-C-Water_Demand_Assessmemt_FINAL.pdf (accessed July 27, 2015).

US Bureau of Reclamation. 2012d. Colorado River Basin Water Supply and Demand Study, Technical Report C—Water Demand Assessment, Appendix C2 Colorado Water Demand Scenario Quantification. Available at http://www.usbr.gov/lc/region/programs/crbstudy/finalreport/Technical%20Report%20C%20-%20Water%20Demand%20Assessment/TR-C_Appendix2_FINAL.pdf (accessed July 27, 2015).

US Bureau of Reclamation. 2012e. Colorado River Basin Water Supply and Demand Study, Technical Report C—Water Demand Assessment, Appendix C3 New Mexico Water Demand Scenario Quantification. Available at http://www.usbr.gov/lc/region/programs/crbstudy/finalreport/Technical%20Report%20C%20-%20Water%20Demand%20Assessment/TR-C_Appendix3_FINAL.pdf (accessed July 27, 2015).

US Bureau of Reclamation. 2012f. Colorado River Basin Water Supply and Demand Study, Technical Report C—Water Demand Assessment, Appendix C4 Utah Water Demand Scenario Quantification. Available at http://www.usbr.gov/lc/region/programs/crbstudy/finalreport/Technical%20Report%20C%20-%20Water%20Demand%20Assessment/TR-C_Appendix4_FINAL.pdf (accessed July 27, 2015).

US Bureau of Reclamation. 2012g. Colorado River Basin Water Supply and Demand Study, Technical Report C—Water Demand Assessment, Appendix C5 Wyoming Water Demand Scenario Quantification. Available at http://www.usbr.gov/lc/region/programs/crbstudy/finalreport/Technical%20Report%20C%20-%20Water%20Demand%20Assessment/TR-C_Appendix5_FINAL.pdf (accessed July 27, 2015).

US Bureau of Reclamation. 2012h. Colorado River Basin Water Supply and Demand Study, Technical Report C—Water Demand Assessment, Appendix C9 Tribal Water Demand Scenario Quantification. Available at http://www.usbr.gov/lc/region/programs/crbstudy/finalreport/Technical%20Report%20C%20-%20Water%20Demand%20Assessment/TR-C_Appendix9_FINAL.pdf (accessed July 27, 2015).

US Bureau of Reclamation. 2012i. Colorado River Basin Water Supply and Demand Study, Technical Report C—Water Demand Assessment, Appendix C11 Modeling of Lower Basin Tributaries in Colorado River Simulation System. Available at http://www.usbr.gov/lc/region/programs/crbstudy/finalreport/Technical%20Report%20C%20-%20Water%20Demand%20Assessment/TR-C_Appendix11_FINAL.pdf (accessed July 27, 2015).

US Bureau of Reclamation. 2012j. Colorado River Basin Water Supply and Demand Study, Technical Report F—Development of Options and Strategies. Available at http://www.usbr.gov/lc/region/programs/crbstudy/finalreport/Technical%20Report%20F%20-%20Development%20of%20Options%20and%20Stategies/TR-F_Development_of_Ops&Strats_FINAL.pdf (accessed July 27, 2015).

US Bureau of Reclamation. 2013. Literature Synthesis on Climate Change, Implications for Reclamation's Water Resources. Available at http://www.usbr.gov/climate/docs/ClimateChangeLiteratureSynthesis3.pdf (accessed July 27, 2015).

US Bureau of Reclamation. 2014. *2015 Lake Powell Water Release to Lake Mead Will Increase*. Available at http://www.usbr.gov/newsroom/newsrelease/detail.cfm?RecordID=47753 (accessed July 27, 2015).

US Bureau of Reclamation. 2015a. Colorado River System Consumptive Uses and Losses Reports. Available at http://www.usbr.gov/uc/library/envdocs/reports/crs/crsul.html (accessed July 27, 2015).

US Bureau of Reclamation. 2015b. Upper Colorado Region, Water Resources Group, Colorado River Basin Annual Operating Plans. Available at http://www.usbr.gov/uc/water/rsvrs/ops/aop/(accessed July 27, 2015).

US Bureau of Reclamation. 2015c. Lower Colorado Region, Lower Colorado River Water Accounting. Available at http://www.usbr.gov/lc/region/g4000/wtracct.html (accessed July 27, 2015).

Vano, J.A. et al. 2014. Understanding Uncertainties in Future Colorado River Streamflow. *Bulletin of the American Meteorological Society* 95:59–78. Available at http://journals.ametsoc.org/doi/pdf/10.1175/BAMS-D-12-00228.1 (accessed July 27, 2015).

17

Using Large-Scale Flow Experiments to Rehabilitate Colorado River Ecosystem Function in Grand Canyon: Basis for an Adaptive Climate-Resilient Strategy

Theodore S. Melis, William E. Pine, III, Josh Korman, Michael D. Yard, Shaleen Jain, and Roger S. Pulwarty

CONTENTS

ABSTRACT Adaptive management of Glen Canyon Dam is improving downstream resources of the Colorado River in Glen Canyon National Recreation Area and Grand Canyon National Park. The Glen Canyon Dam Adaptive Management Program (AMP), a federal advisory committee of 25 members with diverse special interests tasked to advise the US Department of the Interior, was established in 1997 in response to the 1992 Grand Canyon Protection Act. Adaptive management assumes that ecosystem responses to management policies are inherently complex and unpredictable, but that understanding and management can be improved through monitoring. Best known for its high-flow experiments intended to benefit physical and biological resources by simulating one aspect of pre-dam conditions—floods—the AMP promotes collaboration among tribal, recreation, hydropower, environmental, water, and other natural resource management interests. Monitoring has shown that high flow experiments move limited new tributary sand inputs below the dam from the bottom of the Colorado River to shorelines, rebuilding eroded sandbars that support camping areas and other natural and cultural resources. Spring-time high flows have also been shown to stimulate aquatic productivity by disturbing the river bed below the dam in Glen Canyon. Understanding of the responses of nonnative tailwater rainbow trout (*Oncorhynchus mykiss*) and downstream endangered humpback chub (*Gila cypha*) to dam operations has also increased, but this learning has mostly posed *surprise* adaptation opportunities to managers. Since reoperation of the dam to Modified Low Fluctuating Flows in 1996, rainbow trout now benefit from more stable daily flows and high spring releases, but possibly at a risk to humpback chub and other native fishes downstream. In contrast, humpback chub have so far proven robust to all flows, and native fish have increased under the combination of warmer river temperatures associated with reduced storage in Lake Powell and a systemwide reduction in trout from 2000 to 2006, possibly due to several years of natural reproduction under limited food supply. Uncertainties about dam operations and ecosystem responses remain, including how native and nonnative fish will interact and respond to possible increased river temperatures under drier basin conditions. Ongoing assessment of operating policies by the AMP's diverse stakeholders represents a major commitment to the river's valued resources, while surprise learning opportunities can also help identify a resilient climate-change strategy for co-managing nonnative and endangered native fish, sandbar habitats, and other river resources in a region with already complex and ever-increasing water demands.

17.1 Introduction

One of the most prominent ongoing efforts to address environmental impacts using adaptive management (AM) is the Glen Canyon Dam Adaptive Management Program (AMP) (Hamill and Melis 2012). Established in 1997, the AMP was an outgrowth of the Grand Canyon Protection Act of 1992 (GCPA 1992), legislation that provided guidance and legal support to the US secretary of the interior to better operate Glen Canyon Dam (GCD) to protect and improve downstream resources found in Glen Canyon National Recreation Area and Grand Canyon National Park (hereafter, Glen and Grand Canyons; Figure 17.1). Prior to the 1992 law, both the existence and operation of the dam had fueled ongoing debate about tradeoffs centered on societal values associated with *relict* resources of the pre-dam river (native endemic fish, expansive river sandbars, seasonally varied flow, and river temperature), and *artifact* resources (water supply, hydroelectric power, recreational boating, and trout angling) created by the dam and its operation (Schmidt et al. 1998).

One step toward resolving this debate occurred when GCD was re-operated in 1996, following completion of a five-year long environmental impact statement (EIS) (US Department of the Interior 1995). The preferred alternative, termed Modified Low Fluctuating Flows (MLFF, Table 17.1), identified a new operation intended to mitigate downstream river impacts of previously wider ranging daily releases.

The MLFF policy has continued to allow daily hydropeaking releases over a more limited range of flows intended to improve downstream natural and cultural resources, but monitoring revealed that re-operation of GCD achieved mixed results (Lovich and Melis 2007). At the time, AM was identified as a strategy for addressing complex environmental management problems downstream of the dam through scientist–stakeholder collaborations. As its name suggests, AM assumes that ecosystem responses to management actions are highly complex and often unpredictable, but that if policies are monitored appropriately, then they may be evaluated and adaptively modified to improve performance over time. To avoid policy gridlock and enable learning, effective AM leaders also nurture ongoing collaboration among stakeholders, managers, and scientists (Greig et al. 2013).

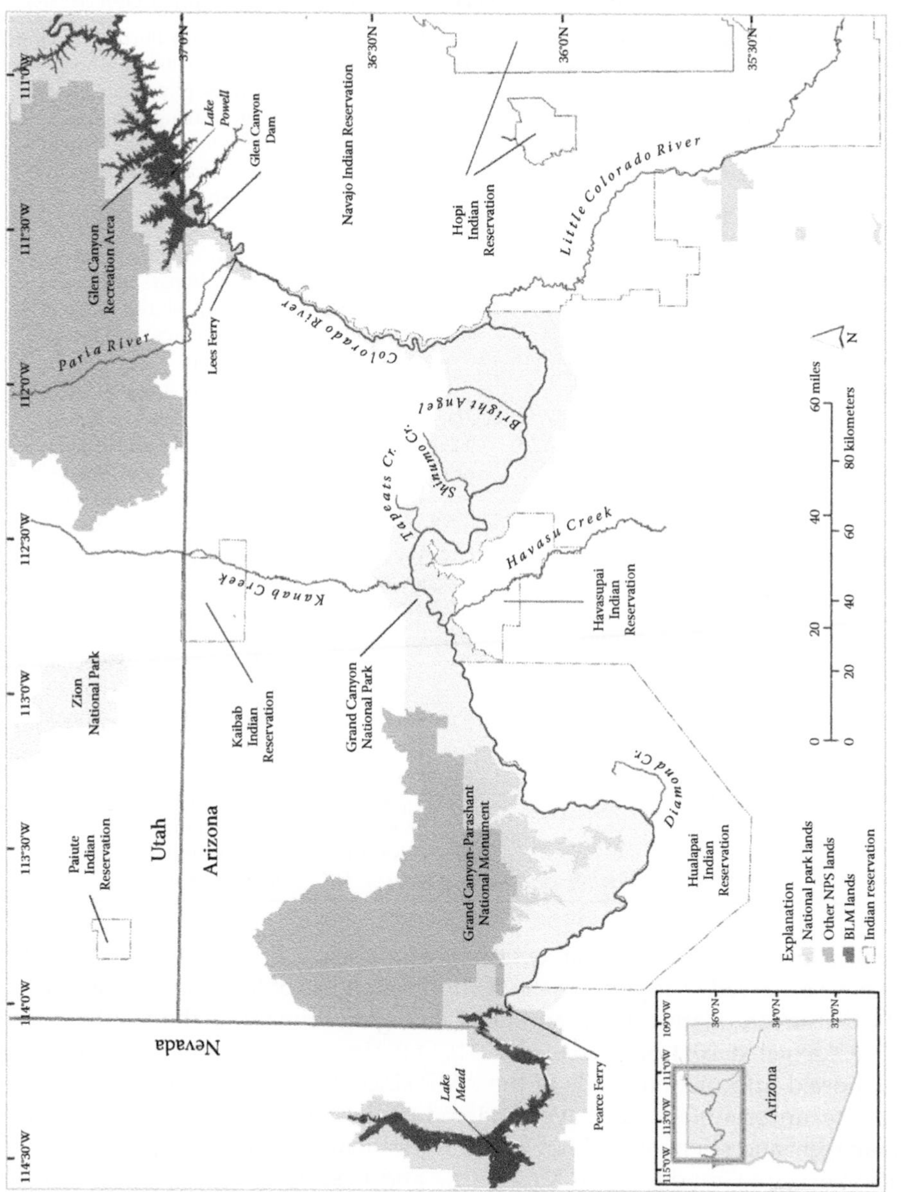

FIGURE 17.1
The Colorado River ecosystem of Glen Canyon National Recreation Area and Grand Canyon National Park. For purposes of the Glen Canyon Dam Adaptive Management Program, the Colorado River ecosystem has been defined as the river segment extending from the forebay of Glen Canyon Dam to the western boundary of Grand Canyon National Park, southwestern United States. Other National Park Service (NPS) and Bureau of Land Management (BLM) areas are also shown.

TABLE 17.1

Ongoing and (or) Previous Flow Experiments Monitored in the Colorado River Ecosystem below Glen Canyon Dam

EXP FLOW Treatment	Seasonal Timing of EXP FLOW	Monitoring Data (Yes/No) and EXP FLOW Frequency	EXP FLOW Treatment Description/Objective(s)
Modified Low Fluctuating Flows (MLFF): 1996 record of decision (preceded by *interim fluctuating flows* with reduced daily ranges, 1991–1996)	**Year-round** with seasonally varied monthly volumes—higher in winter and summer, lower in fall and spring—to follow seasonal peaking energy demands	**Yes** 1991–present	Limited daily fluctuating ranges that vary with monthly release volumes, plus limited hourly hydropeaking ramping rates—to conserve shoreline sandbars and fish and improve navigation
High-flow experiments (HFEs)	**Fall (Oct–Nov) and Spring (Mar–Apr)**—HFEs following summer and winter tributary sand inputs	**Yes** 1996, 1997, 2000 (twice), 2004, 2008, 2012, 2013, and 2014	Max. power plant releases or greater (bypass) for 24–168 h—to rebuild sandbars with tributary sand inputs below dam
High steady flows (HSFs)	**Winter and Spring to Fall**—released for dam safety purposes, and to meet annual downstream water delivery from Upper Colorado River Basin	**Yes** Winter 1997, spring 2000, and late spring–fall 2011	High steady flows in 1997 and 2011 in response to higher upper basin runoff, but also occurred in spring before 2000 low summer steady flows (see below)
Low summer and fall steady flows (LSSF and FSF)	**Summer (Jun–Sep)** and **Fall (Sep–Oct)** timed to coincide with emergence of juvenile chub from tributary spawning habitats	**Yes** 2000 and 2008–2012	Low and steady releases in summer months to warm river and steady flows in fall to stabilize shorelines 125 km below dam where juvenile native humpback chub are found near Little Colorado River (LCR)
Rainbow trout management flows (TMFs)	**Winter to Spring (Jan–Mar)**	**Yes** 2003–2005 (although no data were collected in 2005)	2× increase in daily fluctuations relative to MLFF—to limit juvenile rainbow trout egg viability and manage the tailwater fishery

This chapter is intended to provide an overview of the scientific learning and resource benefits that have occurred in support of the AMP through use of large-scale flow experiments released from the dam during an era of changing climate in the southwestern United States (Olden et al. 2014; Konrad et al. 2011). Our aim is to highlight recent findings that have supported information needs of the AMP since its inception, but more importantly to identify uncertainties that have hindered or delayed decisions, and been discussed by managers and scientists in knowledge assessment workshops organized by the US Geological Survey's Grand Canyon Monitoring and Research Center (GCMRC 2008). We examine key AMP resources, such as endemic native fish, introduced trout, and river sandbars, for which adaptive learning may become increasingly important as climate change influences Upper Colorado River Basin (UCRB) hydrology, water delivery strategies for managing this important basin's reservoirs, and the river's thermal regime.

The Colorado River basin drains 637,000 km^2 of land within seven states of the southwestern United States and Mexico. The majority of the snowmelt runoff that feeds the river originates from relatively few high-elevation watersheds mostly located in Wyoming, Utah, and Colorado. Despite limited high-elevation catchment areas to accumulate snowpack in its headwaters, the Colorado River is relied upon to supply water to some 40 million people of the region, including many of its fastest growing cities. The construction of GCD, intended to meet delivery requirements to downstream water users in the face of repeated UCRB drought, guaranteed higher minimum flows through Glen and Grand Canyons.

Lake Powell, the reservoir created by GCD, stores 32 billion m^3 of water at full pool (elevation, 1128 m) and has a surface area of 658 km^2. Filling of the reservoir began with dam closure in 1963, and was achieved in 1980. Unregulated inflows to the reservoir from 1981 to 2010 are highly variable from April to July but average 13.4 billion m^3 annually (Figure 17.2). The second largest reservoir in the United States (Lake Mead is slightly larger), Lake Powell extends 290 km upstream and has an estimated shoreline length of 3,057 km; with a drainage area above the reservoir in four western states of 279,000 km^2 (Vernieu 2013).

17.1.1 Influence of Glen Canyon Dam Operations on the Colorado River

The main authorized purpose of GCD is water storage; specifically, to ensure that water delivery agreements between the four upper and three lower basin states and Mexico can continue during periods of drought without the need to curtail water use by upstream users. Annual dam releases have varied from ~10 to 26 billion m^3 since 1964 (Figure 17.3).

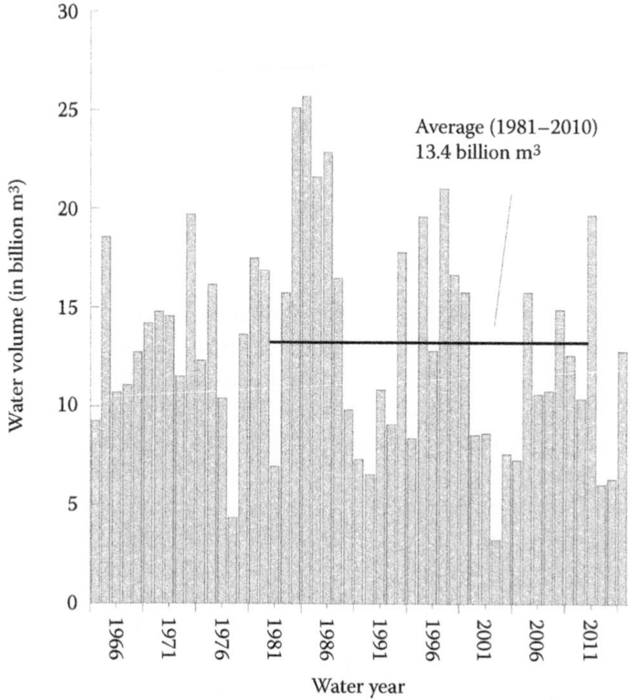

FIGURE 17.2
Unregulated inflow volumes into Lake Powell from Upper Colorado River Basin (1964–2014). (Data courtesy of K. Grantz, written communication, US Department of the Interior Bureau of Reclamation, 2014.)

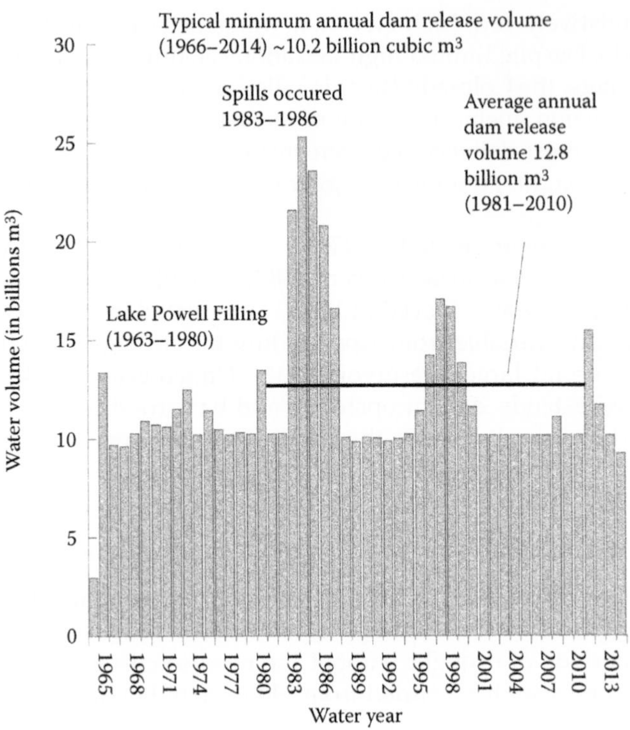

FIGURE 17.3
Annual water volumes released by Glen Canyon Dam from Lake Powell (1964–2014). (Data courtesy of R. Clayton and K. Grantz, written communication, US Department of the Interior Bureau of Reclamation, 2014.)

About two-thirds of the annual releases have been at the historical (1964–2008) minimum volume of 10.2 billion m^3 required by a 1922 interstate compact, and a treaty with Mexico (see Law of the River: http://www.usbr.gov/lc/region/pao/lawofrvr.html). Between 1963 and 2014, three periods of higher than minimum annual GCD releases stand out: the mid-1980s; the late 1990s; and water years (WY, defined as October 1 through September 30) 2011–2012 (Figure 17.3).

In terms of the ecology of pre-dam riverine resources, regulation by GCD has effectively eliminated the influence of UCRB droughts on riverine resources of the Colorado River between Lakes Powell and Mead. Seasonal low flows of the pre-dam era no longer occur due to minimum daily release requirements of 142 cubic meters per second (m^3/s), and releases below 227 m^3/s seldom occur even during most minimum volume release years (Figure 17.3). Topping et al. (2003) report that the median discharge of the Colorado River measured below GCD at Lees Ferry, AZ (226 m^3/s) was increased by 58% to 357 m^3/s as a result of dam operations between 1963 and 2000, while turbidity below the dam was decreased by a factor of 20–2000 because most of the river's fine sediment is retained in Lake Powell (Voichick and Topping 2014). Most of the river's remaining downstream sand supply is now provided by the Paria River and Little Colorado River (Figure 17.1), but constitutes only about 6% to 12% respectively of the sand formerly available to rebuild and maintain sandbars in Grand Canyon (Wright et al. 2005). Also significant to downstream resources, the 1920–1960 pre-dam median daily range in discharge (15 m^3/s) was increased after 1963 by a factor of 15.8–243 m^3/s (Topping et al. 2003). The combined influences of decreased fine-sediment supply below the dam, increased median flow, and increased daily flow range resulted in erosion of river sandbars (Schmidt and Grams 2011), and also influenced opportunities for recreational boating.

Filling of Lake Powell by 1980 also significantly decreased seasonal variation in the river's thermal regime from 0°C to 27°C, down to an annual range of 7°C to 9°C in Glen Canyon, a shift that persisted through 2002. Alteration of the river's thermal variability has long been a topic of discussion among AMP stakeholders, and is an influence that may equal or exceed that of regulated flow effects on aquatic organisms and processes (Ellis and Jones 2013; Olden and Naiman 2010; Kennedy and Gloss 2005; and Gloss and Coggins 2005). Between 1973 and 2002, average dam release temperatures were about 9°C which in summer is about 18°C below pre-dam values (Voichick and Wright 2007). Year-round colder river temperatures and clear water dam releases allowed for establishment of a recreational rainbow trout (*Oncorhynchus mykiss*) tailwater fishery downstream of the dam in Glen Canyon, but also likely limited growth of juvenile native fish (Gloss and Coggins 2005). Stocking was annually required to maintain the Glen Canyon trout fishery, but MLFF operations that increased both minimum daily flows and flow stability in the 1990s, dam releases with a median daily range of 140 m^3/s (Topping et al. 2003), likely supported increased rainbow trout spawning and recruitment, and rainbow trout stocking ended in 1998 (Korman et al. 2011a,b).

Constraints on daily dam operations since 1996, combined with drier UCRB hydrology and reduced Lake Powell storage since 2001, have limited hydropower resources at GCD (Harpman and Douglas 2005), and also increased river temperature since 2002, as a result of warmer dam releases drawing from Lake Powell's epilimnion (Vernieu 2013; Voichick and Wright 2007; Vernieu et al. 2005). In this period, lower dam releases reduced sandbar erosion (Schmidt and Grams 2011), but limited water supply and hydropower energy production, while warmer releases have likely benefitted native fish compared to influences of earlier larger volume and colder hypolimnetic dam releases from 1995–2002 (Walters et al. 2012). Contrasting responses in these Colorado River resource trends tied to UCRB hydrologic variability and MLFF operations over the last two decades provide one example of climate variability and change induced management trade-offs faced by the AMP.

17.1.2 Adaptive Management of the Colorado River through Flow Experiments

The AMP focuses on a segment of the Colorado River corridor extending from the forebay of GCD (lat. 36.937, long. –111.484) within Glen Canyon to the western boundary of Grand Canyon (Figure 17.1). Termed, the Colorado River ecosystem (CRe), the river below the dam is managed by two different National Park Service units within a 470-km long segment of the river; the upstream unit extending from the dam to Lees Ferry within Glen Canyon, and the other unit extending downstream from Lees Ferry through Grand Canyon. In 1956, when the Colorado River Storage Project Act authorized construction of GCD, the goal of federal policy was to control and use the waters of the Colorado River to create wealth and new opportunities through dam building and hydroelectric power generation (US Department of the Interior 1946).

Construction of GCD, which is located just south of the Arizona–Utah border, began in 1956, a period before the enactment of the National Environmental Policy Act of 1969, and the Endangered Species Act of 1973. At the time of GCD construction, little consideration was given to how dam operations might affect downstream river resources. Likewise, early water managers' awareness about the influence of climate variability and changes in UCRB hydrology was mainly derived from a relatively limited period associated with very wet years of the early 20th century followed by southwestern US droughts that reduced flows through Grand Canyon in the 1930s and 50s. Over time, stream flow monitoring since the early 1920s at Lees Ferry, AZ, the point on the river designated to be the boundary between the upper and lower basin states, and research have provided managers with a better perspective about the high variability associated with UCRB water supply (Brekke et al. 2009; Pulwarty 2003; also see Woodhouse et al., Chapter 9, this volume). Recently, Jain et al. (2005) reported increasing year-to-year variance of Colorado River streamflow after the 1970s, which equates to decreasing reliability in the water supply of the southwest during the period following completion of most water projects in the basin.

Following establishment of the AMP, Pulwarty and Melis (2001) identified how the US Department of the Interior's Bureau of Reclamation and other stakeholders were able to use climate forecast information to adaptively manage GCD releases and storage in Lake Powell during the large 1997–1998 ENSO event to meet multiple resource objectives. A key objective achieved by water managers in summer 1998 was refilling Lake Powell without the need for power plant bypass releases (originally designed for use to ensure dam safety during high runoff years when Lake Powell storage is near capacity as occurred in 1983–1986 and 1995–1999). Discussions among AMP stakeholders at meetings in August 2007 (Pulwarty, attachment 8 of meeting minutes) and February 2014 (Grantz, Kuhn, and Harkins; attachments 3a–c of meeting minutes) reflect increasing interest and awareness about climate variability and climate change since 2000, apparently, in response to repeated years of drought in the UCRB (see AMP meeting minutes: http://www.usbr.gov/uc/rm/amp/mtgmin.html). Earlier evidence of concern about how climate change and extended periods of drought might influence CRe resources below GCD is also found in the AMP's science planning documents, such as its 2006 draft monitoring and research plan prepared in collaboration with the Grand Canyon Monitoring and Research Center (GCMRC) (for more about recent drought responses in the UCRB, also see Smith et al., Chapter 7, this volume).

Experimental dam releases have been a mainstay of managing GCD since the AMP was established. Since 1963, most daily GCD operations typically have been diurnal hydropeaking fluctuations, with the first managed high-flow releases studied in 1965 (Grams et al. 2007), and formal reoperation of the dam to MLFF in fall 1996, following completion of the 1995 EIS. In addition to MLFF operations, a variety of experimental flows have also been implemented through additional environmental compliance (Table 17.1). All of the experimental releases have multiple objectives of benefiting downstream resources while also achieving water delivery requirements, providing the greatest practicable level of hydropower, and meeting information needs of US Department of the Interior (DOI) policy decision makers and other stakeholders (Olden et al. 2014; Cook 2013; US Department of the Interior 2012a; Melis et al. 2012; Melis 2011; Gloss et al. 2005). Konrad et al. (2011) define criteria for large-scale flow experiments intended to inform management of freshwater ecosystems and these characteristics describe nearly all GCD flows released since 1996. Dam releases after 2002 have also occurred in coordination with other "non-flow" management experiments, such as nonnative fish removal from limited areas of the CRe in Grand Canyon (Coggins et al. 2011).

17.2 Monitoring Experiments to Reduce Management Uncertainties

As a federal advisory committee composed of 25 members with diverse special interests, the AMP (see: http://www.usbr.gov/uc/rm/amp/) is tasked with evaluating performance of GCD operations and formulating recommendations to the DOI on the basis of monitoring and research information pertaining to at least a dozen resources of concern. This information is provided by the GCMRC and its cooperators (Hamill and Melis 2012; Gloss et al. 2005).

In a 2001 strategic plan, the AMP identified 12 resource goals and 56 objectives (Berkley 2013). This large number of objectives creates an extremely challenging situation for the AMP where attempts to meet resource objectives may pose trade-offs between different resources such as sandbars, fisheries and hydropower within dam operating constraints imposed by prior water management agreements. This challenge is further complicated by a lack of clear prioritization and metrics for the various resources identified by the AMP's collaborative of stakeholders and management agencies (King et al. 2015; Scarlett 2013). This situation improved somewhat in 2011, when the AMP developed, and DOI accepted, desired future conditions that grouped goals into 4 areas: cultural, recreational, hydropower and river ecosystem resources (http://www.usbr.gov/uc/rm/amp/amwg/pdfs/recltr_12April30.pdf).

Three summaries below provide examples of how numerous large-scale flow experiments have helped adaptive managers better understand GCD operating strategies for achieving downstream objectives for sandbars, native humpback chub (*Gila cypha*), and introduced rainbow trout. The order of these summaries reflects lesser-to-greater uncertainty about documented responses to dam operations for these three AMP resources highlighted in recent knowledge assessment workshops convened by the GCMRC in 2005, 2007 and 2011, and annual reporting meetings each year since they are intended to inform stakeholders. We have attempted to capture uncertainties about how dam operations influence these and other CRe resources in an *uncertainty matrix* (Figure 17.4), intended as a road map that some stakeholders may find useful in guiding discussions on monitoring and research needs.

We also intend that Figure 17.4 might provide stakeholders and decision makers with another tool to help focus discussions about planning of future large-scale field experiments under ongoing hydrologic change, evolving water demand, and management strategies for changing water supply throughout the Colorado River basin.

17.2.1 Sandbars (Lower Uncertainty)

Long-term monitoring of the effects of MLFF dam operations combined with research on high-flow experiments (HFEs) since 1996 have greatly reduced managers' uncertainties about dam operations and sandbar conservation strategies in Grand Canyon (Schmidt and Grams 2011). Monitoring and research have resulted in three conclusions related to sandbar resources below the dam that have important implications for designing future HFEs tied to the AMP's desired future conditions for the CRe (Figure 17.4, rows 3–4 and 7). First, it is known that HFEs rebuild sandbars by either eroding existing low-elevation portions of sandbars within eddies, by entraining new tributary-supplied sand stored in the channel outside eddies, or some combination of both of these sand sources (Rubin et al. 2002). Because of this fact, a series of repeated HFEs that are intended to result in larger sandbars in a specific river segment below the dam must be carefully planned and monitored relative to upstream sandbar responses. Monitoring sand budgets between the Paria and Little Colorado Rivers, and river segments further downstream is also required to minimize any net loss of sandbar area (Mueller et al. 2014; Grams 2013; Schmidt and Grams 2011). Such a management experiment is now occurring under a new 2012–2020 HFE Protocol and preliminary results suggest that sandbars are increasing below the dam (US Department of

Resources	↓Hydro	↓FFs	↓HFEs	↑FFs	↑LSFs	↑HSFs	↑TMFs
Subset of Desired Future Resource Conditions Identified by the AMP: (a) Protect and Conserve Humpback Chub (b) Rebuild and Maintain Sandbars (c) Control Nonnative Fishes in GRCA (d) Improve GLCA Tailwater Fishery	**Lower Lake Powell Storage**—from decreased runoff and (or) increased use, leading to increased downstream river temperatures in summer and fall, as in 2003–2011 and 2014, and lower dam releases	**Decreased Daily River Fluctuations**—relative to MLFF, further reduced perhaps by more uniformly distributing monthly release volumes during the year	**Increased Frequency of Artificial Floods**—after fine sediment inputs from downstream tributaries to rebuild and maintain sandbars	**Increased Daily Fluctuating Flows**—relative to MLFF to more closely follow daily peak electrical energy demand	**Increased Sustained Low Steady Flow**—to increase downstream water temperature in warm-season months and conserve tributary sand inputs prior to fall HFEs	**Increased Sustained High Steady Flow**—to meet downstream water deliveries required in years of above-average runoff in upper basin, or for seasonally adjusted steady flow experiments in spring months	**Increased Trout Management Flows**—increase daily fluctuating flows to achieve egg and fry mortality in winter to enhance recreational fishing opportunities in GLCA (as tested in Jan–Mar 2003–2005)
1. Hydropeaking to meet capacity and energy demands	(–)	(–)	(–)	(+)	(–)	(–)	(+)
2. Averaged downstream river temperatures	(+)	(o)	(o)	(o)	(+)	(–)	(o)
3. Sand in lower river channel	(+)	(+)	(–)	(–)	(+)	(–)	(–)
4. River camp area with little to no vegetation	(+)*	(+)*	(+)*	(–)	(+)*	(–)	(–)
5. Riparian vegetation	(+)*	(+)*	(+)*	(–)	(+)*	(+)	(o)
6. Active sand dunes near cultural sites	?*	(+)*	?*	(–)	(+)*	(+)	(–)
7. Higher shore and eddy sandbars	(+)	(+)	(+)	(–)	(+)	(–)	(–)
8. River rafting	(–)	(+)	(o)	(–)	(–)	(+)	(–)
9. GLCA trout	?	(+)*	(+)*	(–)	(+)*	(+)*	(–)

FIGURE 17.4
Uncertainty matrix for Colorado River ecosystem below Glen Canyon Dam in Glen Canyon National Recreation Area (GLCA) and Grand Canyon National Park (GRCA), United States.
(Continued)

Resources	↓Hydro	↓FFs	↓HFEs	↑FFs	↑LSFs	↑HSFs	↑TMFs
10. GRCA trout	?	(+)*	(+)*	?	(+)*	(+)*	(o)
11. Humpback chub in the LCR aggregation	(+)	?*	?*	(o)	?*	?*	(o)
12. Chub and native suckers elsewhere throughout CRe	(+)	?*	?*	?	?*	?*	?
13. Nonnative warm-water fishes throughout CRe	(o)	?	?	?	?	?	?
14. Aquatic food supply	(o)	?	(+)	?	?	?	?

The matrix only generally reflects current status of knowledge about how AMP resources have responded to dam releases on the basis of available information (1996–2014), but more importantly, it is intended to show increasing uncertainty (from top to bottom rows) about influences of basin hydrology and dam operating policies (previously tested or ongoing as recommended by the AMP) on key river resources of concern to managers, including the following: (1) sandbar rebuilding and maintenance through periodic release of HFEs in combination with other dam operations; (2) management of the GLCA rainbow trout tailwater fishery to improve recreational angling opportunities; (3) native fish protection, especially endangered humpback chub; and (4) control of both cold- and warm-water nonnative fish in GRCA. Other important resources depicted in the matrix include hydropower, quality of water (mainly water temperature), aquatic food base, recreational boating and camping areas, riparian vegetation, and windblown sand deposits that may help protect archeological sites. Matrix inputs were derived from literature and expert workshops convened by GCMRC, 2005–2011. (For resource assessment examples, see Gloss et al. 2005; GCMRC 2008.)

Notes: (1) (+), positive response predicted relative to management objective; (o), neutral response; (–), negative response. (2) Responses assume that dam operations are constrained by fixed monthly-to-annual release volumes. (3) Suite of operational elements is contained within column "↑FFs" or "↓FFs," such as hourly release ramp rate, flow range, peak, and minimum flow, for any given monthly volume release, relative to the record of decision implemented in October 1996 for reoperation of the dam's hydropower plant. (4) Asterisk (*) indicates that some uncertainties (?) persist owing to indirect resource responses that may affect other resources. For example, high steady flows in spring and summer are known to increase GLCA trout production (+) but are deemed as neutral (o) to humpback chub. However, increases in GLCA trout may or may not limit GRCA chub population; hence (?), uncertainty persists about high steady flows relative to chub, as they might eventually pose an increased threat to native fish downstream of the GLCA tailwater, but indirectly as a slower variable response. Similarly, exposed sandbar area increases under lower flows, but riparian vegetation expansion may then offset gains in camping areas or areas of desiccated sand for wind transport that supports active dune migration.

Shading code: White—existing models can estimate the direction *and* the magnitude of resource response relative to dam operations (e.g., dam releases related to river stage and inundation estimated for a given camping beach). *Gray*—owing to unresolved uncertainties, scientists can generally estimate the direction, but not the magnitude, of response to dam operations. *Black* ?—uncertainties are so large that a link with dam operations may be suspected, but too little is known to estimate resource response direction *or* magnitude.

FIGURE 17.4 (CONTINUED)
Uncertainty matrix for Colorado River ecosystem below Glen Canyon Dam in Glen Canyon National Recreation Area (GLCA) and Grand Canyon National Park (GRCA), United States.

the Interior 2012a; Grams et al. 2015). Second, monitoring has revealed that under typical MLFF dam operations, only HFEs, conducted soon after new sand has been supplied to the river channel by downstream tributary floods, are effective at increasing both sandbar area and volume and less likely to result in the erosion of low-elevation portions of sandbars (Schmidt and Grams 2011). Third, monitoring data have shown that sandbars erode more quickly as monthly-to-annual dam release volumes and daily fluctuations increase under MLFF and other recent experimental fluctuating (Table 17.1) operations (Schmidt and Grams 2011).

Interdisciplinary learning related to how dam operations influence riparian vegetation and therefore indirectly influence sandbar resources below GCD has provided key information to the AMP and helped highlight trade-offs to be addressed. For example, it is clear that the combined influence of lower and steadier dam releases throughout the year and more frequent tributary sand-enriched HFEs are most likely to promote increases in eddy and shoreline sandbar areas (Wright et al. 2008; Rubin et al. 1994). However, those same dam operations have also promoted increases in riparian vegetation which has reduced open sandbar areas available for recreational camping (Kaplinski et al. 2014), posing a trade-off between values associated with riparian habitat and recreational resources (Figure 17.4, rows 4–5 and 7).

We believe that there is still at least one key long-term uncertainty confronting river managers committed to rebuilding and maintaining CRe sandbars using HFEs. The question remains whether desired sandbar building from potentially more frequent HFEs can occur at a rate faster than sandbars are eroded by GCD operations between HFEs (Wright and Kennedy 2011; Wright et al. 2008). Resolving uncertainty about this outcome is further complicated by predictions for reduced UCRB hydrology (see: http://www.usbr.gov/lc/region/programs/crbstudy.html), current strategies for annual dam releases required to meet water deliveries (Kenney et al. 2011), and variability of future downstream tributary sand supply (Figure 17.5) below the dam.

Although these factors will likely have a significant influence on the fate of CRe sandbars, they cannot all be directly influenced by GCD operations, at least within existing infrastructure and water management policies. Nevertheless, with greater certainty about options for sandbar management downstream of the dam, AMP stakeholders now have more ability to evaluate the various trade-offs associated with using GCD operations during persistent drier periods to rehabilitate one of the river's most recognized and valued *relict* attributes—abundant, expansive and vegetation-free sandbars (Wright and Kennedy 2011).

17.2.2 Nonnative Rainbow Trout (Moderate Uncertainty)

For several decades, the fish community in the CRe has been dominated in terms of abundance by cold water nonnative rainbow trout (Coggins et al. 2011; Makinster et al. 2010; McKinney et al. 2001). Brown trout (*Salmo trutta*) are also found in Glen and Grand Canyons, but in far lower numbers. A large variety of cool and warm water nonnative fish now also inhabit the CRe (Gloss and Coggins 2005) making it possible for some nonnative fish to potentially increase under slightly warmer water as a result of warmer release temperatures from GCD. Nonnative trout are a valued recreational angling resource in Glen Canyon, but in Grand Canyon they pose a threat to native fish because they prey on juvenile humpback chub and also compete with native fish for a limited food supply and habitats (Cross et al. 2013; Yard et al. 2011).

Although by the late 1990s, more information had become available about options for improving the 1995 EIS strategy for increasing sandbars (Wright et al. 2008, 2005; Rubin

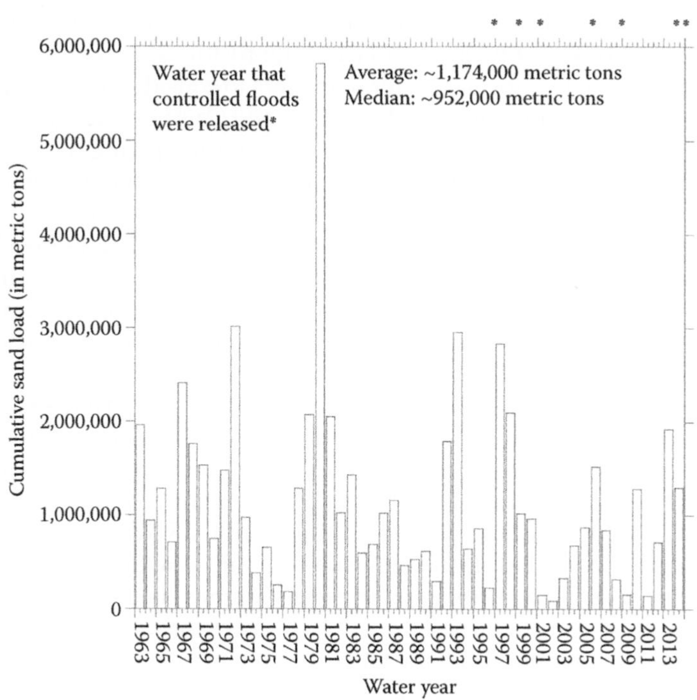

FIGURE 17.5
Annual sand loads delivered to the Colorado River below Glen Canyon Dam by the Paria River (US Geological Survey streamgage #09382000) following dam closure (1963–2014). Asterisks (*) indicate the timing of high-flow experiments (HFEs) released from the dam in Water Years (WY) 1996, 1998, 2000, 2005, 2008, 2013, and 2014 that have occurred under a range of sand-supply conditions in the main channel below this key sand-contributing tributary. Channel sand conditions prior to the HFEs have varied from sand enriched (WYs 2008, 2014, and 2015) to relatively depleted (WYs 1996 and 2000).

et al. 2002), there was much less certainty about factors that promote juvenile humpback chub and other native fish recruitment in Grand Canyon (Melis et al. 2006). As a result, most other non-HFE flow experiments have focused on the uncertainties about co-managing chub and nonnative rainbow trout in the CRe (GCMRC 2008).

In terms of the population dynamics of rainbow trout, recent modeling research strongly suggests that dam operations in 3 years (1997, 2000, and 2008) since MLFF was implemented in WY 1997 have resulted in increased trout recruitment in Glen Canyon (Korman et al. 2012). Monitoring suggests that Glen Canyon trout recruitment was elevated again in 2009 following the 2008 spring HFE (Melis et al. 2012; Korman et al. 2011a), and during 2011 (Avery et al. 2015) in response to high and steady dam releases required for water delivery to Lake Mead. This is another important trade-off confronting the AMP, because increases in rainbow trout production following re-operation of the dam in the 1990s occurred during the same period of time when humpback chub abundance was declining (1989–2000) in Grand Canyon (Coggins et al. 2011). Attempts at reducing natural rainbow trout recruitment in Glen Canyon by experimentally dewatering redds (egg nests in river gravels) through increased daily flow fluctuations in winter months of 2003–2005 (Table 17.1) were not successful, likely owing to increases in trout survival later in life that compensated for the earlier, higher mortality in redds (Figure 17.4, rows 9–10) (Korman et al. 2011a,b).

Our uncertainty matrix (Figure 17.2) identifies knowledge gaps in CRe responses to specific flow policies evaluated by the AMP and indicates that monitoring and research of experimental fluctuations, steady flows and early life-history stages of rainbow trout since 2003 has produced valuable information for managers to consider in planning future long-term experiments (Korman and Melis 2011; Korman and Campana 2009). As with sediment resources, key uncertainties still persist about options for managing the Glen Canyon tail-water fishery, while simultaneously limiting the abundance of trout downstream to protect native fish in Grand Canyon. There are at least three questions that remain to be answered: (1) Are rainbow trout produced in Glen Canyon having a population-level limiting influence on downstream native fish in Grand Canyon, in particular the humpback chub? (2) What factors influence downstream movements of rainbow trout from Glen Canyon into Grand Canyon? and (3) Can the abundance of trout in Glen Canyon and downstream be limited using dam releases (Figure 17.4, rows 9–10)?

Although influences on the CRe's trout population resulting from steadier flows and spring-timed HFEs are now documented (Korman et al. 2012), information available to managers about how dam operations and other factors influence and limit downstream food item abundance and diversity remains highly uncertain. On the basis of monitoring, moderate warming of the CRe since 2003, resulting from UCRB drought and reduced Lake Powell storage, has had a neutral effect on the downstream aquatic food supply (Figure 17.4, row 14) (Kennedy et al. 2013). It remains unclear whether dam operations can be used to manage the downstream food base; possibly to increase diversity and abundance of drifting invertebrates to support both the Glen Canyon tailwater fishery, and native fish in Grand Canyon (Figure 17.4, row 14) (Kennedy 2013). Research has recently characterized the CRe food web as being unstable, and comprised of relatively few invertebrate taxa as drifting prey items available to fish (Cross et al. 2013). This raises key questions about whether native fish populations in Grand Canyon might be limited by direct competition with trout for food and habitat in Grand Canyon, or by trout in Glen Canyon that may consume large quantities of prey resources that might otherwise reach downstream native fish.

17.2.3 Endangered Humpback Chub (Higher Uncertainty)

The Grand Canyon humpback chub population is the largest of the six remaining populations of this endangered species within the Colorado River basin (Gloss and Coggins 2005). Most humpback chub in the CRe are found in the area of the Little Colorado River and its confluence with the Colorado River (Figure 17.1), likely because this major tributary provides suitable spawning and rearing habitat (Coggins et al. 2011). Many management actions in aquatic ecosystems are directed at restoring or improving specific habitats to improve juvenile fish growth and survival. In Grand Canyon, experimental flows as part of the AMP often consider the creation of habitat types (i.e., sandbars and associated near-shore "backwaters"), water temperatures, or flow magnitude and daily fluctuating range as critical for juvenile native fish conservation.

In 2008–2012, a Fall Steady Flow experiment (FSF, Table 17.1) was conducted at GCD to evaluate how steady flow, through its interaction with physical habitat structure, influences the growth, survival rates, and habitat use of juvenile humpback chub in Grand Canyon. The flow experiment was evaluated within a limited CRe study segment (~5 km) located downstream from the Little Colorado River (Figure 17.1) in the same area known to have the largest population of endangered humpback chub. Population recovery of the chub is of keen interest to the AMP and other stakeholders throughout the river basin, and two fundamental research questions related to knowledge of humpback chub population ecology were addressed during FSF testing: (1) Do steadier flows released from GCD during September through October increase survival, abundance, and (or) growth rates of juvenile native and non-native fish? and, (2) do juvenile humpback chub select specific near-shore habitat types and if so, does this selection change under different GCD operations and river flow regimes in Grand Canyon? These questions were motivated by concerns that, as a hydropeaking facility, the availability of certain habitat types (i.e., backwaters), and available area of other habitat types changes with water releases from the dam (Vernieu and Anderson 2013; Behn et al. 2010; Grams et al. 2010; Ralston et al. 2007).

The assessment of juvenile humpback chub population responses to FSF testing yielded several results that addressed key uncertainties related to the relationship between dam operations and native fish population responses. First, by conducting steady flows ranging from 227 m^3/s to 736 m^3/s in September and October, thermal warming 122 km downstream of GCD was minimal due to influences of river volume and travel times under reduced fall solar radiation (Ross and Grams 2013). This suggests that if one objective of future flow experiments is to benefit native fish through stable habitats and warmer water, then future experiments might be better conducted during summer months of higher solar radiation, as scientists had suggested in 2007 (GCMRC 2008). Downstream warming is predicted by modeling (Wright et al. 2009) to be greatest under the lowest summer releases possible, especially when Lake Powell storage is high and dam releases are relatively cold (Figure 17.6), as they were during an earlier 2000 Low Summer Steady Flow (LSSF) test (Table 17.1) (Ralston 2011). The FSF mainly tested hypotheses about the influence of stable shoreline habitats on early life stage success of chub rather than water temperature. Recent research on other CRe native fish species provide no evidence to support the steady-flow habitat hypothesis (Walters et al. 2012), but weight-of-evidence supports the idea that warmer river temperatures (Figure 17.7) and lower abundance of nonnative trout (Coggins et al. 2011) are likely the main factors in juvenile native flannelmouth and bluehead sucker (*Catostomus latipinnis* and *Catostomus discobolus*, respectively) increases after

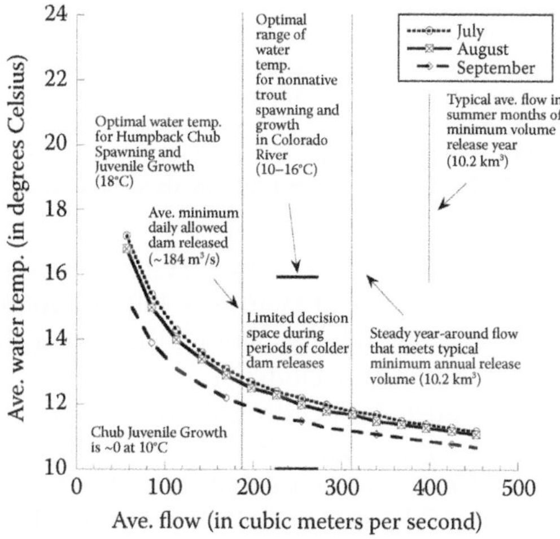

FIGURE 17.6
Average monthly summer water temperature simulations for July, August, and September at a point in the Colorado River 122 km below the dam, near its confluence with the Little Colorado River in Grand Canyon, over average monthly Glen Canyon Dam releases ranging between 57 and 453 m^3/s. Simulations were derived from a model developed by Wright et al. (2009) assuming dam release temperatures of 9°C combined with average monthly air temperatures.

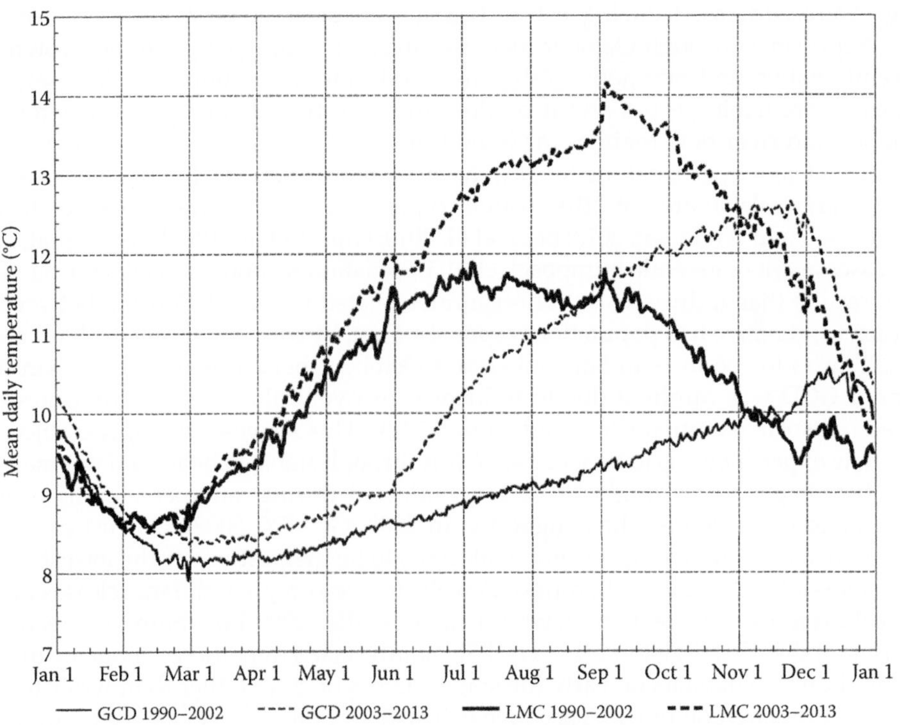

FIGURE 17.7
The mean values of daily (1990–2002) Colorado River water temperature released from Glen Canyon Dam (GCD, thinner solid line) and measured in Lower Marble Canyon (LMC, thicker solid line) near the confluence of the Little Colorado and Colorado Rivers, 122 km below the dam (Colorado River above the Little Colorado River near Desert View, station #09383100). Dashed lines for each location show the magnitude and timing of increased mean daily warming during the era (2003–2013), a period of dam releases during reduced storage in Lake Powell resulting from the combined influences of continued water demand by lower basin states and drier hydrology in the Upper Colorado River Basin. (Plot and Glen Canyon Dam preliminary data courtesy of W. Vernieu, US Geological Survey, written communication, 2014; data on the mainstem above the Little Colorado River confluence from US Geological Survey, after Voichick, N., and Wright, S.A., Water-temperature data for the Colorado River and tributaries between GCD and Spencer Canyon, northern Arizona, 1988–2005, *US Geological Survey Data Series*, 251. 24 pp., http://pubs.usgs.gov/ds/2007/251/, 2007.)

2002 (Walters et al. 2012). Second, no increase in survival or decrease in abundance was found for juvenile humpback chub (size at tagging < 100-mm total length) between summer MLFF fluctuating flows and FSF releases among the varied shoreline habitats where fish were studied (Finch et al. 2015, 2013; Dodrill et al. 2014; Gerig et al. 2014). In contrast to the earlier 2000 LSSF test, it must also be recognized that FSF testing occurred during a unique period of dam releases characterized by summer and fall average water temperatures near the Little Colorado River confluence that were about 1.5°C to 2.5°C warmer than the 1990–2002 period in which MLFF was initially implemented and Lake Powell storage increased (Figure 17.7).

The FSF experiment followed up on the 2000 LSSF, an experiment that also focused on humpback chub responses, and another very limited duration flow experiment in fall 2005 (see Ralston et al. 2007). The LSSF provided key information on how the retention time of water in littoral areas, such as nearshore backwaters and low-angle shorelines, at different discharge volumes and ambient water release temperatures at GCD interact to increase water temperatures throughout the CRe (Vernieu and Anderson 2013). Ralston (2011) summarized the 2000 LSSF results and found that summer (June through September) steady flows at ~227 m^3/s when GCD water releases were at about 9.5°C, resulted in only marginal average downstream warming at the main target location for benefiting native fish near the confluence with the Little Colorado River (~3.5°C), but average June through August warming did reach optimal spawning and rearing temperatures (16°C and 15°C to 25°C, respectively) farther downstream in western Grand Canyon. Fish monitoring data collected in support of the LSSF revealed no evidence that either the moderate downstream warming, or shoreline stability associated with summer 2000 operations produced a population-level response in Grand Canyon humpback chub (Walters et al. 2012; Trammell et al. 2002). Subsequent work on humpback chub energetics and growth rate by Coggins and Pine (2010) estimated that water temperatures in the CRe would need to be closer to 20°C, or about 7°C higher than the average mainstem water temperatures measured near the Little Colorado River confluence during the 2000 LSSF, for significant growth to occur in juvenile chub after they disperse from the Little Colorado River and enter the mainstem.

Combined with the 2000 LSSF findings, results of FSF studies suggest that juvenile humpback chub survival, growth, abundance, and habitat use are robust to fall steady flows. Although data are less abundant, adult chub populations showed no changes in response after low and steady summer dam releases were tested in 2000, or following experimental winter fluctuating flows exceeding MLFF in 2003–2005, that were designed specifically to limit rainbow trout egg survival in Glen Canyon (Korman et al. 2011b). It is likely that more extreme flow treatments (replicates of either average flows below 227 m^3/s in summer, or also higher or lower discharges over wider daily ranges over a longer duration from summer through fall) are required in the segment of the CRe where FSF field studies were focused before changes in these metrics would be observed. These data demonstrating the apparent flexibility of juvenile humpback chub in habitat selection regardless of daily flow patterns as well as the growth, survival, and persistence of juveniles in the main channel of the Colorado River are invaluable additions to the body of knowledge available for managing both the CRe and possibly other segments of the Colorado River affected by hydropeaking dams.

Overall results from the several flow experiments released from GCD since 1996, including MLFF, suggest that responses of humpback chub have been neutral at best or remain highly uncertain, with chub being robust to steady flows, increased and decreased daily fluctuations in different seasons, and spring or fall HFEs. Despite this, the adult humpback chub (age 4+) population in Grand Canyon has perhaps increased by a factor of three to four (Yackulic et al. 2014) since their estimated abundance was first reported by Gloss and Coggins (2005). Besides growing chub numbers (see: appendix S3 of Yackulic et al. 2014; Figure 3 of Coggins and Walters 2009; and Figure 12 of Gloss and Coggins 2005), Walters et al. (2012) also report increases in native suckers, apparently, in response to warmer temperatures and lower trout abundance near the Little Colorado River and in other parts of the CRe (Figure 17.4, rows 11–12). However, Wright and Kennedy (2011) report that Grand Canyon chub might still be negatively influenced, albeit indirectly, by steady and spring-to-summer high flows that benefit trout and possibly lead to increased native–nonnative fish interactions in Grand Canyon (increased competition for habitat and limited food, plus predation of juveniles; Figure 17.4, rows 9–12) (Yard et al. 2011).

17.2.4 Other Downstream Resources

Rows 1 through 4 of Figure 17.4 reflect relatively low remaining uncertainty about how dam operations affect several other important AMP resources, such as the hydroelectric capacity and energy resources at GCD (Harpman and Douglas 2005), general trends in sandbar responses to fine-sediment enriched HFEs (Grams et al. 2015; Mueller et al. 2014; Schmidt and Grams 2011; Wright et al. 2008), sand flux within the main channel in the eastern third of Grand Canyon (Wright et al. 2010; Wright and Grams 2010), availability of shoreline camping areas (Kaplinski et al. 2014; Rubin et al. 1994), and the thermal regime of the CRe (Wright et al. 2009; Voichick and Wright 2007).

In general, periods of lower and steadier releases impose greater limits on the ability of the hydropower plant to meet electrical energy demands of customers in the region where GCD energy is distributed. Reduced UCRB streamflows that lower Lake Powell storage (http://ww2.wapa.gov/sites/western/powerm/Pages/Drought.aspx), as well as environmental rules that constrain daily flows directly influence the value of hydropower resources in a region where demand for energy and water is growing rapidly. Also, some HFEs, those with the highest peak magnitudes allowed (above ~890 m^3/s), require some water to bypass the GCD power plant for periods of hours to a few days and therefore limit hydropower resources (Figure 17.4, row 1) (Harpman and Douglas 2005).

Several modeling research advances now allow managers to estimate sand inputs from the Paria River in near real time, make predictions about how most dam releases transport sand inputs downstream as suspended load in the upper third of the CRe (Figure 17.4,

ROLE OF LONG-TERM EXPERIMENTAL MANAGEMENT PLANNING

Some uncertainty persists even for flow treatments focused on well-understood resources, such as sandbars, in part because no long-term experimental management plans designed to embrace integrated resource responses, climate variability and change were approved until 2012, and conditions under which flow treatments have been repeated cannot be completely controlled. For example, several HFEs evaluated since 1996 have had different characteristics, including antecedent sediment loading from tributaries, release timing, peak-flow magnitude and duration, making it difficult to determine exactly how seasonality of a given HFE or a repeated series of such dam releases in either fall or spring will influence sediment and biological resources of the CRe over multi-year periods. To reduce uncertainties, DOI officials developed and approved a 2012–2020 HFE protocol, a 10-year Nonnative Fish Control plan, and an integrated fisheries management plan for the CRe on the basis of research conducted as part of the AMP; evidence that "learning-by-doing" is occurring in Glen and Grand Canyons. There is hope that a new 20-year long Long-Term Experimental and Management Plan (LTEMP) will produce monitoring information on new operating alternatives to compare with past responses to the MLFF releases documented over the last two decades. The goal being to identify future operating strategies that managers might choose to adopt that more closely reflect what has been learned about how dam operations affect downstream resources. Future LTEMP treatments are likely to focus on reducing some of the uncertainties depicted in Figure 17.4, including climate change influences, more variable and (or) reduced dam releases that affect slower variable resources of concern such as riparian vegetation, aquatic food and fish. Meanwhile, until an LTEMP is implemented, ongoing ecosystem modeling combined with ongoing efforts to better integrate aquatic and terrestrial resource monitoring (King et al. 2015), will likely help ensure that downstream resource trends influenced by climate variability and change are documented over meaningful timescales.

row 3) (Wright et al. 2010; Wright and Grams 2010; Topping 1997), and estimate how average daily dam releases influence downstream water temperatures (Figure 17.4, row 2; Figure 17.6) throughout the CRe (Wright et al. 2009). Models to estimate the evolution of CRe sandbars, and changes in riparian vegetation in response to dam operations remain in development, but monitoring and research of these resources have advanced to the point where the direction of response can be generally predicted for most dam operations (Figure 17.4, rows 4–5 and 7) (Ralston et al. 2014; Ralston 2010, 2005). River stage modeling also informs managers about how existing camping sandbars are inundated over a range of dam releases (Magirl et al. 2008; Hazel et al. 2006), and river stage is a main factor in limiting access and utility of available camping beach area (Figure 17.4, row 4) (Rubin et al. 1994). There remains uncertainty about the interactions of wind, rebuilt sandbars associated with HFEs and vegetation trends on aeolian transport and the preservation of archaeological sites (Figure 17.4, row 6) (Draut 2012). Monitoring and research to resolve this uncertainty continues under GCMRC science support to the AMP (Sankey and Draut 2014).

17.3 Role of Uncertainty Matrix in AMP Planning under Climate Change

Perhaps the greatest uncertainty looming in the AMP is how climate change caused by global warming will influence the Colorado River basin's water supply, management of Lakes Powell and Mead, and GCD operations required to meet growing water demand in the face of reduced upstream supply (Cook et al. 2015; Vano et al. 2014; US Department of the Interior 2012b; Kenney et al. 2011; Brekke et al. 2009; and Seager et al. 2007). A recent report on projections of uses and supply of water in the Colorado River basin identifies that by 2060 average annual runoff in the basin may decline by 9%, while projected future demands will accelerate far beyond the projected water supply in response to population growth. On average, shortfalls of approximately 3.95 billion m^3 annually are estimated over the 50-year study under reduced water supply and increasing use scenarios, but there is apparently high uncertainty about this value and the deficits might be much larger or smaller. In addressing the water supply uncertainties, at least 150 options for addressing the projected 50-year shortfall that have been compiled and slated for further consideration covering four areas: increase supply, reduce demand, modify operations, and governance and implementation (US Department of the Interior 2012b).

The combined potential for reduced water supply and storage in the reservoirs of the Colorado River basin, including Lakes Powell and Mead, and the deep uncertainties associated with how future shortfalls might be addressed, have great implications for the operation of GCD and likely all of the AMP resource goals; a concern included in planning documents since 2000. These uncertainties, when added to those already identified above, make the challenge of estimating future river resource conditions below GCD a daunting task for managers. Realizing and incorporating these uncertainties was particularly relevant in 2012–2015 for managers from the Bureau of Reclamation and the National Park Service tasked with evaluating long-term GCD operating alternatives for a Long Term Experiment and Management Plan (LTEMP) EIS (see: http://ltempeis.anl.gov/) initiated by DOI officials in 2012.

Recent modeling of precipitation, evaporation, surface runoff and soil moisture in the Colorado River headwaters under projections of increased greenhouse gas emissions (anthropogenic climate change) predicts decreased UCRB water supply to Lake Powell relative to 20th century flow records. The projected decrease in Colorado River flow by 2040 is estimated to be equivalent to the most severe, temporary periods of low flow recorded (Seager et al. 2013). Uncertainties about estimates for decreased CRe streamflow vary widely, but a decrease in streamflow at Lees Ferry, AZ of about 6.5% (± 3.5%) per degree °C of atmospheric warming alone is anticipated, resulting in a decrease in water supply ranging from 5% to 35% by 2050 (Vano et al. 2014). An additional 10% to 15% decrease in streamflow might also occur if precipitation in the basin decreases 5% by the middle of the 21st century.

Another regional study recently compared global mean carbon dioxide concentration levels (CO_2) with the magnitude of historical annual river floods recorded by long-term streamflow data collected throughout the coterminous United States. Counter to their hypothesis that increased warming under rising CO_2 levels leads to larger flood peaks, Hirsch and Ryberg (2012) report a significant decreasing trend of flood peaks in southwestern US streamgage records that met their study criteria, including the Paria River. Most sand inputs from the Paria River tend to be delivered to the CRe during warm season floods from July through October. As such, predictions for possible increases in warm-season precipitation in the UCRB (Seager et al. 2013) might also lead to more frequent and (or) larger volume summer to fall sand deliveries from this important tributary below GCD, despite a decreasing trend in annual peak flows. Long-term monitoring of streamflow, sediment-transport and sandbars of the CRe by the GCMRC provides AMP stakeholders ongoing opportunities to evaluate and adaptively respond to evolving trends in UCRB hydrology, experimental dam operations and downstream tributary sand supplies under the current 2012–2020 HFE protocol and beyond through eventual implementation of the LTEMP as learning about the influences of global warming and climate change on CRe resources continues.

Information from the AMP's integrated sandbar and riparian vegetation monitoring program suggests that a future with more frequent minimum annual dam releases through Grand Canyon would likely reduce downstream sand export rates to Lake Mead while also promoting stability of sandbars rebuilt by HFEs, and continued expansion of riparian vegetation throughout open beach camping areas (Figure 17.4, rows 4–5 and 7). As a result, recreational camping areas will be limited without some type of adaptive strategy that limits increasing riparian vegetation. Increased riparian vegetation may benefit other terrestrial organisms and avifauna, but to date, monitoring and modeling of these resources have been quite limited under the AMP (see Table 2 of Walters et al. 2000). In some, but perhaps not all settings, expanding riparian vegetation will likely also limit areas of open, dry sand available for wind transport toward sensitive cultural resources typically found buried in river terraces and sand dunes (Draut 2012).

There is a potential for increased variance in the thermal regime of the Colorado River below GCD. From 1973 to 2002, downstream summer river temperatures at Lees Ferry were about 18°C below pre-dam conditions, and likely limited juvenile native fish growth and survival. Reduced UCRB hydrology under increased temperatures in the southwestern US has already moderately warmed the CRe's thermal regime from 2003–2014 (Figure 17.7); the period in which Lake Powell storage has been consistently reduced. As warmer dam releases persisted after 2002 and scientists reported increases in native fish in Grand Canyon, federal managers deferred plans in 2008 to retrofit the GCD with a selective withdrawal structure; infrastructure needed to manipulate the river's thermal regime, citing possible risks to native fish in Grand Canyon from expansion of nonnative warm water fish. During future periods of reduced UCRB streamflow, variable downstream river warming is likely to continue and may possibly increase beyond recent levels. Increasing year-to-year Colorado River streamflow variance (Jain et al. 2005) could also mean a return to cooler dam releases if Lake Powell storage increases again in coming decades. Warming presents possible risks and benefits to native fish, co-equal or greater benefit to nonnative fish, and might also lead to uncertain changes in the already limited and unstable aquatic food base of the CRe (Stoks et al. 2014; Cross et al. 2013; Hogg and Williams 1996). Increasing variance in streamflow could therefore also present managers with unique future opportunities to observe how aquatic resources respond to multi-year periods of warmer and cooler river conditions under repeated fluctuations in Lake Powell storage. Monitoring of cooler and warmer GCD releases under ongoing HFEs, might also be combined with varied experimental patterns of daily-to-seasonal dam releases to evaluate which combination of releases might best achieve desired downstream resources.

Temperature simulations have been useful for screening dam release scenarios under varied reservoir storage conditions with and without use of previously proposed but never built selective withdrawal structures on the dam's hydroelectric units. Most importantly, modeling revealed the physical limits on downstream river warming under existing water management and dam operating policies. Hourly unsteady flow simulations conducted by GCMRC in 2006 predicted equivalent levels of average downstream river warming under both fluctuating and steady flows for a given monthly release volume (GCMRC 2008). In support of new LTEMP environmental compliance on dam operations, temperature models have informed decision makers that it is possible for the river to approach near-optimal temperatures for native fish (16°C to 18°C) near the Little Colorado River confluence and downstream when Lake Powell storage is high and releases are cold (9°C, Figure 17.6). However, such conditions are only achievable if summer releases are reduced below levels allowed by current operating policies, or if selective withdrawal structures exist. What remains highly uncertain is the frequency at which colder water releases from GCD will occur (Figure 17.8) in the future as a result of increased Lake Powell storage under increased streamflow variability (Jain et al. 2005), evolving water management policies (Kenney et al. 2011) or both. Fish modeling research suggests that future cold GCD releases that likely reduce humpback chub recruitment in the CRe, might limit native fish conservation objectives of the AMP (Yackulic et al. 2014; Pine et al. 2013).

The uncertainty matrix presented here (Figure 17.4) is not intended to be definitive, but we hope it provides a conceptual road map that is useful to decision makers planning future experimental flow treatments for GCD in collaboration with the AMP. Perhaps more importantly, the matrix can be reviewed by stakeholders and scientists during knowledge assessments and updated as new information is available to show where progress in understanding has occurred. This collaborative process can also be used to identify and anticipate complex responses and linkages between CRe resources and long-term variations in UCRB hydrology (Cook et al. 2015; Seager et al. 2013) that will likely influence future dam operations within Law-of-the-River requirements over decade timescales (Kenney et al. 2011). Resilient strategies for adaptively managing the CRe under climate change are most likely to emerge from such frequent assessments of long-term monitoring data (King et al. 2015) and policies for operating GCD and managing water supply throughout the Colorado River basin. This ongoing AM process may be particularly valuable if stakeholders continue to embrace learning opportunities from surprise experimental outcomes like those already reported by scientists (Melis et al. 2015; Walters et al. 2012; GCMRC 2008; Rubin et al. 2002), and they are willing and able to consider trade-offs associated with alternatives to existing reservoir and dam operating policies. The new 2012–2020 HFE Protocol appears to be leading to sandbar increases in Grand Canyon (Grams et al. 2015). We anticipate that new strategies for improving other valued AMP resources may also become apparent, but perhaps over relatively longer periods in which food web and fish resources respond to changes in UCRB hydrology and annual patterns of GCD releases. Stakeholders also need time to evaluate trade-offs posed by new adaptive strategies.

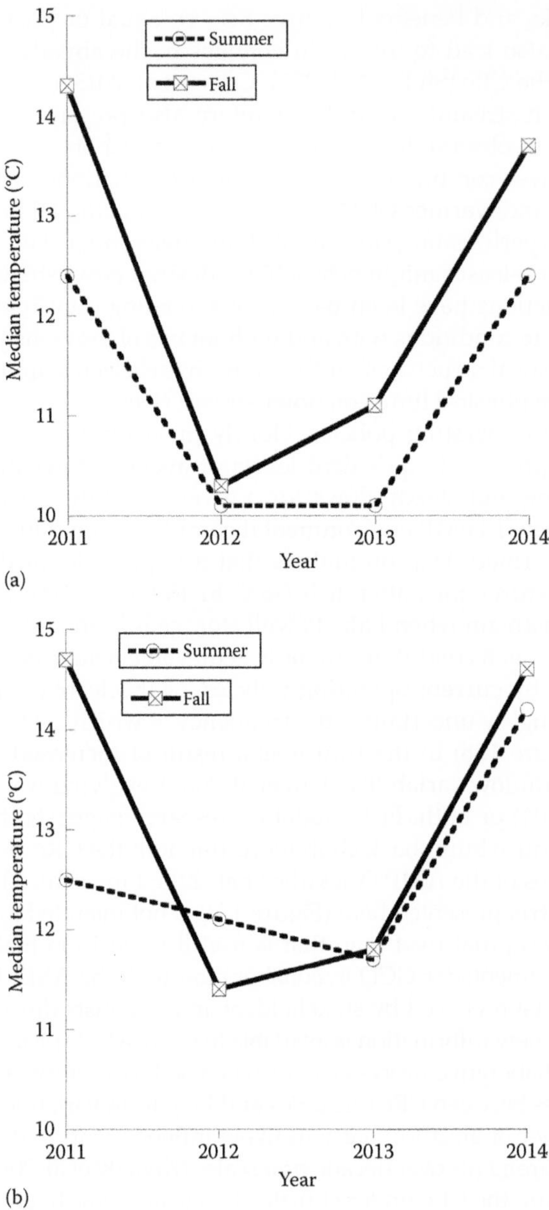

FIGURE 17.8
Median summer and fall (2011–2014) Colorado River water temperature at two points downstream from Glen Canyon Dam: (a) 25 km downstream at Lees Ferry, Arizona (USGS streamgage #0938000) and (b) 122 km downstream near the confluence of the Little Colorado and Colorado Rivers (USGS streamgage #09383100).

(Continued)

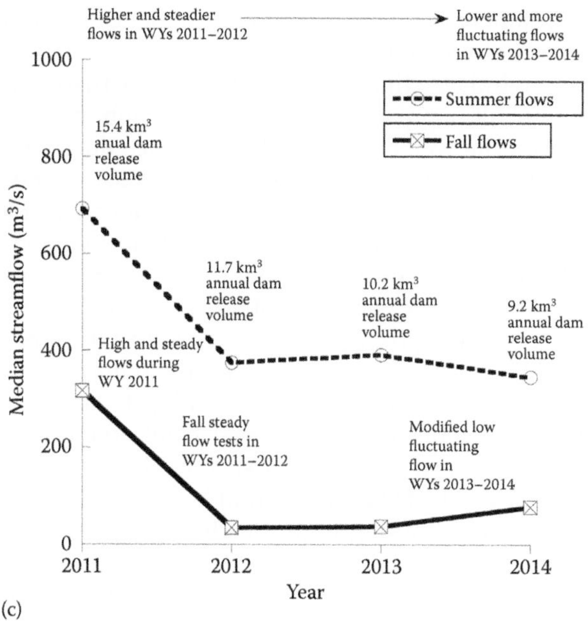

FIGURE 17.8 (CONTINUED)
Median summer and fall (2011–2014) Colorado River water temperature at two points downstream from Glen Canyon Dam: (c) Median summer and fall streamflow at Lees Ferry during a period of declining annual dam release volumes. Dam releases were relatively warm in summer and fall 2011, but downstream warming was limited by high flows associated with above-average annual dam release volume. Dam releases were cooler during 2012–2013, owing to increased storage in Lake Powell. Warmer dam releases returned again in 2014, owing to reduced reservoir elevation that allowed for warmer epilimnetic withdrawals. Dam releases from Lake Powell are typically warmest in late fall. (Data from http://www.gcmrc.gov/discharge_qw_sediment /stations/GCDAMP.)

17.4 Summary

The AMP has overcome some of the barriers to AM identified by Walters (1997) by requiring frequent assessments of ongoing monitoring data, maintaining an ongoing collaborative dialogue between scientists and stakeholders, effectively using models to design monitoring strategies and screen policy options, and willingly adapting learning outcomes. The AMP has also benefitted from strong DOI leadership, innovative technologies, and rapid learning from surprise outcomes. However, AMP stakeholders continue to grapple with other challenges, such as the inability to determine prescriptive treatments from ecosystem modeling alone, confounded experimental results, difficulty in resolving complex trade-offs, and lags in slow variable responses occurring under climate variability and climate change. Integrating indigenous perspectives about traditional ecological knowledge into the AMP has also been challenging, as understanding about native and nonnative fish interactions and the role that dam operations may play in those interactions advances.

Long-term monitoring data suggest that sandbars and rainbow trout responded negatively to the much wider fluctuations associated with hydropeaking prior to MLFF. Reductions in daily fluctuations under MLFF allowed natural production of Glen Canyon rainbow trout, an important objective of recreational anglers, but continue to erode fragile Grand Canyon sandbars between HFEs.

There is no evidence to indicate that steady or diurnal fluctuating flow dam operations either limit or increase adult populations of humpback chub and other native species in Grand Canyon. However, the weight of available evidence does suggest that native fish population increases in Grand Canyon after about 2002, occurred under relatively warmer dam releases from Lake Powell, a limited, but apparently adequate aquatic food availability and during a poorly understood but well documented reduction in trout abundance throughout the CRe. We therefore speculate that if food items remain sufficiently available in Grand Canyon, then humpback chub and other native fish may continue benefitting from warmer GCD hydropeaking releases associated with future periods of drier UCRB hydrology and reduced Lake Powell storage. However, it remains unclear how frequently colder releases might occur, and whether river warming to support juvenile native fish recruitment must also be combined with reduced trout abundance in Grand Canyon, either through avoidance of spring HFEs and steady, high-flow releases in spring and through summer, or other means for limiting trout below Glen Canyon. Avoiding spring HFEs to rebuild sandbars and limiting high, steady spring-to-summer GCD releases to meet downstream water delivery might be a feasible strategy for limiting trout production and downstream movement into segments where most chub are located. However, such an approach requires that HFEs be released only in fall (fortunately, most Paria River sand inputs occur from July through October), and a willingness on the part of water managers to consider revising current schedules for upper to lower basin water transfers from single to perhaps multi-water year periods.

Our review of information on the influence of GCD operations for three key CRe resources, sandbars, rainbow trout, and humpback chub suggests, perhaps somewhat surprisingly, that UCRB hydrology, Lake Powell storage and annual GCD release strategies required to meet downstream water delivery are likely the largest factors influencing these and perhaps other valued downstream resources. If so, then this conclusion suggests that MLFF hydropeaking operations may have only limited influence on long-term trends of downstream resources. Most of the important AMP learning results over the last two decades derived from a commitment to consistent monitoring that has identified serendipitous results rather than from a commitment to any one, long-term, experimental design.

From our collective experiences working with AMP stakeholders, we believe that one of the biggest challenges for the program is to manage and balance what Holling (1998) refers to as the "science-of-parts," or the unambiguous understanding at a small scale, with the science of "integration-of-parts," a process that uses the results of the first to identify gaps, develops hypotheses, and explores uncertainties over large scales. For example, to benefit fully from a long-term experimental and management plan, monitoring data must be collaboratively evaluated by food web and fisheries scientists as well as others such as hydrologists and geomorphologists. This will help develop a more complete interpretation of how the entire 470-km long river corridor responds to repeated HFEs over a multi-decade period; particularly with a backdrop effect of climate change, increasing year-to-year streamflow variance and resulting basin hydrology influences on river temperatures, fish and the food base. With climate change and reduced or perhaps more variable hydrology in the UCRB may come new opportunities to advance the integration of science-of-parts in greater ways than before for CRe managers. Such opportunities might be missed unless future AMP monitoring and research projects are developed with integrated objectives for learning in mind from the start, are implemented over appropriate timescales that allow resilient strategies for adaptation to be identified and embraced by managers, are frequently re-evaluated, and point toward policy options that can be supported by diverse stakeholders following equitable trade-off assessments.

The AMP's rich history of ecosystem modeling and assessment workshops, where stakeholders and scientists have spent substantial time discussing and evaluating uncertainties, information needs, and experimental monitoring options, suggests that such ongoing meetings are the logical public forum in which options for achieving resiliency in the CRe are identified and trade-offs debated relative to relict and artifact river resources. We speculate that the most effective strategy for maintaining resiliency in CRe processes and resources below this large dam will likely be one with the widest and most flexible range of adaptive dam operating possibilities for managers to use as climate change continues to influence hydrology, and water deliveries throughout the Colorado River basin become increasingly more complex.

Acknowledgments

We thank our many science colleagues for their input, particularly their numerous contributions to the available literature that guided our development of the uncertainty matrix presented here, and the many engaging discussions with AMP stakeholders that occurred at numerous assessment meetings and modeling workshops convened by the GCMRC from 1997 through 2011. Documentation of GCMRC's AM-focused assessment workshops was ably accomplished with help from Lara M. Schmit, Kyrie Fry, and Meredith Hartwell, of Northern Arizona University. We thank Barbara Ralston and Mary Freeman for their thoughtful review comments on an earlier version of this chapter; Chris Robinson, Kathleen Miller, and Chad Smith for later reviews; as well as Lara M. Schmit for editorial help in 2013 (Northern Arizona University, Merriam-Powell Center for Environmental Research). We appreciate assistance provided by Bill Vernieu (GCD and downstream CRe water temperature data and plot, US Geological Survey–GCMRC), and Rick Clayton and Katrina Grantz (Lake Powell annual unregulated inflow and GCD annual release volumes, Upper Colorado River Regional office, US Department of the Interior Bureau of Reclamation).

References

Avery, L.A., J. Korman, and W.R. Persons. 2015. Effects of increased discharge on rainbow trout spawning and age-0 recruitment in the Colorado at Lees Ferry, AZ. *North American Journal of Fisheries Management* 35:671–680. http://www.tandfonline.com/doi/pdf/10.1080/02755947.2015.1040560.

Behn, K.E., T.A. Kennedy, and R.O. Hall Jr. 2010. *Basal resources in backwaters of the Colorado River below GCD—Effects of discharge regimes and comparison with mainstem depositional environments.* US Geological Survey Open-File Report 2010-1075. 25 p. http://pubs.usgs.gov/of/2010/1075/.

Berkley, J. 2013. Opportunities for collaborative adaptive management progress: Integrating stakeholder assessments into progress measurement. *Ecology and Society* 18(4):69. http://dx.doi.org/10.5751/ES-05988-180469.

Brekke, L.D., J.E. Kiang, J.R. Olsen et al. 2009. *Climate change and water resources management—A federal perspective.* US Geological Survey Circular 1331. 65 p. http://pubs.usgs.gov/circ/1331/.

Coggins, L.G., Jr., and W.E. Pine, III. 2010. Development of a temperature-dependent growth model for the endangered humpback chub using capture–recapture data. *Open Fish Science Journal* 3:122–131. http://benthamopen.com/contents/pdf/TOFISHSJ/TOFISHSJ-3-122.pdf.

Coggins, L.G., Jr., and C.J. Walters. 2009. *Abundance trends and status of the Little Colorado River population of humpback chub; an update considering data from 1989–2008*. US Geological Survey Open-File Report 2009-1075. 18 p. http://pubs.usgs.gov/of/2009/1075/.

Coggins, L.G., Jr., M.D. Yard, and W.E. Pine, III. 2011. Nonnative fish control in the Colorado River in Grand Canyon, Arizona: An effective program or serendipitous timing? *Transactions of the American Fisheries Society* 140(2):456–470. http://www.tandfonline.com/doi/abs/10.1080/00028487.2011.572009.

Cook T. 2013. A flood of controversy on the Colorado River. *Earth* March: 40–46. http://www.agiweb.org/store/library/imprint.php?id = 2013_03.

Cook, B.I, T.R. Ault, and J.E. Smerdon. 2015. Unprecedented 21st century drought risk in the American Southwest and Central Plains. *Science Advances*. 1:e1400082 12 February 2015. 7 p. http://advances.sciencemag.org/content/advances/1/1/e1400082.full.pdf.

Cross, W.F., C.V. Baxter, E.J. Rosi-Marshall et al. 2013. Food-web dynamics in a large river discontinuum. *Ecological Monographs* 83:311–337. http://dx.doi.org/10.1890/12-1727.1.

Dodrill, M.J., C.B. Yackulic, B. Gerig, W.E. Pine III, J. Korman, and C. Finch. 2014. Do management actions to restore rare habitat benefit native fish conservation? Distribution of juvenile native fish among shoreline habitats of the Colorado River. *River Research and Applications* 1–15. doi: 10:1002/rra.2842. http://onlinelibrary.wiley.com/doi/10.1002/rra.2842/abstract.

Draut, A.E. 2012. Effects of river regulation on aeolian landscapes, Colorado River, southwestern USA. *Journal of Geophysical Research* 117:1–22. http://onlinelibrary.wiley.com/doi/10.1029/2011JF002329/pdf.

Ellis, L.E., and Jones, N.E. 2013. Longitudinal trends in regulated rivers: A review and synthesis within the context of the serial discontinuity concept. *Environmental Reviews* 21(3):136–148. doi: 10.1139/er-2012-0064. http://www.nrcresearchpress.com/doi/abs/10.1139/er-2012-0064#.VRCfLWd01D8.

Finch, C.G., W.E. Pine III, C.B. Yackulic, M.J. Dodrill, M. Yard, B.S. Gerig, L.G. Coggins, and J. Korman. 2015. Assessing juvenile native fish demographic responses to a steady flow experiment in a large regulated river. *River Research and Applications* doi: 10.1002/rra.2893. http://onlinelibrary.wiley.com/doi/10.1002/rra.2893/epdf.

Finch, C.G., W.E. Pine III, and K.E. Limburg. 2013. Do hydropeaking flows alter juvenile fish growth rates? A test with juvenile humpback chub in the Colorado River. *River Research and Applications* 31(2):156–164. doi: 10.1002/rra.2725. http://onlinelibrary.wiley.com/doi/10.1002/rra.2725/abstract.

Gerig, B., M.J. Dodrill, and W.E. Pine III. 2014. Habitat selection and movement of adult humpback chub in the Colorado River in Grand Canyon, Arizona, during an experimental steady flow release. *North American Journal of Fisheries Management* 34(1):39–48. doi: 10.1080/02755947.2013.847880. http://dx.doi.org/10.1080/02755947.2013.847880.

Gloss, S.P., and L.G. Coggins, Jr. 2005. Fishes of Grand Canyon. Pages 33–56. *in* S.P. Gloss, J.E. Lovich, and T.S. Melis eds. *The State of the Colorado River Ecosystem in Grand Canyon*. US Geological Survey Circular 1282. http://pubs.er.usgs.gov/publication/cir1282.

Gloss, S.P., J.E. Lovich, and T.S. Melis eds. 2005. *The State of the Colorado River Ecosystem in Grand Canyon*. US Geological Survey Circular 1282. 220 p. http://pubs.er.usgs.gov/publication/cir1282.

Grams, P.E. 2013. *A Sand Budget for Marble Canyon, Arizona—Implications for Long-Term Monitoring of Sand Storage Change*. US Geological Survey Fact Sheet. 013–3074. 4 p. http://pubs.usgs.gov/fs/2013/3074/.

Grams, P.E., J.C. Schmidt, and M.E. Andersen. 2010, *2008 High-Flow Experiment at Glen Canyon Dam—Morphologic Response of Eddy-Deposited Sandbars and Associated Aquatic Backwater Habitats Along the Colorado River in Grand Canyon National Park*. US Geological Survey Open-File Report 2010-1032. 73 p. http://pubs.er.usgs.gov/publication/ofr20101032.

Grams, P.E., J.C. Schmidt, S.A. Wright, D.J. Topping, T.S. Melis, and D.M. Rubin. 2015. Recent controlled floods contribute to sandbar gains in Grand Canyon. *Eos* 96, June 2015. https://eos.org/features/building-sandbars-in-the-grand-canyon.

Grams, P.E., J.C. Schmidt, and D.J. Topping. 2007. The rate and pattern of bed incision and bank adjustment on the Colorado River in Glen Canyon downstream from GCD, 1956–2000. *Geological Society of America Bulletin* 119(5–6):556–575. doi: 10.1130/B25969.1. http://gsabulletin.gsapubs.org/content/119/5-6/556.full.pdf+html.

Grand Canyon Monitoring and Research Center. 2008. *USGS Workshop on Scientific Aspects of a Long-Term Experimental Plan for GCD, April 10–11, 2007, Flagstaff, Arizona*: US Geological Survey Open-File Report 2008–1153, 79 p. http://pubs.er.usgs.gov/publication/ofr20081153.

Grand Canyon Protection Act of 1992. Pubic Law No. 102-575, §§ 1801–1809, 106 Stat.4600 (1992). http://www.usbr.gov/uc/rm/amp/legal/gcpa1992.html.

Greig, L.A., D.R. Marmorek, C. Murray, and D.C.E. Robinson. 2013. Insight into enabling adaptive management. *Ecology and Society* 18(3): art 24. 1–11. http://dx.doi.org/10.5751/ES-05686-180324.

Hamill, J.F., and T.S. Melis. 2012. The GCD adaptive management program: Progress and immediate challenges. Pages 325–338 *in* P.J. Boon and P.J. Raven eds. *River Conservation and Management*. West Sussex: John Wiley and Sons, Ltd. doi: 10.1002/9781119961819.ch26.http://onlinelibrary.wiley.com/doi/10.1002/9781119961819.ch26/pdf.

Harpman, D.A., and A.J. Douglas. 2005. Status and trends of hydropower production at GCD. Pages 165–176. *in* S.P. Gloss, J.E. Lovich, and T.S. Melis eds. *The State of the Colorado River Ecosystem in Grand Canyon*. US Geological Survey Circular 1282. http://pubs.er.usgs.gov/publication/cir1282.

Hazel, J.E., Jr., M. Kaplinski, R. Parnell, K. Kohl, and D.J. Topping. 2006. *Stage-Discharge Relations for the Colorado River in Glen, Marble, and Grand Canyons, Arizona*. US Geological Survey Open-File Report 2006-1243. 7 p. http://pubs.er.usgs.gov/publication/ofr20061243.

Hirsch, R.M., and K.R. Ryberg. 2012. Has the magnitude of floods across the USA changed with global CO2 levels? *Hydrolological Sciences Journal* 57(1):1–9. http://www.tandfonline.com/doi/suppl/10.1080/02626667.2011.621895#tabModule.

Hogg, I.D., and D.D. Williams. 1996. Response of stream invertebrates to a global-warming thermal regime: An ecosystem-level manipulation. *Ecology* 77(2):395–407. http://dx.doi.org/10.2307/2265617.

Holling, C.S. 1998. Two cultures of ecology. *Conservation Ecology*. 2(2):4. http://www.ecologyandsociety.org/vol2/iss2/art4/.

Jain, S., M. Hoerling, and J. Eischeid. 2005. Decreasing reliability and increasing synchroneity of western North American streamflow. *Journal of Climate* 18:613–618. http://journals.ametsoc.org/doi/pdf/10.1175/JCLI-3311.1.

Kaplinski, M., J.E. Hazel, Jr., R. Parnell et al. 2014. *Colorado River Campsite Monitoring, Grand Canyon National Park, Arizona, 1998–2012*. US Geological Survey Open-File Report 2014-1161, 24 p. plus appendix. http://dx.doi.org/10.3133/ofr20141161.

Kennedy, T.A. 2013. *Identification and Evaluation of Scientific Uncertainties Related to Fish and Aquatic Resources in the Colorado River, Grand Canyon—Summary and Interpretation of an Expert-Elicitation Questionnaire*. US Geological Survey Scientific Investigations Report 2013–5027. 34 p. http://pubs.usgs.gov/sir/2013/5027/.

Kennedy, T.A., W.F. Cross, R.O Hall et al. 2013. *Native and Nonnative Fish Populations of the Colorado River Are Food Limited—Evidence From New Food Web Analyses*. US Geological Survey Fact Sheet 2013-3039. 4 p. http://pubs.usgs.gov/fs/2013/3039.

Kennedy, T.A., and S.P. Gloss. 2005. Ecology: The role of organic matter and invertebrates. Pages 87–101. *in* S.P. Gloss, J.E. Lovich, and T.S. Melis eds. *The State of the Colorado River ecosystem in Grand Canyon*. US Geological Survey Circular 1282. http://pubs.er.usgs.gov/publication/cir1282.

Kenney, D., S. Bates, A. Bensard, and J. Berggren. 2011. The Colorado River and the inevitability of institutional change. *Public Land & Resources Law Review* 32:103–152. http://scholarship.law.umt.edu/cgi/viewcontent.cgi?article = 1025&context = plrlr.

King, A.J., B. Gawne, L. Beesley, J.D. Koehn, D.L. Nielsen, and A. Price. 2015. Improving ecological response monitoring of environmental flows. *Environmental Management* 55(5):991–1005.

Konrad, C.P., J.D. Olden, K.B. Gido et al. 2011. Large-scale flow experiments for managing river systems. *Ecological Applications* 61(12):948–959. http://bioscience.oxfordjournals.org/content/61/12/948.short.

Korman J., S.J.D. Martell, C.J. Walters et al. 2012. Estimating recruitment dynamics and movement of rainbow trout (*Oncorhynchus mykiss*) in the Colorado River in Grand Canyon using an integrated assessment model. *Canadian Journal of Fisheries and Aquatic Science* 69(11):1827–1849. doi: 10.1139/f2012-097. http://www.nrcresearchpress.com/doi/pdf/10.1139/f2012-097.

Korman, J., and T.S. Melis. 2011. *The Effects of GCD Operations on Early Life Stages of Rainbow Trout in the Colorado River*. USGS Fact Sheet 2011-3002. 4p. http://pubs.usgs.gov/fs/2011/3002.

Korman, J., M. Kaplinski, and T.S. Melis. 2011a. Effects of fluctuating flows and a controlled flood on incubation success and early survival rates and growth of age-0 rainbow trout in a large regulated river. *Transactions of the American Fisheries Society* 140(2):487–505. http://www.tandfonline.com/doi/abs/10.1080/00028487.2011.572015.

Korman, J., C.J. Walters, S.J.D. Martell, W.E. Pine, III, and A. Dutterer. 2011b. Effects of flow fluctuations on habitat use and survival of age-0 rainbow trout (*Oncorhynchus mykiss*) in a large, regulated river. *Canadian Journal of Fisheries and Aquatic Sciences* 68:1097–1109. http://www.nrcresearchpress.com/doi/abs/10.1139/f2011-045#.VRCkw2d01D8.

Korman, J., and S.E. Campana. 2009. Effects of hydropeaking on nearshore habitat use and growth of age-0 rainbow trout in a large regulated river. *Transactions of the American Fisheries Society* 138:76–87. http://www.gcmrc.gov/files/2606/w-2606_2011-06-23-18-31-26-337_362402_713582191_936395114.pdf.

Lovich, J.E., and T.S. Melis. 2007. The state of the Colorado River Ecosystem in Grand Canyon: Lessons from 10 years of adaptive ecosystem management. *International Journal of River Basin Management* 5(3):207–221. http://www.tandfonline.com/doi/pdf/10.1080/15715124.2007.9635321.

Magirl, C.F.N., R.H. Webb, and P.E. Griffiths. 2008. *Modeling Water-Surface Elevations and Virtual Shorelines for the Colorado River in Grand Canyon, Arizona*. US Geological Survey. Scientific Investigations Report 2008-5075. 32p. http://pubs.usgs.gov/sir/2008/5075/.

Makinster, A.S., W.R. Persons, L.A. Avery, and A.J. Bunch. 2010. *Colorado River Fish Monitoring in Grand Canyon, Arizona: 2000 to 2009 summary*. US Geological Survey Open-File Report 2010–1246. http://pubs.usgs.gov/of/2010/1246/.

McKinney, T., D.W. Speas, R.S. Rogers, and W.R. Persons. 2001. Rainbow trout in a regulated river below GCD, Arizona, following increased minimum flows and reduced discharge variability. *North American Journal of Fisheries Management* 21(1):216–222. http://www.tandfonline.com/doi/pdf/10.1577/1548-8675%282001%29021%3C0216%3ARTIARR%3E2.0.CO%3B2.

Melis, T.S. (editor). 2011. *Effects of Three High-Flow Experiments on the Colorado River Ecosystem Downstream From Glen Canyon Dam, Arizona*. US Geological Survey Circular 1366. 147 p. http://pubs.usgs.gov/circ/1366/.

Melis, T.S., C.J. Walters, and J. Korman. 2015. Surprise and opportunity for learning in Grand Canyon: The Glen Canyon Dam Adaptive Management Program. *Ecology and Society* 20(3):22. http://dx.doi.org/10.5751/ES-07621-200322.

Melis, T.S., J. Korman, and T.A. Kennedy. 2012. Abiotic and biotic responses of the Colorado River to experimental floods at GCD, Arizona, USA. *River Research and Applications* 28:764–776. http://onlinelibrary.wiley.com/doi/10.1002/rra.1503/.

Melis, T.S., S.J.D. Martell, L.G. Coggins, W.E. Pine III, and M.E. Andersen. 2006. Adaptive management of the Colorado River ecosystem below GCD, Arizona: Using science and modeling to resolve uncertainty in river management. *Proceedings of the Adaptive Management of Water Resources AWRA Summer Specialty Conference*, June 26–28, 2006, Missoula, MT. http://www.gcmrc.gov/publications/library.aspx.

Mueller, E.R., P.E. Grams, J.C. Schmidt, J.E. Hazel, Jr., J.S. Alexander, and M. Kaplinski. 2014. The influence of controlled floods on fine sediment storage in debris fan-affected canyons of the Colorado River basin. *Geomorphology* 226:65–75. doi: 10.1016/j.geomorph.2014.07.029. http://www.sciencedirect.com/science/article/pii/S0169555X14003948.

Olden, J.D., and R.J. Naiman. 2010. Incorporating thermal regimes into environmental flows assessments: Modifying dam operations to restore freshwater ecosystem integrity. *Freshwater Biology* 55:86–107. http://onlinelibrary.wiley.com/doi/10.1111/j.1365-2427.2009.02179.x/epdf.

Olden, J.D., C.P. Konrad, T.S. Melis et al. 2014. Are large-scale flow experiments informing emerging challenges in freshwater management? *Frontiers in Ecology and the Environment* 12(3):176–185. doi:10.1890/130076. http://www.esajournals.org/doi/pdf/10.1890/130076.

Pine, W.E., III, B. Healy, E. Omana-Smith et al. 2013. An individual-based model for population viability analysis of humpback chub in the Colorado River in Grand Canyon, Arizona. *North American Journal of Fisheries Management*. 33(3):626–641, doi: 10.1080/02755947.2013.788587. http://dx.doi.org/10.1080/02755947.2013.788587.

Pulwarty, R.S. 2003. Climate and water in the west: Science, information and decision-making. *Water Resources* 124:4–12. http://www.ucowr.org/files/Achieved_Journal_Issues/V124_A1Climate%20and%20Water%20in%20the%20West%20Science,%20Information%20and%20Decision-Making.pdf.

Pulwarty, R.S., and T.S. Melis. 2001. Climate extremes and adaptive management on the Colorado River: lessons from the 1997–1998 ENSO event. *Journal of Environmental Management* 63:307–324. doi:10.1006/jema.2001.0494. http://www.colorado.edu/geography/geomorph/envs_5810/pulwarty_01.pdf.

Ralston, B.E. 2011. *Summary Report of Responses of Key Resources to the 2000 Low Steady Summer Flow Experiment, Along the Colorado River Downstream From GCD, Arizona*. US Geological Survey Open-File Report 2011–1220. 129 p. http://pubs.er.usgs.gov/publication/ofr20111220/.

Ralston, B.E. 2010. *Riparian Vegetation Response to the March 2008 Short-Duration, High-Flow Experiment—Implications of Timing and Frequency of Flood Disturbance on Nonnative Plant Establishment Along the Colorado River Below GCD*. US Geological Survey Open-File Report 2010–1022. 30 p. http://pubs.usgs.gov/of/2010/1022/.

Ralston, B.E. 2005. Riparian vegetation and associated wildlife. *in* S.P. Gloss, J.E. Lovich, and T.S. Melis eds. *The State of the Colorado River Ecosystem in Grand Canyon*. US Geological Survey Circular 1282. chap. 6, p. 103–122. http://pubs.usgs.gov/circ/1282/.

Ralston, B.E., M.V. Lauretta, and T.A. Kennedy. 2007. *Comparisons of Water Quality and Biological Variables From Colorado River Shoreline Habitats in Grand Canyon, Arizona, Under Steady and Fluctuating Discharges From Glen Canyon Dam*. US Geological Survey Open File Report 2007-1195. 29 p. http://pubs.usgs.gov/of/207/1195/.

Ralston, B.E., A.M. Starfield, R.S. Black, and R.A. Van Lonkhuyzen. 2014. *State-and-Transition Prototype Model of Riparian Vegetation Downstream of Glen Canyon Dam, Arizona*. US Geological Survey Open-File Report 2014-1095. 26 p. http://dx.doi.org/10.3133/ofr20141095.

Ross, R., and P.E. Grams. 2013. *Nearshore Thermal Gradients of the Colorado River Near the Little Colorado River Confluence, Grand Canyon National Park, Arizona, 2010*. US Geological Survey Open-File Report 2013-1013. 65 p. http://pubs.usgs.gov/of/2013/1013/of2013-1013_text.pdf.

Rubin, D.M., D.J. Topping, J.C. Schmidt et al. 2002. Recent sediment studies refute Glen Canyon Dam hypothesis. *Eos, Transactions, American Geophysical Union* 83(25):273, 277–278. http://onlinelibrary.wiley.com/doi/10.1029/2002EO000191/epdf.

Rubin, D.M., J.C. Schmidt, R.A. Anima et al. 1994. *Internal Structure of Bars in Grand Canyon, Arizona, and Evaluation of Proposed Flow Alternatives for Glen Canyon Dam*. US Geological Survey Open-File Report 94-594. 35 p. http://pubs.usgs.gov/of/1994/0594/report.pdf.

Sankey, J.B., and A.E. Draut. 2014. Gully annealing by aeolian sediment: Field and remote-sensing investigation of aeolian–hillslope–fluvial interactions, Colorado River corridor, Arizona, USA. *Geomorphology*. 220:68–80. http://dx.doi.org/10.1016/j.geomorph.2014.05.028.

Scarlett, L. 2013. Collaborative adaptive management: challenges and opportunities. *Ecology and Society* 18(3):26. http://dx.doi.org/10.5751/ES-05762-180326.

Schmidt, J.C., and P.E. Grams. 2011. The high flows: physical science results. Pages 53–91. *in* T.S. Melis ed. *Effects of Three High-Flow Experiments on the Colorado River Ecosystem Downstream From GCD, Arizona.* US Geological Survey Circular 1366. http://pubs.usgs.gov/circ/1366/.

Schmidt, J.C., R.H. Webb, R.A. Valdez, G.R. Marzolf, and L.E. Stevens. 1998. Science and values in river restoration in the Grand Canyon. *BioScience* 48(9):735–747. http://www.jstor.org/stable/1313336.

Seager, R., M. Ting, I. Held et al. 2007. Model projections of an imminent transition to a more arid climate in southwestern North America. *Science* 316(5828):1181–1184. http://www.sciencemag.org/cgi/content/full/316/5828/1181.

Seager, R., M. Ting, C. Li et al. 2013. Projections of declining surface water availability for the southwestern United States. *Nature Climate Change* 3:482–486. doi:10.1038/NCLIMATE1787. http://www.nature.com/nclimate/journal/v3/n5/pdf/nclimate1787.pdf.

Stoks, R., A.N. Geerts, and L. De Meester. 2014. Evolutionary and plastic responses of freshwater invertebrates to climate change: realized patterns and future potential. *Evolutionary Applications* 7(1):42–55. doi: 10.1111/eva.12108PMCID. http://www.ncbi.nlm.nih.gov/pmc/articles/PMC3894897/.

Topping, D.J. 1997. Physics of flow, sediment transport, hydraulic geometry and channel geomorphic adjustment during flash floods in an ephemeral river, the Paria River, Utah and Arizona. Ph.D. dissertation. University of Washington. 205 p. http://www.gcmrc.gov/library/reports/Physical/Fine_Sed/Topping1997V1.pdf.

Topping, D.J., J.C. Schmidt, and L.E. Vierra. 2003. *Computation and Analysis of the Instantaneous-Discharge Record for the Colorado River at Lees Ferry, Arizona—May 8, 1921, Through September 30, 2000.* US Geological Survey Professional Paper 1677. http://pubs.usgs.gov/pp/pp1677/pdf/pp1677.pdf.

Trammell, M., R.A. Valdez, S.W. Carothers, and R. Ryel. 2002. *Effects of a Low Steady Summer Flow Experiment on Native Fishes of the Colorado River in Grand Canyon—Final Report: Flagstaff, Ariz., SWCA, Inc. Environmental Consultants,* submitted to US Geological Survey, Grand Canyon Monitoring and Research Center, cooperative agreement no. 99–FC–40–2260, 77 p. [Available upon request by contacting the Center Director, US Geological Survey, Southwest Biological Science Center, 2255 N. Gemini Drive, Flagstaff, Ariz. 86001.]

US Department of the Interior. 1946. *The Colorado River—A Natural Menace Becomes a National Resource.* Washington, D.C. Bureau of Reclamation. 295 p. http://www.riversimulator.org/Resources/USBR/Menace.pdf.

US Department of the Interior. 1995. *Operation of Glen Canyon Dam—Final Environmental Impact Statement, Colorado River Storage Project, Coconino County, Arizona.* Salt Lake City, Utah, Bureau of Reclamation. Upper Colorado Regional Office. 337 p. http://www.usbr.gov/uc/library/envdocs/eis/gc/pdfs/Cov-con/cov-con.pdf.

US Department of the Interior. 2012a. *Environmental Assessment: Development and Implementation of a Protocol for High-Flow Experimental Releases from Glen Canyon Dam, Arizona, 2011 through 2020.* Bureau of Reclamation, Salt Lake City, UT. 546 p. http://www.usbr.gov/uc/envdocs/ea/gc/HFEProtocol/index.html.

US Department of the Interior. 2012b. *Colorado River Basin Water Supply and Demand Study—Study Report.* Bureau of Reclamation. Boulder City, Nevada. 99p. http://www.usbr.gov/lc/region/programs/crbstudy/finalreport/Study%20Report/StudyReport_FINAL_Dec2012.pdf.

Vano, J.A., B. Udall, D.R. Cayan et al. 2014. Understanding uncertainties in future Colorado River streamflow. *Bulletin of the American Meteorological Society* 95(1):59–78. http://journals.ametsoc.org/doi/abs/10.1175/BAMS-D-12-00228.1.

Vernieu, W.S. 2013. *Historical Physical and Chemical Data for Water in Lake Powell and From GCD Releases, Utah–Arizona, 1964–2012* (ver. 2.0, October 2013). US Geological Survey Data Series 471. 23 p. http://pubs.usgs.gov/ds/471/.

Vernieu, W.S., and C.R. Anderson. 2013. *Water Temperatures in Select Nearshore Environments of the Colorado River in Grand Canyon, Arizona, During the Low Steady Summer Flow Experiment of 2000.* US Geological Survey Open-File Report 2013–1066. 44 p. http://pubs.er.usgs.gov/publication/ofr20131066.

Vernieu, W.S., S.J. Hueftle, and S.P. Gloss. 2005. Water quality in Lake Powell and the Colorado River. Pages 69–86. *in* S.P. Gloss, J.E. Lovich, and T.S. Melis eds. *The State of the Colorado River Ecosystem in Grand Canyon*. US Geological Survey Circular 1282. http://pubs.er.usgs.gov/publication/cir1282.

Voichick, N., and Topping, D.J. 2014. *Extending the Turbidity Record—Making Additional Use of Continuous Data From Turbidity, Acoustic-Doppler, and Laser Diffraction Instruments and Suspended-Sediment Samples in the Colorado River in Grand Canyon*. US Geological Survey Scientific Investigations Report 2014–5097. http://dx.doi.org/10.3133/sir20145097.

Voichick, N., and Wright, S.A. 2007. *Water-Temperature Data for the Colorado River and Tributaries Between GCD and Spencer Canyon, Northern Arizona, 1988–2005*. US Geological Survey Data Series 251. 24 p. http://pubs.usgs.gov/ds/2007/251/.

Walters, C.J. 1997. Challenges in adaptive management of riparian and coastal ecosystems. *Conservation Ecology* 1(2):1. http://www.consecol.org/vol1/iss2/art1/.

Walters, C.J., B.T. van Poorten, and L.G. Coggins. 2012. Bioenergetics and population dynamics of flannelmouth sucker and bluehead sucker in Grand Canyon as evidenced by tag recapture observations. *Transactions of the American Fisheries Society* 141(1):158–173. http://dx.doi.org/10.1080/00028487.2012.654891.

Walters, C.J., J. Korman, L.E. Stevens, and B. Gold. 2000. Ecosystem modeling for evaluation of adaptive management policies in the Grand Canyon. *Conservation Ecology* 4(2):1–38. http://www.consecol.org/vol4/iss2/art1/.

Wright, S.A., and T.A. Kennedy. 2011. Science-based strategies for future high-flow experiments at Glen Canyon Dam. Pages 127–147 *in* T.S. Melis, editor. *Effects of Three High-Flow Experiments on the Colorado River Ecosystem Downstream From Glen Canyon Dam, Arizona.* US Geological Survey Circular 1366. http://pubs.usgs.gov/circ/1366/.

Wright, S.A., and P.E. Grams, 2010. *Evaluation of Water Year 2011 Glen Canyon Dam Flow Release Scenarios on Downstream Sand Storage Along the Colorado River in Arizona*. US Geological Survey Open-File Report 2010-1133. 19 p. http://pubs.usgs.gov/of/2010/1133/.

Wright, S.A., D.J. Topping, D.M. Rubin, and T.S. Melis. 2010. An approach for modeling sediment budgets in supply-limited rivers. *Water Resources Research* 46:1–18. W10538. doi:10.1029/2009WR008600. http://www.agu.org/journals/wr/wr1010/2009WR008600/2009WR008600.pdf.

Wright, S.A., C.R. Anderson, and N. Voichick. 2009. A simplified water temperature model for the Colorado River, below GCD. *River Research and Applications*. 25:675–686. doi: 10.1002/rra.1179. http://onlinelibrary.wiley.com/doi/10.1002/rra.1179/epdf.

Wright, S.A., J.C. Schmidt, T.S. Melis, D.J. Topping, and D.M. Rubin. 2008. Is there enough sand? Evaluating the fate of Grand Canyon sandbars. *GSA Today* 18(8):4–10. doi:10.1130/GSATG12A.1. http://www.geosociety.org/gsatoday/archive/18/8/pdf/i1052-5173-18-8-4.pdf.

Wright, S.A., T.S. Melis, D.J. Topping, and D.M. Rubin. 2005. Influence of Glen Canyon dam operations on downstream sand resources of the Colorado River in Grand Canyon. Pages 17–32. *in* S.P. Gloss, J.E. Lovich, and T.S. Melis eds. *The State of the Colorado River Ecosystem in Grand Canyon*. US Geological Survey Circular 1282. http://pubs.usgs.gov/circ/1282/.

Yackulic, C.B., M.D. Yard, J. Korman, and D.R. Van Haverbeke. 2014. A quantitative life history of endangered humpback chub that spawn in the Little Colorado River: Variation in movement, growth, and survival. *Ecology and Evolution* 4(7):1006–1018. doi: 10.1002/ece3.990. http://onlinelibrary.wiley.com/doi/10.1002/ece3.990/pdf.

Yard, M.D., L.G. Coggins, Jr., C.V. Baxter, G.E. Bennett, and J. Korman. 2011. Trout piscivory in the Colorado River, Grand Canyon. *Transactions of the American Fisheries Society* 140(2):471–486. http://dx.doi.org/10.1080/00028487.2011.572011.

18

Integration of Surface Water and Groundwater Rights: Colorado's Experience

Thomas V. Cech

CONTENTS

ABSTRACT Groundwater and surface water are hydraulically connected in some regions of the world. This type of groundwater system is called a tributary aquifer and can affect flows of adjacent streams. Colorado state law requires all water users—of both surface and tributary groundwater—to follow a strict water-right priority system of "first in time, first in right." This Doctrine of Prior Appropriation, adopted in Colorado in 1876, is rigidly followed today. Since 2003, thousands of tributary groundwater irrigation wells have been curtailed from pumping (legally shut off) due to the negative impacts of pumping depletions that reduce stream flow. Surface and tributary groundwater use conflicts are inevitable in locations where rigid water-allocation systems are followed. Nobel Prize winner Elinor Ostrom points out that common-pool resource management requires collective-choice arrangements if resource users—in this case, surface water and groundwater users—are to develop a stable management plan.

18.1 Introduction

Here is a land where life is written in water.

These words of Thomas Hornsby Ferril, Colorado's poet laureate, are etched in the rotunda of the state capitol in Denver. Colorado's early-day pioneers created their water legacy with an elaborate network of surface water irrigation ditches across the state. Later, second- and third-generation settlers and others constructed wells to pump shallow groundwater—primarily for irrigation. These developments overlapped and intersected, and eventually, tributary groundwater use dramatically altered the surface water flows of many adjacent streams. This led to a dramatic shift in surface water and groundwater laws and management in Colorado, and intense conflict among water users during the first decade of the twenty-first century. Unfortunately, this trend is also developing in other parts of the United States and the world.

Many water users are unaware that surface water and groundwater can be hydraulically connected. This means that the flow of some streams can be enhanced by inflows (accretions) from adjacent shallow groundwater aquifers. Alternatively, some aquifers can be recharged by infiltration from streams. In severe situations, groundwater pumping can deplete the flow of adjacent surface streams.

Aquifers can be found at a variety of depths below the land surface. The depth to groundwater is an important factor that determines the hydraulic connection between a stream and an aquifer. If the groundwater table is only tens of feet below the land surface, and no geologic barrier exists—such as a layer of clay or other impermeable geologic material—groundwater may slowly enter a stream. If the groundwater table declines due to groundwater pumping or drought, water from the stream may move slowly back into the hydraulically connected aquifer.

An *alluvial aquifer* is created when moving surface water deposits sedimentary material (sand, gravel, silt, clay particles, etc.) and later fills with groundwater. This type of stream–aquifer system is also called a *tributary aquifer* and can affect streamflow, as described. Well pumping can significantly affect the direction of movement of groundwater within an alluvial aquifer (Figure 18.1). In turn, this can create conflict between surface water and groundwater users.

By contrast, a *nontributary aquifer* is one that does not affect streamflow—perhaps because there is an impermeable layer of clay, shale, or other geologic material that does not allow groundwater to move into a surface stream. Or, an aquifer might be nontributary because of its significant depth below the land surface. For example, in some US High Plains locations, the Ogallala Aquifer has a groundwater table several hundred feet below the land surface. Thus, no groundwater moves into streams in those regions, because the groundwater table is far below the bottom of riverbeds (Winter et al. 1998).

18.2 Development of Surface Water-Allocation Laws in Colorado

Through the ages, civilizations have developed water-allocation laws to promote economic development and protect public health and safety and recently for environmental protection. In humid regions, surface water-allocation rules are generally based on the concept of *sharing*. By contrast, the concept of *first possession*, or priority-based systems, is generally used in more arid locations. At some locations in ancient times, surface water irrigation was developed and managed by a king, emperor, or pharaoh.

18.2.1 Riparian Doctrine

In some areas of the United States and the world, a *riparian landowner*—one who owns property adjacent to a stream—can make *de minimis* (reasonable) use for milling, domestic,

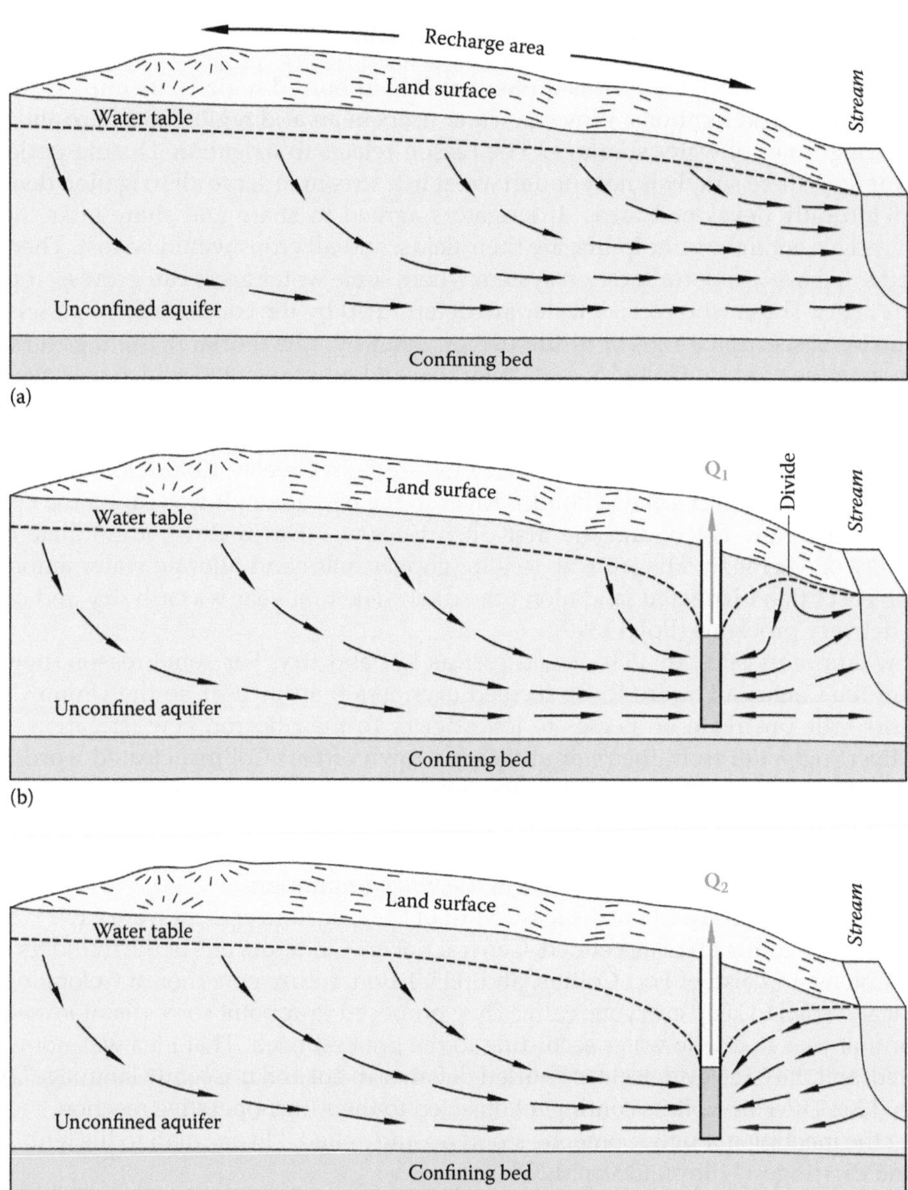

FIGURE 18.1
Effects of groundwater development on groundwater flow to and from surface water bodies. (From Winter, T.C. et al., 1998, Ground Water and Surface Water a Single Resource, *US Geological Survey (USGS) Circular*, 1139, USGS, Denver, CO, 70 pp.)

manufacturing, and agricultural purposes as long as navigation is not injured. A riparian water user must return water to the stream unchanged in water quantity and quality. During times of drought, all water users share limited resources. This concept of riparian water use is important, and in the United States, individual states determine the method of water management used within their borders. States in the more humid eastern portion of the United States generally rely on the riparian doctrine.

18.2.2 Doctrine of Prior Appropriation in Colorado

While the riparian doctrine proves quite workable in humid regions, its implementation is inadequate in arid locations. Why can water users in an arid region not share and share alike during times of water shortage? The reason relates to irrigation. During periods of low river flow, there simply is not enough water in a stream to serve all irrigation demands in agriculturally developed areas. If irrigators agreed to share and share alike, no one would receive enough water to irrigate their fields, and all crops would be lost. Therefore, it is better to have a water-allocation system where some water users can grow a crop during shortages. Those who receive water are determined by the concept of *first possession*—a hierarchy based upon a grant of the use of water by a territorial or state government (common in the western United States, Australia, and other regions), which was generally developed in the 1800s.

In 1870, for example, one of the first large-scale community irrigation systems in America was developed by settlers of the Union Colony—today's Greeley, Colorado. An irrigation ditch was dug from the Cache la Poudre River to the new town. It was called the Greeley Irrigation Company Ditch, and the first ditch rider was David Boyd, a graduate of the University of Michigan. His job was to plug gopher holes and allocate water among the 200 parcels of flood-irrigated land along the ditch. The first year was ash dry and caused many delivery problems (Boyd 1890).

A few summers later, in 1874, it was just as hot and dry. For some reason, however, flows in the Cache la Poudre River seemed even less than in 1870, so the Union Colony irrigators rode upstream on horses to investigate. To their horror, new irrigation canals were diverting water from the river at the new town of Fort Collins. Heated words were exchanged between irrigators because the downstream *senior* irrigators knew the Union Colony settlement was doomed if the upstream *junior* diverters were allowed to take water from the dwindling river. The problem was not resolved, so a meeting was called for later at a neutral schoolhouse midway between the two communities.

The meeting was lively, to say the least. Most of the irrigators present were Civil War veterans, since the conflict had just ended. General Robert Cameron, one of the founders of the Union Colony and also of Fort Collins, and B.H. Eaton, future governor of Colorado, were present and tried to keep everyone calm. They proposed to appoint some disinterested person for that year to divide water according to the greatest need. That idea was not widely accepted, and the Greeley delegates "hurled defiance in hot and unseemly language" (Boyd 1890, p. 120). The Fort Collins contingent objected to their uncooperative reaction.

Then the meeting got ugly. Someone stood up and yelled, "Every man to his tent! To his rifle and cartridges!" (Boyd 1890, p. 120).

Eventually, Mr. Eaton and General Cameron were able to quiet the mob, but no solution was found. Luckily, heavy rains the next few days reduced tensions. Irrigators along the Cache la Poudre River soon adopted the principles of the *priority system* (first in time, first in right), which had been used in the goldfields of California to reduce bloodshed. Since many irrigators in the region were former gold miners, it was natural they would adopt a similar standard. In addition, miners may have known of the tin miners' water-allocation rules in Great Britain. The water rights along the Cache la Poudre River were soon recorded with the territorial court system. Later, in 1876, these irrigators were instrumental in having the Colorado Doctrine written into the state constitution:

> The right to divert the unappropriated waters of any natural stream to beneficial uses shall never be denied. Priority of appropriation shall give the better right as between those using water for the same purpose. (Colorado Constitution 1876)

Surface water development continued throughout Colorado—from the 1860s to the present day. Following the state constitution, the prior person to appropriate, divert, and place water to a beneficial use has a better (more senior) right than someone who diverts later. This system of *first in time, first in right* continues today in Colorado and most Western states for the allocation of surface water (Cech 2010).

18.3 Development of Groundwater-Allocation Laws in Colorado

Surface water conflicts can be relatively straightforward in situations where water is diverted from a stream; a downstream diverter feels shortchanged and protests; facts are assembled regarding diversion points, flow of surface water, patterns of use, etc.; and then the issue is resolved by a state water official, board, or court of law. By contrast, groundwater conflicts are often difficult to understand and resolve because the resource is hidden underground, the movement of groundwater is difficult to substantiate, and impacts to surface water users can be difficult to quantify.

One of the first court cases regarding tributary groundwater pumping and its effect on streamflow was *McClellan v. Hurdle* in northeastern Colorado in 1893. In 1886, the plaintiff owned 400 acres* of land and constructed a diversion structure to irrigate out of nearby Lone Tree Creek. The defendant later sunk a well near the creek and installed a pump. At times, the stream flowed aboveground, but at other times, it went dry. The plaintiff argued that well pumping depleted the flow of the creek and should be stopped. The court ruled that it was an invasion of the rights of the prior appropriator—the landowner who had constructed the diversion structure from Lone Tree Creek—for another to divert from a stream *surface or subterranean* by means of dams, wells, or pumps, whereby the flow of water is diminished. However, in this case, the Court felt that the evidence was vague and indefinite and did approve the claim for damages (*McClellan v. Hurdle* 1893).

Since the late 1960s, Colorado has led the nation, and perhaps the world, in the management of groundwater use as it affects surface streamflow. This is not to say the Colorado's groundwater management methods are the most efficient or reasonable. However, it can be argued that it is the most rigid system in the United States (and perhaps the world) to protect the senior water rights related to surface water.

Recall that the Colorado Constitution of 1876 adopted the Doctrine of Prior Appropriation—first in time, first in right. How has this affected groundwater management in the state? In the South Platte Valley of northeastern Colorado, groundwater wells were generally first dug in the early 1900s, although a few were dug in the late 1800s. This was many decades after the irrigation-ditch construction boom, most of which occurred in the period 1860–1890. During the droughts of the 1930s and 1950s, thousands of wells were drilled and developed. Later, Frank Zybach's invention of the center pivot irrigation system, patented in 1952, led to further groundwater development and depletions throughout Colorado and the West. Prior to 1969, tributary groundwater pumping was not regulated by the Doctrine of Prior Appropriation in Colorado, and tensions between surface water irrigators and well pumpers escalated.

The reason for the coming fight between surface water and groundwater users was simple. First, senior surface water users had developed an elaborate legal and management

* 1 acre = 0.4 ha.

system for allocating scarce water supplies during dry summer months. Surface irrigators knew they had to be *in priority* before water could be diverted from a stream. By contrast, groundwater irrigators could simply flip a switch to pump water from beneath their land. Second, surface water users were convinced that tributary groundwater, located near a stream, *fed* the stream during summer months. When groundwater was pumped, they believed that it had a negative effect on streamflows. As early as 1954, Colorado state senator Ranger Rogers accused well pumpers of "robbing" surface water from the South Platte River (Kryloff 2007).

18.3.1 The Water Rights Determination and Administration Act of 1969

Finally, in 1968, an engineering study was authorized by the Colorado legislature to provide recommendations for integrating the use of alluvial (tributary) groundwater into the surface water system. The following year, lobbying by surface water owners helped secure passage of the Water Rights Determination and Administration Act of 1969, which stated that all tributary irrigation wells had to follow the same priority system of water allocation described in the Colorado Constitution.

The 1969 act created seven water divisions based upon the drainage patterns of various rivers in Colorado (Figure 18.2). Each water division is staffed with a division engineer, appointed by the state engineer; a water court judge, appointed by the Colorado Supreme Court; a water court referee, appointed by the water judge; and a water court clerk, assigned by the district court. Water court judges are district court judges appointed by the Colorado Supreme Court and have jurisdiction in settling litigation regarding water rights, the use and administration of water, and all other water matters within the jurisdiction of the water divisions (Colorado Supreme Court 2015).

The key component of the 1969 act was that a well drilled in 1949 now had a priority date of 1949. That was not good news for tributary-well owners—considering that junior surface water diversions had priority dates in the 1890s and reservoirs had decrees from the early 1900s. By contrast, most well users had priorities junior to the early 1900s. A catastrophe was waiting to happen for 10,000 irrigation-well owners in Colorado, but nature and limited groundwater management intervened during the 1980s and 1990s.

18.3.2 Development of Augmentation Plans in Colorado

In the early 1970s, well owners were required to belong to an augmentation plan such as the Groundwater Appropriators of the South Platte, Inc. (GASP), the Central Colorado Water Conservancy District, or an individual augmentation plan. These augmentation plans were a management tool to replace out-of-priority well-pumping depletions for the South Platte River for senior water users. The amount of augmentation required was determined either by a percentage of the amount of water pumped (initially only 5% of pumping), or through a crop consumptive use analysis. Out-of-priority well-pumping depletions were replaced based upon consumptive use calculations, distance of a well from the South Platte River, and demand for senior water rights downstream (called *river calls*).

Well owners in Central and GASP paid an annual fee to receive augmentation coverage. Fees were used to purchase or rent water for delivery to streams negatively impacted by well pumping, so that streamflows were *augmented*. In this way, downstream senior diverters were protected from injury from out-of-priority well pumping. Cost of water acquisition was generally affordable for the region at US $15–20/acre-foot to annually lease water

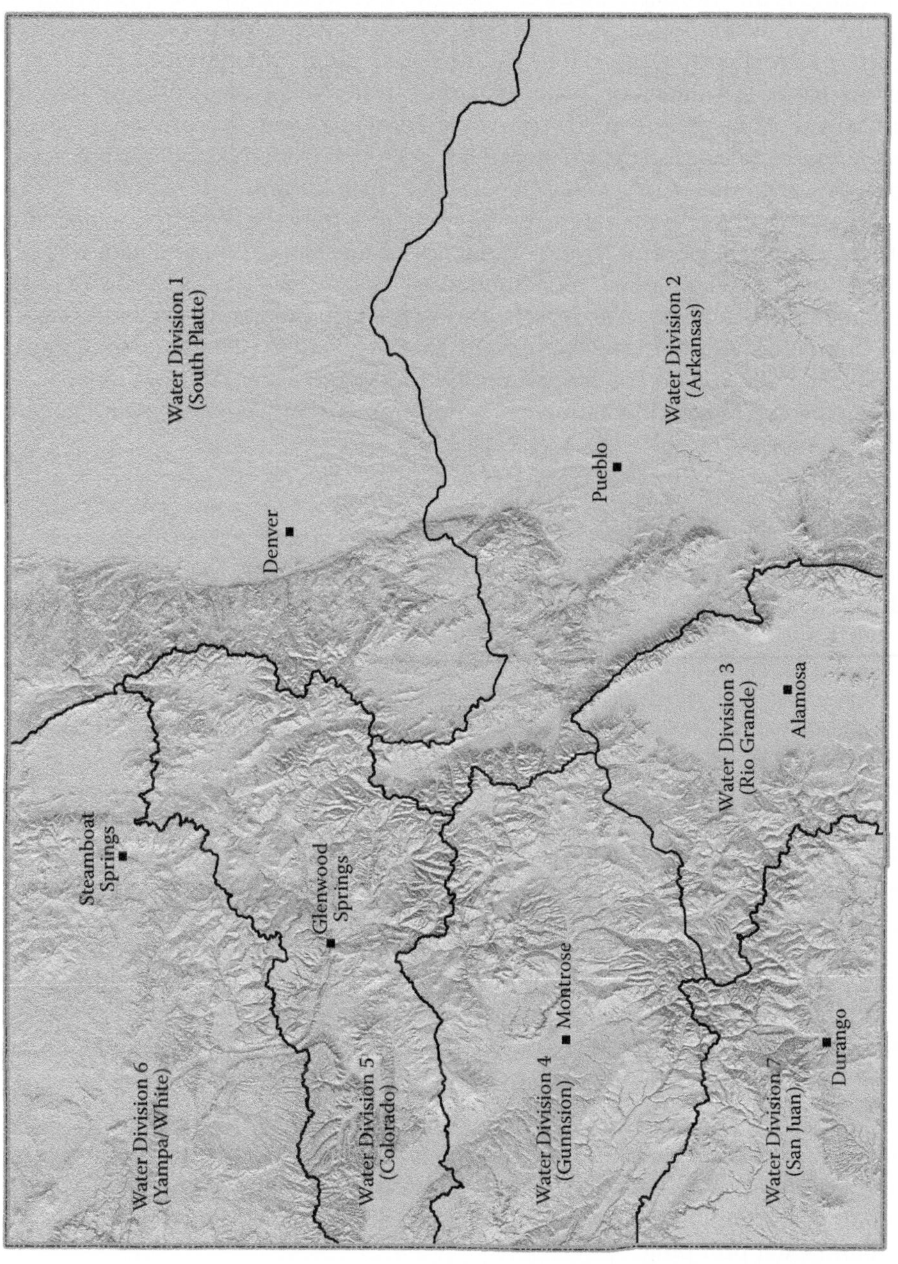

FIGURE 18.2
Seven water divisions of Colorado. (From the Colorado Division of Water Resources.)

usable for augmentation, and around US $3500/acre-foot* of consumption for permanent water use.

Augmentation water sources include water rights within a ditch or reservoir company, reusable municipal effluent from cities along the Front Range, and groundwater recharge projects. *Recharge* is an affordable method whereby river water is diverted, generally during the spring high-water season, into recharge basins or dry creek beds. This recharged water returns to streams later and can be claimed as an augmentation *credit*.

Another source of augmentation water is surface water diverted in priority into lined gravel pits. Mined gravel pits are lined with a bentonite slurry wall to create a belowground impermeable barrier around the perimeter of the gravel pit from the land surface down to the bedrock. The bentonite slurry wall prevents water from escaping from within the lined reservoir and also keeps groundwater from seeping into the lake. Water is delivered into gravel-pit lakes through adjacent irrigation ditches or mechanical pumps near a river. In 1991, the first lined gravel pit for augmentation water storage was completed near Greeley.

Augmentation plans continued to operate during the 1980s and 1990s under the supervision of the Colorado state engineer. These were 20 of the wettest years in recorded history in northeast Colorado and greatly helped provide adequate water supplies for irrigators. Nature supplemented the augmentation water requirements of well owners and generally protected the senior rights of surface water users.

18.3.3 *Empire Lodge v. Moyer*

The record drought of 2002 caused many senior water right holders to legally question the ability of well-augmentation groups to provide adequate augmentation water to protect senior water rights. The state engineer's legal authority to approve these annual augmentation plans was also legally challenged. That same year, the Colorado Supreme Court ruled in *Empire Lodge v. Moyer* that all members of augmentation plans had to obtain a water court–approved decree to operate. This would allow objecting senior water right holders to argue for greater scrutiny of augmentation plans, and to obtain greater accountability of crop consumptive use calculations, groundwater pumping lag depletion analysis, augmentation water delivery timing and location requirements, and the ability to force well-pumping curtailment if a court decree was not properly followed (*Empire Lodge Homeowners Association v. Moyer* 2002).

Central (1000 tributary wells) and GASP (approximately 3500 wells in the augmentation plans) were the two largest well-augmentation groups in the South Platte River Basin. GASP was unable to comply with the supreme court requirements because of inadequate augmentation supplies and ceased operations a few years later.

In 2005, Central obtained a court-approved augmentation decree, which was 70 pages long and included the following major provisions:

- A 6-year postpumping depletion projection tool was required. This prohibits any future pumping unless adequate augmentation water supplies are available to cover all postpumping depletions to the South Platte River from the prior 6 years of groundwater pumping.
- Central was required to set an annual groundwater pumping quota (limitations) by April 1 of each year to meet the 6-year projection tool augmentation requirements.

* 1 acre-foot (af) = 1233.48 cubic meters (m^3).

- Totalizer flowmeters (measuring devices) were required on all wells.
- Stringent water-accounting procedures were required to ensure that groundwater pumping was restricted to augmentation water-replacement availability.

18.3.4 Well-Pumping Quotas

In 2006, Central's pumping quota was set at 50% of historic pumping—in response to the conditions of the new decree. In 2004, a group of 400 wells that had formerly been members of GASP joined a new augmentation subdistrict of the Central Colorado Water Conservancy District. It was called the Well Augmentation Subdistrict (WAS) and was allowed to pump 50% of historic levels in 2005. In 2006, the WAS group anticipated pumping 15% of historic groundwater pumping quantities because of stricter augmentation requirements, but prolonged drought depleted those supplies by late spring. In 2006, these 400 wells were not allowed to pump even though crops had been planted. According to a Colorado State University study, approximately 30,000 acres of irrigated land was fallowed (not irrigated) in 2006, which led to $28 million in economic losses for the region (Thorvaldson and Pritchett 2007). Some losses were covered by crop insurance, but that was generally an exception and not the rule. The well curtailment was necessary under state law, however, to protect senior surface water rights in the river basin (Rocky Mountain News 2006).

How is a pumping quota determined? The key is to align augmentation water supplies with calculated well-pumping depletions—both present and in the future (lagged depletions). Well-pumping depletions are calculated based upon the time, location, and amount of depletions caused by the proposed out-of-priority diversions from each well. This is based upon calculated crop consumptive use coefficients and the number of acres to be irrigated. (Crops such as corn or alfalfa have higher consumptive use figures than, say, irrigated wheat or pinto beans.) These consumptive use depletions are then *tracked* based upon various computer modeling methods to determine where and when streamflows will be impacted.

Next, the quantities of required augmentation water must be calculated so that it is delivered at the correct time, quantity, and location to offset any potential out-of-priority well-pumping depletions to the stream. The aftermath of these groundwater accounting and management changes, from 2002 to today, has changed the agricultural landscape of many regions in Colorado. To date, thousands of wells have been forced to shut down and will not likely pump again because of inadequate augmentation water supplies. Cropland has been lost, as well as some jobs, and tax revenues and agricultural business have declined—particularly in smaller agricultural communities. Some groundwater irrigators are turning to less-consumptive-use crops, such as pinto beans, wheat, and sunflowers, to reduce the depletive effects of groundwater pumping on nearby streams. Some groundwater irrigators moved out of state, or moved a portion of their operations to neighboring Kansas, Nebraska, or New Mexico (Jones and Cech 2009).

18.4 Other Surface Water and Groundwater Use Conflict Situations

Not surprisingly, other locations in the United States and around the world are facing similar challenges. In 2005, Nebraska's Supreme Court ruled in *Spear T. Ranch Inc. v. Melvin G. Knaub* that tributary-well pumping along Pumpkin Creek in western Nebraska caused injury to senior surface water users. There are approximately 500 tributary wells along Pumpkin Creek, and a moratorium on new well drilling was implemented by the local natural resources district (*Spear T. Ranch Inc. v. Melvin G. Knaub et al.* 2005). Long term, Nebraska hopes to use funds from the US Department of Agriculture's Conservation Reserve Enhancement Program (CREP) to retire 100,000 acres of irrigated land from crop production to mitigate effects of tributary groundwater pumping.

Kansas has sued Nebraska for not fulfilling its interstate compact decree on the Republican River—arguing that upstream tributary-well pumping is depleting streamflow in the Republican. This may result in well-pumping curtailments in south-central Nebraska. In 2015, the US Supreme Court ruled that Nebraska must pay the state of Kansas US $5.5 million for damages incurred from violating provisions of the 1943 Republican River Compact.

Additional examples of stream impacts caused by tributary groundwater pumping can be found in Robert Glennon's excellent book, *Water Follies* (2002). His examples include the Santa Cruz River in New Mexico, Edwards Aquifer in Texas, Straight River in Minnesota, and several others.

In Australia, similar trends and conflicts are occurring. According to an Australian Broadcasting Corporation (ABC) report titled "Irrigators Angry at Losing Groundwater Allocation,"

> Groundwater pumpers in the Murray Valley are facing massive cuts to their water licences.... In the lower Murray, irrigators have licences to pump more than 260 000 megalitres, but scientists say only 84 000 megalitres is sustainable. So under the water reform, entitlements will be reduced by two thirds and that's got groundwater pumpers worried. There was clearly some anger at the cutbacks, and concern about how farms will be affected.... Chairman of the Murray CMA, Kel Baxter can understand the frustration. But he says this is the opportunity to have some input on how the cutbacks will be managed.... 100 million dollars will go to groundwater pumpers across the six aquifers but Executive Officer of the Murray Valley Groundwater Users Association, Leigh Chappell, says it's simply not enough. (Australian Broadcasting Corporation 2005)

18.5 Future Trends of Surface Water and Groundwater Use Conflict

> *Institutional arrangements are human creations; they are matters of choice.* (Schlager and Blomquist 2008, p. 23)

The well-pumping curtailments in 2006 were a relatively quick and drastic response to the ongoing drought in Colorado. Many well owners bemoaned this as a draconian policy, but others pointed to the Colorado Constitution and its Doctrine of Prior Appropriation. *First in time, first in right* is the law of Colorado, and no compromises could be forged to ease the economic and social pressures from the well shutdown. The Doctrine of Prior Appropriation served its purpose to protect the interests and property rights of senior water right holders.

Senior water right holders pointed out, and the Colorado Supreme Court repeatedly confirmed after 2005, that the Doctrine of Prior Appropriate governed the allocation of tributary groundwater in the state. The concept of maximizing beneficial use of both surface water and groundwater resources could not be adopted if it injured senior water rights. This rigid water-allocation system is unique to Colorado.

In the past decade, a similar fight occurred in Idaho, but with different results. Tributary-well pumping caused depletions to senior water rights along the Snake River, but the wells continued to pump. This was a great puzzle to well owners in Colorado: why were Idaho groundwater irrigators allowed to utilize a common-pool resource (CPR) management plan while Colorado irrigators could not? The answer lies in the 1890 Constitution of the State of Idaho:

> SECTION 5. PRIORITIES AND LIMITATIONS ON USE … but whenever the supply of such water shall not be sufficient to meet the demands of all those desiring to use the same, *such priority of right shall be subject to such reasonable limitations as to the quantity of water used and times of use as the legislature, having due regard both to such priority of right and the necessities of those subsequent in time of settlement or improvement, may by law prescribe* (emphasis added).
>
> SECTION 7. STATE WATER RESOURCE AGENCY. Additionally, the State Water Resource Agency shall have power to formulate and implement a state water plan for optimum development of water resources in the public interest. (State of Idaho Legislature 1890)

This allowed Idaho water officials to develop management plans, which included economic incentives, to allow well pumping to continue while protecting senior water rights. By contrast, the Colorado Constitution does not allow the state legislature to create "*reasonable* limitations as to the quantity of water used and times of use." In addition, the Colorado Constitution does not authorize a state water agency to "formulate and implement a state water plan for optimum development of water resources in the public interest." These are dramatic differences between the basic constitutional water laws of Colorado and Idaho.

This contrast of resource management between states becomes apparent in the work of Dr. Elinor Ostrom. In 2009, Elinor Ostrom was awarded the Nobel Memorial Prize in Economic Sciences for her work in economic governance of *common-pool resources* (an economic term related to resources that can be used and consumed by many), such as surface water or groundwater. CPRs can be owned by individuals or controlled by public agencies, such as local or state water management agencies. Ostrom identified eight principles of stable local CPR management:

1. The CPRs must have clearly defined management boundaries.
2. The rules for appropriation and provision of CPRs must be adapted to local conditions.
3. Collective-choice arrangements must be used to allow most resource appropriators to participate in the decision-making process.
4. Effective monitoring must be established so that monitors are part of or accountable to appropriators.
5. There should be a scale of graduated sanctions for appropriators who violate community rules.
6. Mechanisms of conflict resolution must be cheap and of easy access.
7. The self-determination of the community is recognized by higher-level authorities.
8. In the case of larger CPRs, various levels of organization should be used, with small local CPRs at the base level (Commons Journal 2010; Ostrom 1990).

Elinor Ostrom's third listed principle of stable local CPR management is not allowed in Colorado: "Collective-choice arrangements must be used to allow most resource appropriators to participate in the decision-making process." This omission allows for the rigid interpretation and administration of the Doctrine of Prior Appropriation in Colorado.

Futilely, many well users in Colorado argued that tributary groundwater and surface water should be managed to maximize the use of both water sources. In essence, well owners desired to treat the South Platte River and underlying aquifer as a CPR and to manage it as a single resource unit. Well users maintained that this would maximize the beneficial use of the entire surface water–tributary groundwater system in the basin, provide for improved economic viability, and encourage maximum utilization of the waters of the state of Colorado. This argument did not gain any traction with legislators.

However, Colorado officials did explore management methods that would allow groundwater pumping during average and wet years but would require curtailment in dry years. In essence, junior well owners wanted to pump during wet years even though their postpumping depletions (cones of depression that would slowly move from a well to an adjacent surface stream) could cause injury to senior water right holders in future years. Senior water users argued that these postpumping (or lagged) depletions caused by tributary groundwater pumping would continue to reduce surface water flows for years into the future: how could a groundwater user predict future dry years so that senior water rights are not injured by prior-year lagged well-pumping depletions? That was a management question left unanswered and resulted in the lack of a collective-choice agreement to allow curtailed Colorado well users to pump their wells as they had done previously.

Conflicts between surface water and tributary groundwater users will continue in Colorado, the western United States, and the world. Increased water demands and drought will lead to greater use of tributary groundwater, which, in turn, will deplete surface water supplies.

Climate change could dramatically change the timing and availability of surface water in some regions of the world. In Colorado, warmer temperatures could lead to an early snowmelt runoff season in the spring. This would reduce the ability of downstream reservoirs to capture and store snowmelt, which could reduce available water supplies for late summer needs. In addition, warmer temperatures would increase crop consumptive use requirements, creating the need for additional irrigation water supplies. Surface water availability changes in timing and quantity could negatively affect groundwater pumping allowances.

In the United States, state constitutions, interstate compacts, and international treaties will dictate the types of surface water and groundwater management systems in alluvial systems. Some locations will be able to use the eight management principles of Nobel Prize winner Elinor Ostrom, while others will be prohibited by state constitutions to maximize surface water and groundwater management benefits.

References

Australian Broadcasting Corporation. 2005, November 11. Irrigators Angry at Losing Groundwater Allocation. *NSW Country Hour*. Available at http://www.abc.net.au/site-archive/rural/nsw/stories/s1508170.htm (accessed May 30, 2015).

Boyd, D. 1890. *A History: Greeley and the Union Colony of Colorado*. Greeley: Greeley Tribune Press. 448 pp.

Cech, T.V. 2010. *Principles of Water Resources: History, Development, Management and Policy*, 3rd Edition. Hoboken: John Wiley & Sons, pp. 252–253.

Colorado Constitution, Article XVI, Section 6. 1876.

Colorado Supreme Court, Colorado Judicial Branch, *Water Courts*. Available at https://www.courts.state.co.us/Courts/Water/Index.cfm (accessed May 30, 2015).

Commons Journal. 2010, February 22. Announcement: Congratulations to Editorial Board Member Elinor Ostrom. Available at http://www.thecommonsjournal.org/index.php/ijc/announcement/view/1 (accessed May 30, 2015).

Empire Lodge Homeowners Association v. Moyer, 39 P.3d 1139 (Colo. 2002).

Glennon, R. 2002. *Water Follies: Groundwater Pumping and the Fate of America's Fresh Waters*. Washington, DC: Island Press, 314 pp.

Jones, A.P. and T. Cech. 2009. *Colorado Water Law for Non-Lawyers*. Boulder: University Press of Colorado, 276 pp.

Kryloff, N.A. 2007. Hole in the River: A Brief History of Groundwater in the South Platte Valley, 1858–1969. *Colorado Water* 24(5):9–11.

McClellan v. Hurdle, 3 Colo. App. 430, 33 P. 280 (1893). Reviewed in the 1970 *Denver University Law Review* 47(2):308–309.

Ostrom, E. 1990. *Governing the Commons: The Evolution of Institutions for Collective Action*. Political Economy of Institutions and Decisions series. Cambridge: Cambridge University Press, 298 pp.

Rocky Mountain News. 2006, May 10. Farms High and Dry: State Shutting 400 Wells to Preserve South Platte; 200 Growers Could Lose Crops. Available at http://www.rockymountainnews.com/ (accessed May 21, 2010).

Schlager, E. and W. Blomquist. 2008. *Embracing Watershed Politics*. Boulder: University Press of Colorado.

Spear T. Ranch Inc. v. Melvin G. Knaub et al. No. S-03-789. 269 Neb. 177. Supreme Court of Nebraska. January 21, 2005.

State of Idaho Legislature. 1890. Article XV Water Rights. *Constitution of the State of Idaho*. Available at http://www.legislature.idaho.gov/idstat/IC/ArtXVSect5.htm (accessed May 30, 2015).

Thorvaldson, J. and J. Pritchett. July 2007. Some Economic Effects of Changing Augmentation Rules in Colorado's Lower South Platte Basin: Producer Survey and Regional Economic Impact Analysis, Completion Report 209, Department of Agricultural and Resource Economics, Colorado State University, Fort Collins, CO.

Winter, T.C., J.W. Harvey, O.L. Franke, and W.M. Alley. 1998. *Ground Water and Surface Water a Single Resource*. US Geological Survey (USGS) Circular 1139. USGS, Denver, CO, 70 pp. Available at http://pubs.usgs.gov/circ/circ1139/.

19

*Floods as Unnatural Disasters: The Role of Law**

Sandra B. Zellmer and Christine A. Klein

CONTENTS

ABSTRACT Flooding is a recurrent hazard, even in the driest of states. Climate change alters flood risks by changing the intensity of precipitation events. Effective policies and laws will need to recognize and respond to the physical and social factors that determine the vulnerability of people and communities to future flood damages. This chapter discusses the major drivers of flood losses, including land use and floodplain management decisions, and considers the distortions caused by public policy choices, focusing primarily on federal flood policies and their unintended consequences. It considers several possible reforms in federal law, including (1) revising the decision-making metrics for the construction of flood control projects; (2) reforming the Fifth Amendment takings doctrine; and (3) recalibrating the National Flood Insurance Program both to keep people out of harm's way and to place the risks and costs of flooding on those who choose to engage in risky behavior.

* Derived from Christine A. Klein and Sandra B. Zellmer, *Mississippi River Tragedies: A Century of Unnatural Disaster*, NYU Press, New York, 2014.

19.1 Introduction

As discussed in previous chapters in this volume (including Chapters 2, 10, 11 and 15), flooding is a recurrent hazard even in the driest of the western states. Climate change is likely to alter flood risks by increasing the intensity of the heaviest precipitation events, a trend that is already evident across the United States (Kunkel et al. 2008; Walsh et al. 2014). Changes in flood frequency and intensity are difficult to predict and will depend on regional differences in the primary weather phenomena that lead to extreme precipitation (Ralph et al. 2014). Research efforts are focusing on extreme weather events and how they might change in a warmer climate (Trenberth 2011). While better understanding of the physical science related to flooding can inform future flood policies, it can play only a small part in averting flood-related losses. It will be far more important to understand and address the social factors that determine the vulnerability of people and communities to flood damages. This chapter discusses the major drivers of flood losses, including long-term land use and floodplain management decisions, and considers the distortions caused by well-intentioned but misguided public policy choices that cause *unnatural disasters.* The analysis focuses especially on federal flood policies, their unintended consequences, and possible reforms.

19.2 Looking Back: The Impetus for Federal Involvement

Our experiment with intensive federal flood management began on the Mississippi River in the early twentieth century, when engineers and government bureaucrats assumed that levees could achieve dual purposes by improving navigation and containing floods (Klein and Zellmer 2007). The US Army Corps of Engineers had long advocated more and bigger levees for navigational purposes, while other federal and state agencies hoped that levees would be an effective means of achieving multiple objectives—thereby providing something for everyone. For its part, Congress was reluctant to get into the flood control business, believing that flooding was a local problem that should be addressed by local responses, and it insisted that federal money could be used for levees *only* insofar as they were related to navigation, *not* to protect land from flooding. After a series of massive floods, however, Congress relented somewhat in the Flood Control Act of 1917, the first federal law that explicitly appropriated money for nonnavigational river infrastructure, i.e., flood control levees, on both the Mississippi and Sacramento Rivers (Pub. L. No. 64-367, 39 Stat. 948). By today's standards, it was a modest measure, and critics of federal power continued to question whether the central government had the authority—much less the responsibility—to shield private property from flooding.

According to historian Donald Pisani, "modern water politics," complete with logrolling, pork-barrel funding, and backroom deal-making, "was born in the 1920s" (Pisani 2002, p. 253). The Great Mississippi River Flood of 1927 had a good deal to do with it. When Mississippi River levees ruptured in 120 places, the flood blanketed nearly 18 million acres in seven states, and at its widest point near Vicksburg, Mississippi, the swollen river formed an inland sea almost 100 miles wide (Barry 1998). After the flood made short work of the federal *levees-only* policy, Congress passed the Federal Flood Control Act of 1928, authorizing far more extensive federal involvement in flood management and placing responsibility for flood control in the Mississippi basin firmly on the Corps of

Engineers (33 U.S.C. §§ 702a–m). Structural responses, including levees, reservoirs, and spillways, were still featured, but by setting a precedent for widespread federal involvement in what had long been perceived as the primarily local affair of flood management, the 1928 act marked a paradigm shift in the division of labor among federal, state, and local governments. Yet even as Congress expanded the federal government's *authority*, it was careful to limit its *responsibility* in case flooding occurred despite of (or even because of) the Corps' efforts. The 1928 act, therefore, immunized the government from liability "of any kind…for any damage from or by floods or flood waters at any place" (33 U.S.C. § 702c). This provision expresses the notion, long embedded in the law, that "the king can do no wrong," and even if he does, the king cannot be sued.

Several pieces of federal flood control legislation followed on the heels of the 1928 act, most notably the Flood Control Act of 1936, which delegates broad discretion to the Corps to construct any flood control project it chooses whenever "the benefits to whomsoever they may accrue are in excess of the estimated costs" (33 U.S.C. § 701a). The Flood Control Act of 1944 stands out as well, in that it authorized six massive main-stem dams and reservoirs on the upper Missouri River in hopes of protecting downstream population centers and farms (58 Stat. 887, codified in various provisions of Titles 16, 33, and 43 of the US Code).

Once Congress committed itself to federal flood control management, it took additional, nonstructural steps by way of institutional involvement. The Disaster Relief Act of 1950, as amended by the Stafford Disaster Relief and Emergency Assistance Amendments of 1988, authorizes the president to deploy federal troops and to distribute federal aid in response to floods and other national disasters (42 U.S.C. §§ 5121–5202). Subsidized flood insurance was added to the mix with the National Flood Insurance Act of 1968 (42 U.S.C. §§ 4001–4129). In establishing the National Flood Insurance Program (NFIP), Congress hoped to defray the expense of after-the-fact disaster relief by encouraging floodplain occupants to pay premiums before disaster struck, but it also hoped that local governments would adopt land-use control measures to prevent improvident occupation of the floodplain. These goals were to be accomplished through a type of quid pro quo arrangement: the federal government would offer insurance to residents (through private insurers) at below-cost rates if their communities adopted certain land-use regulations and other restrictions in so-called *special flood hazard areas*, which are those areas determined to be within the 100-year floodplain—defined as an area that has a 1% chance of flooding in any given year (Klein and Zellmer 2007).

19.3 Treading Water: The Legacy of Our Mistakes

Since the 1920s, the country has spent hundreds of billions of dollars on flood control structures in coastal and floodplain areas, on flood insurance subsidies, and on disaster relief. Despite these massive expenditures, economic losses due to flooding have continued to increase (Singer 1990). According to testimony before the Senate Committee on Environmental and Public Works in 2005, "Because so many Corps flood control projects induce development in harm's way, flood damages have more than tripled in real dollars in the past eighty years—even as the Corps has spent more than $120 billion on flood control projects" (Faber 2005).

By way of example, in both 1993 and 2011, record-breaking, prolonged floods were seen along the Missouri and Mississippi Rivers. In 2005 and 2011, Hurricanes Katrina and Irene devastated communities on the Gulf Coast and in the Northeast, respectively. Total direct flood damages in 2011 alone totaled $8.4 billion (NOAA–National Weather Service 2011). Just 1 year later, in 2012, Superstorm Sandy pummeled the eastern seaboard as a Category 1 hurricane, causing major damage to tens of thousands of residences and businesses and interrupting electricity and fuel supplies for millions (FEMA 2013). The arid western states have not escaped flooding, either.

> Through the first week of September 2013, Colorado was exceptionally warm and dry. By September 12, everything had changed. Flood conditions stretched about 150 miles, from Colorado Springs north to Ft. Collins. Saturated soils left water with no place to go, and puddles turned to ponds throughout the densely populated Colorado Front Range. Rainwater swelled rivers and creeks, overtopped dams, flooded basements, and washed out roads. (NOAA 2013)

None of these events was a wholly isolated incident. In fact, the Front Range city of Boulder "is one of Colorado's most flood-vulnerable communities. The city is situated at the mouth of a canyon, and the Boulder Creek flows through the middle of town" (NOAA 2013).

Over the years, our experiments with federal flood management and disaster relief have highlighted three important lessons: (1) rivers *will* flood; (2) levees *will* fail; and (3) unwise floodplain development *will* happen if we let it.

First, rivers *will* flood. As the US Geological Survey reported, during the twentieth century, floods topped the list of *natural disasters* in terms of death and property damage (Perry 2000). The Federal Emergency Management Agency (FEMA) estimates that flooding accounts for approximately 85% of all disaster declarations (FEMA 2015a). We pour billions of dollars into flood control structures, yet flood damages continue to rise. The Missouri and Mississippi Rivers, in particular, continue to spill over their banks, despite the gargantuan dams and the corset of levees designed to prevent them from occupying the floodplain. Even where levees exist, as the Association of State Floodplain Managers warns, "Building in a floodplain is like pitching your tent on a highway when there are no cars coming" (Wright 2000). We flirted with—in fact, we married ourselves to—the disastrous federal levees-only policy. But the flood of 1927 taught us that the policy was "the most colossal blunder in engineering history" (O'Neill 2006, p. 144). In response, to replace the natural floodplains that had been walled off behind levees, the Corps designed reservoirs and artificial *floodways*. The floodways direct overflow through engineered gaps (*fuse plugs*) where the levee height is lower than in surrounding areas. Dynamite may be used when necessary to blast out an emergency exit for the river. In effect, we are still putting our faith in levees, but we have added a few more tools to the box.

Meanwhile, scientists have taught us that flooding is not such a bad thing after all, ecologically speaking. Floods enrich floodplain soils and deliver sediments to build up river deltas. Just as we have engineered substitutes for natural floodplains, we have also used technology to mimic natural floods and to produce at least some of their benefits. To restore levee-starved riverine habitats, at times, we deliberately manipulate dams and reservoirs to create *managed* floods (Klein and Zellmer 2007). Instead of relying on levees, floodways, spillways, fuse plugs, and other structural responses, in many cases, perhaps it would be more effective to let the river reclaim its floodplain and function more naturally.

Second, levees *will* fail. As one engineer explains, there are two kinds of levees: "Those that have failed and those that will fail!" (Martinsdale and Osman 2007). In fact, levees are *designed* to fail under some circumstances. In engineer-speak, this is dubbed *residual risk*.

The typical 100-year levee is engineered to hold back the level of flooding that has at least a 1% chance of occurring each year (which translates into a 26% chance of flooding over the life of a 30-year mortgage). Beyond that, all bets are off. The American Society of Civil Engineers (ASCE) warns that no levee is flood-proof:

> Levees *reduce* the risk of flooding. But no levee system can *eliminate* all flood risk. A levee is generally designed to control a certain amount of floodwater. If a larger flood occurs, floodwaters will flow over the levee. Flooding also can damage levees, allowing floodwaters to flow through an opening, or breach. (ASCE 2010, p. 2)

Third, unwise floodplain development *will* happen if we let it. And we let it happen, in part, through timid land-use decisions at multiple jurisdictional levels and, at the federal level, by providing subsidized flood insurance and by misguided jurisprudence on landowners' Fifth Amendment takings claims. When Congress created the NFIP in 1968, it did so with trepidation. Our legislators worried that rather than limiting losses, the availability of subsidized insurance policies would cause further development in the floodplains and lead to even greater flood damage. Just 2 years earlier, a national task force had warned, "A flood insurance program is a tool that should be used *expertly or not at all.* Correctly applied, it could promote wise use of floodplains. Incorrectly applied, it could exacerbate the whole problem of flood losses" (Water Science & Technology Board 1995, p. 164). The warning proved apt. Attempting to wield the NFIP tool "expertly," Congress passed significant amendments to the program's legislation after nearly every major flood since the program was first adopted. But communities have been loath to implement floodplain-protective measures for fear of takings liability, as discussed in Section 19.4.2 below.

The 1993 flood, the 2005 hurricane season, and the 2011–2012 storms illustrated critical flaws of the NFIP, and Congress attempted to respond, but for the most part, it did so ineffectively. People had in fact built their homes and businesses in floodplains, not enough of them bought into the insurance pool, and insurance premiums were not high enough to make the program financially sustainable in the long term (King 2012, 2013). For the three and a half decades leading up to Hurricane Katrina, the NFIP was able to support itself with premiums and fees generated by insurance policies, despite satisfying claims for flood after flood, including paying out almost $300 million after the Midwest flood of 1993. But after Hurricane Katrina, over $17 billion worth of claims swamped the NFIP's ability to remain financially self-supporting (King 2013).

These floods have also shown that the goal of steering development away from high-risk areas has not been realized. The case of so-called *repetitive loss* properties illustrates this phenomenon. Tens of thousands of insured properties have sustained flood damage on more than one occasion. In many cases, the cumulative payments for flood losses to these properties have exceeded the value of the property. Rather than move out of the floodplain or take measures to flood-proof their properties (such as elevating them), these landowners have collected insurance benefits and stayed put. Although repetitive loss properties hold only 1% of all NFIP policies, they have consumed one-third of all claims payments. Since 1978, these determined floodplain dwellers have cost the National Flood Insurance Fund a total of $12.1 billion in nominal dollars (King 2013). Repetitive loss properties are damning evidence that the NFIP has failed to reduce flood damage or to dissuade people from occupying hazardous areas. Instead, it has enabled a small (but expensive) group of people to remain in harm's way, perhaps because of their enjoyment of the amenities associated with a riverfront location, or perhaps because they felt trapped by the compensation rules embodied in the NFIP, knowing that their property would be rendered nearly valueless if they failed to accept the insurance payment to rebuild. Moreover, local officials have been reluctant to require them to move. Understanding why these homeowners choose to stay put, and how federal policies can best play a role in informing their decisions and the decision making of local zoning and planning authorities, could inform a more effective insurance policy.

Congress attempted to address some of the NFIP's shortfalls when it passed the Biggert–Waters Flood Insurance Reform Act of 2012. The law was aimed at making the NFIP fiscally sustainable by, among other things, gradually phasing out below-market insurance rates and raising the cap on permissible annual rate increases. No sooner than the legislation passed, however, Hurricane Sandy ravaged portions of the Atlantic coastline, destroying or damaging some 385,000 homes in New Jersey and New York. When it became apparent that the new law would have a severe financial impact on already-devastated homeowners—in some cases requiring dramatic increases in annual insurance premiums—Congress blinked. Shortly after Congress passed the reform legislation, it adopted measures to delay or preclude implementation of the new provisions (Homeowner Flood Insurance Affordability Act 2014).

19.4 Looking Forward: How Law Can Help

Through its flood control policies, the federal government promised more than it could deliver—that it would somehow hold back the mighty rivers, keeping its citizens safe and dry, no matter where they chose to put down stakes. These policies have had, and are continuing to have, significant adverse consequences. If we acknowledge that some tragedies are of our own making—that they are unnatural disasters, not uncontrollable *acts of God*—then we have a fighting chance at accepting the human responsibility for magnifying flood damage and at making better decisions in the future (Klein and Zellmer 2007).

Legal reforms are needed to hold accountable those who endanger themselves or others, both physically and financially. Any type of reform that results in a reallocation of risks and costs will be painful for those who have benefitted from existing flood policies. Choices will need to be made about how far and fast to proceed and how to fairly and effectively manage the conflicts that will arise.

To achieve the goal of reassigning risk to individuals and entities whose decisions cause unwarranted exposure of people and property to flood risks, several reforms might be considered: (1) reforming the decision-making metrics for the construction of flood control projects; (2) reforming the Fifth Amendment takings doctrine; and (3) reforming the NFIP both to keep people out of harm's way and to place the risks, and the costs, of flooding on those who choose to engage in risky behavior.

19.4.1 Reform the Decision-Making Metrics for Construction of Flood Control Projects

The Flood Control Act of 1936 delegates broad discretion to the Corps to construct any flood control project it chooses whenever "the benefits to whomsoever they may accrue are in excess of the estimated costs" (33 U.S.C. § 701a). On its face, the requirement to weigh benefits and costs, in other words, to engage in a cost–benefit analysis (CBA), appears to provide a rational decision-making metric. By requiring that the costs and benefits of a proposed action be quantified, CBA is said to be an unbiased method of evaluating the potential effects of a proposal and exposing bad plans that would impose ruinous costs for relatively minimal benefits. As applied, however, CBA requirements like the one found in the Flood Control Act of 1936 are notoriously easy to manipulate, and excessive reliance on CBA has masked far too many bad policy choices. According to Professor Dan Tarlock, the Corps, in particular, "has a long history of inflated and methodologically unsound benefit–cost analysis techniques" (Tarlock 2004, p. 1315). Since 1936, the Corps has spent billions of dollars on dams, reservoirs, levees, and other structures, many of which have imposed social and ecological costs far in excess of the benefits produced. According to some reports, the benefits of navigational "improvements" on the Missouri to support virtually nonexistent commercial barge traffic were exaggerated nearly tenfold (Upbin 1998), and the use of inflated cost–benefit methodology to justify replacing old locks and dams on the upper Mississippi triggered a congressional investigation and a critical National Research Council report (National Research Council 2001; McCool 2005).

In the aftermath of the 2005 hurricanes, the Corps itself recognized that its approach to CBA was unlikely to justify risk reduction measures for storms like Katrina, in part because its analyses "do not consider such non-economic assets as human life" (Schleifstein 2006). To respond more appropriately to flood risks, Congress should impose more meaningful decision-making metrics that compel careful consideration of the value of human life and the potential for lost lives.

Congress should also require the Corps to weigh the value of ecosystem services—buffering storm surges, capturing and filtering floodwaters, providing habitat for species, and furnishing recreational opportunities for humans—that may be lost as a result of flood control and related structures. The FEMA has a role to play here, too. With federal funds from FEMA's Hazard Mitigation Grant Program and other initiatives, some local governments are engaging in strategic buyouts of properties that are most vulnerable to future flooding. Since the 1993 flood, over 27,000 families have been voluntarily relocated, and title or easements for over a million acres of marginal floodplain farmland has been acquired for natural-use purposes in the upper Mississippi River Basin (Galloway 2005). In the wake of the 2013 Front Range flood, Boulder County is using FEMA funds to purchase damaged properties, demolish them, and create open space. Not only will these initiatives protect lives and prevent structural damages in the future; they will also improve long-term ecological function in the floodplain.

Formal consideration of the likely impacts of climate change on flood risks is also needed. Ideally, Congress would compel such a requirement, but President Obama has taken strides by directing federal agencies to meet a new federal flood risk resilience standard for future federal investments in and affecting floodplains in his executive order on the Federal Flood Risk Management Standard (Executive Order 2015). The new standard will apply whenever federal funds are used to build new structures and facilities or to rebuild those that have been damaged. Implementation is expected to incorporate the "best available, actionable data and methods that integrate current and future changes in flooding based on climate science" (FEMA 2015a).

19.4.2 Reform the Fifth Amendment Takings Doctrine and Empower Governments to Tame Floods Naturally by Giving Nature Some Space

The Fifth Amendment of the US Constitution prevents federal and state governments from taking private property for public use without just compensation. As the supreme court has explained, compensation is required "[w]here the government authorizes a physical occupation of property (or actually takes title)" (*Yee v. Escondido* 522, 1992). In addition, in rare cases where a government action causes a physical occupation of the property or denies all economically beneficial use, a *per se* taking will be found, requiring compensation (*Lucas v. S.C. Coastal Council* 1019, 1992; *Arkansas Game & Fish Comm'n v. United States* 511, 2012). By contrast, when government action impacts private property but does not physically occupy the property or destroy its value, the generally applicable test for determining whether a taking has occurred is whether the action or regulation goes "too far" (*Pennsylvania Coal Co. v. Mahon* 415, 1922). Courts employ a balancing test that considers the effects on the property owner's reasonable investment-backed expectations and the character of the governmental action. As for the latter, "[a] 'taking' may more readily be found when the interference with property can be characterized as a physical invasion by government than when interference arises from some public program adjusting the benefits and burdens of economic life to promote the common good" (*Penn Central Transp. Co. v. New York* 124, 1978).

In its current form, Fifth Amendment takings jurisprudence is unbalanced. Although the doctrine scrutinizes allegations of harm that landowners claim to suffer at the hands of "heavy-handed" government action, it does not call for a systematic inspection of the governmental benefits that landowners have received in order to occupy the floodplain, nor does it account for the damage that these same landowners might otherwise inflict on their neighbors. Economists would call the latter a study of *externalities*—the spillover effects foisted onto others. In lay terms, both themes come down to being a good citizen and taking responsibility for the consequences of one's actions. These could be addressed through several possible legal reforms.

First, an article published in the *Yale Law Journal* developed the concept of *givings*—benefits conferred by the government that increase property values, such as the provision of flood control devices, subsidized federal insurance, sewers, roads, and the like (Bell et al. 2001). Before a court awards compensation in response to a floodplain owner's claim that land-use decisions have taken their property, the article suggests, judges should discount the claimed taking by the value added to the property by government action. This rough accounting of both give and take, the authors conclude, leads to a fairer, more balanced result. Whether we look at it as givings or a more nuanced approach to takings, if we expand our thinking about the relationship between landowners, their neighbors, governments, and the land itself, all sorts of possibilities may come to light. Judges could take steps to improve the doctrine by recognizing that landowners who misuse the land can take from their neighbors just as easily as government regulators can take from landowners.

On a related note, giving greater weight to ecosystem services would help quantify the benefits that landowners take from their neighbors when they use their property destructively. If a landowner builds on a wetland that would otherwise capture and tame floodwaters for the community or places impermeable materials in a floodway that alter the flow and drainage of flood waters, compensation for government actions that diminish the market value of their property should be reduced or, in some cases, denied.

In addition, a doctrine known as the *public trust* should be given greater force as a defense to alleged takings by the government. In *Just v. Marinette County*, the Wisconsin Supreme Court rejected a landowner's challenge to a county shoreline zoning ordinance that protected water quality by preventing landowners from changing the natural character of land adjacent to navigable rivers and lakes. The court held that the ordinance was not unreasonable and that it took nothing from the landowner. The court observed that the state and its counties have an "active public trust duty...not only to promote navigation but also to protect and preserve [navigable] waters for fishing, recreation, and scenic beauty" (*Just v. Marinette County* 768). Similarly, in *National Audubon Society v. Superior Court*, the California Supreme Court concluded that the State Water Resources Control Board had a continuing duty to scrutinize Los Angeles' permit to use water from tributaries of Mono Lake. The court found that the Water Board was obligated to protect the public's interest in the beneficial use of its water and to ensure against violations of the public trust by Los Angeles or other water users (*Nat'l Audubon Society v. Superior Court* 728–732). As in *Just*, Los Angeles' assertion of a takings claim was rejected: The city had no vested property right to prevent the state from carrying out its public trust responsibilities.

Although the parameters of the public trust doctrine have not been clearly defined by the courts, both judges and scholars have recognized that the trust purposes are not set in stone but can be expanded as the public's expectations expand and as our understanding of ecological processes and climate evolves. As the New Jersey Supreme Court stated, "The public trust doctrine...should not be considered fixed or static but should be molded and extended to meet changing conditions and needs of the public it was created to benefit" (*Slocum v. Borough of Belmar* 316). Protecting people and ecosystems from improvident activities that destroy wetlands and shorelines, on the one hand, and preventing people and their structures from inflicting costs on their communities while placing themselves in harm's way in flood-prone areas, on the other, both come within the scope of the public trust doctrine. Governments may be empowered to prevent landowners from degrading trust resources, and courts should protect those governments from takings claims when the landowners turn around and sue. As a result, it will be possible and perhaps even legally necessary to give nature some space for the natural bounty of its overflowing rivers.

19.4.3 Reform NFIP to Move People Out of Harm's Way and to Place the Risk (and the Cost) on Those Who Choose to Engage in Risky Behavior

This proposal involves a number of related reforms.

19.4.3.1 Modify the Levee Loophole

Before purchasing property, buyers (or their lenders) can measure the property's location against flood hazard maps prepared by FEMA. The maps give would-be purchasers information about the potential flood risks they will face and whether they will be required to purchase federal flood insurance. Currently, if an otherwise flood-prone area lies behind a levee certified to protect from the 1%-annual-chance flood (the so-called 100-year flood), the area can be removed from FEMA flood maps. As a consequence, landowners may have a false sense of security and not realize the risks they face if/when the levee fails. Congress should modify this levee loophole. At a minimum, Congress should limit the exclusion so that it better reflects actual risks of levee failure during a flood event. Congress could also consider rescinding the levee loophole altogether in areas that would otherwise be considered a floodplain, since there are only two kinds of levees: those that have failed and those that will fail.

19.4.3.2 Require Accurate, Up-to-Date Federal Maps and State Disclosure Forms

Even if Congress does not modify the levee loophole, FEMA and the states can take measures to improve landowners' understanding of risk. For example, state law could require sellers to disclose to potential purchasers the flood status of the property and the relevance of levees (if any) to that status. To assist sellers in satisfying their disclosure obligation, FEMA maps should make clear whether an area is outside the 100-year floodplain because of natural topography or, conversely, because of the presence of constructed levees. Moreover, while FEMA provides the public with flood risk visualization tools that can be used to explore flooding hazards, such tools should be expanded to allow prospective property buyers and developers to contemplate an area's vulnerability to changing climate conditions (FEMA 2015b). In addition, FEMA could require states that participate in the NFIP to adopt disclosure requirements.

As for the floodplain maps themselves, they should be updated regularly. Existing maps are often inaccurate and outdated (Government Accountability Office 2010). Although there have been efforts to improve mapping by using the best available data to identify a community's flood risk, not enough is being done (Wriggins 2014). Moreover, risks will likely increase as a result of rising sea levels and increases in heavy precipitation and runoff events (Kunkel et al. 2008; Walsh et al. 2014). Although local changes in flood frequency and intensity are difficult to predict (Ralph et al. 2014), Congress should require additional action to ensure up-to-date accuracy of floodplain maps to the greatest extent possible.

19.4.3.3 Increase Participation in the NFIP

To make the program financially sustainable, more floodplain occupants need to purchase and maintain insurance. This reform is a relatively modest one, in that federal officials and lenders must simply enforce existing requirements. A 2006 study found that fewer than 50% of those living in hazardous areas had purchased mandatory flood insurance, and many who purchased it when they first obtained their mortgages subsequently dropped it (Rand Corporation 2006). Banks have routinely failed to enforce the insurance requirement, especially when the initial lenders sell the first mortgage to another bank, which in turn does not enforce the requirement (Wriggins 2014). The penalties for noncompliance are minimal, although in 2012, Congress raised them from $350 per violation to $2000 per violation (Biggert–Waters 2012 § 4012a[f][5]). An alternative means of increasing participation and improving compliance with flood insurance requirements would be to impose the requirements and enforcement mechanisms directly on floodplain property owners, but this proposal has proven highly unpopular, and Congress seems unlikely to embrace it (Lemann 2015).

19.4.3.4 Adjust Insurance Premiums

Insurance premiums should not be subsidized at artificially low rates. This may be accomplished through four approaches:

1. Those who live in floodplains behind levees could pay actuarially sound premiums that recognize the *residual risk* of levee failure.
2. Congress could adopt a "one strike and you're out" policy for repetitive risk properties. That is, floodplain residents would be compensated one time only. The payment should be sufficient to assist the resident in relocating to higher ground. If a second flood occurs, the resident will not be entitled to NFIP benefits. Given the harshness of this option, Congress could instead adopt a rate structure that penalizes repeat claimants according to recalculated loss probability estimates based on actual loss experience.

3. Where older properties (generally less sturdy and less able to withstand flooding) have been *grandfathered* into the NFIP program at below-market rates, Congress could phase out these subsidies over time to ensure that the premiums reflect the risks. Grants and other types of carefully calibrated financial support could be provided to help homeowners who could not otherwise afford moving out of the floodplain or flood-proofing their homes.
4. Congress could exclude or limit voluntary risk takers by making vacation homes and non-water-dependent businesses ineligible for subsidized coverage, and by strengthening the exclusion of barrier island properties from NFIP.

With respect to these and other measures, reformers must take care to ensure that the burdens of their proposals are imposed fairly and manageably. In 2012, the Biggert–Waters Flood Insurance Reform Act adopted a number of reforms to strengthen the future financial solvency and administrative efficiency of the NFIP (Biggert–Waters 2012; King 2013). However, when premiums for floodplain owners went up, Congress rescinded many of the fiscal provisions (Homeowner Flood Insurance Affordability Act 2014). The takeaway lesson from the experiences with these two legislative initiatives is that change does not come easy. But the national dialogue about flood control and risk management is currently underway, in earnest, and one can hope that continued attention to these issues may yield viable and effective reforms.

19.5 Conclusion

For well over a century, our national policies have exploited rivers, often with the best of intentions. Civic-minded engineers squeezed the lower Mississippi River between what they hoped would be an impervious line of levees to protect the delta and its cities and farms from floods. The walled city of Cairo, Illinois, takes advantage of its enviable location near the confluence of the Mississippi and Ohio Rivers but relies on an engineered floodway to replace the natural floodplain. Looking west, an extensive string of levees on the Sacramento–San Joaquin River System in California was constructed to protect Central Valley farmland, cities, and utilities amounting to "billions of dollars in economic production and half of the state's total freshwater supply" (Percy 2007, p. 547).

In the end, though, many of these endeavors have destroyed ecosystem services within the floodplain and have made floodplain communities *more*—not less—vulnerable to future storms and floods. Levees and other engineered structures prevent the nation's rivers from nourishing the floodplain soils with essential nutrients, and they restrict the delivery of land-building sediments to help offset natural subsidence and erosion. In California, for example, the loss of wetlands, along with unstable geologic and soil structure within the Sacramento–San Joaquin Delta, makes the massive system of levees highly vulnerable to failure, which would have far-ranging implications throughout the region (Percy 2007). In fact, in 1986 and 1995, several of the Sacramento River levees failed, "catastrophically inundating some smaller communities" (Brandt et al. 2012, p. 45).

But past need not be future. We have wasted more than a century pursuing a foolish and impossible ideal: *the floodless floodplain.* We struggled to hold back the water from the people. That effort failed time and again, and is more likely to fail in a climate-altered future that poses the likelihood of more frequent or extreme storms. Federal policies, in particular, have given insufficient attention to the opposite notion—holding people back from the waters. The old reliance on levees, insurance, and disaster relief continues to create irresistible incentives to settle in harm's way.

It is time to try something different: *giving rivers room to flood*. At the very least, we should think of sharing floodplains with their rivers. Some communities in the Mississippi River Basin and the Front Range are doing just that, though it took catastrophic flooding to motivate them. Human attitudes shift over time as we build on experiences and learn from both success and failure. Attitudes take time to evolve. But as this chapter suggests, legal reforms can play a role in leading the way.

References

American Society of Civil Engineers (ASCE). 2010. So You Live Behind a Levee! What You Should Know to Protect Your Home and Loved Ones From Floods. Available at http://content.asce.org/files/pdf/SoYouLiveBehindLevee.pdf.

Arkansas Game & Fish Comm'n v. United States, 133 S. Ct. 511. 2012.

Barry, J.M. 1998. *Rising Tide: The Great Mississippi Flood of 1927 and How It Changed America*. New York: Simon & Schuster.

Bell, A., and G. Parchomovsky. 2001. Givings. *Yale Law Journal* 111:547–618.

Biggert–Waters Flood Insurance Reform Act of 2012. P.L. 112-141, 126 Stat. 405, 917-19, 927-30 (amending 42 U.S.C. §§ 4001–4127).

Brandt, A.W., and A.P. Clark. 2012. Preparing for Extreme Floods: California's New Flood Plan. *Natural Resources & Environment* Fall 2012:45–48.

Disaster Relief Act of 1950. Pub. L. No. 81-875, 64 Stat. 1109, 81 Cong. Ch. 1125, Sept. 30, 1950.

Executive Order 13690. 2015. *Establishing a Federal Flood Risk Management Standard and a Process for Further Soliciting and Considering Stakeholder Input*. 80 Fed. Reg. 6425.

Faber, S. 2005. Forming a Comprehensive Approach to Meeting the Water Resources Needs of Coastal Louisiana in the Wake of Hurricanes Katrina and Rita. *Hearing before the Comm. on Env't and Pub. Works*. 109th Cong. 71–72.

Federal Emergency Management Agency (FEMA). 2013. 6 Months Report: Superstorm Sandy from Pre-Disaster to Recovery. Available at http://www.fema.gov/disaster/4086/updates/6-months-report-superstorm-sandy-pre-disaster-recovery.

Federal Emergency Management Agency (FEMA). 2015a. *Fact Sheet: Taking Action to Protect Communities and Reduce the Cost of Future Flood Disasters*. Available at https://www.whitehouse.gov/administration/eop/ceq/Press_Releases/January_30_2015.

Federal Emergency Management Agency (FEMA). 2015b. *Risk MAP Flood Risk Products*. Available at https://www.fema.gov/risk-map-flood-risk-products.

Flood Control Act of Mar. 1, 1917. Pub. L. No. 64-367, 39 Stat. 948.

Flood Control Act of 1928. 45 Stat. 534, 33 U.S.C. §§ 702a–m, May 15, 1928.

Flood Control Act of 1936. Ch. 688, 49 Stat. 1570, codified at 33 U.S.C. § 701(a).

Flood Control Act of 1944. 58 Stat. 887, codified in various provisions of Titles 16, 33, and 43 of the U.S. Code.

Galloway, Jr., G.E. 2005. Corps of Engineers Responses to the Changing National Approach to Floodplain Management since the 1993 Midwest Flood, *Journal of Contemporary Water Research & Education* 130:5–12.

Government Accountability Office. 2010. *National Flood Insurance Program, Continued Actions Needed to Address Financial and Operational Issues*. GAO-10-631T.

Homeowner Flood Insurance Affordability Act of 2014. Pub. L. No. 113-89, §§ 3–4, 128 Stat. 1020, 1021-22 (2014) (amending 42 U.S.C. §§ 4001–4127).

Just v. Marinette County, 56 Wis. 2d 7, 13, 201 N.W.2d 761, 768 (1972).
King, R.O. 2012. National Flood Insurance Program: Background, Challenges, and Financial Status. Congressional Research Service R40650.
King, R.O. 2013. National Flood Insurance Program: Status and Remaining Issues for Congress. Congressional Research Service R42850.
Klein, C.A. and S.B. Zellmer. 2007. Mississippi Stories: A Century of Unnatural Disasters. *Southern Methodist University Law Review* 60:1471–1537.
Klein, C.A. and S.B. Zellmer. 2014. *Mississippi River Tragedies: A Century of Unnatural Disaster.* New York: NYU Press.
Kunkel, K.E., P.D. Bromirski, H.E. Brooks et al. 2008. Ch. 2: Observed changes in weather and climate extremes. *Weather and Climate Extremes in a Changing Climate. Regions of Focus: North America, Hawaii, Caribbean, and U.S. Pacific Islands. A Report by the U.S. Climate Change Science Program and the Subcommittee on Global Change Research*, T.R. Karl, G.A. Meehl, C.D. Miller, S.J. Hassol, A.M. Waple, and W.L. Murray, eds., Washington D.C.: U.S. Climate Change Science Program and the Subcommittee, 35–80.
Lemann, A. 2015. Rolling Back the Tide: Toward an Individual Mandate for Flood Insurance. *Fordham Environmental Law Review* 26:166–206.
Lucas v. S.C. Coastal Council, 505 U.S. 1003, 1019 (1992).
Martinsdale, B., and P. Osman. 2007. Why the Concerns with Levees? They're Safe, Right? *Illinois Association for Floodplain and Stormwater Newsletter Fall 2007.* Available at http://www.illinoisfloods.org/documents/IAFSM_Levee%20Article.pdf.
McCool, D. 2005. The River Commons: A New Era in U.S. Water Policy. *Texas Law Review* 83:1903–1927.
National Audubon Soc'y v. Superior Court, 33 Cal. 3d 419, 658 P.2d 709, 728–732 (1983).
National Flood Insurance Act of 1968. 42 U.S.C. §§ 4001–4129.
National Oceanic and Atmospheric Administration (NOAA). 2013. Historic Rainfall and Floods in Colorado. *Climate.gov.* Available at https://www.climate.gov/news-features/event-tracker/historic-rainfall-and-floods-colorado.
National Oceanic and Atmospheric Administration–National Weather Service. 2011. *United States Flood Loss Report—Water Year 2011.* Available at http://www.nws.noaa.gov/hic/summaries/WY2011.pdf.
National Research Council, Water Sci. & Tech. Bd. 2001. *Inland Navigation System Planning: The Upper Mississippi River–Illinois Waterway.*
O'Neill, K.M. 2006. *Rivers by Design: State Power and the Origins of U.S. Flood Control.* Durham: Duke University Press.
Penn Central Transp. Co. v. City of New York, 438 U.S. 104, 124 (1978).
Pennsylvania Coal Co. v. Mahon, 260 U.S. 393, 415 (1922).
Percy, M.J. 2007. Delta Levees—Tort Immunity vs. Takings Liability. *Real Property, Probate & Trust Journal* 42:547–576.
Perry, C.A. 2000. *Significant Floods in the United States during the 20th Century: USGS Measures a Century of Floods.* USGS Fact Sheet 024-00.
Pisani, D. 2002. *Water and American Government: The Reclamation Bureau, National Water Policy, and the West.* Berkeley and Los Angeles: University of California Press.
Ralph, F.M., M. Dettinger, A. White et al. 2014. A Vision for Future Observations for Western US Extreme Precipitation and Flooding. *Journal of Contemporary Water Research & Education* 153(1):16–32.
Rand Corporation. 2006. *Evaluating National Flood Insurance.* RB-9176, xiii.
Schleifstein, M. 2006. Corps Report Ignores Call for Specifics; Details of Category 5 Protections Left Out. *Times-Picayune (New Orleans)*, July 1, 2006: 1.
Singer, S.J. 1990. Flooding the Fifth Amendment: The National Flood Insurance Program and the Takings Clause. *Boston College Environmental Affairs Law Review* 17:334–370.
Slocum v. Borough of Belmar, 569 A.2d 312, 316 (N.J. Super. L. 1989).
Stafford Disaster Relief and Emergency Assistance Amendments of 1988. Pub. L. No. 100-707, 102 Stat. 4689 (1988) (codified at 42 U.S.C. §§ 5121–5202 [1988]).

Tarlock, A.D. 2004. A First Look at a Modern Legal Regime for a "Post-Modern" United States Army Corps of Engineers. *Kansas Law Review* 52:1285–1325.
Trenberth, K.E. 2011. Changes in precipitation with climate change. *Climate Research*, 47:123–138.
Upbin, B. 1998. A River of Subsidies. *Forbes*. March 23, 1998:86.
Walsh, J., D. Wuebbles, K. Hayhoe et al. 2014. Ch. 2: Our Changing Climate. *Climate Change Impacts in the United States: The Third National Climate Assessment*, J.M. Melillo, T.C. Richmond, and G.W. Yohe, eds., Washington, D.C.: U.S. Global Change Research Program, 19–67.
Water Science & Technology Board. 1995. *Flood Risk Management and the American River Basin: An Evaluation.* Commission on Flood Control Alternatives in the American River Basin.
Wriggins, J. 2014. Flood Money: The Challenge of U.S. Flood Insurance Reform in a Warming World. *Penn State Law Review* 119:361–437.
Wright, J.M. 2000. *The Nation's Responses to Flood Disasters: A Historical Account.* Association of State Floodplain Managers.
Yee v. Escondido, 503 U.S. 519, 522 (1992).

20

Adaptive Management and Governance Lessons from a Semiarid River Basin: A Platte River Case Study

Chadwin B. Smith, Jason M. Farnsworth, David M. Baasch, and Jerry F. Kenny

CONTENTS

ABSTRACT Adaptive management and a unique governance structure are at the center of a large-scale species recovery program on the central Platte River in Nebraska. The Platte River Recovery Implementation Program (Program) began in 2007 as a joint effort between the states of Colorado, Wyoming, and Nebraska; the US Department of the Interior; water users; and conservation groups to address water use and endangered species needs in this semi-arid river basin. The Program, which arose in response to a number of environmental problems and water management challenges in the three basin states in the early 1990s, manages land and water resources in central Nebraska to address habitat loss while preserving existing water uses. Uncertainties related to the response of target species to Program management actions are addressed through the application of adaptive management, in this case defined as a rigorous approach for designing and implementing management actions to maximize learning about critical uncertainties that affect decisions while simultaneously striving to meet multiple management objectives. The Program's Adaptive Management Plan provides the structure for organizing and implementing management actions such as flow releases and sediment augmentation to test priority hypotheses and answer overarching questions related to river form and function and species responses. After nearly 9 years of implementation, a collaborative governance structure that includes stakeholders and clear lines of decision making and communication has the Platte River Program poised to successfully complete one full loop of the six-step adaptive management cycle and actually adjust in response to accumulated learning. This chapter explores those facets of the Platte River Program and how it functions in an area marked by a semiarid climate and intense surface and groundwater use.

20.1 Introduction

As a concept, adaptive management (AM) has long been intriguing for resource managers and decision makers working in complex ecosystems and facing a high degree of uncertainty. AM emerged as an application of the scientific method to resource management, closely tying management to science learning through experimental actions (Holling 1978; Walters 1986). At larger scales, the phrase *learning by doing* seemed to best capture the premise behind developing an experimental management approach that could be applied on the scale of a river system or ecosystem like the Everglades (Walters and Holling 1990). In nearly five decades of discussion and application, however, examples of successful AM implementation at large scales are few, and conflict remains over how to achieve the most essential elements of a true AM approach (Gregory et al. 2006; Walters 2007)—a rigorous process of learning by designing management actions as experiments. In the case of large-scale ecosystem rehabilitation and species recovery program efforts, the application of AM as a guiding framework for science has had mixed success at best (Walters et al. 1992; Lee 1993, 1999; Allan and Curtis 2005; Zellmer and Gunderson 2009; Murray et al. 2015).

AM serves as the science framework for the Platte River Recovery Implementation Program (Program or PRRIP), a large-scale species recovery program on the central Platte River in Nebraska. The Program started in 2007 as the result of a negotiating process that began in 1997 between the states of Colorado, Wyoming, and Nebraska; the US Department of the Interior; waters users; and conservation groups. The Program is intended to address issues related to the Endangered Species Act and loss of river habitat along 90 mi. (145 km) of the Big Bend Reach of the Platte River in central Nebraska (Figure 20.1).

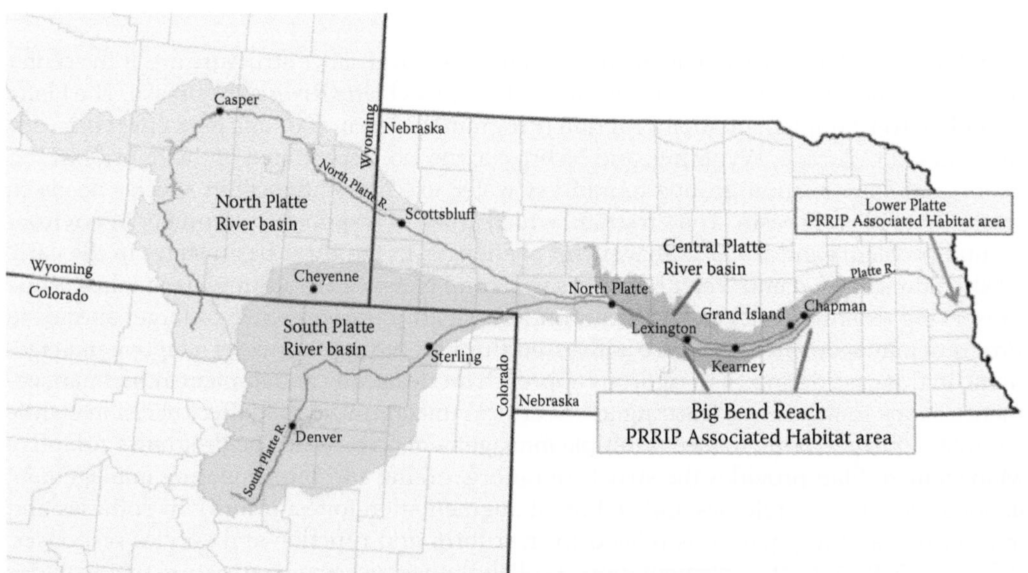

FIGURE 20.1
Platte River Recovery Implementation Program (Program or PRRIP) spatial scale. The Program is focused on species use of habitat and management actions within a 90 mi. (145 km) section of the Big Bend Reach of the Platte River between Lexington and Chapman in central Nebraska called the Associated Habitat Reach. The lower Platte River also includes Associated Habitat for the pallid sturgeon. The main-stem Platte River is formed by the confluence of Colorado's South Platte River and Wyoming's North Platte River near North Platte, Nebraska.

By managing land and water resources and applying science in an AM framework, the Program is to implement of a portion of the recovery plans for four target species: the endangered whooping crane (*Grus americana*), interior least tern (*Sternula antillarum*), and pallid sturgeon (*Scaphirhynchus albus*); and the threatened piping plover (*Charadrius melodus*).

An adaptive management plan (AMP) provides guidance for Program science and offers a systematic process to test priority hypotheses and apply the information learned to improve management on the ground (PRRIP 2006a). The AMP was developed jointly by Program participants, reflects different interpretations of river processes and species' responses to management actions, and represents a shared attempt on the part of Program cooperators and partners to use the best available science in an agreed-upon manner to implement management actions as experiments, learn, and revise management actions to reach a common goal—contribute to the recovery of the target species.

20.2 Impacts of Water Development on the Platte River

20.2.1 Habitat

Water development in the Platte River basin began in the mid-1800s as settlers migrated to the region in search of gold and to homestead after the federal government opened the basin for settlement. The Platte River is now heavily developed, with over 7000 diversion rights and nearly 7 million acre-feet (8.3 BCM)* of storage (Figure 20.2) (Simons & Associates 2000) in Colorado, Wyoming, and Nebraska.

Platte River discharge records begin in 1895, 15 years before the completion of Pathfinder Dam, the first major agricultural storage project in the basin. Mean annual discharge and the magnitude of the mean annual peak discharge in the contemporary river are currently less than 40% of what was observed during the brief period of record prior to reservoir construction (Table 20.1) (Stroup et al. 2006). Predevelopment sediment loads in the central Platte River have been roughly estimated at between 2 and 7.8 million tons (1.8 and 7.1 million metric tons) per year (Simons & Associates 2000; Murphy et al. 2004). Contemporary sediment load estimates are less variable and generally range from 400,000 to 1 million tons (363,000–907,000 metric tons) per year.

One of the most significant changes in sediment dynamics from predevelopment conditions in the central Platte reach where the Program is working is a sediment deficit due to clear water hydropower returns at the Johnson 2 (J-2) return structure on the south channel downstream of Lexington, Nebraska (Figure 20.3).

An average of approximately 73% of remaining Platte River flow, after upstream diversions, is diverted at the Tri-County Diversion Dam downstream of North Platte, Nebraska, and returns to the river at the J-2 return. Once diverted at North Platte, flow travels through several off-channel reservoirs, where almost all of the sediment is trapped. Accordingly, return flows at the J-2 return structure are sediment-starved, resulting in a sediment deficit below the return.

The reduction in active channel width (unvegetated width between permanently vegetated left and right banks) over historical time frames through expansion of woody vegetation was first quantified by Williams (1978) and has been further studied in subsequent

* 1 million acre-feet (MAF) = 1.233 billion cubic meters (BCM).

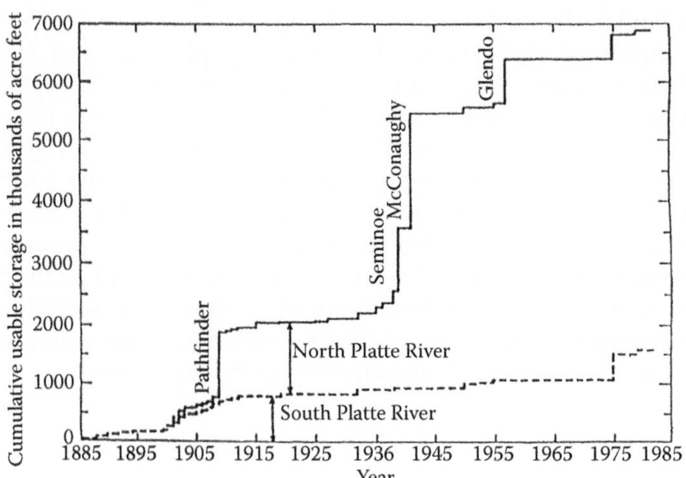

FIGURE 20.2
Cumulative usable storage in reservoirs in the Platte River basin. (From Simons & Associates, Inc., and URS Greiner Woodward Clyde, *Physical History of the Platte River in Nebraska: Focusing upon Flow, Sediment Transport, Geomorphology, and Vegetation*, prepared for the Bureau of Reclamation and Fish and Wildlife Service Platte River EIS Office, August 2000.)

TABLE 20.1
Mean Annual Discharge and Mean Annual Peak Discharge at Overton Gage

	1895–1909	1910–1927	1928–1941	1942–1958	1959–1974	1975–1998	1999–2013
Mean annual discharge in cfs (cms)	4584 (130)	4323 (122)	1845 (52)	1223 (35)	1636 (46)	1938 (55)	1232 (35)
Mean annual peak discharge in cfs (cms)	20,725 (587)	18,218 (516)	11,548 (327)	6685 (189)	7301 (207)	7176 (203)	5056 (143)

Source: Adapted from Stroup, D. et al., *Flow Characterizations for the Platte River Basin in Colorado, Wyoming, and Nebraska*, prepared for the Platte River Recovery Program EIS Office, Lakewood, Colorado, 2006.

analyses (Eschner et al. 1983; Currier et al. 1985; Peake et al. 1985; O'Brien and Currier 1987; Lyons and Randle 1988; Sidle et al. 1989; Johnson 1994; Simons & Associates 2000; Parsons 2003; Murphy et al. 2004; Schumm 2005; Horn et al. 2012). Investigators have generally concluded that the central Platte reach where the Program is working experienced a significant width reduction as a result of the expansion of cottonwood forest into the channel. The change is evident in comparisons of aerial photography. At the earliest aerial photography collection in 1938, unvegetated channel width averaged 2600 ft. (792 m). By 1998, average unvegetated width was 900 ft. (274 m).

Sandbar characteristics in the historical river are not well documented, but early travelers generally characterized the bed of the river as being comprised of innumerable sandbars continually shifting and moving downstream. The first detailed characterization of Platte River sandbar morphology classified Platte River bedforms as transverse bars (Ore 1964). Further attempts to characterize sandbar morphology identified dominant bedforms as transverse/linguoid bars (Smith 1971; Blodgett and Stanley 1980), macroforms (Crowley 1981 and 1983), or a combination of both types (Horn et al. 2012). The historical accounts of Platte

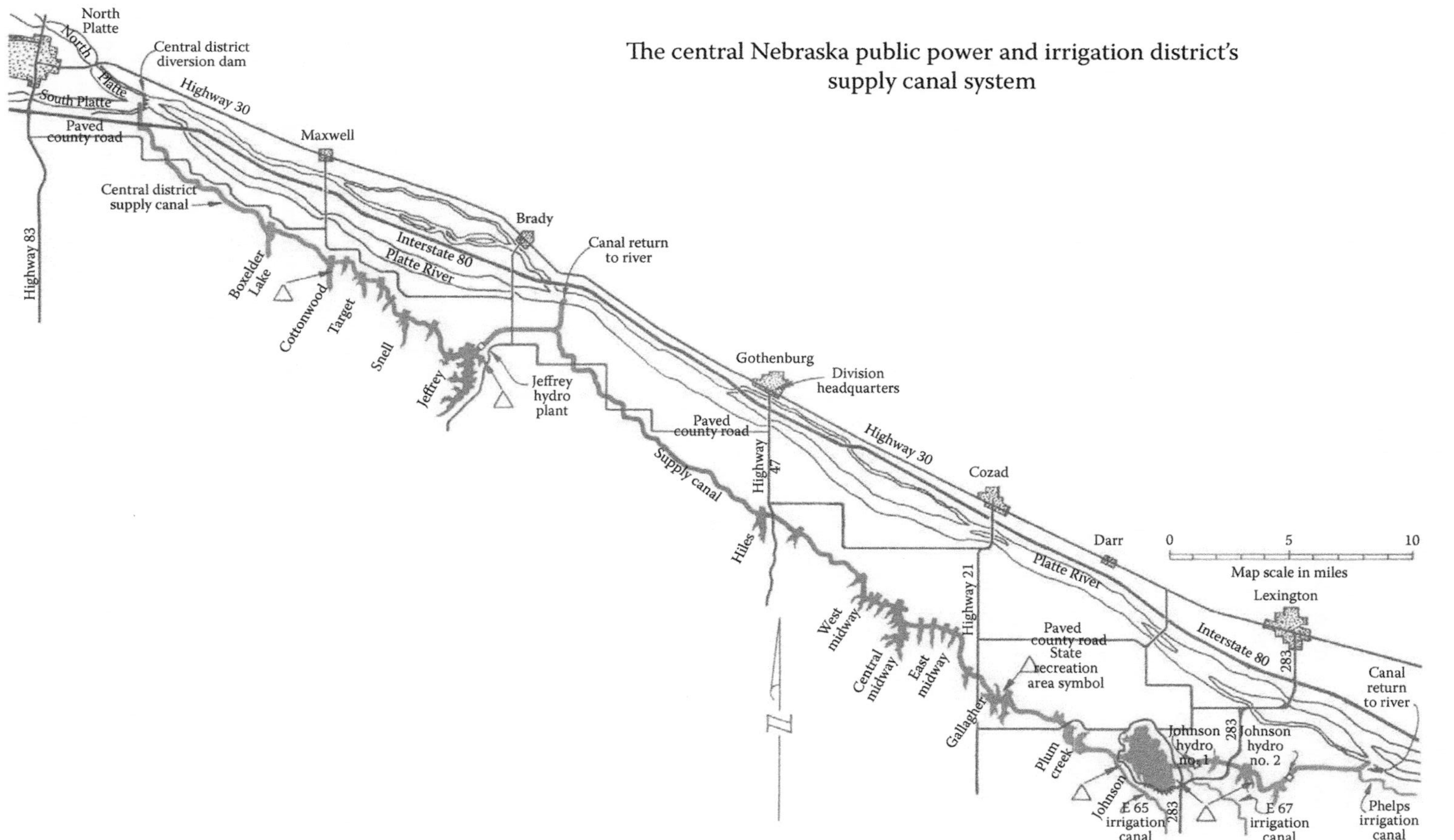

FIGURE 20.3
Map depicting the central Platte River supply canal diversion infrastructure in central Nebraska including the J-2 return near Overton, Nebraska, downstream. (From Central Nebraska Public Power and Irrigation District website, accessed August 12, 2015, http://www.cnppid.com/wp-content/uploads/2014/01/Supply_Canal.pdf.)

River bedforms appear to agree well with contemporary descriptions of transverse/linguoid bars. In all cases, these sandbar types are characterized as being dynamic and mobile, forming, moving, and reforming based on available flow and relationships to channel width. As the characteristics of flow and channel width changed on the central Platte, lower-elevation mobile sandbars started being replaced by larger islands anchored in place by vegetation. The cumulative effects of changes to river flow, channel width, and sandbar morphology led to significant alterations of riverine habitat important to certain imperiled species.

20.2.2 Target Species

The whooping crane (*Grus americana*), considered the rarest species of crane in the world, was federally listed as endangered in 1967 (NRC 2005). Whooping crane stopovers occur throughout the Texas-to-Canada migration corridor and last from one to several days during migrations that can last several weeks. Along the central Platte River, riverine habitat has by far the highest incidence of stopover use for whooping cranes (Austin and Richert 2005; NRC 2005). As such, the US Fish and Wildlife Service (USFWS) designated portions of the central Platte River as critical habitat under the Endangered Species Act (USFWS 1978). The USFWS determined that whooping crane habitat along the central Platte River was threatened by upstream impoundments and diversions that reduce the magnitude of the annual spring runoff credited with historically creating and maintaining open-channel roosting habitat and for sustaining suitable bottomland (wet meadow) habitat deemed to be essential for foraging (DOI 2006).

The interior population of the least tern (*Sternula antillarum*) was federally listed as endangered in 1985, and the piping plover (*Charadrius melodus*) was listed as threatened in 1986. Least terns are colonial nesting birds that breed and nest on barren to sparsely vegetated riverine sandbars, sand and gravel pits, lake and reservoir shorelines, rooftops, ash pits, and salt flats from late April to early August (USFWS 2013). Least terns forage on small fish they capture by diving into shallow riverine habitats and freshwater ponds. The northern Great Plains piping plover population breeds in alkaline wetlands, along lake shorelines of the northern Great Plains, and on the Missouri River and its tributaries in North and South Dakota and Nebraska (USFWS 2009). Piping plovers on the breeding grounds generally forage on insects and spiders. Soon after listing both species, the USFWS made the determination that on the central Platte River, these species were threatened by upstream impoundments and diversions that reduced the magnitude of the annual spring runoff credited with historically creating and maintaining suitable sandbar nesting habitat and associated foraging habitat on a near-annual basis (DOI 2006).

The fourth target species for the Program is the pallid sturgeon (*Scaphirhynchus albus*). The pallid sturgeon was listed as endangered in 1990 (NRC 2005). Pallid sturgeon are long-lived fish typically found in the warm, turbid waters of major river systems. Known to feed on a small fish and insects, the pallid sturgeon is a highly imperiled species and remains at the center of management conflict on the Missouri River, of which the Platte is a significant tributary. While the on-the-ground work of the Program is on the central Platte River, pallid sturgeon are only found in the lower Platte River from its mouth at the Missouri upstream to potentially as far as the Loup River confluence. Given that this geographic location is well beyond the central Platte focus area, the prevailing disagreements about the significance of the lower Platte to the pallid sturgeon population, and the ability of the Program to influence lower Platte flows, the pallid sturgeon remains a controversial aspect of Program implementation. The management objective tied to the pallid sturgeon in the Program's AMP is a *do no harm* objective, and thus far, that has not compelled active

Program management in the lower Platte to benefit pallid sturgeon (PRRIP 2006a). As such, the work of the Program since implementation began in 2007 has focused nearly exclusively on use of the central Platte by the least tern, piping plover, and whooping crane, so the remainder of this chapter will retain a similar focus.

20.3 Water and Species Conflicts in a Semiarid River Basin—A Unique Approach

Like most western rivers, water in the Platte River remains in high demand for diversions related to agriculture, municipal use, and other purposes. Water overappropriation, however, conflicts with an increasing demand for management changes to benefit listed species (NRC 2005). This was true on the Platte River, particularly in the 1980s, 1990s, and early 2000s, as a series of jeopardy opinions from the USFWS on water projects with a federal nexus raised tension over water use and habitat decline on the central Platte River. In particular, the extended relicensing process for the Kingsley Dam hydroelectric project on the North Platte River in western Nebraska, and calls from the USFWS to return 417,000 acre-feet (514.4 million m^3)* of water annually to the central Platte River, pushed parties to find a solution (NRC 2005). Deliberations for the relicensing effort particularly centered on use of the central Platte River by the target species and the alteration of habitat that occurred over the course of several decades associated with water diversions, land-use changes, and other basin alterations.

Beginning in 1997, under a cooperative agreement, the states of Colorado, Wyoming, and Nebraska; the US Department of the Interior; waters users; and conservation groups spent nearly 10 years debating the components of a long-term plan to address endangered species needs while protecting water users. During negotiations, the participants committed to working toward two primary objectives: (1) reduce the shortage of flows in the central Platte River by 130,000–150,000 acre-feet (160–185 million m^3) per year on average and (2) protect or restore 10,000 acres (4100 ha.) of habitat in the central Platte River basin (PRRIP 2006b). Agreement on these two objectives led to the development of the Program, formed the basis for land and water acquisition and management, and provided the context for science-based learning.

The Program is authorized for a 13-year First Increment from 2007 through 2019 and is estimated to cost roughly $325 million, in 2005 dollars, with the monetary portion of that being $187 million and the remainder in commitments of land and water. The federal government and the three states equally share the total cost in terms of cash, water, and land. The focus area of the Program is the Associated Habitat Reach (AHR) in central Nebraska, a 90 mi. reach extending from Lexington downstream to Chapman, and including the Platte River channel and off-channel habitats within 3.5 mi. of the river (see Figure 20.1).

One of the unique aspects of the Program is its governance and management structure. Decisions are the ultimate responsibility of the Governance Committee, which consists of representatives from the Bureau of Reclamation; the USFWS; the states of Colorado, Wyoming, and Nebraska; upstream and downstream water users; and conservation groups. The Governance Committee is assisted by several standing advisory committees made up of technical representatives of Program agencies and institutions and is organized by land, water, and AM issues (Figure 20.4).

* 1 acre-foot (af) = 1233.48 cubic meters (m^3).

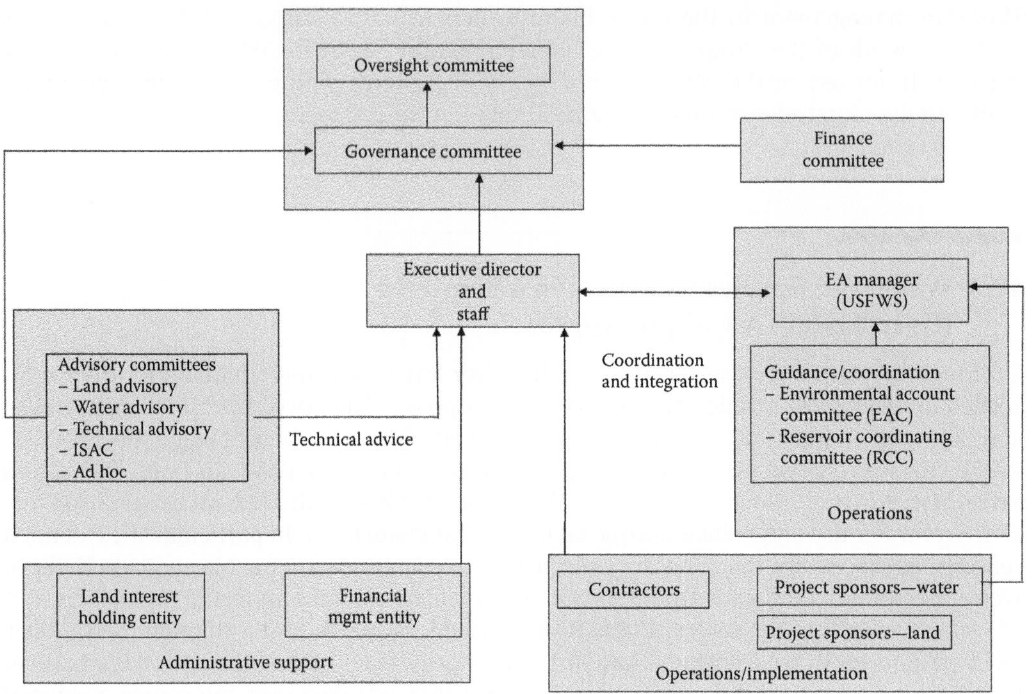

FIGURE 20.4
Platte River Recovery Implementation Program governance structure.

The Governance Committee hired an independent executive director not affiliated with any of its entities. Day-to-day operations of the Program are the responsibility of the executive director and staff. Staff members also operate independently from the partner agencies. The executive director and staff coordinate all Program meetings; provide oversight of the Program budget, lands, and water; and provide a common link between the Governance Committee and the advisory committees.

The Program's approach to governance is much different from other well-known AM programs, such as those in the Everglades and on the Colorado River, where federal agencies are in the lead in terms of both staffing and decision making. In those systems, US Army Corps of Engineers and US Department of the Interior employees staff the programs but are also ultimately in charge of making policy decisions. Within the Platte River Program, all stakeholders including water users and conservation groups are voting members of the policy body and are represented along with state and federal agency representatives. Again, this is a major difference from other programs where stakeholders may be involved in the process at various levels but do not have the opportunity to help make management or policy decisions. On the Platte River, the executive director and staff are independent of the US Department of the Interior, the states, the water users, and the conservation groups. This builds in a considerable level of independence and lack of bias that, so far, has proven useful in moving Program projects forward and building trust across all levels of Program participants. The Program's construct is very much in line with a social learning process that is inherent in AM implementation and engages stakeholders at a decision-making level to build trust and provide a broader context for experimental management actions (Lee 1993).

Independent scientific review is an integral part of implementing AM on the Platte River. The first six members of the Independent Scientific Advisory Committee (ISAC) were appointed by the Governance Committee in 2008, and the ISAC still stands at six members in 2015, with some regular rotation of members as new areas of expertise are identified. The ISAC reports directly to the Governance Committee on issues related to implementation of the AMP; the rigor and robustness of data collection and analysis; and specific topical questions addressed to the ISAC by the Governance Committee, the Program's Technical Advisory Committee, and Program staff. The ISAC meets at least twice per year, including at the annual AMP reporting session, where Program participants, Program staff, contractors, the ISAC, and other independent experts review the pace and results of AMP implementation from the previous year. The ISAC is not a participant in implementing AM; rather, the committee provides independent counsel on experimental design, the overall AM approach, and the methods utilized to integrate science into decision making. In addition, external peer review panels are routinely assembled to review specific Program data analysis and synthesis reports, monitoring protocols, and other AMP documents to ensure scientific rigor. The Program's peer review process is included as part of the AMP (PRRIP 2006a) and, based on guidance from the ISAC, has been updated to conform with peer review guidelines used and accepted by the USFWS and the Office of Management and Budget.

This governance model is unique and serves as an alternative to how most species recovery and ecosystem rehabilitation programs are currently organized and managed. Thus far, this approach has proved effective at consensus-based decision making and helping to internalize science learning from implementation of AM as an input into that decision making. In a semiarid basin characterized by tensions over water and endangered species and differences of opinion over species' responses to river processes and management actions, this structure is providing a path forward. Given that the Program is poised to move further through the six-step AM cycle (assess, design, implement, monitor, evaluate, adjust) than other similarly scaled species recovery and ecosystem restoration programs, this unique governance structure appears to be a key difference facilitating that forward progress.

20.4 Adaptive Management to Reduce Uncertainty

20.4.1 Objectives, Hypotheses, and Big Questions

According to the AMP, the Program defines AM as "a series of scientifically driven management actions (within policy and resource constraints) that use the monitoring and research results provided by the Integrated Monitoring and Research Plan (IMRP) to test priority hypotheses related to management decisions and actions, and apply the resulting information to improve management" (PRRIP 2006a, p. 2). The AMP goes on to identify the common six steps of AM, as noted in Figure 20.5. While many definitions of AM exist, this is the understanding of how AM will be applied within the Program. It also represents how the scientific and technical aspects of the Program have been implemented since 2007.

AM serves as the science framework for the Program. The AMP provides guidance for Program science and offers a systematic process to test priority hypotheses and

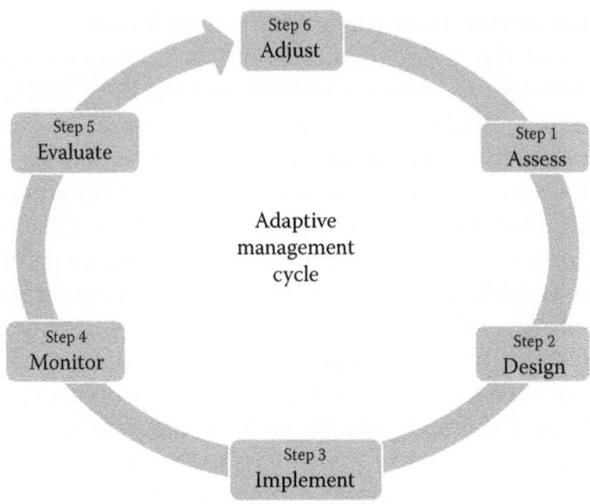

FIGURE 20.5
The six-step adaptive management cycle. (Murray, C. et al. *Middle Rio Grande Endangered Species Collaborative Program Adaptive Management Plan Version 1,* prepared by ESSA Technologies Ltd. (Vancouver, BC) and Headwaters Corporation (Kearney, NE) for the Middle Rio Grande Endangered Species Collaborative Program, Albuquerque, NM, 2011.)

apply the information learned to improve management on the ground (PRRIP 2006a). The AMP was developed jointly by Program participants; reflects different interpretations of river processes and species' responses to management actions; and represents a shared attempt on the part of Program cooperators and partners to use the best available science in an agreed-upon manner to implement management actions as experiments, to learn, and to revise management actions to reach a common goal—contribute to the recovery of the target species. This learning-by-doing approach links monitoring and research to key questions related to species and river process response to management actions. Data collection and analysis are closely tied to assessing Program hypotheses and management actions that feed information into the decision-making process.

Specific priority hypotheses with detailed X–Y graphs were added to the AMP to provide the data evaluation context for exploring the relationships addressed in the broader hypotheses (PRRIP 2006a). As an example, Flow #1 suggests that under a balanced sediment budget, a water discharge volume of 5000–8000 cubic feet per second (cfs; 142–227 cubic meters per second [cms]) for 3 days (roughly 50,000–75,000 acre-feet, or 61.7–92.5 million m^3), the only flow management action prescribed in the AMP, will build sandbars to an elevation that is suitable for least tern and piping plover nesting (Figure 20.6).

One initial challenge in preparing for implementation of the AMP was the set of 42 priority hypotheses that address issues related to each target species or river processes such as sediment transport. Program participants developed hypotheses through the negotiation process, and each includes an alternative hypothesis. The hypotheses are labeled *priority* because they represent the negotiated set of key questions the Program hopes to assess during the First Increment, and they stem from a much larger set of hypotheses filtered out through negotiations over feasibility and relevance to key Program questions and objectives. In the process of developing the Program, negotiators generally agreed to set aside their value systems and thoughts on how to apply management actions, and they built an AMP that incorporates competing views of how

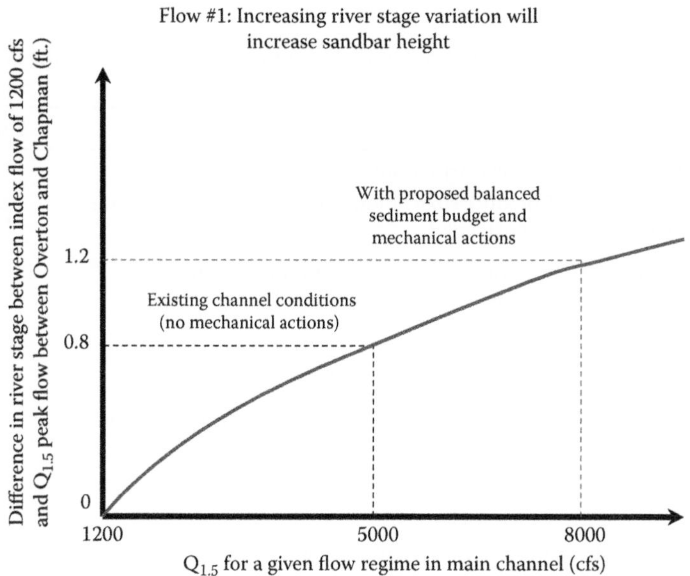

FIGURE 20.6
Priority hypothesis Flow #1, as detailed in the Program's adaptive management plan. Increasing the variation between river stage at peak flow (indexed by $Q_{1.5}$ flow at Overton) and average flows (1200 cfs index flow), by increasing the stage of the peak (1.5-year) flow through program flows, will increase the height of sandbars between Overton and Chapman by 30–50% from existing conditions, assuming balanced sediment budget.

the Platte River works (competing hypotheses) and different methods to assess species response (two management strategies). However, it is almost impossible to test that many hypotheses in normal policy timelines, especially within the 13-year limitation of the First Increment.

In 2012, the Program developed a set of "Big Questions" designed to focus implementation efforts and science learning on the most critical uncertainties related to Program management actions and species responses. Those questions are used as a device to organize Program data synthesis efforts and communicate important science learning to the decision makers on the Governance Committee. Information related to the Big Questions is compiled annually into a State of the Platte Report, which serves "as a synthesis of Program monitoring data, research, analysis, and associated retrospective analyses to provide important information to the Governance Committee regarding key scientific and technical uncertainties" (PRRIP 2014, p. 2). As shown in Table 20.2, each year, the Big Questions are paired with an assessment of progress using icons that quickly show decision makers whether there is a definitive answer, a trend in a positive or negative direction, or a need for more information.

The AMP contains specific management actions grouped collectively into two management strategies (PRRIP 2006a). The first management strategy is the river-centric flow–sediment–mechanical (FSM) approach. Management actions include the following:

- Flow—Use Environmental Account water from Lake McConaughy to generate short-duration near-bank-full flows (short-duration high flows [SDHFs]) of 5000–8000 cfs (142–227 cms) in the habitat reach for 3 days in the springtime or other

TABLE 20.2

The Program's "Big Questions" and Latest Quick-Reference Assessments (in the Form of Icons) from the 2015 State of the Platte Report

PRRIP Big Question	2014 Assessment
Implementation—Program Management Actions and Habitat	
1. Will implementation of SDHF produce suitable tern and plover riverine nesting habitat on an annual or near-annual basis?	
2. Will implementation of SDHF produce and/or maintain suitable whooping crane riverine roosting habitat on an annual or near-annual basis?	
3. Is sediment augmentation necessary for the creation and/or maintenance of suitable riverine tern, plover, and whooping crane habitat?	
4. Are mechanical channel alterations (channel widening and flow consolidation) necessary for the creation and/or maintenance of suitable riverine tern, plover, and whooping crane habitat?	
Effectiveness—Habitat and Target Species Response	
5. Do whooping cranes select suitable riverine roosting habitat in proportions equal to its availability?	
6. Does availability of suitable nesting habitat limit tern and plover use and reproductive success on the central Platte River?	
7. Are both suitable in-channel and off-channel nesting habitats required to maintain central Platte River tern and plover populations?	
8. Does forage availability limit tern and plover productivity on the central Platte River?	
9. Do program flow management actions in the central Platte River avoid adverse impacts to pallid sturgeon in the lower Platte River?	
Larger-Scale Issues—Application of Learning	
10. Do program management actions in the central Platte River contribute to least tern, piping plover, and whooping crane recovery?	
11. What uncertainties exist at the end of the First Increment, and how might the Program address those uncertainties?	

times outside of the main irrigation season. The intent is to achieve these flows on an annual or near-annual basis.

- Sediment—Mechanically place sediment into the river from banks, islands, and out-of-bank areas at a rate that will eliminate the sediment deficiency and restore a balanced sediment budget.
- Mechanical
    a. Consolidate the flow and river channels to maximize stream power and help induce braided channel characteristics.
    b. Mechanically cut banks and lower islands to a level that will be inundated by anticipated annual peak flows.

c. Mechanically clear vegetation from islands and banks in the single channel as needed to aid in the widening process and make sediment available for recruitment to the river. Minimum channel width target is 750 ft. (229 m).

The FSM management strategy is river-centric, as indicated by the management actions and hypothesized beneficial effects. The alternative mechanical creation and maintenance (MCM) approach focuses on mechanical creation and maintenance of both in- and off-channel habitats on an annual or near-annual basis.

20.4.2 Implementation, Analysis, and Synthesis

To date, the Program's focus has largely been on evaluating the ability of the FSM strategy to produce suitable nesting habitat. The most critical least tern and piping plover management uncertainties to be evaluated through implementation of the AMP include the following: (1) the ability of the FSM strategy to produce and maintain riverine sandbars that are suitably high for nesting; (2) whether or not the species will select in-channel habitats over off-channel habitat; and (3) differences in productivity between the two habitat types. The hypothesized beneficial effects of the FSM strategy for the least tern and piping plover include the following:

- Flows of 5000–8000 cfs (142–227 cms) magnitude for a duration of 3 days (defined in the AMP as an SDHF) are needed with both mechanical actions of consolidating flow and river widening to raise sandbars to an elevation suitable for least tern and piping plover nesting habitat.
- Sediment augmentation is required in conjunction with increases in flows and contributes to wider sustainable channels, contributes to increases in occurrence of sandbars, restores streambed elevation, and over time will promote the occurrence of a braided planform in currently anastomosed reaches of the river where multiple channels are separated by vegetated islands.
- The mechanical action of consolidating flows will help shift the river to a braided condition, which widens the river and creates more sandbars. Cutting banks and leveling islands in conjunction with SDHF will widen the river.

In 2012, the Program met the land objective and currently manages over 10,000 acres (4100 ha.) in five complexes with habitat in a sixth complex coming under active management in the latter half of 2015. The Program continues to consider land for acquisition, lease, or easement based on the concept of habitat *complexes* that can be managed to meet important habitat criteria such as channel width, sandbars, and unobstructed width (PRRIP 2006b). As for the water objective, by 2015, the Program had secured about 78–90% of the First Increment target of 130,000–150,000 acre-feet (160–185 million m^3) of water.

The overall status of FSM management action implementation during the first 8 years of the Program is as follows:

- Flow—Peak discharges exceeding minimum SDHF magnitude and duration occurred in 6 out of 8 years (2007, 2008, 2010, 2011, 2013, and 2014) at Grand Island. Most of these were natural flow events, not controlled releases by the Program, and several far exceeded the SHDF criteria.
- Sediment—Augmentation occurred in 5 out of 8 years. Water year augmentation volumes in years when sediment was augmented ranged from 21,875 tons to

182,000 tons (19,845–165,108 metric tons). The upper half of the reach was in sediment deficit, and the lower half was likely aggradational, although that conclusion appears to only be weakly supported by the data.

- Mechanical—Flow consolidation was abandoned as an unimplementable management action. Various combinations of mechanical vegetation removal from banks and islands, island lowering, channel widening, and in-channel disking were ongoing in 47% of the AHR during the first 8 years of Program implementation.

The AMP includes an IMRP that presents the Program's approach to evaluating species and physical process response to Program management actions and natural events on system, reach, and project scales (PRRIP 2006a). The approach consists of monitoring (e.g., baseline data and long-term trend detection); experimental research (e.g., to determine cause-and-effect relationships); simulation modeling (e.g., to provide a tool to design experiments and test scientific understanding); and independent peer review. Program monitoring and data synthesis efforts fit broadly into the categories of implementation, effectiveness, and validation monitoring. Implementation monitoring is conducted to determine if the management actions are being implemented according to design requirements and standards. Effectiveness monitoring of physical habitat performance indicators is conducted to determine if management actions are achieving or moving toward management experiment performance criteria. Validation monitoring of species use and selection determines if species are responding to management actions and/or if the Program is making progress toward achieving species management objectives.

The Program implements system wide least tern and piping plover habitat selection and productivity monitoring in the AHR. The Program also conducts an annual habitat availability analysis to calculate the total acreage of in-channel habitat that conforms to the minimum habitat suitability criteria. The analysis is geographic information system (GIS) based and utilizes annual aerial imagery and topographic (light detection and ranging [LiDAR]) data in conjunction with stage–discharge relationships from the system-scale HEC-GeoRAS model. Table 20.3 provides an overview of key monitoring data sources used to evaluate the relationships between river flow, sandbar creation and maintenance, and tern and plover productivity.

In 2012, the Program developed a set of minimum habitat suitability criteria to define minimum conditions necessary for least tern and piping plover nesting habitat in the AHR. The Program's fundamental sandbar performance criterion is midchannel bars greater than 0.25 acres (0.10 ha.) in size and greater than 1.5 ft. (0.5 m) above the river stage at 1200 cfs (34 cms). The bars must also be less than 25% vegetated, occur in channels greater than 400 ft. (122 m) wide, and be greater than 200 ft. (61 m) from predator perches (PRRIP 2014). Sandbar height is hypothesized to respond to a single flow event (PRRIP 2006a). Therefore, physical channel response data collected by the Program in areas of sediment balance (at a minimum) proved to be useful in testing the ability of the FSM management strategy to provide the hypothesized beneficial effects.

The Program currently has no specific targets or goals to attain for least tern and piping plover abundance or nesting habitat availability. Instead, the Program's objective is to monitor species and system responses to management activities. During the 2007–2014 nesting seasons, total sandbar area in the AHR conforming to the minimum criteria ranged from 0 to approximately 55 acres (0 to approximately 22 ha.). During the same time frame, least tern in-channel nest counts ranged from 0 to 20 nests, and piping plover in-channel nest counts ranged from 0 to 13 nests. With the exception of piping plover in 2010, in-channel nest counts during the period of 2007–2014 generally trended downward in parallel with

TABLE 20.3

Biological and Physical Process Monitoring, Mechanistic Models, and Research Relevant to Evaluating the Ability of the Flow–Sediment–Mechanical Strategy to Create and Maintain Sandbars Suitably High for Least Tern and Piping Plover Nesting

Effort	Frequency	Description
Least tern and piping plover use and productivity monitoring	Annual	Document species use, habitat variables, and productivity in the AHR.
Least tern and piping plover habitat availability analysis	Annual	Document occurrence and amount of habitat in AHR meeting minimum species habitat suitability criteria.
Discharge measurements	Real-time	Real-time Platte River discharge monitoring at six locations in the AHR. Stream gaging conducted in cooperation with the US Geological Survey and Nebraska Department of Natural Resources.
June color-infrared imagery	Annual	Document in-channel and off-channel habitat conditions during least tern and piping plover nest initiation period.
November color-infrared imagery and light detection and ranging	Annual	Document channel morphology and topography under leaf-off and low-discharge conditions.
System-scale geomorphology and vegetation monitoring	Annual	Monitor sediment transport, channel morphology, and in-channel vegetation throughout the AHR. Data include bed and suspended sediment load measurements, repeat channel transect surveys, bed and bank material sampling, and vegetation monitoring.
HEC-GeoRAS hydraulic model of AHR	As Necessary	Segment-scale hydraulic model for evaluation of channel hydraulics and development of water surface profiles across a range of discharges.
HEC-6T sediment transport model of AHR	As Necessary	Segment-scale sediment transport model for evaluation of sediment deficit and augmentation activities.

the reduction in availability of suitable habitat. During the same period (2007–2014), off-channel nest counts in the AHR were stable to increasing. This is an indication that reduction of in-channel nesting incidence is more likely associated with a decrease in habitat availability than other factors that may influence the overall species subpopulations utilizing the central Platte River.

In terms of AM, the FSM strategy was developed in the midst of historic drought conditions in the Platte River basin (Freeman 2010). During the period of 2000–2006, mean annual discharge at Grand Island was 45% of the long-term (1942–2011) mean of 1.15 million acre-feet (1.42 BCM). High flows were also largely absent during most of the Program negotiations. The median annual peak discharge during the 2000–2006 period was 2080 cfs (59 cms), which was less than 30% of the long-term median of 7100 cfs (201 cms). Within the context of drought, the AMP envisioned the need for controlled high-flow releases of at least 5000 cfs (142 cms) on a near-annual basis beginning in the first year of Program implementation to test flow-related hypotheses (PRRIP 2006a). Persistent channel conveyance limitations upstream of the AHR at North Platte, Nebraska, continue to limit the Program's ability to generate flow release magnitudes in the 5000–8000 cfs (142–227 cms) range. However, the easing of basin drought and subsequent river discharge recovery coincident with Program inception in 2007 has provided natural high flows of similar magnitude and greater duration than contemplated in the AMP. During the first 8 years of Program implementation (2007–2014), mean annual discharge more than doubled, and the 3-day mean annual peak discharge at Grand Island exceeded 5000 cfs (142 cms) in 6 out of 8 years and 8000 cfs (227 cms) in 4 out of 8 years (Figure 20.7). Overall, the shift in basin hydrology has resulted in an 8-year period when peak flow frequency, magnitude, and duration substantially exceeded what could have been achieved during the 2000–2006 period under full FSM implementation.

The scale of flow, sediment, and mechanical management actions and natural analogs during 2007–2014 have met or exceeded implementation objectives for the First Increment in at least a portion of the AHR. Overall, the decline of in-channel habitat meeting minimum suitability criteria and associated decline of in-channel nest incidence appear to be strong indicators that the FSM management strategy, as currently conceived, will not produce the suitable nesting habitat necessary to improve in-channel productivity of least tern and piping plover from the AHR.

In 2014, Program staff compiled numerous data analyses related to least terns and piping plovers into a set of synthesis chapters (PRRIP 2015). Those chapters delved into the relationships between channel characteristics, flow, and the results of Program management actions and natural analogs related to these relationships. The chapters were subjected to the Program's rigorous independent peer review process and then used as reference material to write the annual State of the Platte Report for 2014 (PRRIP 2015) for the assessment of Big Question #1—will implementation of SDHF produce suitable tern and plover riverine nesting habitat on an annual or near-annual basis? The peer-reviewed data synthesis led Program staff to answer the question conclusively in the negative, and that answer was affirmed by the Program's six-member ISAC.

In June 2015, the Governance Committee was apprised of the status of Big Question #1 and unanimously voted in support of the conclusive assessment. Program staff asked the Governance Committee about the next steps toward using this information to actually adjust management actions as a result of the information. The Governance Committee directed Program staff to work with the ISAC and technical representatives

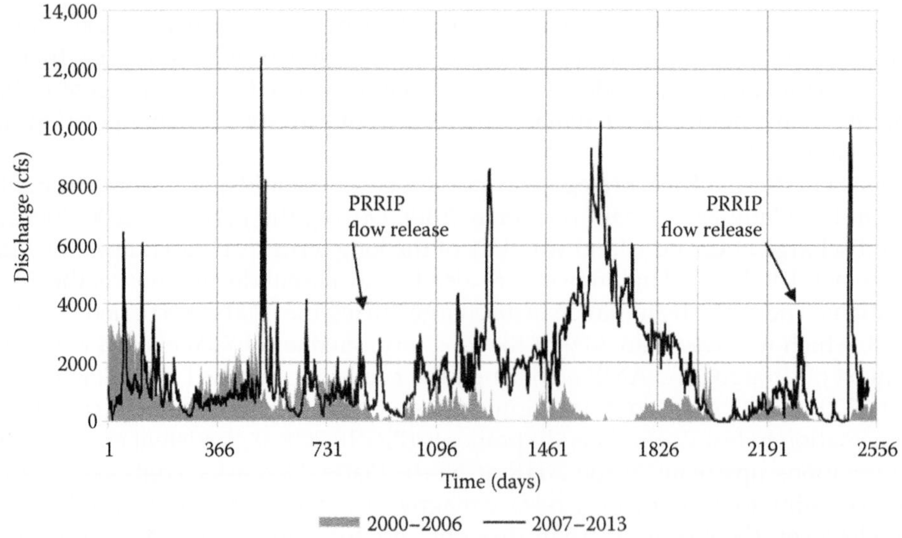

FIGURE 20.7
Comparison of mean daily discharge for periods of 2000–2006 and 2007–2013 at Grand Island (US Geological Survey gage 06770500) including identification of program flow releases in 2009 and 2013.

from Governance Committee members to recommend a path forward. Strategies involving structured decision making as a way to robustly evaluate possible management action options are now underway, and a revised course of management action will most likely be determined by the summer of 2016. At that point, the Program may likely be the first large-scale AM program in the country to successfully complete one full loop of the AM process.

20.5 Summary

Large-scale recovery efforts are complex and demanding, especially in the strained context of low water availability, high water use, and a changing climate. The PRRIP is structured around a collaborative decision-making format that places water use and conservation stakeholders at the decision-making table with federal and state agencies and seeks to find consensus on important management decisions through that collaborative approach. The Program has also found it highly valuable to utilize independent, professional staff to manage data collection, analysis, and synthesis and to provide scientific guidance through the six-step AM cycle to ensure that decision makers are receiving thorough and useful scientific information to consider in the decision-making process. An emphasis on independent science review through a standing independent science committee and the use of a carefully managed independent peer review process ensure that Program science products are developed with the highest level of robustness and rigor. This governance model is unique and serves as an alternative to how most species recovery and ecosystem rehabilitation programs are currently organized and managed. Thus far, this approach has proved effective at consensus-based decision making and helping to internalize science learning from implementation of AM as an input into that decision making. In a semiarid basin characterized by tensions over water and endangered species and differences of opinion over species' responses to river processes and management actions, this structure is providing a path forward. If Program decision makers can successfully use information from management action implementation and data syntheses to adjust management, this Program will have succeeded in completing one full loop of the AM cycle.

The architecture of the Program's independent staffing provides constant, direct implementation action and affords an additional layer of independence from Program governing entities. Collaborative decision making that includes stakeholders and buy-in from all parties on goals, objectives, and what questions need to be answered help to focus the work of the Program and provide the most relevant information to decision makers. A unique governance structure, careful attention to the details of AM implementation, and a rigorous synthesis of multiple lines of evidence are critical pieces of the Program's success. In the case of governance and management, the PRRIP provides a template worth exploring for other systems.

References

Allan, C. and A. Curtis. 2005. Nipped in the bud: Why regional scale adaptive management is not blooming. *Env. Management* 36(3), 414–425.

Austin, J. and A. Reichert. 2005. Patterns of habitat use by whooping cranes during migration: Summary from 1977-199 site evaluation data. *USGS Northern Prairie Wildlife Research Center Paper 6.*

Blodgett, R.H. and K.O. Stanley. 1980. Stratification, bedforms and discharge relations of the Platte braided river system, Nebraska. *Journal of Sedimentary Petrology* 50, 139–148.

Crowley, K.D. 1981. *Large-scale bedforms in the Platte River downstream from Grand Island, Nebraska—Structure, process, and relationship to channel narrowing.* US Geological Survey Open-File Report 81-1059, 33 p.

Crowley, K.D. 1983. Large-scale bed configurations (macroforms), Platte River Basin, Colorado and Nebraska—Primary structures and formative processes. *Geological Society of America Bulletin* 94(1), 117–133.

Currier, P.J., G.R. Lingle, and J.G. VanDerwalker. 1985. *Migratory Bird Habitat on the Platte and North Platte Rivers in Nebraska.* Platte River Whooping Crane Habitat Maintenance Trust, Grand Island, Nebraska.

DOI (Department of the Interior). 2006. *Platte River Recovery Implementation Program Final Environmental Impact Statement.* [Denver, Colo.] Bureau of Reclamation and Fish and Wildlife Service.

Eschner, T.R., R.F. Hadley, and K.D. Crowley. 1983. *Hydrologic and morphologic changes in channels of the Platte River Basin in Colorado, Wyoming and Nebraska: A historical perspective.* U.S. Geological Survey, pp. A1–A39.

Freeman, D.M. 2010. *Implementing the Endangered Species Act on the Plate Basin Water Commons.* University Press of Colorado.

Gregory, R.S., D. Ohlson, and J. Arvai. 2006. Deconstructing adaptive management: Criteria for applications to environmental management. *Ecological Applications* 16, 2411–2425.

Holling, C.S. 1978. *Adaptive Environmental Assessment and Management*, The Blackburn Press, New Jersey.

Horn, J.D., R.M. Joeckel, and C.R. Fielding. 2012. Progressive abandonment and planform changes of the central Platte River in Nebraska, central USA, over historical timeframes. *Geomorphology* 139–140(2012), 372–383.

Johnson, W.C. 1994. Woodland expansion in the Platte River, Nebraska: Patterns and causes. *Ecological Monographs* 64(1), 45–84.

Lee, K.N., 1993. *Compass and Gyroscope: Integrating Science and Politics for the Environment*, Island Press, Washington, D.C.

Lee, K.N. 1999. Appraising adaptive management. *Ecology and Society* 3(2), 3. http://www.consecol.org/vol3/iss2/art3/.

Lyons, J.K. and T.J. Randle. 1988. *Platte River channel characteristics in the Big Bend Reach, Prairie Bend Project.* US Department of the Interior, Bureau of Reclamation, Denver, Colorado.

Murray, C., C. Smith, and D. Marmorek. 2011. *Middle Rio Grande Endangered Species Collaborative Program Adaptive Management Plan Version 1.* Prepared by ESSA Technologies Ltd. (Vancouver, BC) and Headwaters Corporation (Kearney, NE) for the Middle Rio Grande Endangered Species Collaborative Program, Albuquerque, NM.

Murray, C., D. Marmorek, and L. Greig. 2015. Adaptive management today: A practitioner's perspective. In: Allen, C., Garmestani, A., (Eds.), *Adaptive Management of Social–Ecological Systems,* Springer.

Murphy, P.J., T.J. Randle, L.M. Fotherby, and J.A. Daraio. 2004. *Platte River Channel: History and Restoration.* Bureau of Reclamation, Technical Service Center, Sedimentation and River Hydraulics Group, Denver, Colorado.

NRC (National Research Council). 2005. *Endangered and Threatened Species of the Platte River.* Committee on Endangered and Threatened Species in the Platte River Basin, National Research Council, National Academy of Sciences. The National Academies Press, Washington, D.C.

O'Brien, J.S. and P.J. Currier. 1987. *Channel Morphology, Channel Maintenance, and Riparian Vegetation Changes in the Big Bend Reach of the Platte River in Nebraska.* Unpublished report.

Ore, H.T. 1964. Some criteria for recognition of braided stream deposits: Laramie, Wyo., University of Wyoming. *Contributions to Geology* 3, 1–14.

Parsons. 2003. *Platte River Channel Dynamics Investigation*. Prepared for States of Colorado, Nebraska and Wyoming.

Peake, J.S. 1985. *Interpretation of Vegetation Encroachment and Flow Relationships in the Platte River by Use of Remote Sensing Techniques*. Department of Geography–Geology, University of Nebraska at Omaha, Nebraska Water Resources Center.

PRRIP (Platte River Recovery Implementation Program). 2006a. *Adaptive Management Plan*. US Department of the Interior, State of Wyoming, State of Nebraska, State of Colorado.

PRRIP (Platte River Recovery Implementation Program). 2006b. *Final Program Document*. US Department of the Interior, State of Wyoming, State of Nebraska, State of Colorado.

PRRIP (Platte River Recovery Implementation Program). 2014. *2013 State of the Platte Report*. Prepared by the Executive Director's Office of the PRRIP. Kearney, NE.

PRRIP (Platte River Recovery Implementation Program). 2015. *2014 State of the Platte Report*. Prepared by the Executive Director's Office of the PRRIP. Kearney, NE.

Schumm, S.A. 2005. *River Variability and Complexity*. Cambridge University Press, Cambridge.

Sidle, J.G., E.D. Miller, and P.J. Currier. 1989. Changing habitats in the Platte River Valley of Nebraska. *Prairie Naturalist* 21, 91–104.

Simons & Associates, Inc. and URS Greiner Woodward Clyde. 2000. *Physical History of the Platte River in Nebraska: Focusing Upon Flow, Sediment Transport, Geomorphology, and Vegetation*. Prepared for Bureau of Reclamation and Fish and Wildlife Service Platte River EIS Office, dated August 2000.

Smith, N.D. 1971. Transverse bars and braiding in the lower Platte River Nebraska. *Geological Society of America Bulletin* 82, 3407–3420.

Stroup, D., M. Rodney, and D. Anderson. 2006. *Flow Characterizations for the Platte River Basin in Colorado, Wyoming, and Nebraska*. Prepared for Platte River Recovery Program EIS Office, Lakewood, Colorado.

USFWS (US Fish and Wildlife Service). 1978. *Determination of Critical Habitat for the Whooping Crane*. Federal Register 43:20938–20942.

USFWS (US Fish and Wildlife Service). 2009. *Piping Plover* (Charadrius melodus) *5-Year Review: Summary and Evaluation*.

USFWS (US Fish and Wildlife Service). 2013. *Interior Least Tern* (Sternula antillarum) *5-Year Review: Summary and Evaluation*.

Walters, C.J. 1986. *Adaptive Management of Renewable Resources*, Macmillan, New York.

Walters, C.J. 1997. Challenges in adaptive management of riparian and coastal systems. *Ecology & Society* 1, 1.

Walters, C.J. 2007. Is adaptive management helping to solve fisheries problems? *Ambio* 36, 304–307.

Walters, C.J. and C. S. Holling. 1990. Large-scale management experiments and learning by doing. *Ecology* 71, 2060–2068.

Walters, C.J., L.H. Gunderson, and C.S. Holling. 1992. Experimental policies for water management in the Everglades. *Ecol. Appl.* 2, 189–202.

Zellmer, S. and L.H. Gunderson. 2009. Why resilience may not always be a good thing: Lessons in ecosystem restoration from Glen Canyon and the Everglades. *Neb. Law Rev.* 87, 893–949.

21

Drought as an Opportunity for Legal and Institutional Change in Texas

Ronald Kaiser

CONTENTS

ABSTRACT Drought, in conjunction with a rapidly growing urban population and new environmental, industrial, and municipal water demands, has pushed Texas to a crossroads in its water resource management. The traditional approach of building more reservoirs to satisfy new demands is limited by economic, environmental, fiscal, political, and physical constraints. Texas water plans continue to emphasize a strategy of reservoir development for providing water to meet new municipal growth demands, but there is increasing emphasis on water conservation, reuse of treated effluent, aquifer storage, desalination, and transfer of water from agricultural uses to municipal uses. New state funding priorities require local expenditures for conservation projects.

21.1 Introduction

Drought is a recurrent condition in Texas, and it has played a key role in driving the evolution of the state's water law, infrastructure, and management practices. Droughts are distinctive among climate conditions in that they develop slowly but can ultimately have social, economic, and environmental consequences as devastating as floods, tornadoes, and hurricanes. The consequences of drought result in crop, livestock and wildlife losses, damage to aquatic and upland ecosystems, increases in fire threats, reduced surface-water supplies, and increases in water demands. Although drought harms many water users and resources, its damages fall heavily on agriculture. The 2011–2012 drought affected over 80% of the south-central United States, resulting in estimated damages and costs over $31 billion (NOAA 2013). In Texas, the 2011 drought was one of the most intense 1-year droughts since 1895 when statewide records began (Neilson-Gammon 2011). By September of 2011, nearly all of the state was in exceptional drought (see Figure 21.1). Direct and indirect agricultural losses exceeded $8 billion (Texas Comptroller 2012). Other estimates put losses at over $13 billion (StateImpact Texas 2012). Wildfires resulting from the 2011 drought burned some 4 million acres (1.6 million ha.)* and destroyed almost 3000 homes (StateImpact Texas 2012). The Bastrop County Complex fire that started in September of 2011 was the most destructive in Texas history. It burned some 32,000 acres, destroyed 1660 homes and caused two deaths (Texas A&M Forest Service 2011).

Drought definitions vary, as do the methods of assessment. Scientists group drought into meteorological, hydrological, agricultural, and socioeconomic categories (Wilhite and Glantz 1985). Different methods for quantifying drought severity include the Standardized Precipitation Index, Palmer Drought Severity Index (PDSI), US Drought Monitor Surface Water Supply Index, Crop Moisture Index (National Drought Mitigation Center 2015). For planning purposes, Texas uses the Palmer Index (Texas Water Development Board 2012; hereinafter, references are cited as TWDB).

* 1 acre = 0.4 ha.

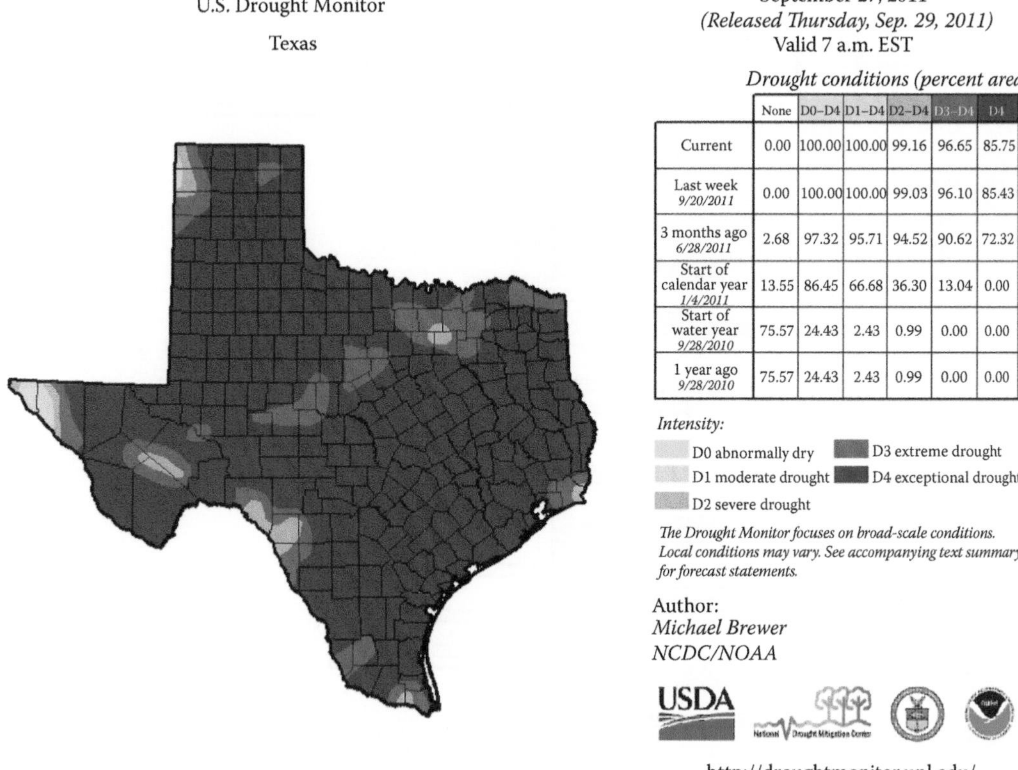

	None	D0–D4	D1–D4	D2–D4	D3–D4	D4
Current	0.00	100.00	100.00	99.16	96.65	85.75
Last week 9/20/2011	0.00	100.00	100.00	99.03	96.10	85.43
3 months ago 6/28/2011	2.68	97.32	95.71	94.52	90.62	72.32
Start of calendar year 1/4/2011	13.55	86.45	66.68	36.30	13.04	0.00
Start of water year 9/28/2010	75.57	24.43	2.43	0.99	0.00	0.00
1 year ago 9/28/2010	75.57	24.43	2.43	0.99	0.00	0.00

FIGURE 21.1
Extent of Texas drought by end of September 2011.

Drought is a recurring phenomenon in Texas occurring at least once every decade since the 1880s (see Figure 21.2). More detailed analysis indicates that at least 10 significant drought events have occurred in each of the 10 Texas climate regions over the last 150 years (South Central Climate Science Center 2013). Information on pre-1880 droughts is incomplete but some insight about their presence can be gleaned from early newspaper accounts, settler diaries, memoirs, and letters. A central Texas drought in 1756 dried up the San Gabriel River, north of present-day Austin, forcing the abandonment of a Spanish missionary settlement. Likewise an 1822 drought caused crop failures for Stephen Austin's first colonists. In 1883, Texas opened its western school lands for settlement and thousands of farmers moved to West Texas. Ironically, an 1884–1886 drought caused massive crop failures and the exodus of farmers from the region (Texas Historical Society 2015).

Instrumental precipitation records indicate that the 1950s drought was the worst statewide since the 1890s (Lowry 1959). The 1950s drought is the benchmark for state water planning (TWDB 2012). However, scientists using tree-ring proxy analysis have identified drought periods predating instrumental records. These studies indicate that decadal-scale droughts have occurred in Texas at least once a century since the 1500s and that some may have been longer and more intense than the 50s drought (Acuna Soto et al. 2002; Stahle et al. 1988, 2009; Cleaveland et al. 2011).

Historically, droughts and floods have spurred evolutionary changes in Texas surface water laws and institutions. Four legal and institutional eras can be traced to drought:

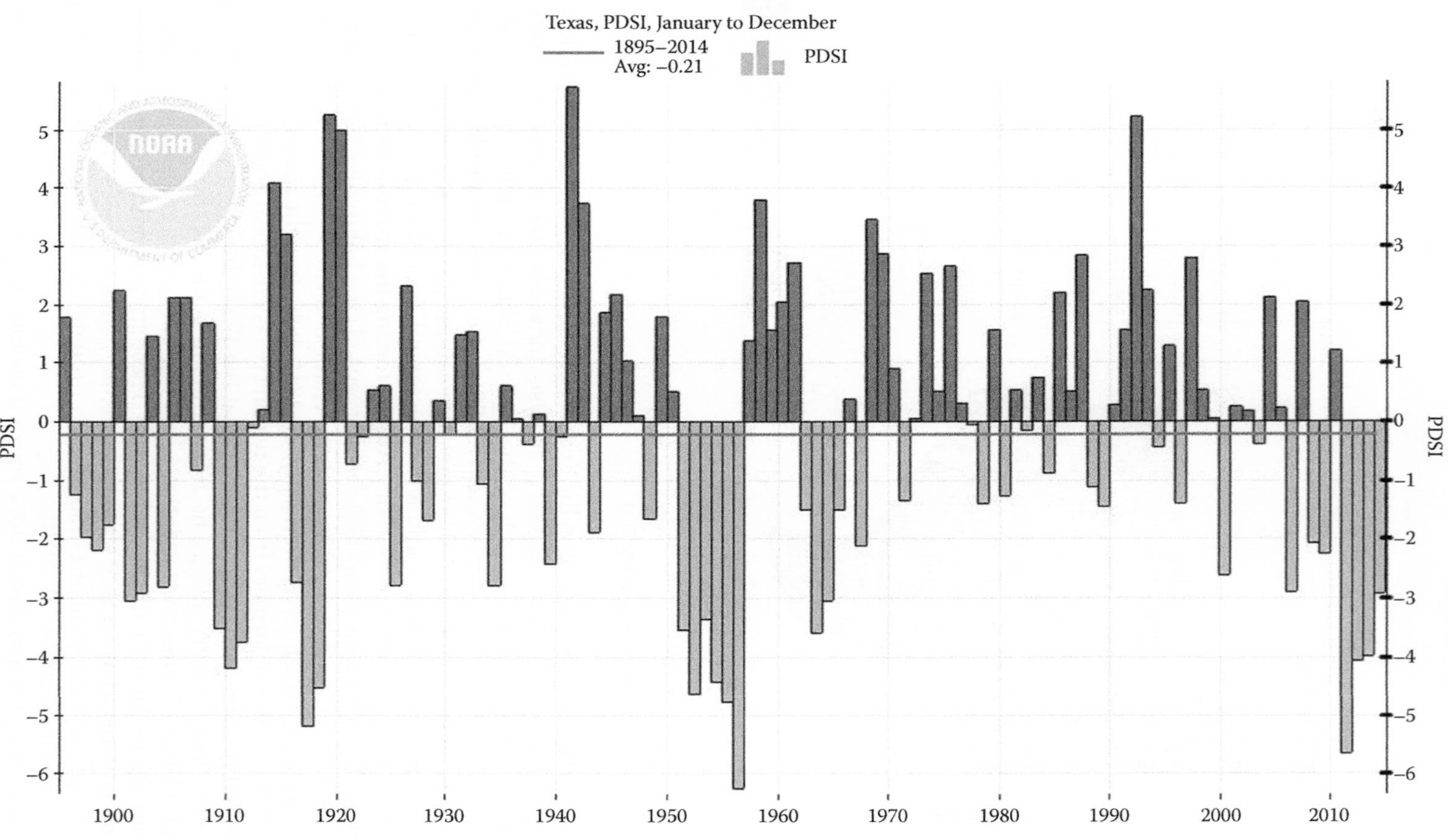

FIGURE 21.2
Statewide average Palmer Drought Severity Index: 1895–2014.

TABLE 21.1

Texas Drought Since 2011

Calendar Year	Statewide Palmer Drought Severity Index (PDSI)	Severity Rank within the 1895–2014 Record
2011	−5.66	2
2012	−4.07	9
2013	−3.98	10
2014	−2.92	18

Source: NOAA National Climate Data Center, http://www.ncdc.noaa.gov/cag/time-series/us/41/00/pdsi/ytd/12/1895–2015?base_prd=true&firstbaseyear=1895&lastbaseyear=2014.

(1) Adoption of Prior Appropriation and Rise of Water Institutions; (2) Top-Down Planning and Big Dam Construction; (3) Conservation Awakening; and (4) Bottom-Up Water Planning. However, the very severe 2011 drought and continuing dry conditions during the subsequent three years may be the harbinger of a new normal for Texas (see Table 21.1). This drought, in conjunction with significant population growth, highlights the reality that the state has reached the crossroads in water resource management. In addition to supply development, a new era of water management is one of increased conservation, innovation, and diversification. Adjusting to this new reality will be a challenge for planners and policy makers accustomed to reservoir construction and groundwater development. In order to better understand the differential impacts of drought on legal and institutional change, it is helpful to provide a brief overview of Texas water laws, water use patterns and supplies.

21.1.1 Texas Water Law

Water law is central to allocating water rights, resolving conflicts over water shortages, and guiding water-management institutions. Texas water law reflects a confluence of geology, hydrology, history, and politics. It has been influenced by the Spanish and Mexican civil law, the English common law, and the customs and practices in other western states (Kaiser 1986). While it is generally accepted that all water is interconnected, Texas uses geology to define its water-rights framework. Three geomorphic categories of water are recognized: (1) surface water, (2) diffused surface water, and (3) groundwater. Different rules and institutions govern each.

21.1.1.1 Surface Water

In 1889, Texas began transitioning from riparian water law to the prior appropriation system. By 1913, conversion to prior appropriation was extended to the entire state. Today, surface water is owned by the state and allocated pursuant to the statutory appropriation process. The Texas Water Code (§11.021) takes an expansive view of state water, defining it as the ordinary flow, underflow, floodwater, storm water, rainwater, and tidewater flowing in natural watercourses defined by a bed and banks. Procedures for acquiring water rights, or amending existing rights are outlined in Chapter 11 of the Water Code (Texas Water Code 2005, hereinafter TWC). The Texas Commission on Environmental Quality (TCEQ), a state agency, administers and enforces surface-water rights for the state.

21.1.1.2 Diffused Surface Water

Diffused surface water is rainwater, or snowmelt, flowing in unpatterned ways across the land without being confined to a channel. Diffused surface waters are the private property of the landowner, until they enter a natural watercourse (*Hoefs v. Short 1925*). Once surface drainage becomes confined to a channel or a natural watercourse, it becomes state water governed by the Water Code (*Domel v. City of Georgetown 1999*).

21.1.1.3 Groundwater

In contrast to surface water, which is owned and managed by the state, groundwater is treated as the private property of the landowner and is allocated by the rule of capture (*Houston and T.C. Ry v. East 1904 Edwards Aquifer Authority v. Day 2012*). Under this rule, the overlying landowners can pump an unlimited amount of groundwater from beneath their property, regardless of the impact on adjacent or more distant landowners, as long as the pumping is not done to maliciously harm a neighbor, is not wasteful, and does not cause subsidence (*Sipriano v. Great Spring Waters of American, Inc. 1999*).

The capture rule is not absolute and is subject to modification by local groundwater conservation districts (Chapter 36 of TWC 2005). Unlike many western states, Texas does manage groundwater at a state agency level. It has deferred regulation and management to some 98 local groundwater districts. According to the Texas legislature, these districts are the state's preferred method for managing groundwater (TWC §36.0015).

21.1.2 Texas Water Uses

On average, Texans consumptively use between 16 and 17 million acre-feet (MAF) (~20 BCM)* of water annually (TWDB 2007). This amount varies based on annual drought and rainfall conditions. In wet years, water use declines and in dry years it increases. Water use data are compiled by the TWDB from annual surveys of municipal and industrial users and from estimates of annual crop acreage compiled by the Farm Service Administration (See http://www.twdb.texas.gov/waterplanning/waterusesurvey/estimates/). TWDB and Regional Water Planning groups use this information and it becomes part of the Texas Water Plan.

Dramatic shifts in Texas water-use patterns are occurring and are projected to continue for the next three decades. Agricultural water use is on the decline and municipal and industrial use is on the increase. In the 1970s, agricultural irrigation accounted for 75% of total water use, and by 2012, that percentage dropped to about 60%. According to the 2002, 2007, and 2012 Texas Water Plans, the state's population of 25 million people is expected to increase to nearly 40 million by 2040 resulting in municipal and industrial water uses exceeding agricultural water uses. Most of the decline in irrigation use is not the result of urbanization but is related to the declining aquifer availability, increased pumping costs, improved irrigation efficiencies, shifts in market demand for agricultural commodities, voluntary transfers of water from irrigation to municipal use, and the decline in cheap water for agriculture (TWDB 2012).

In spite of the decline in irrigation water use, irrigated acreage has remained relatively steady for the last quarter century. Texas has just over 6 million irrigated acres comprising more than 10% of the irrigated acres in the United States (TWDB 2011). Most of the state's irrigated acres are concentrated in the sparsely populated Texas Panhandle, along the middle and lower Gulf coast, in the more populated Lower Rio Grande Valley, and in

* 1 million acre-feet (MAF) = 1.233 billion cubic meters (BCM).

the Winter Gardens area near San Antonio. Groundwater is the sole source of irrigation water in the Texas High Plains, while the Lower Rio Grande Valley and the Gulf Coast rely mostly on surface water.

Another water-use trend is the increasing municipal per capita water use. Statewide per capita water use over the last 50 years increased from an average of 100 gal. to over 190 gal. per person per day (Hardberger 2008). While per capita trends vary based on climate regions and other factors, it is a useful metric for gauging the efficacy of municipal water conservation practices. Per capita water use also provides the basis for approximating municipal demand projections in Texas water plans (TWDB 2012). These demand projections, with few exceptions, indicate that per capita water use will increase for the 40 largest Texas cities (TWDB 2012). It is ironic that state water plans posit that conservation is an imperative, yet planners see little change in practices over time. One consequence of maintaining high per capita water use is that demand projections in the state water plan are elevated thereby justifying additional supply development. See discussion in Section 21.3.5.2 of this chapter.

The Dallas suburban area has some of the highest per capita use in the state, whereas the cities of San Antonio and El Paso some of the lowest (TWDB 2007, 2012). This Texas trend is at variance with declining per capita water-use trends in other American southwest cities (World Water Forum 2012). Attempting to reverse this trend, Texas now requires that new public water-supply conservation plans include savings-based per capita targets (TWC §11.1271[c]). The results from this legislation should be reflected in the next version of the Texas water plan.

21.1.2.1 Groundwater Resources

Texas is a groundwater-dependent state relying on nine major aquifers to provide about 60% of the estimated 17 MAF annual usage (see Table 21.2). Texas aquifers are geologically and hydrologically diverse. The Texas Water Development Board recognizes nine major and 20 minor aquifers that are capable of producing water for agricultural, industrial and municipal purposes (TWDB 2012). These aquifers underlie approximately 80% of the state. Some aquifers like the Edwards, located along the Interstate 35 (I-35) corridor near San Antonio, are highly rechargeable limestone formations while others, such as the sand, gravel, silt Texas Panhandle Ogallala, have little recharge but store large volumes of water. The Ogallala Aquifer supplies over 6 MAF of water, which is about two-thirds of all the groundwater and more than one third (38%) of all the water used in Texas (see Table 21.3). Due to limited rainfall and geology, the Texas portion of the Ogallala receives very little natural recharge.

TABLE 21.2

Texas Water Sources, Uses, and Allocation Rules

Sources and Uses		Allocation Rule
Groundwater	(10 MAF)	Private ownership: capture rule as modified by groundwater districts
Amount used		
Agriculture	80%	
Municipal	15%	
Industrial/other	5%	
Surface water	(7 MAF)	State owned: prior appropriation
Amount used		
Municipal	65%	
Agriculture	35%	

TABLE 21.3
Aquifer Use and Availability in Texas

Aquifer	Estimated Pumping (Acre-Feet/Year)[a]	Estimated Available (Acre-Feet/Year)[b]
Ogallala	6,300,000	4,200,000
Gulf Coast	1,100,000	1,400,000
Carrizo–Wilcox	450,000	625,000
Edwards	390,000	340,000
Edwards–Trinity	150,000	230,000
Trinity	170,000	255,000
Seymour	190,000	145,000
Hueco Bolson	110,000	130,000
Pecos Valley	55,000	120,000
Total Majors	8,915,000	7,445,000
Minor aquifers	775,000	655,000
All aquifers	9,690,000	8,100,000

[a] Data from 2007 Texas Water Plan (Texas Water Development Board 2007, pp. 176–238).
[b] Desired pumping limits set by local groundwater conservation districts. Not yet achieved.

Statewide, about one-third of the water used to meet municipal demands is from groundwater. Most of western Texas and a significant part of East Texas rely on groundwater for municipal and industrial uses. San Antonio relies almost exclusively on the Edwards Aquifer, while Houston and El Paso rely on a combination of surface and groundwater. Because of land subsidence from groundwater pumping, the Houston area is converting from ground to surface water.

21.1.2.2 Surface-Water Resources

Texas has some 23 river basins, including 15 major basins and eight coastal basins. Because of its size, spanning some 800 mi. (1280 km)* from east to west and north to south, the state encompasses an extremely varied range of rainfall and aridity regimes. Except for the wetter eastern portion of the state, potential evaporation exceeds precipitation in the central and western parts of the state yielding a semiarid climate that becomes arid in far west Texas (TWDB 2012). East Texas watersheds receive the greatest amount of rainfall averaging nearly 60 in. (1524 mm),† whereas central Texas rivers receive between 20 and 35 in. per year while west Texas rivers flow through arid parts of the state with rainfall averaging less than 10 in. per year.

On average, just over 40 MAF of water annually flows through Texas rivers and streams (TWDB 1968, 2007). Texas has 196 major reservoirs, 175 of which are designated for water supply. These reservoirs have a conservation storage capacity of just less than 40 MAF. However, surface-water availability during the 50s drought of record is estimated at just over 13 MAF (TWDB 2007, 2012). About half of this amount is stored in the Sabine, Neches, and Trinity River reservoirs east of I-45 that runs between Dallas and Houston while the majority of Texas' population lives west of I-45. Planning data indicate that East Texas rivers have the greatest potential to provide additional surface-water supplies that could be exported to the Dallas and Houston metropolitan growth centers of the state (TWDB 2012).

* 1 mile (mi.) = 1.6 kilometers (km).
† 1 in. = 25.4 mm.

Loss of reservoir storage capacity from sedimentation is a common reservoir problem. It wasn't until 1991, after most reservoirs were completed, that the TWDB began a sediment survey program to determine how quickly the state's reservoirs were filling with sediment. The 2007 Texas Water Plan indicated that Texas' major reservoirs are losing about 90,000 acre-feet of storage per year from sedimentation. Based on this sedimentation rate over the next 50 years, even with the construction of 14 new reservoirs, Texas will lose more storage capacity than it will gain (TWDB 2007).

Most of Texas' surface water, about 65%, is used by cities and industry (see Table 21.2). The remaining 35% is used for irrigation, steam-electric power generation, mining, and livestock production. Cities and industry in the Dallas area, in central Texas, along the Gulf Coast and in the Lower Rio Grande Valley rely primarily on surface-water resources. Because of subsidence from groundwater pumping, Houston is converting to surface-water supplies from the San Jacinto and Trinity rivers.

21.2 Drought and Institutional Change

The story of Texas water law is also a story of drought. While no two droughts were alike in extent, duration, and impact, each significant drought period resulted in legal and institutional changes. Four distinct eras of change in Texas water law and institutions can be traced to drought. See Figure 21.2 for significant drought periods.

21.2.1 Era 1: Adoption of Prior Appropriation System and Rise of Local Water Districts

A series of single and multiyear droughts during the 1880s until 1920 resulted in the adoption of the prior appropriation system (Kaiser 1986; Jarvis 2014). These droughts also led to the formation of local irrigation districts and other specialized water organizations.

21.2.1.1 Prior Appropriation

From 1840 until 1895, Texas followed riparian water law in allocating surface water (*Haas v. Choussard 1856*). Under this rule, riparian landowners have a right to make a reasonable use of the normal river flows. The 1880s drought, which dried many of the state's western rivers, revealed the limitations of riparian water law and in 1889, Texas adopted the prior appropriation system for the western part of the state. Following droughts in 1909 and 1910 the state completed the conversion by establishing a Board of Water Engineers to keep records and produce a set of rules and regulations for water rights (Jarvis 2014). In making this conversion the Texas legislature allowed riparian landowners to retain some riparian water rights. This duel water rights system remained in place until 1967 when the legislature began the water rights adjudication process.

21.2.1.2 Groundwater Law

During this era, the Texas Supreme Court adopted the capture rule. The capture rule did not evolve from drought conditions, but from a landowner dispute. In 1904, the Texas Supreme Court had to confront groundwater ownership and liability rules in a conflict between a railroad and a rancher. The conflict arose when the railroad drilled a deeper well on its property that caused a nearby rancher's well to go dry. Ruling in favor of the railroad, the Texas Supreme Court adopted the capture rule refusing to limit the use of groundwater, even though the railroad company's use clearly deprived "East" of his historical use (*Houston and T.C. Ry v. East 1904*). In 1999 and again in 2012, absent malice, waste, or subsidence, the Texas Supreme Court has retained the capture rule in holding no liability for groundwater pumping (*Sipriano v. Great Spring Waters of American Inc. 1999, Edwards Aquifer Authority v. Day 2012*).

21.2.1.3 Rise of Water Organizations

Water laws and organizations are interdependent. Water organizations have been and continue to be central to agricultural development. Ditch companies, mutual water companies, or water districts constitute about 85% of the organizations providing irrigation water to farmers (Thompson 1993). Because cities have the legal authority and economic resources to supply water to their residential and industrial customers, they do not need special water organizations. However, in areas outside city limits, water organizations often furnish water for domestic uses.

Local water organizations in Texas have existed in various forms since the early 1900s. The Legislature established irrigation districts in 1904, drainage districts in 1905 and levee improvement and navigation districts in 1909 (Jarvis 2014). Over the years, the Legislature has been especially creative in establishing local water agencies (see Table 21.4). Literally a *cast of thousands* is involved in water supply and management. Many districts were initially used to develop the state's agricultural resources, but they have evolved to provide water and wastewater services for urban and rural areas (Stepherson 2014).

TABLE 21.4

Water Districts in Texas

District Type	Active	Inactive	Dissolved
Municipal utility	650	290	449
Water control and improvement	173	44	500
Groundwater conservation	92	1	19
Freshwater supply	54	12	89
Drainage	43	5	58
Special utility	38	16	2
Levee improvement	31	15	78
River authority	31	0	1
Navigation	24	2	8
Irrigation	24	1	3
Water improvement	17	0	40
Municipal management	14	27	4
Others	50	29	53
Total	1241	442	1304

Source: Data from House Committee on Natural Resources, Texas House of Representatives, Interim Report 2006, A Report to the House of Representatives, 80th Texas Legislature, November 29, 2006.

21.2.1.4 Rise of Regional River Authorities

The 1930s Dust Bowl (see Figure 21.2) brought legislative recognition that regional water and watershed problems were best managed by regional rather than local water districts. River authorities were established to engage in watershed planning, reservoir development, water supply, wastewater treatment and hydropower. Most were established during the 1930 to 1950 time period. Many river authorities have substantial water rights and play a major role as wholesale providers of water to cities and water utilities.

21.2.2 Era 2: Top-Down Planning and Big Dams, 1950s–1990s

> It crept up out of Mexico, touching first along the brackish Pecos and spreading then in all directions, a cancerous blight burning a scar on upon the land. Just another dry spell men said at first. * * * Men grumbled, but you learned to live with the dry spells if you stayed in west Texas; there were more dry spells than wet ones. No one expected another drought like that of 33, and the really big dries like 1918 came once in a lifetime. Why worry they said. It would rain this fall. It always had. But it didn't and many a boy would become a man before the land was green again. (Kelton 1973, p. xiii)

The Time it Never Rained, a historical novel by Elmer Kelton, poignantly describes the devastating impacts of the 1950s drought of record on Texas farmers, cities, and industry. Agricultural towns turned into ghost towns, and it raised the specter of municipal and industrial water shortages. In 1952, Lake Dallas held only 11% of its capacity and by the end of 1956, 244 of Texas' 254 counties were classified as disaster areas (TWDB 2007). It is generally accepted that this statewide megadrought ushered in the modern era of statewide water planning, state funding and dam building.

21.2.2.1 Statewide Planning

One of the first actions the Legislature took in 1957 was to establish the Texas Water Development Board (TWDB) and task it with the responsibility to prepare a statewide water plan. An initial plan, prepared by agency staff and released in 1961, was the precursor to modern state water planning. Since that initial effort the TWDB has prepared eight additional plans. (Plans are available on the web: http://www.twdb.texas.gov.) While the Plans have changed in scope and coverage, their central tenet emphasized reservoir development. The 1968 Plan proposed 62 new reservoirs and a canal system to import water from the Mississippi river (TWDB 1968). Many of the reservoirs were built but the canal system idea was abandoned. An additional 44 reservoirs were proposed in the 1984 Plan. Following the recommendations in these Plans, the state embarked on a massive reservoir construction program. Each subsequent plan, including the 2012 Plan, advocated additional reservoir development. However after 1990, Plans began a subtle shift identifying conservation, water transfers, desalination, reuse, and aquifer storage as additional ways to address growing water needs.

21.2.2.2 Funding for State and Local Water Projects

Texas voters have not been shy in approving funding for water projects. Starting in 1957 with approval of the first $200 million, Texas voters approved an additional 13 constitutional amendments authorizing the state to borrow money through the issuance of general obligation bonds (Schoolmaster 1992). These bond funds have provided loans to local units of government enabling dam/reservoir development, water infrastructure and wastewater projects. Through 2010 Texas voters authorized just over $4.7 billion in bonded indebtedness (Texas Legislative Budget Board 2011).

21.2.2.3 The Big Dam Era

Dams and reservoirs have been the dominant water supply and flood control approach of federal and state water management. At the federal level the big dam era began in 1930 and lasted to 1980. During this time more than 1000 large dams were built by the federal government (Gleick 1998). Reservoirs store about 60% of the annual river flow of the United States and in the Colorado basin storage is equal to nearly five years of average runoff (Hirsch et al. 1990; Gleick 1998).

Texas' big dam era began after the drought of the '50s and continued to 2000. Since 1950 a total of 149 reservoirs were built (see Figure 21.3). After 1990 the construction pace slowed considerably as only 8 major dams were completed during the decade (TWDB 2002). Since 2000 no major reservoirs have been completed in the state. In total, Texas has an inventory of 196 major reservoirs; defined as having a conservation storage capacity of more than 5000 acre-feet of water per year. However less than 20 of these 196 reservoirs store about 75% of the 2010 total reservoir firm yield for the state (TWDB 2012).

21.2.2.4 Resolving Dueling Water Law Systems

In adopting prior appropriation system statewide in 1913, the Texas legislature did not explicitly repeal the riparian doctrine. A dual system of riparian and prior appropriation water rights presented few problems when sufficient water was available. However, drought revealed the incompatibility of these legal regimes especially on the Rio Grande (*State v. Hidalgo County 1969*). Water rights claims based on civil law, riparian law and the prior appropriation system exceeded the amount of water available in the Rio Grande resulting in litigation adjudicating competing water rights claims. Litigation dragged on for 13 years, involved about 3000 parties, and generated an estimated $10 million in court costs and attorneys' fees (Caroom and Elliot 1981). This case illustrated the futility of judicial attempts to manage water resources in Texas. Finally in 1967 the Texas legislature

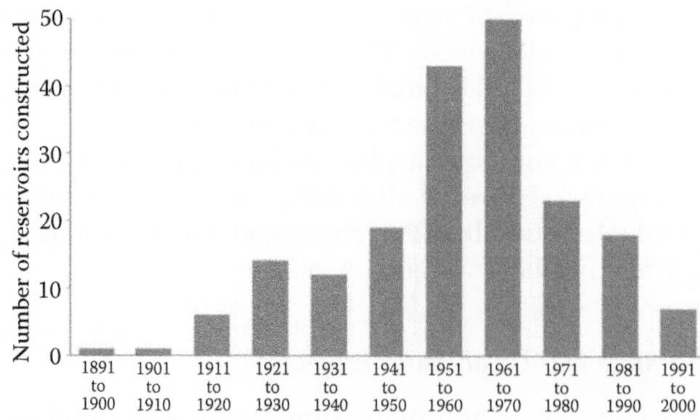

FIGURE 21.3
Texas reservoir construction by decade. Since 2000, no new reservoirs have been completed in Texas. Source: 2007 Texas Water Plan (Texas Water Development Board 2007, p. 114).

merged the civil law, riparian law into the prior appropriation system with the passage of the Water Rights Adjudication Act (TWC §§11.301–11.341). Under terms of this Act, all water rights holders were required to file claims with the Texas Water Commission (now the TCEQ). The Texas Supreme Court approved this conversion in 1982 (in re Adjudication 1982). By the late 1990s the surface water adjudication process was completed. More than 6000 surface water rights permits were issued. Today, any party seeking to use the little remaining surface water must comply with the permit procedures of the TCEQ.

21.2.3 Era 3: Conservation and Environmental Awakening, 1980s–1990s

After an early 1980s drought, conservation and environmental water needs became featured political and policy issues for the Legislature and voters. Legislation in the 1980s and 1990s required state agencies and cities to prepare drought contingency and water conservation plans. Also a modicum of protection for environmental and instream flows became part of water management.

At the state level, a drought preparedness council was established to develop a state drought preparedness plan that is separate from the state water plan. The council is a collection of state agencies that coordinate activities on drought, prepare a plan, monitor conditions, officially declare a drought and advise the governor on disaster declarations (TWC §16.055). Basically this effort involves developing a drought information network and coordinating agency responses.

Concerns that a number of small communities were perilously close to exhausting their water supplies compelled the legislature to require public retail water suppliers to prepare and update every five years a drought contingency plan (TWC §11.1272). Because many retail suppliers were not familiar with this type of plan, the TCEQ developed a model plan that served as a framework to help retail suppliers meet the contingency plan requirements (TCEQ 2005). Today most communities follow this planning framework.

During this era, water conservation plan requirements became part of the surface water rights systems (TWC §11.1271). A conservation plan requirement now applies to retail water suppliers using more than 1000 acre-feet per year. Agricultural irrigators, if they use more than 10,000 acre-feet per year, must also prepare a plan (TWC §11.1271[b]). For public water suppliers the process has evolved to that point that all water conservation plans must now include specific, quantified 5- and 10-year per capita targets for savings. These plans must be updated every five years. However, there are no penalties for failure to achieve per capita target reductions.

Environmental water science finally began to influence state water regulations and management. Biologists as early as the 1960s documented environmental flows as essential factors in maintaining the biological productivity of rivers, lakes, bays and estuaries (Gunter and Hildebrand 1954). It wasn't until 1985 that the Texas Legislature finally provided limited recognition for environmental flows (Kaiser and Kelly 1987). First, the 1985 legislation provided that 5% of the firm yield of reservoirs located within 200 river miles of the coast was to be reserved for environmental flows. However, since 1985 not one reservoir has been constructed that meets this criteria. A second method for protecting environmental flows is through the water permitting process. New permits, or permit amendments must consider environmental flows as part of the review process. If the review indicates real or potential negative impacts on flows, mitigation will be required.

In reality, environmental flow protection is a story of too little too late. Twelve of the 15 major river basins in Texas are fully appropriated and there is no water remaining for the environment (TNRCC 1995). The only long-term option is to reallocate water from some other preexisting uses (National Research Council 1992; Kaiser 1998).

21.2.4 Era 4: Bottom-Up Planning and Growth of Groundwater Districts, 1997–2012

21.2.4.1 Regional Planning

Although state water planning for Texas began in the 1960s the current process for regional planning resulted in part from an intense drought in 1995 and 1996 (see Figure 21.2). Championed by Lt. Governor Bob Bullock in 1997, Senate Bill 1 changed Texas water planning from a top down to a bottom-up regional approach (Brown 1998). This new process mandates extensive stakeholders' involvement in Plan preparation on a 5-year cycle. The state is divided into 16 planning regions with each regional planning stakeholder group composed of a minimum of 11 members. Stakeholder groups must include a representative from agriculture, industry, environmental organizations, the public, cities, businesses, water districts, river authorities, counties, groundwater management districts, and power generation interests (TWC §16.053). Planning groups may add additional members in order to reflect diverse stakeholder interests in that region. State agency staff from Parks and Wildlife and Agriculture serve as ex officio members in each of the 16 Regional Planning groups.

Planning groups evaluate population and water-use projections, compare them with existing water supplies to determine if demands exceed available supplies. If demands exceed existing supplies, each planning group recommends water-management strategies and costs to fill the demand–supply gap. Regional plans are sent to the Texas Water Development Board for approval and incorporation into the state water plan. The state water plan is essentially a compilation of all 16 regional plans. Since 2002, the Texas Water Development Board has followed this practice in publishing the 2002, 2007 and 2012 state water plans.

21.2.4.2 Groundwater Conservation Districts

In contrast to the unified regulatory system for surface water, the Texas Legislature adopted a decentralized approach to groundwater regulation deferring management to local groundwater management districts. First authorized in 1949, these districts are the "state's preferred way to manage groundwater" and have the ability to regulate groundwater through well location, spacing and pumping limitations (TWC §36.0015). Districts may change the capture rule by limiting and enforcing the amount of water pumped from any given well. While the Texas Supreme Court has sustained the authority of districts to regulate groundwater, they have not precluded landowners from challenging district rules for the taking of private property (*Barshop v. Medina County Underground Water Conservation District 1996*).

Over the past two decades there has been a major increase in the number of districts. From 1949 to 2000, Texas had 50 confirmed districts organized predominately along county lines and not aquifer boundaries (Kaiser and Skillern 2001). In the 15 years since 2000, an additional 48 districts were established resulting in a patchwork of 98 different districts. This piecemeal approach has led to many single county districts over one aquifer. Of the 98 districts, 58 are single county jurisdiction.

Critics suggest that problems of self-interest, limited funding, local politics, and the self-limiting nature of these districts prevent meaningful management and protection of groundwater resources (Behrens 1991; Kinkade 1996). Problems of limited funding plague many districts, particularly when facing legal challenges. Districts are funded through property taxes or well production fees. A significant majority of the districts have less than a $100,000 annual operating budget. This lack of funding is a major deterrent to meaningful management and regulation of groundwater.

21.3 The New Water Reality

The 2011 drought highlighted the vulnerability of agriculture and small communities to water shortages. Shortages were most acute in the central, southern and western portions of the state as a number of small cities were within 60 days of losing their water supply. Statewide, some 1000 out of 4700 public water suppliers issued mandatory lawn irrigation restrictions (Texas Comptroller 2012). Declining reservoirs levels, dry rivers and failed wells became the 2011 reality and news stories were not lost on politicians and the public. The 2012 Texas Water Plan reported future water shortages of 8.3 MAF and an infrastructure price tag of $53 billion. Legislators and voters responded by approving more money for water development projects.

21.3.1 The New Face of Water Funding

Billions of dollars in new funding, an emphasis on conservation and designated support for rural communities is the new water reality. During the height of the 2011 drought, voters approved a constitutional amendment authorizing an additional $6 billion of so-called evergreen bonding authority. The unique feature of this constitutional amendment is the ability of the state to continually reissue bonds up to the $6 billion limit. In the past, once bonds were paid off voters' approval was needed to issue a new set of bonds. Under the evergreen concept, voter approval is not needed to continually reissue up to $6 billion in bonds. Funds from the sale of state bonds are made available by the Texas Water Development Board to communities through low-interest loans, extended repayment terms, and deferral of loan repayments to develop local and regional water infrastructure.

In 2014, voters approved transferring $2 billion from the Texas rainy-day fund to the State Water Implementation Fund for Texas (SWIFT). This is unique in two ways. First, it did not require that Texas issue more debt through the sale of bonds, rather it transferred surplus revenue in the rainy-day fund to SWIFT to help communities develop or optimize water supplies. The program provides low interest loans and deferred payment loans to communities. Approximately $800 million will be allocated each year for project funding. Second, the SWIFT legislation directed that during a 5-year period at least 10% of the fund must be allocated to rural communities and at least 20% must be used for conservation and reuse projects (TWC §15.434). Funding can also be used for desalinating groundwater, building new pipelines, and developing well fields and reservoirs. To be eligible for funding, projects must be recommended in the 2012 Sate Water Plan.

21.3.2 Emphasis on Conservation and Reuse

Municipal water conservation has undergone a paradigm shift in state water planning, transitioning from a demand management tool to a supply strategy (TWDB 2012). It is now in the vernacular of water planning that the cheapest water is that water saved through conservation. Given this new shift, SWIFT funding can now be use to replace leaky pipes and aging infrastructure. More problematic and yet to be resolved is the use of SWIFT funds for private customer-side rebate programs such as retrofitting plumbing for gray water reuse and replacement of water-intensive landscaping. Texas has a constitutional prohibition (gift clause) against using public funds for private benefit and unless these customer-side rebate programs can be construed as a "public purpose and benefit," communities will not be able to use these funds for rebates (Texas Constitution, Art. III, §52[a]).

Water reuse, defined as municipal wastewater treated to a quality suitable for beneficial use, has been used in Texas agricultural irrigation for more than a century (TWDB 2011). Municipalities are now increasingly considering it as a new source of water. Twelve of the 16 regional water planning groups have recommended water reuse as a source of additional supply in their regional water plans. However, there are legal challenges in reusing water.

Texas law makes important conveyance distinctions that affect water reuse. Direct reuse occurs when effluent is piped directly from a wastewater plant back to a drinking water treatment plant or to a golf course or city park. Generally, municipalities do not need new authorizations for direct reuse, provided the original permit contains no provision to the contrary, and it meets water quality standards for the intended reuse. Greater complexity exists for indirect reuse where treated effluent is discharged into a watercourse for diversion and use downstream. Where the discharger seeks to retain ownership of the treated effluent, they must seek a new permit, called a *bed-and-banks permit* to convey this water downstream (TWC §11.042). These permits require the TCEQ to consider the impact of the bed-and-banks conveyance on other water right holders and the environment.

There are a number of challenges to further development of reuse projects in Texas. Most notably, striking a balance between environmental and human needs and impacts on downstream water rights holders. The base flow in many rivers comes from discharges from municipal wastewater treatment plants. If these flows are significantly reduced from direct reuse projects there is less water in the river for the environment. Absent legislative or judicial guidance, indirect reuse authorizations to enhance water supplies are uncertain and challenging (*City of San Marcos v. Texas Commission on Environmental Quality 2004*; Smith 2014). In spite of these legal uncertainties, treated effluent now provides a valuable source of nonpotable water for oil and gas development, power plant cooling, irrigating city landscapes and parks, enhancing environmental flows, and as a potable use suitable for drinking. Nearly 200 water entities have applied for 1.6 MAF of reuse water (TWDB 2011).

21.3.3 Desalination

There is good news and bad news regarding the amount of water available and cost of desalination. The good news is that the state has some 2.7 billion acre-feet of brackish groundwater in storage (LBG-Guyton 2003). Most of this water contains dissolved minerals in the range of 1000 to 15,000 milligrams per liter (mg/L). By comparison, seawater contains about 35,000 mg/L of dissolved solids. The bad news is the capital and operating cost to desalinate this water is a major barrier to greater use of this strategy. However, in areas with a shortage of freshwater, brackish groundwater is a viable alternative. According to the TWDB there are 44 municipal brackish water desalination and two seawater desalination plants operating in Texas. These plants can produce up to 137,000 acre-feet of fresh water per year, provided they are operating at full capacity (TWDB Desalination Data Base 2015). The 2012 State Water Plan recommends funding for 39 new desalination plants in south and west Texas that by 2060 would produce over 180,000 acre-feet of water per year (see Table 21.5). So long as there is fresh water available in these areas, these desalination plants will remain in the planning stages.

21.3.4 Other Strategies

21.3.4.1 Aquifer Storage and Recovery

Opposition to surface reservoir projects has encouraged Texas officials to explore aquifer storage and recovery (ASR). These projects are designed to create a bubble of water around

TABLE 21.5

Potential Water Supply Volumes in Acre-Feet by Type of Strategy

Type of Water Management Strategy	2010	2020	2030	2040	2050	2060
Municipal conservation	137,847	264,885	353,620	436,632	538,997	647,361
Irrigation conservation	624,151	1,125,494	1,351,175	1,415,814	1,463,846	1,505,465
Other conservation[a]	4660	9242	15,977	18,469	21,371	23,432
New major reservoir	19,672	432,291	918,391	948,355	1,230,573	1,499,671
Other surface water	742,447	1,510,997	1,815,624	2,031,532	2,700,690	3,050,049
Groundwater	254,057	443,614	599,151	668,690	738,484	800,795
Reuse	100,592	428,263	487,795	637,089	766,402	915,589
Groundwater desalination	56,553	81,156	103,435	133,278	163,083	181,568
Conjunctive use	26,505	88,001	87,496	113,035	136,351	135,846
Aquifer storage and recovery	22,181	61,743	61,743	72,243	72,243	80,869
Weather modification	0	15,206	15,206	15,206	15,206	15,206
Drought management	41,701	461	461	461	461	1912
Brush control	18,862	18,862	18,862	18,862	18,862	18,862
Seawater desalination	125	125	143	6049	40,021	125,514
Surface water desalination	0	2700	2700	2700	2700	2700
Total supply volumes	**2,049,353**	**4,483,040**	**5,831,779**	**6,518,415**	**7,909,290**	**9,004,839**

Source: 2012 Texas Water Plan, p. 189.
[a] Associated with manufacturing, mining and steam-electric industries.

the aquifer well injection site that can be extracted from that same site when water is needed. ASR projects have certain advantages over reservoir storage because they avoid surface damages, do not have evaporation losses, and they are substantially cheaper (Malcolm Pirnie 2011). Three Texas cities; El Paso, Kerrville and San Antonio, have operating ASR systems. Each has a different source of supply. El Paso uses treated wastewater, Kerrville, treated river water and San Antonio groundwater. The San Antonio Twin Oaks ASR project is designed to store more than 100,000 acre-feet of water (SAWS 2015) while the El Paso and Kerrville projects are smaller in size. During wet periods and winter months when there are no restrictions on the Edwards Aquifer pumping, San Antonio extracts Edwards water and injects it into the Carrizo Aquifer. Then during the hot, dry months this water is pumped from storage back into the San Antonio water distribution system (SAWS 2015).

In spite of the promise of ASR, technical, geologic and regulatory issues remain and may be barriers to wider adoption of the practice (Malcolm Pirnie 2011). The 2012 Texas Water Plan indicates it as making only a minimal contribution to improving water supplies in the state (see Table 21.5).

21.3.4.2 Brush Control

By removing salt cedar and other phreatophytes, the hope is that more surface and groundwater will be available to meet future demands. From 2000 to 2014, Texas spent $62 million for brush control, believing it would increase water supply (House Committee on Agriculture and Livestock 2015). Allocations were made through the State Soil and Water Conservation Board (TSSWCB) to landowners paying up to 70% of the cost to remove brush. The TSSWCB has never conducted field tests to determine if water supplies have been enhanced through brush control (House Committee on Agriculture and Livestock 2015).

Critics and range scientists posit that there is little evidence that brush control increases water supply (Environmental Defense Fund 2003; Wilcox 2015). In spite of these findings, the state continues to fund landowner brush control activities and the 2012 Texas Water Plan suggests that brush control will supply 19,000 acre-feet of water per year (see Table 21.5).

21.3.5 Challenges and Options

In spite of more than 50 years of water planning experience, including the last 15 years of stakeholder-based regional planning, some $8 billion in new spending authority, and a newly reconstituted water planning agency, Texas must address a number of water policy issues. The following are the most notable in need of attention.

21.3.5.1 Removing Interbasin Surface Water Transfer Barriers

Surface water resources in Texas do not align with its population. East Texas is water rich and sparsely populated. Central Texas from Dallas to Houston and along the I-35 corridor extending to San Antonio is experiencing the greatest population growth and has limited surface water resources. Interbasin transfers have been used to move water for critical municipal and industrial needs as more than 100 transfers have been authorized (TWDB 2002). Until 1997, the policy and practice was to allow transfers from basins with surplus water to basins with a shortage. Legislation enacted in that year discourages interbasin transfers by changing the seniority date for new transfers (TWC §11.085[s]). Seniority is important because Texas follows the rule in allocating water during times of shortage. Before the rule change, any entity could acquire a senior water right in the originating basin and retained the seniority date. The new provision cancels the seniority date and reduces it to a junior right, greatly reducing the reliability of a transferred surface water right.

Advocates claim that the rule has protected water in rural areas from the "thirst of growing cities," but it might have fostered just the opposite effect. In order to provide a reliable source of water, a number of cities are turning to groundwater as a replacement source. Because groundwater transfers are not subject to the junior rights provisions, a number of cities and water supply agencies have proposals to move groundwater from rural to urban areas thereby exacerbating rural and urban conflicts. Fifteen of the 44 recommended water transfer projects proposed in the 2012 State Water Plan involve interbasin transfers. Water planners suggest that the junior rights rule has served as a significant obstacle to solving the Texas water supply puzzle (TWDB 2012).

21.3.5.2 Mandated Targets for Conservation

Texas has made significant strides in recognizing the importance of water conservation. However major municipal and irrigation conservation improvements can still be made. In the Dallas area the 2012 State Water Plan has per capita use at 207 gal. declining to 198 by 2060. Adding population growth to this 198 per capita target results in a need for 2.9 MAF in new water to be supplied by new reservoirs. The Texas Water Conservation Task Force recommended a goal of 140 gal. per capita (GPC) for potable water supplied to municipal customers (2004). However, it is only a goal and is lacking a mandate. Critics suggest that applying a 140 GPS standard to the Dallas area would mean that municipal demand in 2060 would be over 1 MAF less than projected in the 2012 State Water Plan, thereby lessening the need for new reservoirs (Texas Center for Policy Studies 2014). While the 140 GPC may be too stringent for Texans accustomed to large green lawns, serious consideration must be given to linking any new state funding to targeted reductions in agricultural and municipal water use.

21.3.5.3 Extending the Watermaster Program to All the Major River Basins

Mayordomos* and watermasters encourage compliance with water rights systems. As a result of limited staffing, enforcing Texas water rights during drought is based on the honor system. Fortunately, Texas began a watermaster program after the 1950s drought and with each successive drought has further refined the system. Watermasters have the legal authority to ensure compliance with water rights by monitoring stream flows, reservoir levels, and limiting diversions based on seniority to prevent the wasting of water or its being used in quantities beyond a user's right (TWC §11.327).

Watermaster programs have been established on six (Brazos, Guadalupe, Lavaca, Nueces, San Antonio and Rio Grande) of the 15 major river basins in the state and on four sub basins. Since most Texas rivers are over appropriated and new demands will outstrip existing supplies, watermaster programs will be increasingly important to ensuring compliance with the Texas water rights system (Hubert and Bullock 1999). In addition to enforcing water rights, watermaster programs encourage water marketing and conservation (Texas Senate Committee 2004). Research suggests that water marketing and serious conservation efforts could meet the water needs of cities for the next decade (Kaiser 1996; De Laughter 2000; Tarlock 2002; Texas Center for Policy Studies 2014). Legal authority exists to establish watermaster programs on all major basins, political will and policymaker leadership is in short supply.

21.3.5.4 Critique of State Water Planning

The state water plan and the regional planning process are not without critics. A recent report questions the demand forecasting methods for the Dallas–Fort Worth planning region and asserts that it underestimates the potential for water conservation savings (Texas Center for Policy Studies 2014). Other criticisms include the lack of prioritization of water projects, unrealistic wish lists of projects, exaggeration of water needs and failure to promote drought contingency plans. Recommendations for change include changing the planning cycle from five to 10 years, improving the diversity of interests on regional planning groups, incorporating conservation as the first priority for meeting unmet water needs and adopting a project prioritization process (Texas Leadership Roundtable on Water 2014).

21.3.5.5 Rethinking Climate Change Impacts

Climate change consequences have received scant discussion in Texas Water Plans. Former Governor Rick Perry, a climate change skeptic, wrote in his 2010 book *Fed Up,*

> …the complexities of the global atmosphere have eluded the most sophisticated scientist and that draconian policies with dire economic effects based on so-called science may not stand the test of time. Quite frankly when science gets hijacked by the political left we should all be concerned. (Perry 2010, p. 92)

* These are ditch masters in community irrigation ditches (acequias) that operate under a patchwork of state and Spanish water laws and customs. John Wesley Powell proposed water masters to the Montana Constitutional Convention as early as 1889. See Wallace Stegner, *Beyond the Hundredth Meridian: John Wesley Powell and the Second Opening of the West*, University of Nebraska Press, 1954.

It should be noted that the governor appoints the director of the Texas Water Development Board, the agency responsible for preparing state water plans.

The 2012 Texas Water Plan acknowledges that if temperatures rise and precipitation decreases as predicted by climate models, Texas would experience droughts as bad as or worse than the drought of the 1950s. In response to these concerns, the Texas Water Development Board held a Climate Change Conference in 2010 and funded a statistical analysis of uncertainty in water resources management. However, water planners continue advocating use of the 50s drought case for planning purposes and plan to continue to monitor climate policy and science and incorporate new developments when appropriate (TWDB 2012).

21.4 Conclusions

The story of Texas water planning, reservoir development, conservation, environmental awareness, and funding is also a story of drought. Each significant drought has resulted in incremental changes in surface and groundwater laws, institutions, and management approaches. Texas faces an impending water crisis that is not one of absolute water scarcity but it is one of readily available cheap water. The cheap and easy water supply fixes have been exhausted. While most rivers are fully appropriated, and the Ogallala Aquifer is being rapidly depleted, an estimated 2.7 billion acre-feet of brackish groundwater are available throughout the state. Desalination technology exists to tap and use this new source but the cost will be higher than most Texans are accustomed to paying. It is clear that conservation, reuse, desalination, interbasin transfers, and water marketing should receive greater emphasis. Perhaps the next drought will bring about greater use of these practices.

References

Acuna-Soto, R., D.K. Stahle, M.K. Cleaveland, M.D. Therrell. 2002. Megadrought and Megadeath in 16th Century Mexico. *Emerging Infectious Diseases* 8(4):360–362.

Barshop v. Medina County Underground Water Conservation District et al., 925 S.W.2d 618, 619, (Tex. 1996).

Behrens, E., M. Dore. 1991. Rights of landowners to percolating groundwater in Texas. *Texas Law Review* 325:185–202.

Brown, B. 1998. Senate bill 1: We've never changed Texas Water Law this way before. *Texas Environmental Law Journal* 18:152–165.

Caroom, D., J. Elliot. 1981. Water rights adjudication—Texas style. *Texas Bar Journal* November 1981:1183–1186.

City of San Marcos v. Texas Commission on Environmental Quality (TCEQ), 128 S.W.3rd 264 (Tex. App—Austin 2004).

Cleaveland, M.K., T.H. Vottler, D.K. Stahle, R.C. Casteel, J.L. Banner. 2011. Extended chronology of drought in South Central, Southeastern and West Texas. *Texas Water Journal* 2(1):54–96.

DeLaughter, C. 2000. Comment: Priming the water industry pump. 37 *Houston Law Review* 37:1465–1493.

Domel v. City of Georgetown, 6 S.W.3d 349 (Tx App—Austin, 1999, pet. denied).
Edwards Aquifer Authority v. Day, 369 S.W.3d 814 (Tex. 2012).
Environmental Defense Fund. 2003. Brush Management: Myths and Facts. http://texaslivingwaters.org/wp-content/uploads/2013/04/tlw-brush_management-03.pdf (accessed April 1, 2015).
Gleick, P. 1998. *The World's Water: The Biennial Report on Freshwater Resources 1998–1999*. Washington D.C.: Island Press.
Gunter, G., H. Hildebrand. 1954. The relation of rainfall of the state and catch of marine shrimp in Texas waters. *Bulletin of Marine Science* 4:95–103.
Hardberger, A. 2008. *From Policy to Reality: Maximizing Urban Water Conservation in Texas*. Environmental Defense Fund Report. Austin, TX.
Haas v. Choussard, 17 Tex.588 (Tex1856).
Hirsch, R.M., J.F. Walker, J.C. Day, R. Kallio.1990. The influence of man on hydrologic systems. *The Geology of North American, Vol. 1 Surface Water Hydrology*. Boulder, CO: Geological Society of America.
Hoefs v. Short, 273 S.W. 785 (Tex. 1925).
House Committee on Agriculture and Livestock. 2015. Interim Report to the 84th Legislature. http://www.house.state.tx.us/_media/pdf/committees/reports/83interim/House-Committee-on-Agriculture-and-Livestock-Interim-Report-2014.pdf (accessed April 1, 2015).
Houston & T.C. Ry. v. East, 98 Tex. 146, 81 S.W. 279 (1904).
Hubert, M., B. Bullock. 1999. Senate bill 1: The first big and bold step toward meeting Texas's future water needs. *Texas Tech Law Review* 30:53–70.
In re Adjudication of Water Rights of Upper Guadalupe Segment, 642 S.W.2d 438 (Tex. 1982).
Jarvis, G. 2014. Historical development of Texas Surface Water Law. *Essentials of Texas Water Resources*, Mary Sahs ed., 3rd ed. Austin, TX: State Bar of Texas.
Kaiser, R. 1986. *Handbook of Texas Water Law*. Texas Water Resources Institute. EIS 86-236 I/87-2m.
Kaiser, R., S. Kelly. 1987. Water rights for Texas estuaries. *Texas Tech University Law Review* 18:1121–1156.
Kaiser, R. 1996. Water marketing in the next millennium: A conceptual and legal framework. *Texas Tech Law Review* 27:181–262.
Kaiser, R. 1998. Untying the Gordian Knot: Negotiated strategies for protecting instream flows in Texas. *Natural Resources Journal* 38:157–196.
Kaiser, R., F. Skillern. 2001. Deep trouble: Options for managing the hidden threat of aquifer depletion in Texas. *Texas Tech Law Review* 32:249–304.
Kelton, E. 1973. *The Time It Never Rained*. Garden City, NY: Doubleday & Co.
Kinkade, J. 1996. Compromise and groundwater conservation. *State Bar of Texas Environmental Law Journal* 26:230–238.
LBG-Guyton Associates. 2003. Brackish Groundwater Manual for Texas Regional Water Planning Groups. http://www.twdb.texas.gov/publications/reports/contracted_reports/doc/2001483395.pdf (accessed April 1, 2015).
Lowry, R. 1959. A study of droughts in Texas, Texas Board of Water Engineers. Bulletin 5914, Austin, TX.
Malcolm Pirnie Inc. 2011. An Assessment of Aquifer Storage and Recovery in Texas. Report to the Texas Water Development Board #0904830940. http://www.twdb.texas.gov/innovativewater/asr/projects/pirnie/doc/2011_03_asr_final_rpt.pdf (accessed April 1, 2105).
National Drought Mitigation Center. 2015. Drought Monitoring in the U.S. http://drought.unl.edu/MonitoringTools/DroughtMonitoringintheUS.aspx (accessed April 1, 2015).
National Research Council. 1992. *Water Transfers in the West: Efficiency, Equity and the Environment*. Washington D.C.: National Academies Press.
Neilson-Gammon, J. 2011. The 2011 Texas Drought: A Briefing Packet fort the Texas Legislature, Office of the State Climatologist. http://climatexas.tamu.edu/files/2011_drought.pdf, (accessed April 1, 2015).
National Oceanic and Atmospheric Administration. 2013. National Centers for Environmental Information, https://www.ncdc.noaa.gov/billions/events (accessed April 1, 2015).
Perry, R. 2010. *Fed Up: Our Fight to Save America From Washington*. New York: Little, Brown & Co.

San Antonio Water System (SAWS). 2015. Twin Oaks-Aquifer Storage and Recovery. http://saws.org/Your_Water/WaterResources/projects/asr.cfm (accessed April 1, 2015).
Schoolmaster, A. 1992. A geographical analysis of water related constitutional amendments in Texas, 1998–1991. *Water Resources Bulletin* 28(3):495–505.
Smith, R. 2014. Special issues in water rights permitting. In *Essentials of Texas Water Resources.* 3rd edition, Mary Sahs ed. Austin, TX: State Bar of Texas.
Sipriano v. Great Spring Waters of American, Inc. 1 S.W.3d 75 (Tex 1999).
Stahle, D.W., M.K. Cleaveland. 1988. Texas drought history reconstructed and analyzed from 1698 to 1980. *Journal of Climate* 1(1):59–74.
Stahle, D.W., M.K. Cleaveland, H.D. Grissino-Mayer, R.D. Griffin, F.K. Fye, M.D. Therrell, D.J. Burnette, D.M. Meko, J. Villanueva Diaz. 2009. Cool- and warm-season precipitation reconstructions over western New Mexico. *Journal of Climate* 22(13):3729–3750. DOI:10.1175/2008JCLI2752.1.
Stepherson, A. 2014. Historical development of Texas Surface Water Law. In *Essentials of Texas Water Resources,* 3rd edition, Mary Sahs (ed). Austin, TX: State Bar of Texas.
State v. Hidalgo County WCID No 18, 443 S.W.2d 728 (Tex. Civ. App.—Corpus Christi 1969, writ ref'd n.r.e.
StateImpact-Texas, National Public Radio, 2012. http://stateimpact.npr.org/texas/2012/05/03/eyes-of-the-fires-a-look-back-at-the-2011-texas-wildfires/(accessed on April 1, 2015).
Tarlock, A.D. 2002. *Water Resource Management: A Casebook in Law and Public Policy.* Minneapolis, MN: West Publish Co.
Texas Administrative Code, Title 30, Chapter 210.
Texas Center for Policy Studies. 2014. A Report on Learning from Drought: The Next Generation of Water Planning for Texas. www.texascenter.org (accessed April 1, 2015).
Texas Commission on Environmental Quality. 2005. Handbook for Drought Contingency Planning for Retail Public Water Suppliers. RG-424. https://www.tceq.texas.gov/assets/public/comm_exec/pubs/archive/rg424.pdf (accessed April 1, 2015).
Texas Constitution, Art III, §52[a].
Texas Comptroller of Public Accounts. 2012. The Impact of the 2011 Drought and Beyond, February 2012, Publication 96-1704. http://www.window.state.tx.us/specialrpt/drought/(accessed April 1, 2015).
Texas Comptroller of Public Accounts. 2014. Texas Water Report: Going Deeper for the Solution, Publication 96-1746. http://www.window.state.tx.us/specialrpt/water/96-1746.pdf#page= 20 (accessed April 1, 2015).
Texas A&M Forest Service. 2011. 2011 Texas Wildfires: Common Denominators of Home Destructions. http://texasforestservice.tamu.edu/uploadedFiles/TFSMain/Preparing_for_Wildfires/Prepare_Your_Home_for_Wildfires/Contact_Us/2011%20Texas%20Wildfires.pdf (accessed April 1, 2015).
Texas Historical Society. The Handbook of Texas Online. http://www.tshaonline.org/handbook/online/articles/ybd01, (accessed April 1, 2015).
Texas Leadership Roundtable on Water: Core Principles and Recommendations, October 2014. http://wrgh.org/docs/TLR%20Water%20final%20core.pdf (accessed April 1, 2015).
Texas Legislative Budget Board. 2011. State Funding for Water Program (2nd Ed), January 2011. Report to the Texas Legislature. Austin, TX.
Texas Senate Select Committee on Water Policy. 2004. Interim Report to the 79th Legislature.
Texas Natural Resource Conservation Commission. 1995. A Regulatory Guidance Document for Applications to Divert, Store or Use State Water. 26, Austin Texas.
Texas Water Code. 2005. Chapters 11 and 36.
Texas Water Development Board. 1968. Water for Texas: 1968 State Water Plan. http://www.twdb.texas.gov/waterplanning/index.asp (accessed April 1, 2015).
Texas Department of Water Resources. 1984. Water for Texas: 1984 State Water Plan. http://www.twdb.texas.gov/waterplanning/index.asp (accessed April 1, 2015).
Texas Department of Water Resources. 1990. Water for Texas: 1990 State Water Plan. http://www.twdb.texas.gov/waterplanning/index.asp (accessed April 1, 2015).

Texas Water Development Board. 1997. Water For Texas: 1997 State Water Plan. http://www.twdb.texas.gov/waterplanning/index.asp (accessed April 1, 2015).

Texas Water Development Board. 2002. Water For Texas: 2002 State Water Plan. http://www.twdb.texas.gov/waterplanning/index.asp (accessed April 1, 2015).

Texas Water Development Board. 2007. Water for Texas: 2007 State Water Plan. http://www.twdb.texas.gov/waterplanning/index.asp (accessed April 1, 2015).

Texas Water Development Board. 2011. A Report on the History of Water Reuse in Texas. www.twdb.texas.gov/innovativewater/reuse/projects/reuseadvance/doc/component_a_final.pdf (accessed April 1, 2015).

Texas Water Development Board. 2012. Water for Texas: 2012 State Water Plan. http://www.twdb.texas.gov/waterplanning/index.asp (accessed April 1, 2015).

Texas Water Development Board. 2014. Water for Texas: Desalination of Brackish Groundwater. http://www.twdb.texas.gov/innovativewater/desal/doc/2014_TheFutureofDesalinationinTexas_Final.pdf (accessed April 1, 2015).

Texas Water Development Board. 2015. Desalination Plant Data Base. http://www2.twdb.texas.gov/apps/desal/default.aspx (accessed March 20, 2015).

The Texas Water Conservation Implementation Task Force, Report to the 79th Legislature, November 2004. http://www.allianceforwaterefficiency.org/ (accessed April 1, 2015).

Texas Leadership Roundtable on Water. 2014. A Report of the Wye River Group. http://wrgh.org/TLR_Water.asp (accessed April 1, 2015).

Texas Water Rights Adjudication Act. 1967. Texas Water Code §§ 11.301–11.341.

Thompson, B. 1993. Institutional perspectives on water policy and markets. *California Law Review* 81:673–764.

Wilcox, B. 2014. Can brush control enhance water supplies? Austin American Statesman. http://essm.tamu.edu/news/2014/water-saving-program-a-point-of-contention-can-brush-control-program-enhance-water-supplies/#.VQ3xIpPF990 (accessed April 1, 2015).

Wilhite, D.A., M.H. Glantz. 1985. Understanding the drought phenomenon: The role of definitions. *Water International* 10(3):111–120.

World Water Forum. 2012. Water in the U.S. American West: 150 Years of Adaptive Strategies, Policy Report for the 6th World Water Forum, March 2012. http://www.building-collaboration-for-water.org/documents/wwfh20amwest%20full2.28lr.pdf (accessed April 1, 2015).

Index

Page numbers followed by f and t indicate figures and tables, respectively.

F

G

R

T

Printed and bound by CPI Group (UK) Ltd, Croydon, CR0 4YY
22/10/2024
01777614-0013